Lecture Notes in Mathematics

1991

Editors:
J.-M. Morel, Cachan
F. Takens, Groningen
B. Teissier, Paris

Filippo Gazzola · Hans-Christoph Grunau
Guido Sweers

Polyharmonic Boundary Value Problems

Positivity Preserving and Nonlinear Higher
Order Elliptic Equations in Bounded Domains

 Springer

Filippo Gazzola
Dipartimento di Matematica
Politecnico di Milano
Piazza Leonardo da Vinci 32
20133 Milano, Italy
filippo.gazzola@polimi.it

Guido Sweers
Mathematisches Institut
Universität zu Köln
Weyertal 86-90
50931 Köln, Germany
gsweers@math.uni-koeln.de

Hans-Christoph Grunau
Institut für Analysis und Numerik
Otto von Guericke-Universität
Postfach 4120
39016 Magdeburg, Germany
hans-christoph.grunau@ovgu.de

ISBN: 978-3-642-12244-6   e-ISBN: 978-3-642-12245-3
DOI: 10.1007/978-3-642-12245-3
Springer Heidelberg Dordrecht London New York

Lecture Notes in Mathematics ISSN print edition: 0075-8434
                    ISSN electronic edition: 1617-9692

Library of Congress Control Number: 2010927754

Mathematics Subject Classification (2000): 35J40, 35J66, 35J91, 35B50, 35B45, 35J35, 35J62, 46E35, 53C42, 74K20

Dedicated to our wives Chiara, Brigitte and Barbara.

The cover figure displays the solution of $\Delta^2 u = f$ in a rectangle with homogeneous Dirichlet boundary condition for a nonnegative function f with its support concentrated near a point on the left hand side. The dark part shows the region where $u < 0$.

Preface

Linear elliptic equations arise in several models describing various phenomena in the applied sciences, the most famous being the second order stationary heat equation or, equivalently, the membrane equation. For this intensively well-studied linear problem there are two main lines of results. The first line consists of existence and regularity results. Usually the solution exists and "gains two orders of differentiation" with respect to the source term. The second line contains comparison type results, namely the property that a positive source term implies that the solution is positive under suitable side constraints such as homogeneous Dirichlet boundary conditions. This property is often also called positivity preserving or, simply, maximum principle. These kinds of results hold for general second order elliptic problems, see the books by Gilbarg-Trudinger [198] and Protter-Weinberger [347]. For linear higher order elliptic problems the existence and regularity type results remain, as one may say, in their full generality whereas comparison type results may fail. Here and in the sequel "higher order" means order at least four.

Most interesting models, however, are nonlinear. By now, the theory of second order elliptic problems is quite well developed for semilinear, quasilinear and even for some fully nonlinear problems. If one looks closely at the tools being used in the proofs, then one finds that many results benefit in some way from the positivity preserving property. Techniques based on Harnack's inequality, De Giorgi-Nash-Moser's iteration, viscosity solutions etc., all use suitable versions of a maximum principle. This is a crucial distinction from higher order problems for which there is no obvious positivity preserving property. A further crucial tool related to the maximum principle and intensively used for second order problems is the truncation method, introduced by Stampacchia. This method is helpful in regularity theory, in properties of first order Sobolev spaces and in several geometric arguments, such as the moving planes technique which proves symmetry of solutions by reflection. Also the truncation (or reflection) method fails for higher order problems. For instance, the modulus of a function belonging to a second order Sobolev space may not belong to the same space. The failure of maximum principles and of truncation methods, one could say, are the main reasons why the theory of nonlinear higher order elliptic equations is by far less developed than the theory of analogous second order equations. On the other hand, in view of many applications and increasing interest especially in the last twenty years, one should try to develop new tools suitable for higher order problems involving polyharmonic operators.

The simple example of the two functions $x \mapsto \pm |x|^2$ shows that already for the biharmonic operator the standard maximum principle fails. Nevertheless, taking also boundary conditions into account could yield comparison or positivity preserving properties and indeed, in certain special situations, such behaviour can be observed. It is one goal of the present exposition to describe situations where positivity preserving properties hold true or fail, respectively, and to explain how we have tackled the main difficulties related to the lack of a general comparison principle. In the present book we also show that in many higher order problems positivity preserving "almost" occurs. By this we mean that the solution to a problem inherits the sign of the data, except for some small contribution. By the experience from the present work, we hope that suitable techniques may be developed in order to obtain results quite analogous to the second order situation. Many recent higher order results give support to this hope.

A further goal of the present book is to collect some of those problems, where the authors were particularly involved, and to explain by which new methods one can replace second order techniques. In particular, to overcome the failure of the maximum principle and of the truncation method several ad hoc ideas will be introduced.

Let us now explain in some detail the subjects we address within this book.

Linear Higher Order Elliptic Problems

The polyharmonic operator $(-\Delta)^m$ is the prototype of an elliptic operator L of order $2m$, but with respect to linear questions, much more general operators can be considered. A general theory for boundary value problems for linear elliptic operators L of order $2m$ was developed by Agmon-Douglis-Nirenberg [4–6, 148]. Although the material is quite technical, it turns out that the Schauder theory as well as the L^p-theory can be developed to a large extent analogously to second order equations. The only exception are maximum modulus estimates which, for linear higher order problems, are much more restrictive than for second order problems. We provide a summary of the main results which hopefully will prove to be sufficiently wide to be useful for anybody who needs to refer to linear estimates or existence results.

The main properties of higher – at least second – order Sobolev spaces will be recalled. Since more orders of differentiation are involved, several different equivalent norms are available in these spaces. A crucial role in the choice of the norm is played by the regularity of the boundary. For the second order Dirichlet problem for the Poisson equation a nonsmooth boundary leads to technical difficulties but, due to the maximum principle, there is an inherent stability so that, when approximating nonsmooth domains by smooth domains, one recovers most of the features for domains with smooth boundary, see [46]. For Neumann boundary conditions the situation is more complicated in domains with rather wild boundaries, although even for polygonal boundaries they do not show spectacular changes. For higher order boundary value problems some peculiar phenomena occur. For instance, the so-called Babuška and Sapondžyan paradoxes [28, 358] forces one to be very careful in the choice of the norm in second order Sobolev spaces since some boundary value problems strongly depend on the regularity of the boundary. This phenomenon and its consequences will be studied in some detail.

Positivity in Higher Order Elliptic Problems

As long as existence and regularity results are concerned, the theory of linear higher order problems is already quite well developed as explained above. This is no longer true as soon as qualitative properties of the solution related to the source term are investigated. For instance, if we consider the clamped plate equation

$$\begin{cases} \Delta^2 u = f & \text{in } \Omega, \\ u = \frac{\partial u}{\partial \nu} = 0 & \text{on } \partial\Omega, \end{cases} \tag{0.1}$$

the "simplest question" seems to find out whether the positivity of the datum implies the positivity of the solution, Or, physically speaking,

does upwards pushing of a clamped plate yield upwards bending?

Equivalently, one may ask whether the corresponding Green function G is positive. In some special cases, the answer is "yes", while it is "no" in general. However, in numerical experiments, it appears very difficult to display the negative part and heuristically, one feels that the negative part of G – if present at all – is small in a suitable sense compared with the "dominating" positive part. We discuss not only the cases where one has positive Green functions and develop a perturbation theory of positivity, but we shall also discuss systematically under which conditions one may expect the negative part of the Green function to be small. We expect such smallness results to have some impact on future developments in the theory of non-linear higher order elliptic boundary value problems.

Boundary Conditions

For second order elliptic equations one usually extensively studies the case of Dirichlet boundary conditions because other boundary conditions do not exhibit too different behaviours. For the biharmonic equation $\Delta^2 u = f$ in a bounded domain of $\mathbb{R}^n$ it is not at all obvious which boundary condition would serve as a role model. Then a good approach is to focus on some boundary conditions that describe physically relevant situations. We consider a simplified energy functional and derive its Euler-Lagrange equation including the corresponding natural boundary conditions. We start with the linearised model for the beam. From a physical point of view, as long as the fourth order planar equation is considered, the most interesting seem to be not only the Dirichlet boundary conditions but also the Navier or

Steklov boundary conditions. The Dirichlet conditions correspond to the clamped plate model whereas Navier and Steklov conditions correspond to the hinged plate model, either by neglecting or considering the contribution of the curvature of the boundary. Each one of these boundary conditions requires the unknown function to vanish on the boundary, the difference being on the second boundary condition. These three boundary conditions have their own features and none of them may be thought to play the model role. We discuss all of them and emphasise their own peculiarities with respect to the comparison principles, to their variational formulation and to solvability of related nonlinear problems.

Eigenvalue Problems

For second order problems, such as the Dirichlet problem for the Laplace operator, one has not only the existence of infinitely many eigenvalues but also the simplicity and the one sign property of the first eigenfunction. For the biharmonic Dirichlet problem, this property is true in a ball but it is false in general. Again, a crucial role is played by the sign of the corresponding Green function. Concerning the isoperimetric properties of the first eigenvalue of the Dirichlet-Laplacian, the Faber-Krahn [162, 254, 255] result states that, among domains having the same finite volume it attains its minimum when the domain is a ball. A similar result was conjectured to hold for the biharmonic operator under homogeneous Dirichlet boundary conditions by Lord Rayleigh [351] in 1894. Although this statement has been proved only in domains of dimensions $n = 2, 3$, it is the common feeling that it should be true in any dimension. The minimisation of the first Steklov eigenvalue appears to be less obvious. And, indeed, we will see that a Faber-Krahn type result does not hold in this case.

Semilinear Equations

Among nonlinear problems for higher order elliptic equations one may just mention models for thin elastic plates, stationary surface diffusion flow, the Paneitz-Branson equation and the Willmore equation as frequently studied. In membrane biophysics the Willmore equation is also known as Helfrich model [228]. Moreover, several results concerning semilinear equations with power type nonlinear sources are also extremely useful in order to understand interesting phenomena in functional analysis such as the failure of compactness in the critical Sobolev embedding and in related inequalities.

One further motivation to study nonlinear higher order elliptic reaction-diffusion type equations like

$$(-\Delta)^m u = f(u) \tag{$*$}$$

in bounded domains is to understand whether the results available in the simplest case $m = 1$ can also be proved for any m, or whether the results for $m = 1$ are special, in particular as far as positivity and the use of maximum principles are concerned. The differential equation $(*)$ is complemented with suitable boundary conditions. As already mentioned above, if $m = n = 2$, equation $(*)$ may be considered as a nonlinear plate equation for plates subject to nonlinear feedback forces, one may think e.g. of suspension bridges. In this case, $(*)$ may also be interpreted as a reaction-diffusion equation, where the diffusion operator Δ^2 refers to (linearised) surface diffusion.

The first part of Chapter 7 is devoted to the proof of symmetry results for positive solutions to $(*)$ in the ball under Dirichlet boundary conditions. As already mentioned, truncation and reflection methods do not apply to higher order problems so that a suitable generalisation of the moving planes technique is needed here.

Equation $(*)$ deserves a particular attention when $f(u)$ has a power-type behaviour. In this case, a crucial role is played by the critical power $s = (n+2m)/(n-2m)$ which corresponds to the critical (Sobolev) exponent which appears whenever $n > 2m$. Indeed, subcritical problems in bounded domains enjoy compactness properties as a consequence of the Rellich-Kondrachov embedding theorem. But compactness is lacking when the critical growth is attained and by means of Pohožaev-type identities, this gives rise to many interesting phenomena. The existence theory can be developed similarly to the second order case $m = 1$ while it becomes immediately quite difficult to prove positivity or nonexistence of certain solutions. Nonexistence phenomena are related to so-called critical dimensions introduced by Pucci-Serrin [348, 349]. They formulated an interesting conjecture concerning these critical dimensions. We give a proof of a relaxed form of it in Chapter 7. We also give a functional analytic interpretation of these nonexistence results, which is reflected in the possibility of adding L^2–remainder terms in Sobolev inequalities with critical exponent and optimal constants. Moreover, the influence of topological and geometrical properties of Ω on the solvability of the equation is investigated. Also applications to conformal geometry, such as the Paneitz-Branson equation, involve the critical Sobolev exponent since the corresponding semilinear equation enjoys a conformal covariance property. In this context a key role is played by a fourth order curvature invariant, the so-called Q-curvature. Our book does not aim at giving an overview of this rapidly developing subject. For this purpose we refer to the monographs of Chang [89] and Druet-Hebey-Robert [149]. We want to put a spot on some special aspects of such kind of equations. First, we consider the question whether in suitable domains in euclidean space it is possible to change the euclidean background metric conformally into a metric which has strictly positive constant Q-curvature, while at the same time, certain geometric quantities vanish on the boundary. Secondly, we study a phenomenon of nonuniqueness of complete metrics in hyperbolic space, all being conformal to the Poincaré-metric and all having the same constant Q-curvature. This result is in strict contrast with the corresponding problem involving constant negative scalar curvature.

We conclude the discussion of semilinear elliptic problems with some observations on fourth order problems with supercritical growth. Corresponding second order results heavily rely on the use of maximum principles and constructions of many refined auxiliary functions having some sub- or supersolution property. Such techniques are not available at all for the fourth order problems. In symmetric situations, however, they could be replaced by different tools so that many of the results being well established for second order equations do indeed carry over to the fourth order ones.

A Dirichlet Problem for Willmore Surfaces of Revolution

A frame invariant modeling of elastic deformations of surfaces like thin plates or biological membranes gives rise to variational integrals involving curvature and area terms. A special case is the Willmore functional

$$\int_{\Gamma} \mathrm{H}^2 \, d\omega,$$

which up to a boundary term is conformally invariant. Here H denotes the mean curvature of the surface Γ in $\mathbb{R}^3$. Critical points of this functional are called Willmore surfaces, the corresponding Euler-Lagrange equation is the so-called Willmore equation. It is quasilinear, of fourth order and elliptic. While a number of beautiful results have been recently found for closed surfaces, see e.g. [35, 156, 263–265, 372], only little is known so far about boundary value problems since the difficulties mentioned earlier being typical for fourth order problems due to a lack of maximum principles add here to the difficulty that the ellipticity of the equation is not uniform. The latter reflects the geometric nature of the equation and gives rise e.g. to the problem that minimising sequences for the Willmore functional are in general not bounded in the Sobolev space H^2. In this book we confine ourselves to a very special situation, namely the Dirichlet problem for symmetric Willmore surfaces of revolution. Here, by means of some refined geometric constructions, we succeed in considering minimising sequences of the Willmore functional subject to Dirichlet boundary conditions and with suitable additional C^1-properties thereby gaining weak H^2- and strong C^1-compactness. We expect the theory of boundary value problems for Willmore surfaces to develop rapidly and consider this chapter as one contribution to outline directions of possible future research in quasilinear geometric fourth order equations.

Acknowledgements

The authors are grateful to their colleagues, friends, and collaborators Gianni Arioli, Elvise Berchio, Francisco Bernis, Isabeau Birindelli, Dorin Bucur, Daniele Cassani, Philippe Clément, Anna Dall'Acqua, Cesare Davini, Klaus Deckelnick, Alberto Ferrero, Steffen Fröhlich, Maurizio Grasselli, Gerhard Huisken, Paschalis Karageorgis, Bernd Kawohl, Marco Kühnel, Ernst Kuwert, Monica Lazzo, Andrea Malchiodi, Joseph McKenna, Christian Meister, Erich Miersemann, Enzo Mitidieri, Serguei Nazarov, Mohameden Ould Ahmedou, Vittorino Pata, Dario Pierotti, Wolfgang Reichel, Frédéric Robert, Bernhard Ruf, Edoardo Sassone, Reiner Schätzle, Friedhelm Schieweck, Paul G. Schmidt, Marco Squassina, Franco Tomarelli, Wolf von Wahl, and Tobias Weth for inspiring discussions and sharing ideas.

Essential parts of the book are based on work published in different form before. We thank our numerous collaborators for their contributions. Detailed information on the sources are given in the bibliographical notes at the end of each chapter.

And last but not least, we thank the referees for their helpful and encouraging comments.

Contents

Chapter 1
Models of Higher Order

The goal of this chapter is to explain in some detail which models and equations are considered in this book and to provide some background information and comments on the interplay between the various problems. Our motivation arises on the one hand from equations in continuum mechanics, biophysics or differential geometry and on the other hand from basic questions in the theory of partial differential equations.

In Section 1.1, after providing a few historical and bibliographical facts, we recall the derivation of several linear boundary value problems for the plate equation. In Section 1.8 we come back to this issue of modeling thin elastic plates where the full nonlinear differential geometric expressions are taken into account. As a particular case we concentrate on the Willmore functional, which models the pure bending energy in terms of the squared mean curvature of the elastic surface. The other sections are mainly devoted to outlining the contents of the present book. In Sections 1.2-1.4 we introduce some basic and still partially open questions concerning qualitative properties of solutions of various linear boundary value problems for the linear plate equation and related eigenvalue problems. Particular emphasis is laid on positivity and – more generally – "almost positivity" issues. A significant part of the present book is devoted to semilinear problems involving the biharmonic or polyharmonic operator as principal part. Section 1.5 gives some geometric background and motivation, while in Sections 1.6 and 1.7 semilinear problems are put into a context of contributing to a theory of nonlinear higher order problems.

1.1 Classical Problems from Elasticity

Around 1800 the physicist Chladni was touring Europe and showing, among other things, the nodal line patterns of vibrating plates. Jacob Bernoulli II tried to model these vibrations by the fourth order operator $\frac{\partial^4}{\partial x^4} + \frac{\partial^4}{\partial y^4}$ [54]. His model was not accepted, since it is not rotationally symmetric and it failed to reproduce the nodal line patterns of Chladni. The first use of Δ^2 for the modeling of an elastic plate is attributed to a correction of Lagrange of a manuscript by Sophie Germain from 1811.

F. Gazzola et al., *Polyharmonic Boundary Value Problems*,
Lecture Notes in Mathematics 1991, DOI 10.1007/978-3-642-12245-3_1,
© Springer-Verlag Berlin Heidelberg 2010

For historical details we refer to [79, 250, 325, 398]. For a more elaborate history of the biharmonic problem and the relation with elasticity from an engineering point of view one may consult a survey of Meleshko [300]. This last paper also contains a large bibliography so far as the mechanical engineers are interested. Mathematically interesting questions came up around 1900 when Almansi [8,9], Boggio [62,63] and Hadamard [222, 223] addressed existence and positivity questions.

In order to have physically meaningful and mathematically well-posed problems the plate equation $\Delta^2 u = f$ has to be complemented with prescribing a suitable set of boundary data. The most commonly studied boundary value problems for second order elliptic equations are named Dirichlet, Neumann and Robin. These three types appear since they have a physical meaning. For fourth order differential equations such as the plate equation the variety of possible boundary conditions is much larger. We will shortly address some of those that are physically relevant. Most of this book will be focussed on the so-called clamped case which is again referred to by the name of Dirichlet. An early derivation of appropriate boundary conditions can be found in a paper by Friedrichs [173]. See also [58, 141]. The following derivation is taken from [388].

1.1.1 The Static Loading of a Slender Beam

If $u(x)$ denotes the deviation from the equilibrium of the idealised one-dimensional beam at the point x and $p(x)$ is the density of the lateral load at x, then the elastic energy stored in the bending beam due to the deformation consists of terms that can be described by bending and by stretching. This stretching occurs when the horizontal position of the beam is fixed at both endpoints. Assuming that the elastic force is proportional to the increase of length, the potential energy density for the beam fixed at height 0 at the endpoints a and b would be

$$J_{st}(u) = \int_a^b \left(\sqrt{1 + u'(x)^2} - 1 \right) dx.$$

For a string one neglects the bending and, by adding a force density p, one finds

$$J(u) = \int_a^b \left(\sqrt{1 + u'(x)^2} - 1 - p(x)u(x) \right) dx.$$

For a thin beam one assumes that the energy density stored by bending the beam is proportional to the square of the curvature:

$$J_{sb}(u) = \int_a^b \frac{u''(x)^2}{(1 + u'(x)^2)^3} \sqrt{1 + u'(x)^2}\, dx. \tag{1.1}$$

Formula (1.1) for J_{sb} highlights the curvature and the arclength. A two-dimensional analogue of this functional is the Willmore functional, which is discussed below in

Section 1.8. Note that the functional J_{sb} does not include a term that corresponds to an increase in the length of the beam which would occur if the ends are fixed and the beam would bend. That is, the function in $H^2 \cap H_0^1(a,b)$ minimising $J_{sb}(u) - \int_a^b pu\ dx$ should be an approximation for the so-called supported beam which is free to move in horizontal directions at its endpoints.

For small deformations of a beam an approximation that takes care of stretching, bending and a force density would be

$$J(u) = \int_a^b \left(\tfrac{1}{2}u''(x)^2 + \tfrac{c}{2}u'(x)^2 - p(x)u(x) \right)\ dx,$$

where $c > 0$ represents the initial tension of the beam which is also fixed horizontally at the endpoints.

The linear Euler-Lagrange equation that arises from this situation contains both second and fourth order terms:

$$u'''' - cu'' = p. \tag{1.2}$$

If one lets the beam move freely at the boundary points (and in the case of zero initial tension), one arrives at the simplest fourth order equation $u'''' = p$. This differential equation may be complemented with several boundary conditions.

The mathematical formulation that corresponds to the boundary conditions in Figure 1.1 are as follows:

- Clamped: $u(a) = 0 = u'(a)$, also known as homogeneous Dirichlet boundary conditions.
- Hinged: $u(b) = 0 = u''(b)$, also known as homogeneous Navier boundary conditions. This is not a real hinged situation since the vertical position is fixed but the beam is allowed to slide in the hinge itself.
- Simply supported: $\max(u(b),0)\,u'''(b) = 0 = u''(b)$. In applications, when the force is directed downwards, this boundary condition simplifies to the hinged one $u(b) = 0 = u''(b)$. However, when upward forces are present it might happen that $u(b) > 0$ and then the natural boundary condition $u'''(b) = 0$ appears.
- Free: $u'''(b) = 0 = u''(b)$.
- Free vertical sliding but with fixed derivative: $u'(b) = u'''(b) = 0$.

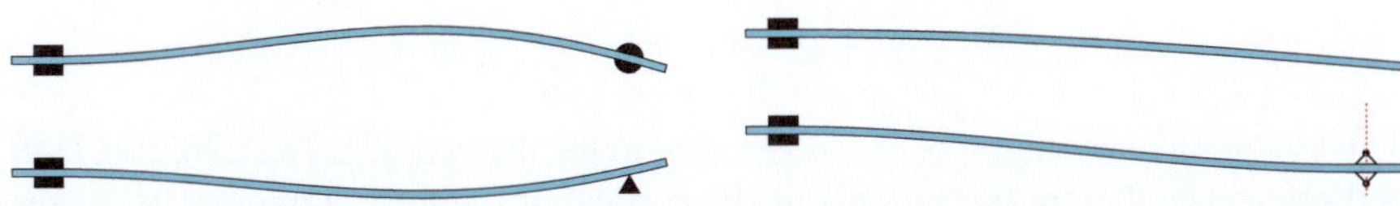

Fig. 1.1 The depicted boundary condition for the left endpoints of these four beams is "clamped". The boundary conditions for the right endpoints are respectively "hinged" and "simply supported" on the left; on the right one finds "free" and one that allows vertical displacement but fixes the derivative by a sliding mechanism.

The second and third order derivatives appear as natural boundary conditions by the derivation of the strong Euler-Lagrange equations.

If the beam would be moving in an elastic medium, then, again for small deviations one adds a further term to J and finds

$$J(u) = \int_a^b \left(\tfrac{1}{2}(u'')^2 + \tfrac{\gamma}{2}\, u^2 - pu \right) dx.$$

This leads to the Euler-Lagrange equation $u'''' + \gamma u = p$.

Also a suspension bridge may be seen as a beam of given length L, with hinged ends and whose downward deflection is measured by a function $u(x,t)$ subject to three forces. These forces can be summarised as the stays holding the bridge up as nonlinear springs with spring constant k, the constant weight per unit length of the bridge W pushing it down, and the external forcing term $f(x,t)$. This leads to the equation

$$\begin{cases} u_{tt} + \gamma u_{xxxx} = -ku^+ + W + f(x,t), \\ u(0,t) = u(L,t) = u_{xx}(0,t) = u_{xx}(L,t) = 0, \end{cases} \tag{1.3}$$

where γ is a physical constant depending on the beam, Young's modulus, and the second moment of inertia. The model leading to (1.3) is taken from the survey papers [271, 296].

The famous collapse of the Tacoma Narrows Bridge, see [16, 61], was the consequence of a torsional oscillation. McKenna [296, p. 106] explains this fact as follows.

> *A large vertical motion had built up, there was a small push in the torsional direction to break symmetry, the instability occurred, and small aerodynamic torsional periodic forces were sufficient to maintain the large periodic torsional motions.*

For this reason, a major role is played by *travelling waves*. If one neglects the effect of external forces and normalises all the constants, then (1.3) becomes

$$u_{tt} + u_{xxxx} = -u^+ + 1. \tag{1.4}$$

In order to find travelling waves, one seeks solutions of (1.4) for $(x,t) \in \mathbb{R}^2$ of the kind $u(x,t) = 1 + y(x - ct)$ where $c > 0$ denotes the speed of propagation. Hence, the function y satisfies the fourth order ordinary differential equation

$$y'''' + c^2 y'' + (y+1)^+ - 1 = 0 \qquad \text{in } \mathbb{R}.$$

This is a nonlinear version of (1.2). We refer to the papers [271, 272, 296, 298, 299] and references therein for variants of these equations and for a number of results and open problems related to suspension bridges.

1.1.2   The Kirchhoff-Love Model for a Thin Plate

As for the beam we assume that the plate, the vertical projection of which is the planar region $\Omega \subset \mathbb{R}^2$, is free to move horizontally at the boundary. Then a simple model for the elastic energy is

$$J(u) = \int_{\Omega} \left(\tfrac{1}{2} (\Delta u)^2 + (1 - \sigma) \left(u_{xy}^2 - u_{xx} u_{yy} \right) - f\, u \right) dxdy, \qquad (1.5)$$

where f is the external vertical load. Again u is the deflection of the plate in vertical direction and, as above for the beam, first order derivatives are left out which indicates that the plate is free to move horizontally.

This modern variational formulation appears already in [173], while a discussion for a boundary value problem for a thin elastic plate in a somehow old fashioned notation is made already by Kirchhoff [250]. See also the two papers of Birman [57, 58], the books by Mikhlin [304, §30], Destuynder-Salaun [141], Ciarlet [102], or the article [103] for the clamped case.

In (1.5) σ is the Poisson ratio, which is defined by $\sigma = \frac{\lambda}{2(\lambda+\mu)}$ with the so-called Lamé constants λ, μ that depend on the material. For physical reasons it holds that $\mu > 0$ and *usually* $\lambda \geq 0$ so that $0 \leq \sigma < \tfrac{1}{2}$. Moreover, it *always* holds true that $\sigma > -1$ although some exotic materials have a negative Poisson ratio, see [266]. For metals the value σ lies around 0.3 (see [281, p. 105]). One should observe that for $\sigma > -1$, the quadratic part of the functional (1.5) is always positive.

For small deformations the terms in (1.5) are taken as approximations being purely quadratic with respect to the second derivatives of u of respectively twice the squared mean curvature and the Gaussian curvature supplied with the factor $\sigma - 1$. For those small deformations one finds

$$\begin{aligned} \tfrac{1}{2} (\Delta u)^2 + (1 - \sigma) \left(u_{xy}^2 - u_{xx} u_{yy} \right) &\approx \tfrac{1}{2} (\kappa_1 + \kappa_2)^2 - (1 - \sigma)\, \kappa_1 \kappa_2 \\ &= \tfrac{1}{2} \kappa_1^2 + \sigma \kappa_1 \kappa_2 + \tfrac{1}{2} \kappa_2^2, \end{aligned}$$

where κ_1, κ_2 are the principal curvatures of the graph of u. Variational integrals avoiding such approximations and involving the original expressions for the mean and the Gaussian curvature are considered in Section 1.8 and lead as a special case to the Willmore functional.

Which are the appropriate boundary conditions? For the clamped and hinged boundary condition the natural settings, that is the Hilbert spaces for these two situations, are respectively $H = H_0^2(\Omega)$ and $H = H^2 \cap H_0^1(\Omega)$. Minimising the energy functional leads to the weak Euler-Lagrange equation $\langle dJ(u), v \rangle = 0$, that is

$$\int_{\Omega} \left(\Delta u \Delta v + (1 - \sigma) \left(2 u_{xy} v_{xy} - u_{xx} v_{yy} - u_{yy} v_{xx} \right) - f\, v \right) dxdy = 0 \qquad (1.6)$$

for all $v \in H$. Let us assume both that minimisers u lie in $H^4(\Omega)$ and that the exterior normal $v = (v_1, v_2)$ and the corresponding tangential $\tau = (\tau_1, \tau_2) = (-v_2, v_1)$ are well-defined. Then an integration by parts of (1.6) leads to

$$
0 = \int_\Omega \left(\Delta^2 u - f \right) v \, dxdy + \int_{\partial\Omega} \left(\frac{\partial}{\partial v} \Delta u \right) v \, ds
$$
$$
+ (1 - \sigma) \int_{\partial\Omega} \left(\left(v_1^2 - v_2^2 \right) u_{xy} - v_1 v_2 \left(u_{xx} - u_{yy} \right) \right) \frac{\partial}{\partial \tau} v \, ds
$$
$$
+ \int_{\partial\Omega} \left(\Delta u + (1 - \sigma) \left(2 v_1 v_2 u_{xy} - v_2^2 u_{xx} - v_1^2 u_{yy} \right) \right) \frac{\partial}{\partial v} v \, ds. \tag{1.7}
$$

- Following [141] let us split the boundary $\partial\Omega$ in a clamped part Γ_0, a hinged part Γ_1 and a free part $\Gamma_2 = \partial\Omega \setminus (\Gamma_0 \cup \Gamma_1)$, which are all assumed to be smooth. Moreover, to keep our derivation simple, we assume that Γ_2 has empty relative boundary in $\partial\Omega$, i.e. it is a union of connected components of $\partial\Omega$.
  On Γ_0 one has $u = u_v = 0$. The type of boundary conditions on Γ_0 are generally referred to as homogeneous Dirichlet.
  On Γ_1 one has $u = 0$ and may rewrite the second boundary condition that appears from (1.7) as

$$
\Delta u + (1 - \sigma) \left(2 u_{xy} v_1 v_2 - u_{xx} v_2^2 - u_{yy} v_1^2 \right)
$$
$$
= \sigma \Delta u + (1 - \sigma) \left(2 u_{xy} v_1 v_2 + u_{xx} v_1^2 + u_{yy} v_2^2 \right)
$$
$$
= \sigma \Delta u + (1 - \sigma) u_{vv} = \sigma \left(u_{vv} + \kappa u_v \right) + (1 - \sigma) u_{vv}
$$
$$
= u_{vv} + \sigma \kappa u_v = \Delta u - (1 - \sigma) \kappa u_v. \tag{1.8}
$$

Here κ is the curvature of the boundary. We use the sign convention that $\kappa \geq 0$ for convex boundary parts and $\kappa \leq 0$ for concave boundary parts.
On Γ_2, which we recall to have empty relative boundary in $\partial\Omega$, an integration by parts along the boundary shows

$$
\int_{\Gamma_2} \left(\frac{\partial}{\partial v} \Delta u \right) v \, ds + (1 - \sigma) \int_{\Gamma_2} \left(\left(v_1^2 - v_2^2 \right) u_{xy} - v_1 v_2 \left(u_{xx} - u_{yy} \right) \right) \frac{\partial}{\partial \tau} v \, ds
$$
$$
= - \int_{\Gamma_2} (1 - \sigma) \left(u_{\tau\tau v} + \frac{\partial}{\partial v} \Delta u \right) v \, ds.
$$

Summarising, on domains with smooth $\Gamma_0, \Gamma_1, \Gamma_2$ one finds the following boundary value problem:

$$
\begin{cases}
\Delta^2 u = f & \text{in} \quad \Omega, \\
u = \frac{\partial u}{\partial v} = 0 & \text{on} \quad \Gamma_0, \\
u = \Delta u - (1 - \sigma) \kappa \frac{\partial u}{\partial v} = 0 & \text{on} \quad \Gamma_1, \\
\sigma \Delta u + (1 - \sigma) u_{vv} = (1 - \sigma) u_{\tau\tau v} + \frac{\partial}{\partial v} \Delta u = 0 & \text{on} \quad \Gamma_2.
\end{cases}
$$

The differential equation $\Delta^2 u = f$ is called the Kirchhoff-Love model for the vertical deflection of a thin elastic plate.

- The clamped plate equation, i.e. the pure Dirichlet case when $\partial\Omega = \Gamma_0$, is as follows:

$$\begin{cases} \Delta^2 u = f & \text{in } \Omega, \\ u = \frac{\partial u}{\partial v} = 0 & \text{on } \partial\Omega. \end{cases} \tag{1.9}$$

Notice that σ does not play any role for clamped boundary conditions. In this case, after an integration by parts like in (1.7), the elastic energy (1.5) becomes

$$J(u) = \int_\Omega \left(\tfrac{1}{2}(\Delta u)^2 - f\,u \right) dx$$

and this functional has to be minimised over the space $H_0^2(\Omega)$.

- The physically relevant boundary value problem for the pure hinged case when $\partial\Omega = \Gamma_1$ reads as

$$\begin{cases} \Delta^2 u = f & \text{in } \Omega, \\ u = \Delta u - (1 - \sigma)\,\kappa \frac{\partial u}{\partial v} = 0 & \text{on } \partial\Omega. \end{cases} \tag{1.10}$$

See [141, II.18 on p. 42]. These boundary conditions are named after Steklov due the first appearance in [380]. In this case, with an integration by parts like in (1.7) and arguing as in (1.8), the elastic energy (1.5) becomes

$$J(u) = \int_\Omega \left(\tfrac{1}{2}(\Delta u)^2 - f\,u \right) dx - \frac{1 - \sigma}{2} \int_{\partial\Omega} \kappa\, u_v^2\, d\omega; \tag{1.11}$$

for details see the proof of Corollary 5.23. This functional has to be minimised over the space $H^2 \cap H_0^1(\Omega)$.

- On straight boundary parts $\kappa = 0$ holds and the second boundary condition in (1.10) simplifies to $\Delta u = 0$ on $\partial\Omega$. The corresponding boundary value problem

$$\begin{cases} \Delta^2 u = f & \text{in } \Omega, \\ u = \Delta u = 0 & \text{on } \partial\Omega, \end{cases} \tag{1.12}$$

is in general referred to as the one with homogeneous Navier boundary conditions, see [141, II.15 on p. 41]. On polygonal domains one might naively expect that (1.10) simplifies to (1.12). Unless $\sigma = 1$ this is an erroneous conclusion and instead of $\kappa \frac{\partial u}{\partial v}$ one should introduce a Dirac-δ-type contribution at the corners. See Section 2.7 and [294].

1.1.3 Decomposition into Second Order Systems

Note that the combination of the boundary conditions in (1.12) or (1.10) allows for
rewriting these fourth order problems as a second order system

$$\begin{cases} -\Delta u = w \ \text{ and } \ -\Delta w = f \ \text{ in } \Omega, \\ \quad u = 0 \quad \text{ and } \quad w = 0 \quad \text{ on } \partial\Omega, \end{cases} \tag{1.13}$$

respectively

$$\begin{cases} -\Delta u = w \ \text{ and } \quad -\Delta w = f \qquad \text{ in } \Omega, \\ \quad u = 0 \quad \text{ and } \qquad w = -(1-\sigma)\,\kappa \frac{\partial u}{\partial \nu} \text{ on } \partial\Omega. \end{cases} \tag{1.14}$$

The boundary value problems in (1.13) can be solved consecutively. Indeed, for
smooth domains the solution u coincides with the minimiser in $H^2 \cap H_0^1(\Omega)$ of

$$J(u) = \int_\Omega \left(\tfrac{1}{2}(\Delta u)^2 - f\,u \right) dx. \tag{1.15}$$

For domains with corners this is not necessarily true. For a reentrant corner a phenomenon may occur that was first noticed by Sapondžyan, see Section 1.4.1 and
Example 2.33.

A splitting into a system of two consecutively solvable second order boundary
value problems is not possible for (1.14). Nevertheless, for convex domains we have
$\kappa \geq 0$ and this fact turns (1.14) into a cooperative second order system for which
some of the techniques for second order equations apply. "Cooperative" means that
the coupling supports the sign properties of the single equations. Cooperative systems of second order boundary value problems are well-studied in the literature and
will not be addressed in this monograph.

A more intricate situation occurs for the clamped case where a similar approach
to split the fourth order problem into a system of second order equations results in

$$\begin{cases} \quad -\Delta u = w \quad \text{ and } \ -\Delta w = f \ \text{ in } \Omega, \\ u = \frac{\partial}{\partial \nu}u = 0 \ \text{ and } \qquad - \qquad \text{ on } \partial\Omega. \end{cases} \tag{1.16}$$

For most questions such a splitting has not yet appeared to be very helpful. The first
boundary value problem has too many boundary conditions, the second one none at
all. Techniques for second order equations, however, can be used e.g. in numerical
approximations, when the problem is put as follows. Find stationary points $(u,w) \in
H_0^1(\Omega) \times H^1(\Omega)$ of

$$F(u,w) = \int_\Omega \left(\nabla u \cdot \nabla w - f\,u - \frac{1}{2}w^2 \right) dx. \tag{1.17}$$

The weak Euler-Lagrange equation becomes

$$\langle dF(u,w),(\varphi,\psi)\rangle = \int_\Omega (\nabla u \cdot \nabla \psi + \nabla \varphi \cdot \nabla w - f\,\varphi - w\,\psi)\,dx = 0 \qquad (1.18)$$

for all $(\varphi, \psi) \in H_0^1(\Omega) \times H^1(\Omega)$. Assuming $u, w \in H^2(\Omega)$, an integration by parts gives

$$\int_{\partial\Omega} \frac{\partial}{\partial v} u\,\psi\,d\omega + \int_\Omega (-\Delta u - w)\,\psi\,dx + \int_\Omega (-\Delta w - f)\,\varphi\,dx = 0.$$

Testing with $(\varphi, \psi) \in H_0^1(\Omega) \times H^1(\Omega)$ we find $u \in H_0^2(\Omega)$, $-\Delta u = w$ and $-\Delta w = f$, thereby recovering (1.16) as Euler-Lagrange-equation for the functional F in (1.17).

The formulation in (1.18) can be used to construct approximate solutions using piecewise linear finite elements instead of the $C^{1,1}$ elements that are necessary for functionals containing second order derivatives. For smooth domains one may show that the stationary points of (1.15) and (1.17) coincide. For nonsmooth domains similar phenomena like the Babuška paradox might appear, which is described below in Section 1.4.2, see also Section 2.7.

1.2 The Boggio-Hadamard Conjecture for a Clamped Plate

Since maximum principles do not only allow for proving nice results on geometric properties of solutions of *second* order elliptic problems but are also extremely important technical tools in this field, one might wonder in how far such results still hold in higher order boundary value problems. First of all it is an obvious remark that a general maximum principle can no longer be true. The biharmonic functions $x \mapsto \pm|x|^2$ have a strict global minimum or maximum respectively in any domain containing the origin. On the other hand, it may be reasonable to ask for positivity preserving properties of *boundary value problems*, i.e. whether positive data yield positive solutions. In physical terms this question may be rephrased as follows:

Does upwards pushing of a plate yield upwards bending?

The answer, of course, depends on the model considered and on the imposed boundary conditions. For instance, in the Dirichlet problem for the plate equation

$$\begin{cases} \Delta^2 u = f & \text{in } \Omega, \\ u = \dfrac{\partial u}{\partial v} = 0 & \text{on } \partial\Omega, \end{cases} \qquad (1.19)$$

there is – at least no obvious way – to take advantage of second order comparison principles and in this sense, it may be considered as the prototype of a "real" fourth order boundary value problem. On the other hand, the plate equation complemented with Navier boundary conditions (1.12) can be written as a system of two second order boundary value problems and enjoys a sort of comparison principle. In particular, under these conditions it is obvious that $f \geq 0$ implies that $u \geq 0$. However, when adding lower order perturbations, the case of a so-called noncooperative coupling may occur and this simple argument breaks down. In this case, the positivity issue becomes quite involved also under Navier boundary conditions, see e.g. [310].

A significant part of the present book will be devoted to discussing the following mathematical question.

What remains true of: "$f \geq 0$ in the clamped plate boundary value problem (1.19) implies positivity of the solution $u \geq 0$"?

In view of the representation formula

$$u(x) = \int_B G_{\Delta^2,\Omega}(x,y)f(y)\,dy,$$

one equivalently may wonder whether the corresponding Green function is positive or even strictly positive, i.e. $G_{\Delta^2,\Omega} > 0$? Lauricella ([269], 1896) found an explicit formula for $G_{\Delta^2,\Omega}$ in the special case of the unit disk $\Omega = B := B_1(0) \subset \mathbb{R}^2$. Boggio ([63, p. 126], 1905) generalised this formula to the Dirichlet problem for any polyharmonic operator $(-\Delta)^m$ in any ball in any $\mathbb{R}^n$ and found a particularly elegant expression for the Green function, see Lemma 2.27. In case of the biharmonic operator in the two-dimensional disk $B \subset \mathbb{R}^2$, this formula reads:

$$G_{\Delta^2,B}(x,y) = \frac{1}{8\pi}|x-y|^2 \int_1^{\left|\,|x|y-\frac{x}{|x|}\right|/|x-y|} \frac{(v^2-1)}{v}\,dv > 0. \qquad (1.20)$$

Positivity is here quite obvious since

$$\left|\,|x|y - \frac{x}{|x|}\right|^2 - |x-y|^2 = (1-|x|^2)(1-|y|^2) > 0.$$

Almansi ([8], 1899) found an explicit construction for solving $\Delta^2 u = 0$ with prescribed boundary data for u and u_v on domains $\Omega \subset \mathbb{R}^2$ with $\Omega = p(B)$ and $p : B \to \Omega$ being a conformal polynomial mapping. Probably inspired by Almansi's result and supported by physically plausible behaviour of plates, Boggio conjectured (see [222, 223]) that for the clamped plate boundary value problem (1.19), the Green function is always positive.

In 1908, Hadamard [223] already knew that this conjecture fails e.g. in annuli with small inner radius (see also [317]). He writes that Boggio had mentioned to him that the conjecture was meant for simply connected domains. In [223] he also writes:

Malgré l'absence de démonstration rigoureuse, l'exactitude de cette proposition ne paraît pas douteuse pour les aires convexes.

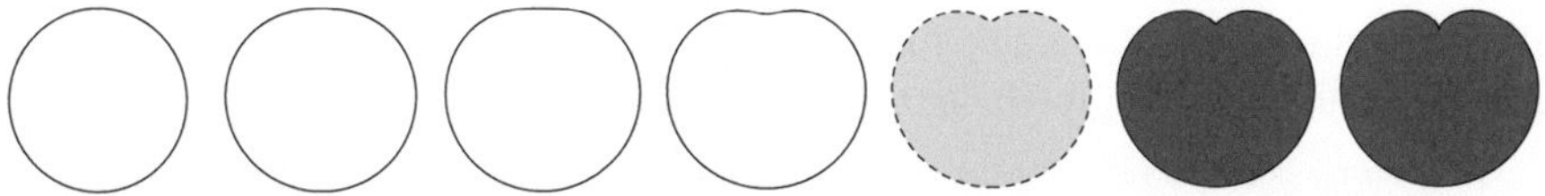

Fig. 1.2 Limaçons vary from circle to cardioid. The fifth limaçon from the left is critical for a positive Green function.

Accordingly the conjecture of Boggio and Hadamard may be formulated as follows:

> *The Green function $G_{\Delta^2,\Omega}$ for the clamped plate boundary value problem on convex domains is positive.*

Using the explicit formula from [8] for the "limaçons de Pascal", see Figure 1.2, Hadamard in [223] even claimed to have proven positivity of the Green function $G_{\Delta^2,\Omega}$ when Ω is such a limaçon.

However, after 1949 numerous counterexamples ([107, 108, 150, 176, 253, 279, 327, 368, 371, 390]) disproved the positivity conjecture of Boggio and Hadamard. The first result in this direction came by Duffin ([150, 152]), who showed that the Green function changes sign on a long rectangle. A most striking example was found by Garabedian. He could show change of sign of the Green function in ellipses with ratio of half axes ≈ 1.6 ([176], [177, p. 275]). For an elementary proof of a slightly weaker result see [371]. Hedenmalm, Jakobsson and Shimorin [227] mention that sign change occurs already in ellipses with ratio of half axes ≈ 1.2. Nakai and Sario [318] give a construction how to extend Garabedian's example also to higher dimensions. Sign change is also proven by Coffman-Duffin [108] in any bounded domain containing a corner, the angle of which is not too large. Their arguments are based on previous results by Osher and Seif [327, 368] and cover, in particular, squares. This means that neither in arbitrarily smooth uniformly convex nor in rather symmetric domains the Green function needs to be positive. Moreover, in [120] it has been proved that Hadamard's claim for the limaçons is not correct. Limaçons are a one-parameter family with circle and cardioid as extreme cases. For domains close enough to the cardioid, the Green function is no longer positive. Surprisingly, the extreme case for positivity is not convex. Hence convexity is neither sufficient nor necessary for a positive Green function. One should observe that in one dimension any bounded interval is a ball and so, one always has positivity there thanks to Boggio's formula.

For the history of the Boggio-Hadamard conjecture one may also see Maz'ya's and Shaposhnikova's biography [295] of Hadamard.

Despite the fact that the Green function is usually sign changing, it is very hard to find real world experiments where loss of positivity preserving can be observed. Moreover, in all numerical experiments in smooth domains, it is very difficult to display the negative part and heuristically, one feels that the negative part of $G_{\Delta^2,\Omega}$ – if present at all – is small in a suitable sense compared with the "dominating" positive part. We refine the Boggio-Hadamard conjecture as follows:

> In arbitrary domains $\Omega \subset \mathbb{R}^n$, the negative part of the biharmonic Green's function $G_{\Delta^2,\Omega}$ is small relative to the singular positive part. In the investigation of nonlinear problems, the negative part is technically disturbing but it does not give rise to any substantial additional assumption in order to have existence, regularity, etc. when compared with analogous second order problems.

The present book may be considered as a first contribution to the discussion of this conjecture and Chapters 5 and 6 are devoted to it. Chapter 4 provides the necessary kernel estimates. Let us mention some of those results which we have obtained so far to give support to this conjecture. For any smooth domain $\Omega \subset \mathbb{R}^n$ $(n \geq 2)$ we show that there exists a constant $C = C(\Omega)$ such that for the biharmonic Green's function $G_{\Delta^2,\Omega}$ under Dirichlet boundary conditions one has the following estimate from below:

$$G_{\Delta^2,\Omega}(x,y) \geq -C \operatorname{dist}(x,\partial\Omega)^2 \operatorname{dist}(y,\partial\Omega)^2.$$

This means that although in general, $G_{\Delta^2,\Omega}$ has a nontrivial negative part, this behaves completely regular and is in this respect not affected by the singularity of the Green's function. Qualitatively, only its positive part is affected by its singularity. See Theorem 6.24 and the subsequent remarks. Moreover, in Theorems 6.3 and 6.29 we show that positivity in the Dirichlet problem for the biharmonic operator does hold true not only in balls but also in smooth domains which are close to balls in a suitably strong sense. Although being a perturbation result it *is not* just a consequence of continuous dependence on data. The problem in proving positivity for Green's functions consists in gaining uniformity when their singularities approach the boundaries.

Finally, in Section 5.4 positivity issues for the biharmonic operator under Steklov boundary conditions are addressed. With respect to positivity it may be considered, at least in some cases, to be intermediate between Dirichlet conditions on the one hand and Navier boundary conditions on the other hand, see Theorems 5.26 and 5.27.

1.3 The First Eigenvalue

It is well-known that for general second order elliptic Dirichlet problems the eigenfunction φ_1 that corresponds to the first eigenvalue is of one sign. In case of the Laplacian such a result can be proven directly sticking to the variational characterisation of the first eigenvalue

$$\Lambda_{1,1} := \min_{v \in H_0^1 \setminus \{0\}} \frac{\int |\nabla v|^2\,dx}{\int |v|^2\,dx} = \frac{\int |\nabla \varphi_1|^2\,dx}{\int |\varphi_1|^2\,dx}$$

by comparing $|\varphi_1|$ with φ_1. For quite general and even non-selfadjoint second order Dirichlet problems the same result is proven by using more abstract results such as the Kreĭn-Rutman theorem. The first approach uses the truncation method and so, a version of the maximum principle, while the Kreĭn-Rutman theorem requires the presence of a comparison principle. A simple alternative is provided by the dual cone method of Moreau [312]. This approach, which is explained in Section 3.1.2, is on one hand restricted to a symmetric setting in a Hilbert space but on the other hand, can also be applied in semilinear problems.

Considering $\Omega \mapsto \Lambda_{1,1}(\Omega)$ in dependence of the domains Ω being subject to having all the same volume as the unit ball $B \subset \mathbb{R}^n$ one may wonder whether this map is minimised for $\Omega = B$. Indeed, this was proved by Faber-Krahn [162,254,255] and, moreover, balls of radius 1 are the only minimisers.

1.3.1 The Dirichlet Eigenvalue Problem

Whenever the biharmonic operator under Dirichlet boundary conditions has a strictly positive Green's function, the first eigenvalue $\Lambda_{2,1}$ is simple and the corresponding first eigenfunction is of fixed sign, see Section 3.1.3. Related to the first eigenvalue is a question posed by Lord Rayleigh in 1894 in his celebrated monograph [351]. He studied the vibration of (planar) plates and conjectured that among domains of given area, *when the edges are clamped, the form of gravest pitch is doubtless the circle*, see [351, p. 382]. This corresponds to saying that

$$\Lambda_{2,1}(B) \leq \Lambda_{2,1}(\Omega) \qquad \text{whenever } |\Omega| = \pi \tag{1.21}$$

for planar domains ($n = 2$). Szegö [389] assumed that in any domain the first eigenfunction for the clamped plate has always a fixed sign and proved that this hypothesis would imply the isoperimetric inequality (1.21). The assumption that the first eigenfunction is of fixed sign, however, is not true as Duffin pointed out. In [152], where he explains some counterexamples, he referred to this assumption as Szegö's conjecture on the clamped plate. Details of these counterexamples can be found in [153–155].

Subsequently, concerning Rayleigh's conjecture, Mohr [311] showed in 1975 that if among all domains of given area there exists a smooth minimiser for $\Lambda_{2,1}$ then the domain is a disk. However, he left open the question of existence. In 1981, Talenti [393] extended Szegö's result in two directions. He showed that the statement remains true under the weaker assumption that the nodal set of the first eigenfunction φ_1 of (3.1) is empty or is included in $\{x \in \Omega; \nabla \varphi_1 = 0\}$. This result holds in any space dimension $n \geq 2$. Moreover, for general domains, instead of (1.21) he showed that

$$C_n \Lambda_{2,1}(B) \leq \Lambda_{2,1}(\Omega) \qquad \text{whenever } |\Omega| = e_n$$

where $0.5 < C_n < 1$ is a constant depending on the dimension n. These constants were increased by Ashbaugh-Laugesen [24] who also showed that $C_n \to 1$ as $n \to \infty$.

A complete proof of Rayleigh's conjecture was finally obtained one century later than the conjecture itself in a celebrated paper by Nadirashvili [316]. This result was immediately extended by Ashbaugh-Benguria [22] to the case of domains in $\mathbb{R}^3$.

More results about the positivity of the first eigenfunction in general domains and a proof of Rayleigh's conjecture can be found in Chapter 3.

1.3.2 An Eigenvalue Problem for a Buckled Plate

In 1910, Th. von Kármán [404] described the large deflections and stresses produced in a thin elastic plate subject to compressive forces along its edge by means of a system of two fourth order elliptic quasilinear equations. For a derivation of this model from three dimensional elasticity one may also see [174] and references therein. An interesting phenomenon associated with this nonlinear model is the appearance of "buckling", namely the plate may deflect out of its plane when these forces reach a certain magnitude. We also refer to more recent work in [48, 101].

The linearisation of the von Kármán equations for an elastic plate over planar domains $\Omega \subset \mathbb{R}^2$ under pressure leads to the following eigenvalue problem

$$\begin{cases} \Delta^2 u = -\mu \Delta u & \text{in } \Omega, \\ u = \Delta u - (1-\sigma)\kappa u_\nu = 0 & \text{on } \partial\Omega. \end{cases} \tag{1.22}$$

Miersemann [302] studied this eigenvalue problem and he was one of the first to apply the dual cone setting of Moreau [312] to a fourth order boundary value problem. He could show that on convex $C^{2,\gamma}$-domains the first eigenvalue for (1.22) is simple and that the corresponding eigenfunction is of fixed sign. The setting introduced by Moreau will be also most convenient for a number of nonlinear problems as we shall outline in Chapters 3 and 7, see in particular Sections 7.2.3 and 7.3.

We also consider the Dirichlet eigenvalue problem

$$\begin{cases} \Delta^2 u = -\mu \Delta u & \text{in } \Omega, \\ u = u_\nu = 0 & \text{on } \partial\Omega, \end{cases} \tag{1.23}$$

related to (1.22) and where the least eigenvalue $\mu_1(\Omega)$ represents the buckling load of a clamped plate. Inspired by Rayleigh's conjecture (1.21), Pólya-Szegö [344, Note F] conjectured that

$$\mu_1(B) \leq \mu_1(\Omega) \qquad \text{whenever } |\Omega| = \pi \tag{1.24}$$

for any bounded planar domain $\Omega \subset \mathbb{R}^2$. And again, using rearrangement techniques they proved (1.24) under the assumption that the solution u to (1.23) is positive, see [344, 389]. Unfortunately, as for the clamped plate eigenvalue, this property fails in general, for instance in the square $(0,1)^2$, see Wieners [413]. Without imposing this sign assumption on the first eigenfunction, Ashbaugh-Laugesen [24] proved the bound $\gamma\mu_1(B) \leq \mu_1(\Omega)$ whenever $|\Omega| = \pi$ for $\gamma = 0.78\ldots$ which is, of course, much weaker than (1.24).

A complete proof of (1.24) is not yet known. A quite well established strategy which could be used to prove (1.24) involves shape derivatives, see e.g. [229]. It mainly consists in three steps.

1. In a suitable class of domains, prove the existence of a minimiser Ω_o for the map $\Omega \mapsto \mu_1(\Omega)$.
2. Prove that $\partial\Omega_o$ is smooth, for instance $\partial\Omega_o \in C^{2,\gamma}$, in order to be able to compute the derivative of $\Omega \mapsto \mu_1(\Omega)$ and to impose that it vanishes when $\Omega = \Omega_o$.
3. Exploit the just obtained stationarity condition, which usually gives an overdetermined condition on $\partial\Omega_o$, to prove that Ω_o is a ball.

In Section 3.2 we show how Item 1 has been successfully settled by Ashbaugh-Bucur [23] and how Item 3 has been achieved by Weinberger-Willms [416], see also [245, Proposition 4.4]. Therefore, for a complete proof of (1.24), "only" Item 2 is missing!

1.3.3 A Steklov Eigenvalue Problem

Usually, eigenvalue problems arise when one studies oscillation modes in the respective time dependent problem in order to have a physically well motivated theory and representation of solutions.

However, in what follows, a most natural motivation for considering a further eigenvalue problem comes from a seemingly quite different mathematical question. We explain how L^2-estimates for the Dirichlet problem for harmonic functions link with the Steklov eigenvalue problem for biharmonic functions.

Let $\Omega \subset \mathbb{R}^n$ be a bounded smooth domain and consider the problem

$$\begin{cases} \Delta u = 0 & \text{in } \Omega, \\ u = g & \text{on } \partial\Omega, \end{cases} \tag{1.25}$$

where $g \in L^2(\partial\Omega)$. It is well-known that (1.25) admits a unique solution $u \in H^{1/2}(\Omega) \subset L^2(\Omega)$, see e.g. [276, Remarque 7.2, p. 202] and also [238, 239] for an extension to nonsmooth domains. One is then interested in a priori estimates, namely in determining the sharp constant C_Ω such that

$$\|u\|_{L^2(\Omega)} \leq C_\Omega \|g\|_{L^2(\partial\Omega)}.$$

By Fichera's principle of duality [170] (see also Section 3.3.2) one sees that C_Ω coincides with the inverse of the first Steklov eigenvalue $\delta_1 = \delta_1(\Omega)$, namely the smallest constant a such that the problem

$$\begin{cases} \Delta^2 u = 0 & \text{in } \Omega, \\ u = \Delta u - au_\nu = 0 & \text{on } \partial\Omega, \end{cases} \tag{1.26}$$

admits a nontrivial solution. Notice that the "true" eigenvalue problem for the hinged plate equation should include the curvature in the second boundary condition, see (1.8). The map $\Omega \mapsto \delta_1(\Omega)$ has several surprising properties which we establish in Section 3.3.2. By rescaling, one sees that $\delta_1(k\Omega) = k^{-1}\delta_1(\Omega)$ for any bounded domain Ω and any $k > 0$ so that $\delta_1(k\Omega) \to 0$ as $k \to \infty$. One is then led to seek domains which minimise δ_1 under suitable constraints, the most natural one being the *volume constraint*. Smith [374] stated that, analogously to the Faber-Krahn result [162, 254, 255], the minimiser for δ_1 should exist and be a ball, at least for planar domains. But, as noticed by Kuttler and Sigillito, the argument in [374] contains a gap. In the "Note added in proof" in [375, p. 111], Smith writes:

Although the result is probably true, a correct proof has not yet been found.

A few years later, Kuttler [259] proved that a (planar) square has a first Steklov eigenvalue $\delta_1(\Omega)$ which is strictly smaller than the one of the disk having the same measure. The estimate by Kuttler was subsequently improved in [165]. Therefore, it is *not true* that $\delta_1(\Omega^*) \leq \delta_1(\Omega)$ where Ω^* denotes the spherical rearrangement of Ω. For this reason, Kuttler [259] suggested a different minimisation problem with a *perimeter constraint*; in [259, Formula (11)] he conjectures that a planar disk minimises δ_1 among all domains having fixed perimeter. He provides numerical evidence that on rectangles his conjecture seems true, see also [260, 262]. In Theorem 3.24 we show that also this conjecture is false and that an optimal shape for δ_1 does not exist under a perimeter constraint in any space dimension $n \geq 2$. In fact, under such a constraint, the infimum of δ_1 is zero.

The spectrum of (1.26) has a nice application in functional analysis. In Section 3.3.1 we show that the closure of the space spanned by the Steklov eigenfunctions is the orthogonal complement of $H_0^2(\Omega)$ in $H^2 \cap H_0^1(\Omega)$.

1.4 Paradoxes for the Hinged Plate

The most common domains for plate problems that appear in engineering are polygonal ones. On the straight boundary parts of a polygonal domain the hinged boundary conditions (1.10) lead to Navier boundary conditions (1.12). Without taking care of a possible singularity due to "$\kappa = \infty$" in the corners it would mean that the solution no longer depends on the Poisson ratio σ. Sapondžyan [358] noticed that the solution one obtains by solving (1.12) iteratively does not necessarily have a bounded energy. Babuška noticed in [28] that the difference between (1.10) and (1.12) would mean that by approximating a curvilinear domain by polygons, as is done in most finite elements methods, the approximating solutions would not converge to the solution on the curvilinear domain.

Although both paradoxes are usually referred to by the name Babuška, they do cover different phenomena as we will explain in more detail.

1.4.1 Sapondžyan's Paradox by Concave Corners

One might expect that the problem that appeared in these paradoxes is due to a boundary condition not being well-defined in corners. Indeed, the curvature that appears in the boundary condition is singular and apparently leads to a δ-distribution type contribution. By adding appropriate extra terms in the corners there is some hope to find the real solution. The situation for reentrant corners can be 'worse'. Due to Kondratiev [65, 252], Maz'ya et al. [289, 290], Grisvard [200] and many others, it is well-known that corners may lead to a loss of regularity. It is less known that a corner may lead to multiple solutions, that is, the solution depends crucially on the space that one chooses.

An example where two different solutions appear naturally from two straightforward settings goes as follows. Both fourth order boundary value problems, hinged or Steklov (1.10) as well as Navier (1.12) boundary conditions, allow a reformulation as a coupled system, see (1.14) and (1.13), respectively. In the latter case, one tends to solve by an iteration of the Green operator for the second order Poisson problem. This approach works fine for bounded smooth domains, but whenever the domain has a nonconvex corner, one does not necessarily get the solution one is looking for. Indeed, for the fourth order problem the natural setting for a weak solution to the Navier boundary value problem would be $H^2 \cap H_0^1(\Omega)$. The second Navier boundary condition $\Delta u = 0$ would follow naturally on smooth boundary parts from the weak formulation where u satisfies

$$\int_\Omega (\Delta u \Delta \varphi - f\varphi)\, dx = 0 \text{ for all } \varphi \in H^2 \cap H_0^1(\Omega). \tag{1.27}$$

However, for the system in (1.13) the natural setting is that one looks for function pairs $(u, v) \in H_0^1(\Omega) \times H_0^1(\Omega)$. In [321] it is shown that for domains with a reentrant corner both problems have a unique solution but the solutions u_1 to (1.13) and u_2 to (1.27) are different. Indeed, there exist a constant c_f and a nontrivial biharmonic function b that satisfies (1.13) with zero Navier boundary condition except in the corner such that $u_1 = u_2 + c_f b$. The related problem for domains with edges is considered in [320]. We refer to Section 2.7 for more details and an explicit example.

1.4.2 The Babuška Paradox

In the original Babuška or polygon-circle paradox one considers problem (1.10) for $f = 1$ and when $\Omega = P_m \subset B$ $(m \geq 3)$ is the interior of the regular polygon with corners $e^{2k\pi i/m}$ for $k \in \mathbb{N}$, namely

$$\begin{cases} \Delta^2 u = 1 & \text{in } P_m, \\ u = \Delta u = 0 & \text{on } \partial P_m. \end{cases}$$

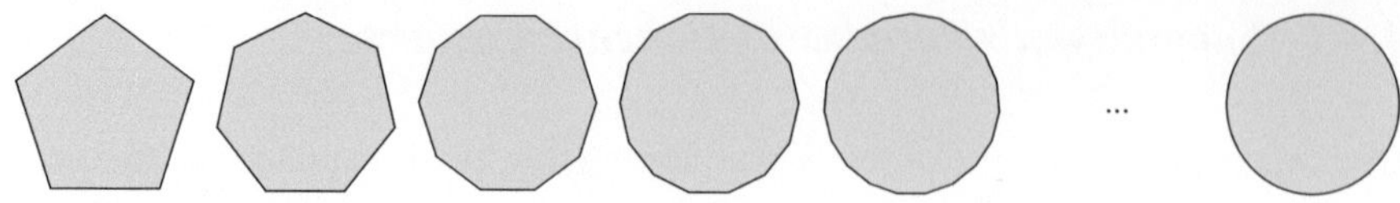

Fig. 1.3 The Babuška or polygon-circle paradox. On polygonal domains (1.10) = (1.12); on curvilinear domains (1.10) ≠ (1.12). Approximating curvilinear domains by polygonal ones does not give the correct limit solution to the hinged plate problem.

If u_m denotes the solution of this problem extended by 0 in $B \setminus P_m$, it can be shown that the sequence (u_m) converges uniformly to

$$u_\infty(x) := \frac{3}{64} - \frac{1}{16} |x|^2 + \frac{1}{64} |x|^4$$

which *is not* the solution to the "limit problem" (where $\kappa = 1$), namely

$$\begin{cases} \Delta^2 u = 1 & \text{in } B, \\ u = \Delta u - (1 - \sigma) \kappa \frac{\partial u}{\partial \nu} = 0 & \text{on } \partial B \end{cases}$$

unless $\sigma = 1$, see Figure 1.3.

For more details on this Babuška paradox see Section 2.7.

1.5 Paneitz-Branson Type Equations

Let $(\mathcal{M}, g)$ be an n–dimensional Riemannian manifold with $n > 4$. The conformal Laplacian is frequently studied and well understood and one may be interested in higher order analogues. Again, the biharmonic case is particularly interesting. The metric g is subject to a conformal change $g_u := u^{\frac{4}{n-4}} g$, $u > 0$, and one wonders about the existence of a fourth order differential operator enjoying a conformal covariance property such that for all $\varphi \in C^\infty(\mathcal{M})$ one has

$$(P_4^n)_u(\varphi) = u^{-\frac{n+4}{n-4}} (P_4^n)(u\varphi).$$

Here, P_4^n denotes the desired operator with respect to the background metric g, while $(P_4^n)_u$ refers to the conformal metric g_u. Indeed, Paneitz [330, 331] and Branson [66, 67] found the following conformal covariant fourth order elliptic operator

$$P_4^n := \Delta^2 - \sum_{i,j=1}^n \nabla^i \left(\frac{(n-2)^2 + 4}{2(n-1)(n-2)} R g_{ij} - \frac{4}{n-2} R_{ij} \right) \nabla^j + \frac{n-4}{2} Q_4^n$$

on $\mathcal{M}$, where $\Delta = \frac{1}{\sqrt{g}}\partial_i\left(\sqrt{g}g^{ij}\partial_j\right)$ denotes the Laplace-Beltrami operator with respect to g in local coordinates, R_{ij} the Ricci-tensor and R the scalar curvature. Moreover, $\nabla^j\varphi = \sum_{k=1}^n g^{jk}\partial_k\varphi$ gives the gradient of a function and

$$\sum_{i=1}^n \nabla^i Z_i = \sum_{i,j=1}^n \frac{1}{\sqrt{g}}\partial_i\left(\sqrt{g}g^{ij}Z_j\right)$$

the divergence of a covector field. A key role is played by the following fourth order curvature invariant

$$Q_4^n := -\frac{2}{(n-2)^2}|(R_{ij})|^2 + \frac{n^3 - 4n^2 + 16n - 16}{8(n-1)^2(n-2)^2}R^2 - \frac{1}{2(n-1)}\Delta R,$$

the so-called Q-curvature. Here $|(R_{ij})|^2 = \sum_{i,j,k,\ell}^n g^{ij}g^{k\ell}R_{ik}R_{j\ell}$. The transformation of the corresponding Q_4^n-curvature under this conformal change of metrics is governed by the Paneitz equation

$$P_4^n u = \frac{n-4}{2}(Q_4^n)_u u^{\frac{n+4}{n-4}}. \tag{1.28}$$

In analogy to the second order Yamabe problem (for an overview see [382, Section III.4]), obvious questions here concern the existence of conformal metrics with constant or prescribed Q-curvature. Huge work has so far been done by research groups around Chang-Yang-Gursky and Hebey, as well as many others. For a survey and references see the books by Chang [89] and by Druet-Hebey-Robert [149]. Difficult problems arise from ensuring the positivity requirement of the conformal factor $u > 0$ and from the necessity to know about the kernel of the Paneitz operator. These problems have only been solved partly yet.

In order to explain the geometrical importance of the Q-curvature, we assume now for a moment that the manifold $(\mathcal{M}, g)$ is four-dimensional. Then, the Paneitz operator is defined by

$$P_4^4 := \Delta^2 - \sum_{i,j=1}^4 \nabla^i\left(\frac{2}{3}Rg_{ij} - 2R_{ij}\right)\nabla^j$$

in such a way that under the conformal change of metrics $g_u = e^{2u}g$ one has

$$(P_4^4)_u(\varphi) = e^{-4u}P_4^4(\varphi).$$

In order to achieve a prescribed Q-curvature on the four-dimensional manifold $(\mathcal{M}, g_u)$, one has to find u solving

$$P_4^4 u + 2Q_4^4 = 2Qe^{4u},$$

where Q_4^4 is the curvature invariant

$$12Q_4^4 = -\Delta R + R^2 - 3|(R_{ij})|^2.$$

In this situation, one has the following Gauss-Bonnet-formula

$$\int_{\mathcal{M}} \left(Q + \frac{1}{8}|W|^2 \right) dS = 4\pi^2 \chi(\mathcal{M}),$$

where W is the Weyl tensor and $\chi(\mathcal{M})$ is the Euler characteristic. Since $\chi(\mathcal{M})$ is a topological and $|W|^2 dS$ is a pointwise conformal invariant, this shows that $\int_{\mathcal{M}} Q\, dS$ is a conformal invariant, which governs e.g. the existence of conformal Ricci positive metrics (see e.g. Chang-Gursky-Yang [90, 91]) and eigenvalue estimates for Dirac operators (see Guofang Wang [408]). All these facts show that the Q-curvature in the context of fourth order conformally covariant operators takes a role quite analogous to the scalar curvature with respect to second order operators.

Getting back to the general case $n > 4$, let us outline what we are going to prove in the present book. We do not aim at giving an overview – not even of parts – of the theory of Paneitz operators but at giving a spot on some aspects of this issue. Namely, in Section 7.9 we address the question whether in specific bounded smooth domains $\Omega \subset \mathbb{R}^n$ $(n > 4)$ there exists a metric $g_u = u^{4/(n-4)}(\delta_{ij})$ being conformal to the flat euclidean metric and subject to certain homogeneous boundary conditions such that it has strictly positive constant Q-curvature. In view of the nonexistence results in Section 7.5.1 one expects that for generic domains the corresponding boundary value problems do not have a positive solution. Hence, in geometrically or topologically simple domains, such a conformal metric does in general not exist. Nevertheless, the boundary value problems have nontrivial solutions in topologically or specific geometrically complicated domains (see Section 7.9). For the Navier problem, i.e. $u = \Delta u = 0$ on $\partial\Omega$, one can also show positivity of u so that it may be considered as a conformal factor and one has such a nontrivial conformal metric as described above. Under Dirichlet boundary conditions, which could be interpreted as vanishing of length and normal curvature of the conformal metric on $\partial\Omega$, the positivity question has so far to be left open. The same difficulty prevents Esposito and Robert [161] from solving the Q-curvature analogue of the Yamabe problem.

In Section 7.10 the starting point is the hyperbolic ball $B = B_1(0) \subset \mathbb{R}^n$ which is equipped with the Poincaré metric $g_{ij} = 4\delta_{ij}/(1 - |x|^2)^2$. This metric has constant Q-curvature $Q \equiv \frac{1}{8}n(n^2 - 4)$ and we address the question, whether there are further conformal metrics $g_u = u^{4/(n-4)}g$ having the same constant Q-curvature such that the resulting manifold is complete. Somehow surprisingly there exists infinitely many such metrics and even infinitely many among them have negative scalar curvature. This high degree of nonuniqueness is in sharp contrast with the corresponding question for the scalar curvature. There is no further conformal complete metric having the same constant negative curvature as g, see [280].

1.6 Critical Growth Polyharmonic Model Problems

The prototype to be studied is the semilinear polyharmonic eigenvalue problem

$$\begin{cases} (-\Delta)^m u = \lambda u + |u|^{s-1} u, & u \not\equiv 0 \text{ in } \Omega, \\ D^\alpha u|_{\partial\Omega} = 0 & \text{for } |\alpha| \leq m-1. \end{cases} \tag{1.29}$$

Here $\Omega \subset \mathbb{R}^n$ is a bounded smooth domain, $n > 2m$, $\lambda \in \mathbb{R}$; $s = (n+2m)/(n-2m)$ is the critical Sobolev exponent. If $m = 2$ and $\lambda = 0$ we are back in the situation discussed in the previous section with a euclidean background metric. The existence theory for (1.29) can be developed similarly to the second order case $m = 1$ while it becomes immediately quite difficult or even impossible to prove positivity or nonexistence of certain solutions. In particular, thanks to a Pohožaev identity [340, 341] one can exclude the existence of solutions to (1.29) in starshaped domains whenever $\lambda < 0$ but as far as the limit case $\lambda = 0$ is considered, things change dramatically in the two situations where $m = 1$ and $m \geq 2$. With a suitable application of the unique continuation principle (see e.g. [248, 346]), one can exclude when $m = 1$ the existence of any solution to (1.29) in starshaped domains even for $\lambda = 0$. In order to apply the same principle to (1.29) when $m \geq 2$ one would need to know the boundary behaviour of more derivatives than those already included in the Dirichlet boundary conditions and provided by the Pohožaev identity. Therefore, when $m \geq 2$ one can try to prove nonexistence of *positive* solutions in *strictly* starshaped domains, see Theorem 7.33 for the case $m = 2$. Unfortunately, this result is not satisfactory since positivity is not ensured in general domains, see also the discussion in the next section. So far, only in balls a more satisfactory discussion can be given. We refer to Section 7.5.1 for an up-to-date state of the art.

A first natural question is then to find out whether the nonexistence result for $\lambda = 0$ really depends on the geometry of the domain and starshapedness is not just a technical assumption. The answer is positive. For instance, problem (1.29) with $m = 2$ and $\lambda = 0$ admits a solution in domains with small holes and in some contractible non-starshaped domains, see Section 7.9. A second natural question then arises. Do the nonexistence results also depend on the boundary conditions considered? It is known that (1.29) admits no *positive* solution if $m = 2$, $\lambda = 0$, Ω is starshaped and *Navier* boundary conditions are considered, see [308, 399] and also Section 7.6. Moreover, in Section 7.7 we address the same problem under Steklov boundary conditions when $m = 2$ and Ω is a ball. We find all the values of the boundary parameter a in (1.26) for which the critical growth equation in (1.29) admits a positive solution.

Problem (1.29) in the case $m = 1$ has been studied extensively by Brezis-Nirenberg [72] who also discovered an interesting phenomenon when Ω is the unit ball. There exists a positive radial solution to (1.29) for every $\lambda \in (0, \Lambda_{1,1})$ if $n \geq 4$ and for every $\lambda \in (\frac{1}{4}\Lambda_{1,1}, \Lambda_{1,1})$ if $n = 3$. Moreover, they could show that in the latter case problem (1.29) has no nontrivial radial solution if $\lambda \leq \frac{1}{4}\Lambda_{1,1}$. Here and in the sequel $\Lambda_{m,1}$ denotes the first eigenvalue of $(-\Delta)^m$ in B under homogeneous Dirichlet boundary conditions.

Pucci and Serrin [349] raised the question in which way this critical behaviour of certain dimensions depends on the order $2m$ of the semilinear polyharmonic eigenvalue problem (1.29). They introduced the name critical dimensions.

Definition 1.1. Let $\Omega \subset \mathbb{R}^n$ be a ball. The dimension n is called *critical* if there is a positive bound $\Lambda > 0$ such that a necessary condition for the existence of a nontrivial radial solution to (1.29) is $\lambda > \Lambda$.

Pucci and Serrin [349] showed that for any m the dimension $n = 2m + 1$ is critical and, moreover, that $n = 5, 6, 7$ are critical in the fourth order problem, $m = 2$. They suggested

Conjecture 1.2 (Pucci–Serrin).
The critical dimensions are precisely $n = 2m + 1, \ldots, 4m - 1$.

In Section 7.5.2 we prove a weakened version of this conjecture. This nonexistence phenomenon has a functional analytic interpretation, which is reflected in the possibility of adding L^2–remainder terms in Sobolev inequalities with critical exponent and optimal constants in any bounded domain Ω, see Section 7.8.

1.7 Qualitative Properties of Solutions to Semilinear Problems

Radial symmetry of positive solutions to suitable semilinear higher order Dirichlet problems in the ball is obtained thanks to a suitable implementation of the moving planes procedure, see Section 7.1.2. One of the crucial steps in the moving planes procedure consists in comparing the solution u in a segment of the ball with its reflection u^r across the hyperplane which bounds the segment, see e.g. [195, Lemma 2.2]. For second order problems the comparison follows from suitable versions of the maximum principle since $u^r \geq u$ holds a priori on the boundary of this segment. This information however is not enough for higher order problems, and therefore the classical moving planes method fails. We employ a different technique to carry out the moving planes mechanism, using the integral representation of u in terms of the Green function of the polyharmonic operator $(-\Delta)^m$ in B under Dirichlet boundary conditions.

As repeatedly emphasised, linear higher order boundary value problems in general do not enjoy a positivity preserving property. This feature may also be observed in nonlinear problems. Let us illustrate this situation for the subcritical model problem corresponding to (1.29), namely

$$\begin{cases} (-\Delta)^m u = \lambda u + |u|^{p-1} u, & u \not\equiv 0 \text{ in } \Omega, \\ D^\alpha u|_{\partial\Omega} = 0 & \text{for } |\alpha| \leq m - 1, \end{cases} \tag{1.30}$$

where $1 < p < s$. Thanks to some compactness, which is not available for (1.29), one may find a nontrivial solution to (1.30) as a suitable constrained minimum provided that $\lambda < \Lambda_{m,1}$. If $m = 1$ one can easily prove that such a minimum is positive just by replacing it with its modulus and by applying the maximum principle. This procedure fails in general if $m \geq 2$, even if Ω is a ball. This problem is discussed in detail in Section 7.2.

Bifurcation branches of solutions to nonlinear problems depending on some parameter λ are often quite complicated to be figured out. The case where only positive solutions are considered is much simpler. This situation is well illustrated by the so-called (second order) Gelfand problem [194, 240] where the nonlinearity is of exponential type, namely λe^u. A similar behaviour can be observed for the "approximate problem" where the nonlinearity is $\lambda (1+u)^p$. For this power-type nonlinearity, the bifurcation branch for the second order problem appears particularly interesting in the supercritical case $p > \frac{n+2}{n-2}$. In order to find out whether a similar behaviour can also be observed in higher order problems, one has to face the possible lack of positivity of the solution. As already discussed in Section 1.2 this can be overcome so far only in some particular situations, such as the case where Ω is a ball. In Section 7.11 we carefully study the branch of solutions to this biharmonic supercritical growth problem with the help of a suitable Lyapunov functional. Our study also takes advantage of the radial symmetry of positive solutions in the ball.

1.8 Willmore Surfaces

At the beginning of this chapter the modeling of thin elastic plates was explained in some detail. There, curvature expressions were somehow "linearised" in order to have a purely quadratic behaviour of the leading terms of the energy functionals. This simplification results in linear Euler-Lagrange equations, which are justified for small deviations from a horizontal equilibrium shape. As soon as large deflections occur or a coordinate system is chosen in such a way that the equilibrium shape is not the x-y-plane, one has to stick to the frame invariant modeling of the bending energy in terms of differential geometric curvature expressions. When compared with the "linearised" energy integral (1.5) in Section 1.1, the integral

$$\int_{\Gamma} \left(\alpha + \beta (\mathrm{H} - \mathrm{H}_0)^2 - \gamma \mathrm{K} \right) d\omega \tag{1.31}$$

with suitable constants $\alpha, \beta, \gamma, \mathrm{H}_0$ may serve as a more realistic model for the bending and stretching energy of a thin elastic plate, which is described by a two-dimensional manifold $\Gamma \subset \mathbb{R}^3$. Here, H denotes its mean and K its Gaussian curvature. According to [325], α is related to the surface tension, β and γ are elastic moduli, while one may think of H_0 as some preferred "intrinsic" curvature due to particular properties of the material under consideration. Physically reasonable assumptions on the coefficients are $\alpha \geq 0$, $0 \leq \gamma \leq \beta$, $\beta \gamma \mathrm{H}_0^2 \leq \alpha(\beta - \gamma)$, which ensure the functional to be positive definite. For modeling aspects and a thorough explanation of the meaning of each term we refer again to the survey article [325] by Nitsche. A discussion of the full model (1.31), however, seems to be out of reach at the moment, and for this reason one usually confines the investigation to the most important and dominant term, i.e. the contribution of H^2.

Given a smooth immersed surface Γ, the Willmore functional is defined by

$$W(\Gamma) := \int_\Gamma \mathrm{H}^2 d\omega.$$

Apart from its meaning as a model for the elastic energy of thin shells or biological membranes, it is also of great geometric interest, see e.g [414, 415]. Furthermore, it is used in image processing for problems of surface restoration and image in-painting, see e.g. [105] and references therein. In these applications one is usually concerned with minima, or more generally with critical points of the Willmore functional. It is well-known that the corresponding surface Γ has to satisfy the Willmore equation

$$\Delta_\Gamma \mathrm{H} + 2\mathrm{H}(\mathrm{H}^2 - \mathrm{K}) = 0 \qquad \text{on } \Gamma, \tag{1.32}$$

where Δ_Γ denotes the Laplace-Beltrami operator on Γ with respect to the induced metric. A solution of (1.32) is called a Willmore surface. An additional difficulty here arises from the fact that Δ_Γ depends on the unknown surface so that the equation is quasilinear. Moreover, the ellipticity is not uniform which, in the variational framework, is reflected by the fact that minimising sequences may in general be unbounded in H^2. A difficult step is to pass to suitable minimising sequences enjoying sufficient compactness in H^2 and C^1.

In the past years a lot of very interesting work has been done, mainly on closed Willmore surfaces, see e.g. [35, 60, 156, 263–265, 288, 356, 362, 372, 373]. For instance, one knows about minimisers of the Willmore energy of prescribed genus and about global existence and convergence of the Willmore flow to the sphere under explicit smallness assumptions which, by means of counterexamples, have been proved to be sharp.

The situation changes if one considers boundary value problems. Except for small data results, our knowledge is still somehow limited, see e.g. [50, 115, 116, 138, 361] and references therein.

Possible boundary value problems for the linear plate equation were discussed in Section 1.1 above to some extent. In the nonlinear context here, one could discuss the same issue, but now considering the geometric terms instead of their linearisations. For details again we refer to [325]. Here we will be concerned with a Dirichlet problem for Willmore surfaces where, in some particularly symmetric situations, results are available. These are not just small data results or application of linear theory combined with the implicit function theorem. Let us mention an important recent contribution by Schätzle [361]. He proved a general result concerning existence of branched Willmore immersions in $\mathbb{S}^n$ with boundary which satisfy Dirichlet boundary conditions. Assuming the boundary data to obey some explicit geometrically motivated smallness condition these immersions can even be shown to be embedded. By working in $\mathbb{S}^n$, some compactness problems could be overcome; on the other hand, when pulling pack these immersions to $\mathbb{R}^n$ it cannot be excluded that they contain the point ∞. Moreover, in general, the existence of branch points cannot be ruled out, and due to the generality of the approach, it seems to us that only little topological information about the solutions can be extracted from the existence proof. We think that it is quite interesting to identify situations where it is possible to work with a priori bounded minimising sequences or where solutions with additional properties like e.g. being a graph or enjoying certain symmetry properties can be found. In view of the lack of general comparison principles and of the highly nonlinear character of (1.32) this is a rather difficult task. In order to outline directions of future research we think that it is a good strategy to investigate first relatively special situations which e.g. enjoy some symmetry.

This is exactly the subject of Section 8. We restrict ourselves to surfaces of revolution satisfying Dirichlet boundary conditions. In this class we can find minimising sequences enjoying sufficient compactness properties thereby constructing a classical solution where a number of additional qualitative properties are obtained. While the underlying differential equation is one-dimensional the geometry is already two-dimensional. The interplay between mean and Gaussian curvature in (1.32) already causes great difficulties.

Chapter 2
Linear Problems

Linear polyharmonic problems and their features are essential in order to achieve the
main tasks of this monograph, namely the study of positivity and nonlinear problems. With no hope of being exhaustive, in this chapter we outline the main tools
and results, which will be needed subsequently. We start by introducing higher order
Sobolev spaces and relevant boundary conditions for polyharmonic problems. Then
using a suitable Hilbert space, we show solvability of a wide class of boundary value
problems. The subsequent part of the chapter is devoted to regularity results and a
priori estimates both in Schauder and L^p setting, including also maximum modulus
estimates. These regularity results are particularly meaningful when writing explicitly the solution of the boundary value problem in terms of the data by means of a
suitable kernel. Focusing on the Dirichlet problem for the polyharmonic operator,
we introduce Green's functions and the fundamental formula by Boggio in balls.
We conclude with a study of a biharmonic problem in nonsmooth domains explain-
ing two paradoxes which are important in particular when approximating solutions
numerically.

2.1 Polyharmonic Operators

Unless otherwise specified, throughout this monograph Ω denotes a bounded domain (open and connected) of $\mathbb{R}^n$ ($n \geq 2$). The smoothness assumptions on the
boundary $\partial\Omega$ will be made precise in each situation considered. However, we shall
always assume that $\partial\Omega$ is Lipschitzian so that the tangent hyperplane and the unit
outward normal $v = v(x)$ are well-defined for a.e. $x \in \partial\Omega$, where a.e. means here
with respect to the $(n-1)$-dimensional Hausdorff measure. When it is clear from
the context, in the sequel we omit writing "a.e."

The Laplacian Δu of a smooth function $u : \Omega \to \mathbb{R}$ is the trace of its Hessian
matrix, namely

$$\Delta u := \sum_{i=1}^{n} \frac{\partial^2 u}{\partial x_i^2} .$$

F. Gazzola et al., *Polyharmonic Boundary Value Problems*,
Lecture Notes in Mathematics 1991, DOI 10.1007/978-3-642-12245-3_2,
© Springer-Verlag Berlin Heidelberg 2010

We are interested in iterations of the Laplace operator, namely *polyharmonic operators* defined inductively by

$$\Delta^m u = \Delta(\Delta^{m-1} u) \qquad \text{for } m = 2, 3, \dots .$$

Arguing by induction on m, it is straightforward to verify that

$$\Delta^m u = \sum_{\ell_1 + \dots + \ell_n = m} \frac{m!}{\ell_1! \dots \ell_n!} \frac{\partial^{2m} u}{\partial x_1^{2\ell_1} \dots \partial x_n^{2\ell_n}} .$$

The polyharmonic operator Δ^m may also be seen in an abstract way through the polynomial $L_m : \mathbb{R}^n \to \mathbb{R}$ defined by

$$L_m(\xi) = \sum_{\ell_1 + \dots + \ell_n = m} \frac{m!}{\ell_1! \dots \ell_n!} \left(\prod_{i=1}^{n} \xi_i^{2\ell_i} \right) = |\xi|^{2m} \text{ for } \xi \in \mathbb{R}^n.$$

Formally, $\Delta^m = L_m(\nabla)$. In particular, this shows that $L_m(\xi) > 0$ for all $\xi \neq 0$ so that Δ^m is an *elliptic operator*, see [5, p. 625] or [276, p. 121]. Ellipticity is a property of the principal part (containing the highest order partial derivatives) of the differential operator.

In this chapter, we study linear differential elliptic operators of the kind

$$u \mapsto Au = (-\Delta)^m u + \mathscr{A}(x; D)u, \tag{2.1}$$

where

$$\mathscr{A} : \Omega \times \mathbb{R}^n \times \mathbb{R}^{n^2} \times \dots \times \mathbb{R}^{n^{2m-1}} \to \mathbb{R}$$

is a linear operator containing all the lower order partial derivatives of the function u. The coefficients of the derivatives are measurable functions of x in Ω. For elliptic differential operators A of the form (2.1) and under suitable assumptions on f, we shall consider solutions $u = u(x)$ to the equation

$$(-\Delta)^m u + \mathscr{A}(x; D)u = f \qquad \text{in } \Omega, \tag{2.2}$$

which satisfy some boundary conditions on $\partial\Omega$. We discuss the class of "admissible" boundary conditions in Section 2.3. What we mean by *solution* to (2.2) will be made clear in each situation considered.

Finally, let us mention that our statements also hold if we replace $(-\Delta)^m$ with the m-th power of any other second order elliptic operator L; for instance, in Section 6.1 we consider powers of

$$Lu = -\sum_{i,j=1}^{2} \tilde{a}_{ij}(x) \frac{\partial^2 u}{\partial x_i \partial x_j} \qquad \text{with the matrix } \{\tilde{a}_{ij}\} \text{ being positive definite,}$$

or

$$Lu = -\frac{2}{|\nabla h|^2} \Delta u \qquad \text{with } \nabla h \neq 0.$$

2.2 Higher Order Sobolev Spaces

Before introducing the boundary conditions to be associated to (2.2), we briefly recall the definition and basic properties of higher order Sobolev spaces and of their embedding into L^q spaces. In particular, we need to define the traces in order to give some meaning to the boundary conditions. We restrict our attention to those statements which will be frequently used in this book. Except in this section, Ω is assumed to be bounded throughout the whole Chapter 2.

2.2.1 Definitions and Basic Properties

Given a domain $\Omega \subset \mathbb{R}^n$, $\|.\|_{L^p}$ denotes the standard $L^p(\Omega)$-norm for $1 \le p \le \infty$. For all $m \in \mathbb{N}^+$ let us define the norm

$$u \mapsto \mathbf{N}(u) := \left(\sum_{k=0}^{m} \|D^k u\|_{L^p}^p \right)^{1/p}, \tag{2.3}$$

where $D^0 u = u$,

$$D^k u \cdot D^k v = \sum_{i_1,\dots,i_k=1}^{n} \frac{\partial^k u}{\partial x_{i_1} \dots \partial x_{i_k}} \frac{\partial^k v}{\partial x_{i_1} \dots \partial x_{i_k}} \quad \text{and} \quad |D^k u| = \left(D^k u \cdot D^k u \right)^{1/2}.$$

Note that we will specify the domain Ω in $\|.\|_{L^p}$ only when it is not clear from the context. Next, we define the space

$$W^{m,p}(\Omega) := \overline{\{u \in C^m(\Omega); \mathbf{N}(u) < \infty\}}^{\mathbf{N}},$$

that is, the completion with respect to the norm (2.3). Alternatively, $W^{m,p}(\Omega)$ may be defined as the subspace of $L^p(\Omega)$ of functions having generalised derivatives up to order m in $L^p(\Omega)$, see [301].

If $\Omega \neq \mathbb{R}^n$ and its boundary $\partial\Omega$ is smooth, then a function $u \in W^{m,p}(\Omega)$ admits some traces on $\partial\Omega$ where, for our purposes, it is enough to restrict the attention to the case $p \in (1,\infty)$. More precisely, if ν denotes the unit outer normal to $\partial\Omega$, then for any $u \in C^m(\overline{\Omega})$ and any $j = 0,\dots,m$ we define the traces

$$\gamma_j u := \frac{\partial^j u}{\partial \nu^j}\bigg|_{\partial\Omega}. \tag{2.4}$$

By [276, Théorème 8.3], these linear operators may be extended continuously to the larger space $W^{m,p}(\Omega)$. We set

$$W^{m-j-1/p,p}(\partial\Omega) := \gamma_j[W^{m,p}(\Omega)] \quad \text{for } j = 0,\dots,m-1. \tag{2.5}$$

In particular, $W^{1/p',p}(\partial\Omega) = \gamma_{m-1}[W^{m,p}(\Omega)]$, where p' is the conjugate of p (that is, $p + p' = pp'$). We also put

$$\gamma_m[W^{m,p}(\Omega)] = W^{-1/p,p}(\partial\Omega) := [W^{1/p,p'}(\partial\Omega)]'$$
$$= \text{the dual space of } W^{1/p,p'}(\partial\Omega), \qquad (2.6)$$

so that (2.5) makes sense for all $j = 0, \ldots, m$. With an abuse of notation, in the sequel we simply write u (respectively $\frac{\partial^j u}{\partial \nu^j}$) instead of $\gamma_0 u$ (respectively $\gamma_j u$ for $j = 1, \ldots, m$).

When $p = 2$, we put $H^m(\Omega) := W^{m,2}(\Omega)$. Moreover, when $p = 2$ and $m \geq 1$ we write $H^{m-1/2}(\partial\Omega) = W^{m-1/2,2}(\partial\Omega)$ and

$$H^{-m+\frac{1}{2}}(\partial\Omega) = [H^{m-\frac{1}{2}}(\partial\Omega)]' = \text{the dual space of } H^{m-\frac{1}{2}}(\partial\Omega). \qquad (2.7)$$

The space $H^m(\Omega)$ becomes a Hilbert space when endowed with the scalar product

$$(u,v) \mapsto \sum_{k=0}^{m} \int_{\Omega} D^k u \cdot D^k v\, dx \qquad \text{for all } u, v \in H^m(\Omega).$$

In some cases one may simplify the just defined norms and scalar products. As a first step, we mention that thanks to interpolation theory, see [1, Theorem 4.14], one can neglect intermediate derivatives in (2.3). More precisely, $W^{m,p}(\Omega)$ is a Banach space also when endowed with the following norm, which is equivalent to (2.3):

$$\|u\|_{W^{m,p}} = \left(\|u\|_{L^p}^p + \|D^m u\|_{L^p}^p\right)^{1/p} \qquad \text{for all } u \in W^{m,p}(\Omega), \qquad (2.8)$$

whereas $H^m(\Omega)$ is a Hilbert space also with the scalar product

$$(u,v)_{H^m} := \int_{\Omega} (uv + D^m u \cdot D^m v)\, dx \qquad \text{for all } u, v \in H^m(\Omega).$$

Of particular interest is the closed subspace of $W^{m,p}$ defined as the intersection of the kernels of the trace operators in (2.4), that is for any bounded domain Ω we consider

$$W_0^{m,p}(\Omega) := \bigcap_{j=0}^{m-1} \ker \gamma_j.$$

Moreover, for bounded domains Ω and for $1 < p < \infty$, if p' is the conjugate of p we write

$$W^{-m,p'}(\Omega) := [W_0^{m,p}(\Omega)]' = \text{the dual space of } W_0^{m,p}(\Omega) \qquad (2.9)$$

and, for $p = 2$,

$$H^{-m}(\Omega) := [H_0^m(\Omega)]' = [W_0^{m,2}(\Omega)]'.$$

Consider the bilinear form

$$(u,v)_{H_0^m} := \begin{cases} \displaystyle\int_\Omega \Delta^k u\, \Delta^k v\, dx & \text{if } m = 2k, \\[2mm] \displaystyle\int_\Omega \nabla(\Delta^k u) \cdot \nabla(\Delta^k v)\, dx & \text{if } m = 2k+1, \end{cases} \tag{2.10}$$

and the corresponding norm

$$\|u\|_{H_0^m} := \begin{cases} \|\Delta^k u\|_{L^2} & \text{if } m = 2k, \\[2mm] \|\nabla(\Delta^k u)\|_{L^2} & \text{if } m = 2k+1. \end{cases} \tag{2.11}$$

For general $p \in (1, \infty)$, one has the choice of taking the L^p-version of (2.11) or the equivalent norm

$$\|u\|_{W_0^{m,p}} := \|D^m u\|_{L^p}.$$

Thanks to these norms, one may define the above spaces in a different way.

Theorem 2.1. *If $\Omega \subset \mathbb{R}^n$ is a bounded domain, then*

$$\begin{aligned} W_0^{m,p}(\Omega) &= \text{ the closure of } C_c^\infty(\Omega) \text{ with respect to the norm } \|\,.\,\|_{W^{m,p}} \\ &= \text{ the closure of } C_c^\infty(\Omega) \text{ with respect to the norm } \|\,.\,\|_{W_0^{m,p}}. \end{aligned}$$

Theorem 2.1 follows by combining interpolation inequalities (see [1, Theorem 4.14]) with the classical Poincaré inequality $\|\nabla u\|_{L^p} \geq c\|u\|_{L^p}$ for all $u \in W_0^{1,p}(\Omega)$. If Ω is unbounded, including the case where $\Omega = \mathbb{R}^n$, we define

$$\|u\|_{\mathscr{D}^{m,p}(\Omega)} := \|D^m u\|_{L^p(\Omega)},$$

$$\mathscr{D}^{m,p}(\Omega) := \text{ the closure of } C_c^\infty(\Omega) \text{ with respect to the norm } \|\,.\,\|_{\mathscr{D}^{m,p}},$$

and, again, let $W_0^{m,p}(\Omega)$ denote the closure of $C_c^\infty(\Omega)$ with respect to the norm $\|\,.\,\|_{W^{m,p}}$. In this unbounded case, a similar result as in Theorem 2.1 is no longer true since although $W_0^{m,p}(\Omega) \subset \mathscr{D}^{m,p}(\Omega)$, the converse inclusion fails. For instance, if $\Omega = \mathbb{R}^n$, then $W_0^{m,p}(\mathbb{R}^n) = W^{m,p}(\mathbb{R}^n)$, whereas the function $u(x) = (1+|x|^2)^{(1-n)/4}$ belongs to $\mathscr{D}^{1,2}(\mathbb{R}^n)$ but not to $H_0^1(\mathbb{R}^n) = H^1(\mathbb{R}^n)$.

Theorem 2.1 states that, when Ω is bounded, the space $H_0^m(\Omega)$ is a Hilbert space when endowed with the scalar product (2.10). The striking fact is that not only all lower order derivatives (including the derivative of order 0!) are neglected but also that some of the highest order derivatives are dropped. This fact has a simple explanation since

$$(u,v)_{H_0^m} = \int_\Omega D^m u \cdot D^m v\, dx \qquad \text{for all } u, v \in H_0^m(\Omega). \tag{2.12}$$

One can verify (2.12) by using a density argument, namely for all $u, v \in C_c^\infty(\Omega)$. And with this restriction, one can integrate by parts several times in order to obtain (2.12). The bilinear form (2.10) also defines a scalar product on the space $\mathscr{D}^{m,2}(\Omega)$ whenever Ω is an unbounded domain. We summarise all these facts in

Theorem 2.2. *Let $\Omega \subset \mathbb{R}^n$ be a smooth domain. Then the bilinear form*

$$(u, v) \mapsto \begin{cases} \displaystyle\int_\Omega \Delta^k u \, \Delta^k v \, dx & \text{if } m = 2k, \\ \displaystyle\int_\Omega \nabla(\Delta^k u) \cdot \nabla(\Delta^k v) \, dx & \text{if } m = 2k+1, \end{cases} \tag{2.13}$$

defines a scalar product on $H_0^m(\Omega)$ (respectively $\mathscr{D}^{m,2}(\Omega)$) if Ω is bounded (respectively unbounded). If Ω is bounded, then this scalar product induces a norm equivalent to (2.3).

2.2.2 Embedding Theorems

Consider first the case of unbounded domains.

Theorem 2.3. *Let $m \in \mathbb{N}^+$, $1 \le p < \infty$, with $n > mp$. Assume that $\Omega \subset \mathbb{R}^n$ is an unbounded domain with uniformly Lipschitzian boundary $\partial\Omega$, then:*

1. $\mathscr{D}^{m,p}(\Omega) \subset L^{np/(n-mp)}(\Omega)$;

2. $W^{m,p}(\Omega) \subset L^q(\Omega)$ for all $p \le q \le \frac{np}{n-mp}$.

On the other hand, in bounded domains subcritical embeddings become compact.

Theorem 2.4 (Rellich-Kondrachov). *Let $m \in \mathbb{N}^+$, $1 \le p < \infty$. Assume that $\Omega \subset \mathbb{R}^n$ is a bounded Lipschitzian domain, then for any $1 \le q < \frac{np}{n-mp}$ there exists a compact embedding $W^{m,p}(\Omega) \subset L^q(\Omega)$. Here we make the convention that $\frac{np}{n-mp} = +\infty$ if $n \le mp$.*

Remark 2.5. The optimal constants of the compact embeddings in Theorem 2.4 are attained on functions solving corresponding Euler-Lagrange equations. We refer to Section 7.2 for a discussion of these problems where, for simplicity, we restrict again our attention to the case $m = 2$.

In fact, if $n < mp$, Theorem 2.4 may be improved by the following statement.

Theorem 2.6. *Let $m \in \mathbb{N}^+$ and let $\Omega \subset \mathbb{R}^n$ be a bounded domain with Lipschitzian boundary. Assume that there exists $k \in \mathbb{N}$ such that $n < (m-k)p$. Then*

$$W^{m,p}(\Omega) \subset C^{k,\gamma}(\overline{\Omega}) \qquad \text{for all } \gamma \in \left(0, m-k-\frac{n}{p}\right] \cap (0,1)$$

with compact embedding if $\gamma < m-k-\frac{n}{p}$.

The statements of Theorems 2.4 and 2.6 also hold if we replace $W^{m,p}(\Omega)$ with its proper subspace $W_0^{m,p}(\Omega)$. In this case, no regularity assumption on the boundary $\partial\Omega$ is needed. Let us also mention that there is a simple way to remember the embeddings in Theorem 2.6. It is based on the so-called *regularity index*, see [11, Section 8.7]. In n-dimensional bounded domains Ω, the regularity index for $W^{m,p}(\Omega)$ is $m - n/p$ whereas for $C^{k,\gamma}(\overline{\Omega})$ it is $k + \gamma$. A Sobolev space is embedded into any other space with a smaller regularity index. For instance, $W^{m,p}(\Omega) \subset W^{\mu,q}(\Omega)$ provided $m - n/p \geq \mu - n/q$ (and $m \geq \mu$). Also $W^{m,p}(\Omega) \subset C^{k,\gamma}(\overline{\Omega})$ whenever $m - n/p \geq k + \gamma$ and $\gamma \in (0,1)$, which is precisely the statement in Theorem 2.6. A similar rule is also available for trace operators, namely if $m - n/p \geq \mu - (n-1)/q$ (and $m > \mu$) then the trace operator on $W^{m,p}(\Omega)$ is continuous into $W^{\mu,q}(\partial\Omega)$.

We conclude this section with the multiplicative properties of functions in Sobolev spaces.

Theorem 2.7. *Assume that $\Omega \subset \mathbb{R}^n$ is a Lipschitzian domain. Let $m \in \mathbb{N}^+$ and $p \in [1,\infty)$ be such that $mp > n$. Then $W^{m,p}(\Omega)$ is a commutative Banach algebra.*

Remark 2.8. Theorem 2.7 can be generalised by considering multiplications of functions in possibly different Sobolev spaces. For instance, if $m_1, m_2 \in \mathbb{N}^+$ and $\mu = \min\{m_1, m_2, m_1 + m_2 - [\frac{n}{2}] - 1\}$, then $H^{m_1}(\Omega)H^{m_2}(\Omega) \subset H^{\mu}(\Omega)$.

We postpone further properties of the Hilbertian critical embedding, that is, $H^m \subset L^{2n/(n-2m)}$ with $n > 2m$, to Sections 7.3 and 7.8. The reasons are both that we need further tools and that these properties have a natural application to nonexistence results for semilinear polyharmonic equations at critical growth.

2.3 Boundary Conditions

For the rest of Chapter 2, we assume the domain Ω to be bounded. Under suitable assumptions on $\partial\Omega$, to equation (2.2) we may associate m boundary conditions. These conditions will be expressed by linear differential operators $B_j(x;D)$, namely

$$B_j(x;D)u = h_j \text{ for } j = 1,\dots,m \text{ on } \partial\Omega, \tag{2.14}$$

where the functions h_j belong to suitable functional spaces. Each B_j has a maximal order of derivatives $m_j \in \mathbb{N}$ and the coefficients of the derivatives are sufficiently smooth functions on $\partial\Omega$. The regularity assumptions on these coefficients and on $\partial\Omega$ will be made precise in each statement.

For the problems considered in this monograph, it always appears that

$$m_j \leq 2m - 1 \qquad \text{for all } j = 1,\dots,m. \tag{2.15}$$

Therefore, we shall always assume that (2.15) holds, although some of our statements remain true under less restrictive assumptions. The meaning of (2.14) will remain unclear until the precise definition of solution to (2.2) will be given; in most cases, they should be seen as traces, namely satisfied in a generalised sense given by the operators (2.4).

The choice of the B_j's is not completely free, we need to impose a certain algebraic constraint, the so-called *complementing condition*. For any j, let B_j' denote the highest order part of B_j which is precisely of order m_j, then for equations (2.2) which have the polyharmonic operator as principal part, we have the following.

Definition 2.9. For every point $x \in \partial\Omega$, let $\nu(x)$ denote the normal unit vector. We say that the *complementing condition* holds for (2.14) if, for any nontrivial tangential vector $\tau(x)$, the polynomials in t $B_j'(x; \tau + t\nu)$ are linearly independent modulo the polynomial $(t - i|\tau|)^m$.

As explained in [5, Section 10], the complementing condition is crucial in order to obtain a priori estimates for solutions to (2.2)-(2.14) and, in turn, existence and uniqueness results.

Clearly, the solvability of (2.2)-(2.14) depends on the assumptions made on $\mathscr{A}$, f, B_j and h_j. We are here interested in *structural assumptions*, namely properties of the problem and not of its data.

Assumptions on the homogeneous problem. If we assume that $f = 0$ in Ω and that $h_j = 0$ on $\partial\Omega$ for all $j = 1, ..., m$, then (2.2)-(2.14) admits the trivial solution $u = 0$, in whatever sense this is intended. The natural question is then to find out whether this is the only solution. The answer depends on the structure of the problem. In fact, for any "reasonable" $\mathscr{A}$ and B_j's there exists a discrete set $\Sigma \subset \mathbb{R}$ such that, if $\sigma \notin \Sigma$, then the problem

$$\begin{cases} (-\Delta)^m u + \sigma\mathscr{A}(x;D)u = 0 & \text{in } \Omega, \\ B_j(x;D)u = 0 & \text{with } j = 1, ..., m \text{ on } \partial\Omega, \end{cases} \tag{2.16}$$

only admits the trivial solution. If $\sigma \in \Sigma$, then the solutions of (2.16) form a nontrivial linear space; if $\mathscr{A}$ and the B_j's are well-behaved (in the sense specified below) this space has finite dimension. Therefore, we shall assume that

the associated homogeneous problem only admits the trivial solution $u = 0$.
$$\tag{2.17}$$

Assumption (2.17) is a structural assumption which only depends on $\mathscr{A}$ and the B_j's. Thanks to the Fredholm alternative (see e.g. [69, Theorem VI.6]), we know that if (2.17) fails, then for any possible choice of the data f and h_j problem (2.2)-(2.14) fails to have either existence or uniqueness of the solution.

Assumptions on $\mathscr{A}$. Assume that $\mathscr{A}$ has the following form

$$\mathscr{A}(x;D)u = \sum_{|\beta| \leq 2m-1} a_\beta(x)D^\beta u, \qquad a_\beta \in C^{|\beta|}(\overline{\Omega}). \tag{2.18}$$

Actually, for some of our results, less regularity is needed on the coefficients a_β but we will not go deep into this. We just mention that, for instance, if $n \geq 5$ then in order to obtain existence of a weak solution (according to Theorem 2.16 below) it is enough to assume $a_0 \in L^{n/4}(\Omega)$.

Assumptions on the boundary conditions. Assume that, according to Definition 2.9,

$$\text{the linear boundary operators } B_j\text{'s satisfy the complementing condition.} \qquad (2.19)$$

We now discuss the main boundary conditions considered in this monograph.

Dirichlet boundary conditions. In this case, $B_j(x,D)u = B'_j(x,D)u = \frac{\partial^{j-1}u}{\partial \nu^{j-1}}$ for $j = 1,\ldots,m$ so that $m_j = j-1$ and (2.14) become

$$u = h_1, \quad \ldots, \quad \frac{\partial^{m-1}u}{\partial \nu^{m-1}} = h_m \qquad \text{on } \partial\Omega. \qquad (2.20)$$

Hence, $B'_j(x; \tau + t\nu) = t^{j-1}$ and, as mentioned in [5, p. 627], the complementing condition is satisfied for (2.20).

Navier boundary conditions. In this case, $B_j(x,D)u = B'_j(x,D)u = \Delta^{j-1}u$ for $j = 1,...,m$ so that $m_j = 2(j-1)$ and (2.14) become

$$u = h_1, \quad \ldots, \quad \Delta^{m-1}u = h_m \qquad \text{on } \partial\Omega. \qquad (2.21)$$

Under these conditions, if $\mathscr{A}$ has a suitable form then (2.2) may be written as a system of m Poisson equations, each one of the unknown functions satisfying Dirichlet boundary conditions. Therefore, the complementing condition follows by the theory of elliptic systems [6].

Mixed Dirichlet-Navier boundary conditions. We make use of these conditions in Section 5.2. They are a suitable combination of (2.20)-(2.21). For instance, if m is odd, they read $B_j(x,D)u = \frac{\partial^{j-1}u}{\partial \nu^{j-1}}$ for $j = 1,...,m-1$ and $B_m(x,D)u = \Delta^{(m-1)/2}u$. Again, the complementing condition is satisfied.

Steklov boundary conditions. We consider these conditions only for the biharmonic operator. Let $a \in C^0(\partial\Omega)$ and to the equation $\Delta^2 u = f$ in Ω we associate the boundary operators $B_1(x,D)u = u$ and $B_2(x,D)u = \Delta u - a\frac{\partial u}{\partial \nu}$. Then (2.14) become

$$u = h_1 \qquad \text{and} \qquad \Delta u - a\frac{\partial u}{\partial \nu} = h_2 \qquad \text{on } \partial\Omega. \qquad (2.22)$$

Since B'_j (for $j = 1,2$) is the same as for (2.21), also (2.22) satisfy the complementing condition.

More generally, Hörmander [231] characterises all the sets of boundary operators B_j which satisfy the complementing condition.

We conclude this section by giving an example of boundary conditions which do not satisfy the complementing condition. Consider the fourth order problem

$$\begin{cases} \Delta^2 u = 0 & \text{in } \Omega, \\ \Delta u = 0 & \text{on } \partial\Omega, \\ \frac{\partial(\Delta u)}{\partial \nu} = 0 & \text{on } \partial\Omega. \end{cases} \qquad (2.23)$$

For any unit vector τ tangential to $\partial\Omega$ we have $B_1(\tau+t\nu) = B_1'(\tau+t\nu) = t^2+1$ and $B_2(\tau+t\nu) = B_2'(\tau+t\nu) = t^3+t$. These polynomials *are not* linearly independent modulo $(t-i)^2$ so that the complementing condition is not satisfied. Note also that any harmonic function solves (2.23) so that the space of solutions *does not* have finite dimension. In particular, if we take any point $x_0 \in \mathbb{R}^n \setminus \overline{\Omega}$, the fundamental solution u_0 of $-\Delta$ having pole in x_0 (namely, $u_0(x) = \log|x-x_0|$ if $n=2$ and $u_0(x) = |x-x_0|^{2-n}$ if $n \geq 3$) solves (2.23). This shows that it is not possible to obtain uniform a priori bounds in any norm. Indeed, as x_0 approaches the boundary $\partial\Omega$ it is clear that (for instance!) the H^1-norm of the solution cannot be bounded uniformly in terms of its L^2-norm.

2.4 Hilbert Space Theory

2.4.1 Normal Boundary Conditions and Green's Formula

In this section we study the solvability of the polyharmonic equation

$$(-\Delta)^m u + \sum_\circ D^\beta \left[a_{\beta,\mu}(x) D^\mu u \right] = f \qquad \text{in } \Omega \qquad (2.24)$$

complemented with the linear boundary conditions

$$\sum_{|\alpha|\leq m_j} b_{j,\alpha}(x) D^\alpha u = h_j \quad \text{on } \partial\Omega \text{ with } j = 1,\dots,m, \qquad (2.25)$$

where $m_j \leq 2m-1$, see (2.15), and $\sum_\circ$ means summation over all multi-indices β and μ such that

$$|\beta| \leq m, \qquad |\mu| \leq m, \qquad |\beta|+|\mu| \leq 2m-1. \qquad (2.26)$$

With the notations of (2.2) and (2.14), we have

$$\mathscr{A}(x;D)u = \sum_\circ D^\beta[a_{\beta,\mu}(x)D^\mu u], \qquad B_j(x;D)u = \sum_{|\alpha|\leq m_j} b_{j,\alpha}(x)D^\alpha u.$$

Assume that

$$a_{\beta,\mu} \in C^{|\beta|}(\overline{\Omega}) \qquad \text{for all } \beta \text{ and } \mu \text{ satisfying (2.26).} \tag{2.27}$$

To the linear differential operator A defined by

$$Au := (-\Delta)^m u + \sum_{\circ} D^{\beta}[a_{\beta,\mu}(x)D^{\mu}u] \tag{2.28}$$

we associate the bilinear form

$$\Psi(u,v) = (u,v) + \sum_{\circ}(-1)^{|\beta|}\int_{\Omega} a_{\beta,\mu}(x)D^{\mu}u D^{\beta}v\,dx \text{ for all } u,v \in H^m(\Omega), \tag{2.29}$$

where $(.,.)$ is defined in (2.13). Formally, Ψ is obtained by integrating by parts $\int Au\,v$ and by neglecting the boundary integrals. We point out that, in view of (2.27), $\Psi(u,v)$ is well-defined for all $u,v \in H^m(\Omega)$.

Let us recall that m_j denotes the highest order derivatives of u appearing in B_j. With no loss of generality, we may assume that the boundary conditions (2.25) are ordered for increasing m_j's so that

$$m_j \leq m_{j+1} \qquad \text{for all } j = 1,\dots,m-1. \tag{2.30}$$

Moreover, we assume that the coefficients in (2.25) satisfy

$$b_{j,\alpha} \in C^{2m-m_j}(\overline{\Omega}) \qquad \text{for all } j = 1,\dots,m \text{ and } |\alpha| \leq m_j; \tag{2.31}$$

by this, we mean that the functions $b_{j,\alpha}$ are restrictions to the boundary $\partial\Omega$ of functions in $C^{2m-m_j}(\overline{\Omega})$.

We also need to define well-behaved systems of boundary operators.

Definition 2.10. Let $k \in \mathbb{N}^+$. We say that the boundary value operators $\{F_j(x;D)\}_{j=1}^{k}$ satisfying (2.30) form a *normal system* on $\partial\Omega$ if $m_i < m_j$ whenever $i < j$ and if $F_j(x;D)$ contains the term $\partial^{m_j}/\partial v^{m_j}$ with a coefficient different from 0 on $\partial\Omega$. Moreover, we say that $\{F_j(x;D)\}_{j=1}^{k}$ is a *Dirichlet system* if, in addition to the above conditions, we have $m_j = j - 1$ for $j = 1,\dots,k$; the number k is then called the *order* of the Dirichlet system.

Remark 2.11. The assumption "F_j contains the term $\partial^{m_j}/\partial v^{m_j}$ with a coefficient different from 0 on $\partial\Omega$" requires some explanations since it may happen that the term $\partial^{m_j}/\partial v^{m_j}$ does not appear explicitly in F_j. One should then rewrite the boundary conditions on $\partial\Omega$ in local coordinates; the system of coordinates should contain the $n-1$ tangential directions and the normal direction v. Then the assumption is that in this new system of coordinates the term $\partial^{m_j}/\partial v^{m_j}$ indeed appears with a coefficient different from 0. For instance, imagine that $m_j = 2$ and that Δu represents the terms of order 2 in F_j; it is known that if $\partial\Omega$ and u are smooth, then

$\Delta u = \frac{\partial^2 u}{\partial v^2} + (n-1)H\frac{\partial u}{\partial v} + \Delta_\tau u$ on $\partial\Omega$, where H denotes the mean curvature at the boundary and $\Delta_\tau u$ denotes the tangential Laplacian of u. Therefore, any boundary operator which contains Δ as principal part satisfies this condition.

It is clear that if a normal system of boundary value operators $\{F_j(x;D)\}_{j=1}^k$ is such that $m_k = k-1$, then it is a Dirichlet system.

Proposition 2.12. *Let $\Omega \subset \mathbb{R}^n$ be a bounded smooth domain. Let $k \in \mathbb{N}^+$ and assume that the boundary value operators $\{B_j(x;D)\}_{j=1}^k$ form a normal system on $\partial\Omega$. If $m \geq m_k$, then there exists a (non-unique) system $\{S_j(x;D)\}_{j=k+1}^m$ such that $\{B_1,...,B_k,S_{k+1},...,S_m\}$ forms a Dirichlet system of order m. Here, all the boundary operators are supposed to have smooth coefficients.*

We can now give a suitable version of Green's formula.

Theorem 2.13. *Let*

$$\partial\Omega \in C^{2m,1} \tag{2.32}$$

and suppose that the differential operator A in (2.28) has coefficients satisfying (2.27). Assume also that $\{F_j(x;D)\}_{j=1}^m$ forms a Dirichlet system of order m (so that $m_j = j-1$) with coefficients satisfying (2.31). Then there exists a normal system of boundary operators $\{\Phi_j(x;D)\}_{j=1}^m$ with coefficients satisfying (2.31) (and with Φ_j of order $2m-j$) such that

$$\Psi(u,v) = \int_\Omega Auv\,dx + \sum_{j=1}^m \int_{\partial\Omega} \Phi_j(x;D)u\,F_j(x;D)v\,d\omega \ \text{ for all } u,v \in H^{2m}(\Omega).$$

The operators $\{\Phi_j(x;D)\}_{j=1}^m$ given by Theorem 2.13 are called *Green adjoint boundary operators* of $\{F_j(x;D)\}_{j=1}^m$.

2.4.2 Homogeneous Boundary Value Problems

In this section we study the solvability of (2.24) in the case of vanishing boundary data h_j in conditions (2.25), namely

$$\begin{cases} (-\Delta)^m w + \Sigma_\circ D^\beta \left(a_{\beta,\mu}(x)D^\mu w\right) = g & \text{in } \Omega, \\ B_j(x;D)w = 0 & \text{for } j=1,...,m \quad \text{on } \partial\Omega. \end{cases} \tag{2.33}$$

The solvability of (2.33) is studied in the framework of Hilbertian Sobolev spaces. To this end, let us explain what is meant by a Hilbert triple.

Definition 2.14. Let V and H be Hilbert spaces such that $V \subset H$ with injective, dense and continuous embedding. Let V' denote the dual space of V; a scheme of this type (namely $V \subset H \subset V'$) is called a *Hilbert triple*.

For a Hilbert triple $V \subset H \subset V'$ also the embedding $H \subset V'$ is necessarily injective, dense and continuous, see [417, Theorem 17.1]. Notice also that, although there exists the Riesz isomorphism between V and V' (see [69, Theorem V.5]), we will represent functionals from V' with the scalar product in H and not with the scalar product in V.

We proceed in several steps in order to simplify problem (2.33) and to give the correct assumptions for its solvability.

Introduction of a suitable Hilbert triple. Divide the boundary operators in (2.33) into two classes. If $m_j < m$ we say that the boundary operator $B_j(x;D)$ is *stable* while if $m_j \geq m$ we say that it is *natural*. Assume that there are p stable boundary operators, with p being an integer between 0 and m. If $p = 0$ all the boundary operators are natural, whereas if $p = m$ all boundary operators are stable. In view of (2.30) the stable operators correspond to indices $j \leq p$. Then we define the space

$$V := \{v \in H^m(\Omega); \ B_j(x,D)v = 0 \text{ on } \partial\Omega \text{ for } j = 1,...,p\}. \qquad (2.34)$$

Clearly, if $p = 0$ we have $V = H^m(\Omega)$ while if $p = m$ we have $V = H_0^m(\Omega)$ (provided the assumption (2.36) below holds). In particular, in the case of Dirichlet boundary conditions (2.20) we have

$$V = H_0^m(\Omega),$$

in the case of Navier boundary conditions (2.21) we have

$$V = H_\vartheta^m(\Omega) := \left\{v \in H^m(\Omega); \ \Delta^j v = 0 \text{ on } \partial\Omega \text{ for } j < \frac{m}{2}\right\}, \qquad (2.35)$$

in the case of Steklov boundary conditions (2.22) we have

$$V = H^2 \cap H_0^1(\Omega) = H_\vartheta^2(\Omega).$$

In any case, the space V is well-defined since each B_j contains trace operators of maximal order $m_j < m$. Moreover, V is a closed subspace of $H^m(\Omega)$ which satisfies $H_0^m(\Omega) \subset V \subset H^m(\Omega)$ with continuous embedding. Therefore, V inherits the scalar product and the Hilbert space structure from $H^m(\Omega)$. If we put $H = L^2(\Omega)$, then $V \subset H \subset V'$ forms a Hilbert triple with compact embeddings.

Assumptions on the boundary operators. Assume that

$$\{B_j(x;D)\}_{j=1}^m \quad \text{forms a normal system} \qquad (2.36)$$

and that the orders of the B_j's satisfy

$$m_i + m_j \neq 2m - 1 \quad \text{for all } i,j = 1,...,m. \qquad (2.37)$$

This assumption is needed since we are not free to choose the orders of the B_j's. For every $k = 0,...,m-1$ there must be exactly one m_j in the set $\{k, 2m-k-1\}$.

Let p denote the number of stable boundary operators. In view of (2.30) we know that these operators are precisely $\{B_j\}_{j=1}^{p}$ and, of course, they also form a normal system of boundary operators. By Proposition 2.12, there exists a family of normal operators $\{S_j\}_{j=p+1}^{m}$ such that $\{B_1,...,B_p,S_{p+1},...,S_m\}$ forms a Dirichlet system of order m. We relabel this system and define

$$\{F_j\}_{j=1}^{m} \equiv \{B_1,...,B_p,S_{p+1},...,S_m\} \tag{2.38}$$

the re-ordered system in such a way that the order of F_j equals $j-1$. The indices $j=1,...,m$ are so divided into two subsets J_1 and J_2 according to the following rule: $j \in J_1$ if $F_j \in \{B_j\}_{j=1}^{p}$ whereas $j \in J_2$ if $F_j \in \{S_j\}_{j=p+1}^{m}$.

Let $\{\Phi_j\}_{j=1}^{m}$ denote the Green adjoint boundary operators of $\{F_j\}_{j=1}^{m}$ according to Theorem 2.13. We finally assume that the S_j's and the Φ_j's may be chosen in a such a way that

$$\{B_j\}_{j=p+1}^{m} \subset \{\Phi_j\}_{j=1}^{m} . \tag{2.39}$$

The condition in (2.39) is quite delicate since it requires the construction of the S_j's and the Φ_j's before being checked. Note that if $p = m$ (Dirichlet boundary conditions) or $p = 0$, then (2.37) and (2.39) are automatically fulfilled.

Assumption on g. Assume that

$$g \in V' . \tag{2.40}$$

If $V = H_0^m(\Omega)$, then $V' = H^{-m}(\Omega)$ and V' has a fairly simple representation, see [417, Theorem 17.6]. If $V = H^m(\Omega)$, then elements of V' have a more difficult characterisation, see [417, Theorem 17.5]. In all the other cases, V' has even more complicated forms but we always have $[H^m(\Omega)]' \subset V' \subset H^{-m}(\Omega)$ with continuous embeddings.

Coercivity of the bilinear form. In order to ensure solvability of (2.42) we need a crucial assumption on the bilinear form Ψ. By (2.27) we know that there exists $c_1 > 0$ such that $\Psi(u,v) \leq c_1\|u\|_{H^m(\Omega)}\|v\|_{H^m(\Omega)}$ for all $u,v \in H^m(\Omega)$. Assume that there exists $c_2 \in (0,c_1)$ such that

$$\Psi(u,u) \geq c_2\|u\|_{H^m(\Omega)}^2 \qquad \text{for all } u \in V. \tag{2.41}$$

In fact, (2.41) is nothing else but a strengthened ellipticity assumption for the operator A; it gives a quadratic lower bound behaviour for Ψ (in terms of the H^m norm) but only on the subspace V. One is then interested in finding sufficient conditions which ensure that (2.41) holds. The most general such condition is due to Agmon [3] and is quite technical to state; since it is beyond the scope of this book, we will not discuss it here. We just limit ourselves to verify (2.41) in some simple cases. If $Au = (-\Delta)^m u$ for some $m \geq 2$ then $\Psi(u,v) = (u,v)_{H_0^m}$ and (2.41) holds with $c_1 = c_2 = 1$ and $V = H_0^m(\Omega)$; hence, Dirichlet boundary conditions (2.20) are allowed with $Au = (-\Delta)^m u$. If $Au = \Delta^2 u$ then $\Psi(u,v) = (u,v)_{H_0^2}$ and (2.41) holds again with $c_1 = c_2 = 1$ but now for both the cases $V = H_0^2(\Omega)$ and $V = H^2 \cap H_0^1(\Omega)$ so that Dirichlet (2.20) and Navier (2.21) boundary conditions are allowed, see also

Theorem 2.31 below. As we shall see in Section 3.3.1 and in Theorem 5.22, if $Au = \Delta^2 u$ and $V = H^2 \cap H_0^1(\Omega)$, also Steklov boundary conditions (2.22) are allowed but now with the bilinear form $\Psi(u,v) = (u,v)_{H_0^2} - \int_{\partial\Omega} a u_\nu v_\nu \, d\omega$ provided a satisfies suitable assumptions which ensure (2.17).

Finally, we say that $w \in V$ is a weak solution to (2.33) if

$$\Psi(w,\varphi) = \langle g, \varphi \rangle \qquad \text{for all } \varphi \in V. \tag{2.42}$$

Thanks to the Lax-Milgram theorem we may now state the existence and uniqueness result for weak solutions to the homogeneous problem (2.33).

Theorem 2.15. *Let $\Omega \subset \mathbb{R}^n$ be a bounded domain satisfying (2.32). Assume that:*

- *the operator A in (2.28) and the bilinear form Ψ in (2.29) satisfy (2.27) and (2.41);*
- *the operators B_j satisfy (2.30), (2.31), (2.36), (2.37), (2.39);*
- *g satisfies (2.40).*

Then problem (2.42) admits a unique weak solution $w \in V$; moreover, there exists a constant $C = C(\Omega, m, \mathscr{A}, B_j) > 0$ independent of g, such that

$$\|w\|_{H^m(\Omega)} \leq C \|g\|_{V'}.$$

To conclude, let us highlight the existing connection between (2.42) and (2.33). It is clear that any solution $w \in H^{2m}(\Omega)$ to (2.33) is also a solution to (2.42). On the other hand, any $w \in V$ satisfying (2.42) automatically satisfies the stable boundary conditions since these are contained in the definition of V. We show that if g and w are smooth then w also satisfies the natural boundary conditions and solves (2.33). To see this, let $\{F_j\}$ be as in (2.38) and let $\{\Phi_j\}$ denote the normal system of boundary operators associated to $\{F_j\}$ through Theorem 2.13. Then if we assume that $g \in L^2(\Omega)$ and $w \in V \cap H^{2m}(\Omega)$, Theorem 2.13 combined with (2.42) gives

$$\int_\Omega Aw\,\varphi\,dx + \sum_{j=1}^m \int_{\partial\Omega} \Phi_j(x;D)w\,F_j(x;D)\varphi\,d\omega = \int_\Omega g\,\varphi\,dx \tag{2.43}$$

for all $\varphi \in V$. Taking arbitrary $\varphi \in C_c^\infty(\Omega)$ in (2.43) shows that $Aw = g$ a.e. in Ω so that the equation in (2.33) is satisfied (recall the definition of A in (2.28)). Once this is established, (2.43) yields

$$\sum_{j \in J_2} \int_{\partial\Omega} \Phi_j(x;D)w\,F_j(x;D)\varphi\,d\omega = \sum_{j=1}^m \int_{\partial\Omega} \Phi_j(x;D)w\,F_j(x;D)\varphi\,d\omega = 0 \tag{2.44}$$

for all $\varphi \in V$, where the first equality is a consequence of the fact that $\varphi \in V$, namely $F_j\varphi = 0$ on $\partial\Omega$ for all $j \in J_1$. Again by arbitrariness of $\varphi \in V$, (2.44) shows that

$$\Phi_j(x;D)w = 0 \quad \text{on } \partial\Omega \text{ for all } j \in J_2.$$

By assumptions (2.37) and (2.39) we know that $\Phi_j = B_{2m-m_j-1}$ for all $j \in J_2$, therefore the latter is equivalent to

$$B_j(x;D)w = 0 \quad \text{on } \partial\Omega \text{ for all } j = p+1,...,m$$

and w also satisfies the natural boundary conditions in (2.33).

2.4.3 Inhomogeneous Boundary Value Problems

In this section we study weak solvability of (2.24)-(2.25) without assuming that the boundary data h_j vanish. After requiring suitable regularity on the data h_j, we explain what is meant by weak solution and we reduce the inhomogeneous problem to an homogeneous one.

Regularity assumptions on the data. Let V be as in (2.34) and assume that

$$f \in V'. \tag{2.45}$$

Weak solutions to (2.24)-(2.25) will be sought in a suitable convex subset of $H^m(\Omega)$. According to Theorem 8.3 in Chapter 1 in [276], it is then necessary to assume that

$$h_j \in H^{m-m_j-\frac{1}{2}}(\partial\Omega) \quad \text{for all } j = 1,...,m. \tag{2.46}$$

We have $m - m_j - \frac{1}{2} > 0$ for all $m_j < m$, namely for all $j = 1,...,p$ where p is the number of stable boundary operators. If $j = p+1,...,m$, we have $m - m_j - \frac{1}{2} < 0$ and we recall the definition in (2.7).

If we assume (2.31), (2.32), (2.36) and (2.46), we may apply [417, Theorem 14.1] to infer that

$$\text{there exists } v \in H^m(\Omega) \text{ such that } B_j(x;D)v = h_j \text{ on } \partial\Omega \tag{2.47}$$

for all $j = 1,...,p$. Then consider the set

$$K := \{w \in H^m(\Omega); w - v \in V\} ;$$

it is straightforward to verify that K is a closed convex nonempty subset of $H^m(\Omega)$. If $p = 0$, then no v needs to be determined by (2.47) and K becomes the whole space $V = H^m(\Omega)$. Let us define the (ordered) family of boundary operators $\{F_j\}_{j=1}^m$ as in (2.38) and let J_1 and J_2 denote the subsets defined there. We say that $u \in K$ is a weak solution to (2.24)-(2.25) if

$$\Psi(u,\varphi) = \langle f, \varphi \rangle + \sum_{j\in J_2} \langle h_{2m-m_j-1}, F_j(x;D)\varphi \rangle_{\partial\Omega,j} \quad \text{for all } \varphi \in V, \tag{2.48}$$

where Ψ is defined in (2.29), $\langle .,. \rangle$ denotes the duality between V' and V and $\langle .,. \rangle_{\partial\Omega,j}$ denotes the duality between $H^{m-m_j-\frac{1}{2}}(\partial\Omega)$ and $H^{-m+m_j+\frac{1}{2}}(\partial\Omega)$.

Reduction to an homogeneous boundary value problem. Let $v \in H^m(\Omega)$ be defined by (2.47) and let $u \in K$ be a weak solution to (2.24)-(2.25), according to (2.48). Subtract $\Psi(v, \varphi)$ from the equations in (2.48) to obtain

$$\Psi(u - v, \varphi) = \langle f, \varphi \rangle + \sum_{j \in J_2} \langle h_{2m-m_j-1}, F_j(x;D)\varphi \rangle_{\partial\Omega, j} - \Psi(v, \varphi)$$

for all $\varphi \in V$. By (2.45), the linear functional g defined by

$$g : \varphi \mapsto \langle f, \varphi \rangle + \sum_{j \in J_2} \langle h_{2m-m_j-1}, F_j(x;D)\varphi \rangle_{\partial\Omega, j} - \Psi(v, \varphi) \quad \text{for } \varphi \in V$$

is continuous on V so that (2.40) holds. Now put $w := u - v$; then $w \in V$ satisfies (2.42). Therefore, we shall proceed as follows. We first determine a function v as in (2.47), then we solve problem (2.33) (whose variational formulation is (2.42)) and find $w \in V$. Putting $u = v + w$ we obtain a solution $u \in K$ to (2.24)-(2.25) (whose variational formulation is (2.48)).

With these arguments, Theorem 2.15 immediately gives

Theorem 2.16. *Let $\Omega \subset \mathbb{R}^n$ be a bounded domain satisfying (2.32). Assume that:*

- *the operator A in (2.28) and the bilinear form Ψ in (2.29) satisfy (2.27) and (2.41);*
- *the operators B_j satisfy (2.30), (2.31), (2.36), (2.37), (2.39);*
- *f satisfies (2.45) and the h_j's satisfy (2.46).*

Then problem (2.48) admits a unique weak solution $u \in K$; moreover, there exists a constant $C = C(\Omega, m, \mathscr{A}, B_j) > 0$ independent of f and of the h_j's, such that

$$\|u\|_{H^m(\Omega)} \leq C \left(\|f\|_{V'} + \sum_{j=1}^{p} \|h_j\|_{H^{m-m_j-\frac{1}{2}}(\partial\Omega)} \right).$$

As for the homogeneous problem, let us explain the link between weak and strong solutions. Again, any strong solution $u \in H^{2m}(\Omega)$ to (2.24)-(2.25) certainly satisfies (2.48); note that a strong solution may exist only if

$$h_j \in H^{2m-m_j-\frac{1}{2}}(\partial\Omega) \text{ for } j = 1, ..., m \text{ and } f \in L^2(\Omega). \tag{2.49}$$

Conversely, assume that (2.49) holds and let $u \in K \cap H^{2m}(\Omega)$ be a solution to (2.48). Let $\{F_j\}$ be as in (2.38) and let $\{\Phi_j\}$ denote the normal system of boundary operators associated to $\{F_j\}$ through Theorem 2.13. Then (2.48) gives

$$\int_\Omega Au\, \varphi\, dx + \sum_{j=1}^{m} \int_{\partial\Omega} \Phi_j(x;D)u\, F_j(x;D)\varphi\, d\omega$$

$$= \int_\Omega f\, \varphi\, dx + \sum_{j \in J_2} \int_{\partial\Omega} h_{2m-m_j-1} F_j(x;D)\varphi\, d\omega \quad \text{for all } \varphi \in V. \tag{2.50}$$

Taking arbitrary $\varphi \in C_c^\infty(\Omega)$ in (2.50) shows that $Au = f$ a.e. in Ω so that (2.24) is satisfied. Once this is established, (2.50) yields

$$\sum_{j \in J_2} \int_{\partial\Omega} \Phi_j(x;D)u\, F_j(x;D)\varphi\, d\omega = \sum_{j \in J_2} \int_{\partial\Omega} h_{2m-m_j-1}\, F_j(x;D)\varphi\, d\omega$$

for all $\varphi \in V$. Then the same arguments used after (2.44) show that

$$B_j(x;D)u = h_j \quad \text{on } \partial\Omega \text{ for all } j = p+1,\ldots,m,$$

which proves that u also satisfies the natural boundary conditions in (2.25).

Remark 2.17. Although (2.17) and the complementing condition (see Definition 2.9) do not explicitly appear in Theorem 2.16, they are hidden in the assumptions. The coercivity assumption (2.41) ensures that (2.17) is satisfied, see Theorem 2.15. On the other hand, assumptions (2.36) and (2.39) ensure that the complementing condition holds, see [276, Section 2.4].

If the boundary $\partial\Omega$ and the data f and h_j are more regular, elliptic theory applies and also the solution u given in Theorem 2.16 is more regular, see the next section.

2.5 Regularity Results and A Priori Estimates

2.5.1 Schauder Theory

Here we consider classical solutions to (2.2)-(2.14). To do so, we need the Schauder theory and a good knowledge of Hölder continuity.

First fix an integer ℓ such that $\max\{m_j\} \le \ell \le 2m$. Then slightly modify the problem and consider the equation

$$(-\Delta)^m u + \sum_* D^\beta \left[a_{\beta,\mu}(x)D^\mu u\right] = \sum_{|\beta| \le 2m-\ell} D^\beta f_\beta \qquad \text{in } \Omega, \tag{2.51}$$

complemented with the boundary conditions

$$\sum_{|\alpha| \le m_j} b_{j,\alpha}(x)D^\alpha u = h_j \quad \text{on } \partial\Omega \text{ with } j = 1,\ldots,m, \tag{2.52}$$

where $\sum_*$ means summation over all multi-indices β and μ such that

$$|\beta| \le 2m-\ell\,, \qquad |\mu| \le \ell\,, \qquad |\beta| + |\mu| \le 2m-1\,. \tag{2.53}$$

With the notations of (2.2) and (2.14), we have now

$$\mathscr{A}(x;D)u = \sum_{*} D^{\beta}\left[a_{\beta,\mu}(x)D^{\mu}u\right], \qquad B_j(x;D)u = \sum_{|\alpha|\leq m_j} b_{j,\alpha}(x)D^{\alpha}u. \qquad (2.54)$$

Fix a second integer $k \geq \ell$ and put $\overline{\ell} = \max\{2m,k\}$. Then assume that for some $0 < \gamma < 1$ we have

$$\begin{cases} a_{\beta,\mu} \in C^{k-\ell,\gamma}(\overline{\Omega}) & \text{for all } \beta,\mu \text{ satisfying } (2.53), \\ f_{\beta} \in C^{k-\ell,\gamma}(\overline{\Omega}) & \text{for all } |\beta| \leq 2m-\ell, \\ b_{j,\alpha} \in C^{k-m_j,\gamma}(\partial\Omega) & \text{for all } j=1,...,m \text{ and } |\alpha| \leq m_j, \\ h_j \in C^{k-m_j,\gamma}(\partial\Omega) & \text{for all } j=1,...,m. \end{cases} \qquad (2.55)$$

Note that under assumptions (2.55), problem (2.51)-(2.52) needs not to make sense in a classical way. Therefore, we first need to introduce a different kind of solution.

Definition 2.18. We say that $u \in C^{k,\gamma}(\overline{\Omega})$ is a *mild solution* to (2.51)-(2.52) if

$$\int_{\Omega} u(-\Delta)^m \varphi\, dx + \sum_{*}(-1)^{|\beta|}\int_{\Omega} a_{\beta,\mu}(x)D^{\mu}u D^{\beta}\varphi\, dx$$

$$= \sum_{|\beta|\leq 2m-\ell}(-1)^{|\beta|}\int_{\Omega} f_{\beta}(x)D^{\beta}\varphi\, dx$$

for all $\varphi \in C_c^{\infty}(\Omega)$ and if u satisfies pointwise the boundary conditions in (2.52).

Hence, for any mild solution the boundary conditions (2.52) are well-defined since $k \geq \ell \geq m_j$ for all j.

We are now ready to state

Theorem 2.19. *Let $k \geq \ell \in [\max\{m_j\},2m]\cap\mathbb{N}$ and $\overline{\ell} = \max\{2m,k\}$. Assume that (2.55) holds and that $\mathscr{A}$ and the B_j's satisfy (2.54)-(2.19). Assume (2.15) and (2.17). Assume moreover that $\partial\Omega \in C^{\overline{\ell},\gamma}$. Then (2.51)-(2.52) admits a unique mild solution $u \in C^{k,\gamma}(\overline{\Omega})$. Moreover, there exists a constant $C = C(\Omega,k,m,a_{\beta,\mu},b_{j,\alpha}) > 0$ independent of the f_{β}'s and of the h_j's, such that the following a priori estimate holds*

$$\|u\|_{C^{k,\gamma}(\overline{\Omega})} \leq C\left(\sum_{|\beta|\leq 2m-\ell}\|f_{\beta}\|_{C^{k-\ell,\gamma}(\overline{\Omega})} + \sum_{j=1}^{m}\|h_j\|_{C^{k-m_j,\gamma}(\partial\Omega)}\right).$$

The constant C depends on Ω only through its measure $|\Omega|$ and the $C^{\overline{\ell},\gamma}$-norms of the local maps which define the boundary $\partial\Omega$. If $k \geq 2m$ then the solution u is classical.

Finally, if (2.17) is dropped, then for any solution u to (2.51)-(2.52) one has the following local variant of the estimate

$$\|u\|_{C^{k,\gamma}(\overline{\Omega}\cap B_R(x_0))} \le C\left(\sum_{|\beta|\le 2m-\ell} \|f_\beta\|_{C^{k-\ell,\gamma}(\overline{\Omega}\cap B_{2R}(x_0))} \right.$$

$$\left. + \sum_{j=1}^{m} \|h_j\|_{C^{k-m_j,\gamma}(\partial\Omega\cap B_{2R}(x_0))} + \|u\|_{L^1(\Omega\cap B_{2R}(x_0))} \right)$$

for any $R > 0$ and any $x_0 \in \Omega$. Here, C also depends on R.

Roughly speaking, equation (2.51) says that $2m$ derivatives of the solution u belong to $C^{k-2m,\gamma}(\overline{\Omega})$; if $k \ge 2m$ this has an obvious meaning while if $k < 2m$ this should be intended in a generalised sense. In any case, Theorem 2.19 states that the solution gains $2m$ derivatives on the datum $\sum_\beta D^\beta f_\beta$.

2.5.2 L^p-Theory

In this section we give an existence result for (2.2)-(2.14) in the framework of L^p spaces. Under suitable assumptions on the parameters involved in the problem, we show that the solution has at least $2m$ derivatives in $L^p(\Omega)$. In this case, the equation (2.2) is satisfied a.e. in Ω and we say that u is a *strong solution*.

The following statement should also be seen as a regularity complement to Theorem 2.16.

Theorem 2.20. *Let $1 < p < \infty$ and take an integer $k \ge 2m$. Assume that $\partial\Omega \in C^k$ and that*

$$\begin{cases} a_\beta \in C^{k-2m}(\overline{\Omega}) & \text{for all } |\beta| \le 2m-1, \\ b_{j,\alpha} \in C^{k-m_j}(\partial\Omega) & \text{for all } j = 1,...,m,\ |\alpha| \le m_j. \end{cases} \tag{2.56}$$

Assume also that (2.15), (2.17) hold and that $\mathscr{A}$ and the B_j's satisfy (2.18)-(2.19). Then for all $f \in W^{k-2m,p}(\Omega)$ and all $h_j \in W^{k-m_j-\frac{1}{p},p}(\partial\Omega)$ with $j = 1,...,m$, the problem (2.2)-(2.25) admits a unique strong solution $u \in W^{k,p}(\Omega)$. Moreover, there exists a constant $C = C(\Omega,k,m,\mathscr{A},B_j) > 0$ independent of f and of the h_j's, such that the following a priori estimate holds

$$\|u\|_{W^{k,p}(\Omega)} \le C\left(\|f\|_{W^{k-2m,p}(\Omega)} + \sum_{j=1}^{m} \|h_j\|_{W^{k-m_j-\frac{1}{p},p}(\partial\Omega)} \right).$$

The constant C depends on Ω only through its measure $|\Omega|$ and the C^k-norms of the local maps which define the boundary $\partial\Omega$. If $k > 2m + \frac{n}{p}$ then u is a classical solution.

Finally, if (2.17) is dropped, then for any solution u to (2.2)-(2.25) one has the following local variant of the estimate

$$\|u\|_{W^{k,p}(\Omega\cap B_R(x_0))} \le C\Bigg(\|f\|_{W^{k-2m,p}(\Omega\cap B_{2R}(x_0))}$$

$$+ \sum_{j=1}^{m} \|h_j\|_{W^{k-m_j-\frac{1}{p},p}(\partial\Omega\cap B_{2R}(x_0))} + \|u\|_{L^1(\Omega\cap B_{2R}(x_0))}\Bigg)$$

for any $R > 0$ and any $x_0 \in \Omega$. Here, C also depends on R.

The proof of this general result is quite involved, especially if $p \ne 2$. It requires the representation of the solution u in terms of the fundamental solution and the Calderon-Zygmund theory [83] on singular integrals in L^p.

In the case of Dirichlet boundary conditions Theorem 2.20 reads

Corollary 2.21. *Let $1 < p < \infty$ and take an integer $k \ge 2m$. Assume that $\partial\Omega \in C^k$ and that (2.56) holds. Assume moreover that (2.17) holds and that $\mathscr{A}$ satisfies (2.18). Then for all $f \in W^{k-2m,p}(\Omega)$ equation (2.2) admits a unique strong solution $u \in W^{k,p} \cap W_0^{m,p}(\Omega)$; moreover, there exists a constant $C = C(\Omega,k,m,\mathscr{A}) > 0$ independent of f, such that*

$$\|u\|_{W^{k,p}(\Omega)} \le C\|f\|_{W^{k-2m,p}(\Omega)}.$$

For equations in variational form such as (2.51), L^p-estimates are available under weaker regularity assumptions. For our purposes we just consider the following special situation.

Theorem 2.22. *For $\partial\Omega \in C^2$, $p \in (1,\infty)$, and $f \in L^p(\Omega)$ there exists a unique solution $u \in W_0^{2,p}(\Omega)$ of*

$$\begin{cases} \Delta^2 u = \nabla^2 f & \text{in } \Omega, \\ u = u_v = 0 & \text{on } \partial\Omega, \end{cases}$$

where ∇^2 means any second derivative. Moreover, the following L^p-estimate holds

$$\|u\|_{W^{2,p}(\Omega)} \le C\|f\|_{L^p(\Omega)}$$

with $C = C(p,\Omega) > 0$.

For Steklov boundary conditions (2.22) associated to the biharmonic operator, Theorem 2.20 reads as follows.

Corollary 2.23. *Let $1 < p < \infty$ and take an integer $k \ge 4$. Assume that $\partial\Omega \in C^k$ and $a \in C^{k-2}(\partial\Omega)$, then there exists $C = C(k,p,\alpha,\Omega) > 0$ such that*

$$\|u\|_{W^{k,p}(\Omega)} \le C\Bigg(\|u\|_{L^p(\Omega)} + \|\Delta^2 u\|_{W^{k-4,p}(\Omega)} + \|u\|_{W^{k-\frac{1}{p},p}(\partial\Omega)}$$

$$+ \|\Delta u - au_v\|_{W^{k-2-\frac{1}{p},p}(\partial\Omega)}\Bigg)$$

for all $u \in W^{k,p}(\Omega)$. The same statement holds for any $k \geq 2$, provided the norms in the right hand side are suitably interpreted, see (2.5), (2.6), and (2.9).

Remark 2.24. In the estimates of Theorems 2.19 and 2.20 and of Corollaries 2.21 and 2.23, the constants depend in an indirect and nonconstructive way on the particular differential and boundary operators. As soon as one puts (for instance) the L^1-norm of the solution on the right hand side, the constants become explicit and depend only on bounds for the data (k,m, domain, and coefficients) of the problem. This kind of uniformity will be needed in the proof of positivity for Green's functions in perturbed domains, see Section 6.5. There we have uniformly coercive problems which yield an explicit estimate for some lower order norms, so that L^p or Schauder estimates depending on the specific operator would be useless.

2.5.3 *The Miranda-Agmon Maximum Modulus Estimates*

We start by recalling that it is in general false that $\Delta u \in C^0$ implies $u \in C^2$ even if u satisfies homogeneous Dirichlet boundary conditions. Therefore, this lack of regularity is a local problem, irrespective of how smooth the boundary data are. To see why the implication fails, consider the function

$$u(x_1,x_2) = \begin{cases} x_1 x_2 \log |\log(x_1^2 + x_2^2)| & \text{if } (x_1,x_2) \neq (0,0), \\ 0 & \text{if } (x_1,x_2) = (0,0), \end{cases}$$

which is well-defined for $|x| < 1$. Some computations show that u solves the problem

$$\begin{cases} -\Delta u = f & \text{in } B_r(0), \\ u = 0 & \text{on } \partial B_r(0), \end{cases}$$

where $r = 1/\sqrt{e}$ and

$$f(x_1,x_2) = \begin{cases} \dfrac{4x_1 x_2 (1 - 2\log(x_1^2 + x_2^2))}{(x_1^2 + x_2^2)\log^2(x_1^2 + x_2^2)} & \text{if } (x_1,x_2) \neq (0,0), \\ 0 & \text{if } (x_1,x_2) = (0,0), . \end{cases}$$

One can check that $f \in C^0(\overline{B_r(0)})$. On the other hand, for $(x_1,x_2) \neq (0,0)$ we have

$$u_{x_1 x_2}(x_1,x_2) = \log |\log(x_1^2 + x_2^2)|$$
$$+ \frac{2(x_1^4 + x_2^4)}{(x_1^2 + x_2^2)^2 \log(x_1^2 + x_2^2)} - \frac{4x_1^2 x_2^2}{(x_1^2 + x_2^2)^2 \log^2(x_1^2 + x_2^2)}$$

which is unbounded for $(x_1,x_2) \to (0,0)$. Therefore, $u \notin C^2(B_r(0))$.

This example shows that a version of Theorem 2.19 in the framework of spaces C^k of continuously differentiable functions is not available. On the other hand, the well-known Poisson integral formula shows that for continuous Dirichlet boundary data any harmonic function in the ball B is of class $C^0(\overline{B})$, see [198, Theorem 2.6]. In other words, the solution inherits continuity properties from its trace. We state below the corresponding result for polyharmonic equations in a particular situation which is, however, general enough for our purposes. We consider boundary conditions (2.14) with *constant coefficients* and the problem

$$
\begin{cases}
(-\Delta)^m u + \mathscr{A}(x;D)u = f & \text{in } \Omega \\[2ex]
B_j(D)u = \displaystyle\sum_{|\alpha| \le m_j} b_{j,\alpha}D^\alpha u = h_j & \text{on } \partial\Omega \text{ with } j = 1,\dots,m
\end{cases}
\tag{2.57}
$$

for some constants $b_{j,\alpha} \in \mathbb{R} \setminus \{0\}$. Then we have the following a priori estimates for the maximum modulus of solutions and some of their derivatives.

Theorem 2.25. *Assume (2.15), (2.17) and that $\mathscr{A}$ and the B_j's satisfy (2.18)-(2.19). Assume also that $\partial\Omega \in C^{2m}$ and let $\mu = \max_j m_j$. Finally, assume that $f \in C^0(\overline{\Omega})$ and that $h_j \in C^{\mu - m_j}(\partial\Omega)$ for any $j = 1,\dots,m$. Then (2.57) admits a unique strong solution $u \in C^\mu(\overline{\Omega}) \cap W^{2m,p}_{loc}(\Omega)$ for any $p \in (1,\infty)$. Moreover, there exists $C > 0$ independent of f, h_j such that*

$$
\max_{0 \le k \le \mu} \|D^k u\|_{L^\infty} \le C \left(\sum_{j=1}^m \|h_j\|_{C^{\mu - m_j}(\partial\Omega)} + \|f\|_{L^\infty} + \|u\|_{L^1} \right).
$$

Proof. We split problem (2.57) into the two subproblems

$$
\begin{cases}
(-\Delta)^m v + \mathscr{A}(x;D)v = f & \text{in } \Omega, \\
B_j(D)v = 0 & \text{on } \partial\Omega, \ j = 1,\dots,m,
\end{cases}
\tag{2.58}
$$

$$
\begin{cases}
(-\Delta)^m w + \mathscr{A}(x;D)w = 0 & \text{in } \Omega, \\
B_j(D)w = h_j & \text{on } \partial\Omega, \ j = 1,\dots,m.
\end{cases}
\tag{2.59}
$$

Since $f \in C^0(\overline{\Omega}) \subset L^p(\Omega)$ for any $p \ge 1$, by Theorem 2.20 (with $k = 2m$) we know that there exists a unique solution $v \in W^{2m,p}(\Omega)$ to (2.58). By Theorem 2.6 we infer that $v \in C^{2m-1,\gamma}(\overline{\Omega})$ for all $\gamma \in (0,1)$. Moreover, there exist constants $c_1, c_2 > 0$ such that

$$
\|v\|_{C^{2m-1,\gamma}(\overline{\Omega})} \le c_1 \|f\|_{L^\infty} + c_2 \|v\|_{L^1},
$$

see again Theorem 2.20.

On the other hand, by generalising the Miranda-Agmon procedure [4, 305, 306] one shows that (2.59) admits a unique solution $w \in C^\mu(\overline{\Omega})$ satisfying

$$
\max_{0 \le k \le \mu} \|D^k w\|_{L^\infty} \le c_3 \sum_{j=1}^m \|h_j\|_{C^{m_j}(\partial\Omega)} + c_4 \|w\|_{L^1}
\tag{2.60}
$$

for some $c_3, c_4 > 0$. This procedure consists in constructing a suitable approximate solution w_0 to (2.59). To this end one uses the explicit Poisson kernels which solve a related boundary value problem in the half space. These Poisson kernels are determined in [5] and, since (2.19) holds, it makes no difference to consider the Dirichlet problem as in [4, 305, 306] or the general boundary value problem in (2.59). Once this approximate solution w_0 is constructed, one shows that it satisfies (2.60) with $c_4 = 0$. Then one uses again L^p elliptic estimates from Theorem 2.20 and embedding arguments in order to show that the solution w to (2.59) satisfies (2.60).

Once the solutions v and w to (2.58) and (2.59) are obtained, the solution u of (2.57) is determined by adding, $u = v + w$, so that also the estimate of the C^μ-norm follows. $\qquad\square$

2.6 Green's Function and Boggio's Formula

The regularity results of the previous sections are somehow directly visible when writing explicitly the solution of the boundary value problem in terms of the data by means of a suitable kernel. Let us focus on the polyharmonic analogue of the clamped plate boundary value problem

$$\begin{cases} (-\Delta)^m u = f & \text{in } \Omega, \\ D^\alpha u|_{\partial\Omega} = 0 & \text{for } |\alpha| \leq m-1. \end{cases} \tag{2.61}$$

Here $\Omega \subset \mathbb{R}^n$ is a bounded smooth domain, f a datum in a suitable functional space and u denotes the unknown solution.

In order to give an explicit formula for solving (2.61), the first step is to define the fundamental solution of the polyharmonic operator $(-\Delta)^m$ in $\mathbb{R}^n$. We put

$$F_{m,n}(x) = \begin{cases} \dfrac{2\Gamma(n/2-m)}{ne_n 4^m \Gamma(n/2)(m-1)!} |x|^{2m-n} & \text{if } n > 2m \text{ or } n \text{ is odd,} \\[2ex] \dfrac{(-1)^{m-n/2}}{ne_n 4^{m-1}\Gamma(n/2)(m-n/2)!(m-1)!} |x|^{2m-n}(-\log|x|) & \text{if } n \leq 2m \text{ is even,} \end{cases}$$

so that, in distributional sense

$$(-\Delta)^m F_{m,n} = \delta_0, \tag{2.62}$$

where δ_0 is the Dirac mass at the origin. Of course, one may add any m-polyharmonic function to $F_{m,n}$ and still find (2.62). For $n > 2m$ there is a unique fundamental solution when one adds the "boundary condition"

$$\lim_{|x|\to\infty} F_{m,n}(x) = 0. \tag{2.63}$$

For $n \leq 2m$ no fundamental solution satisfies (2.63) and there does not seem to be a natural restriction to fix a unique fundamental solution.

Thanks to the fundamental solution, we may introduce the notion of Green function.

Definition 2.26. A Green function for (2.61) is a function $(x,y) \mapsto G(x,y) : \overline{\Omega} \times \overline{\Omega} \to \mathbb{R} \cup \{\infty\}$ satisfying:

1. $x \mapsto G(x,y) - F_{m,n}(x-y) \in C^{2m}(\Omega) \cap C^{m-1}(\overline{\Omega})$ for all $y \in \Omega$ if defined suitably for $x = y$;
2. $(-\Delta_x)^m \left(G(x,y) - F_{m,n}(x-y)\right) = 0$ for all $(x,y) \in \Omega^2$ if defined suitably for $x = y$;
3. $D_x^{\alpha} G(x,y) = 0$ for all $(x,y) \in \partial\Omega \times \Omega$ and $|\alpha| \leq m-1$.

Formally, the Green function enables one to write the unique solution to (2.61) as

$$u(x) = \int_{\Omega} G(x,y) f(y) dy. \tag{2.64}$$

Provided f belongs to a suitable functional space, this formula makes sense and gives the solution u.

Clearly, the exact form of the Green function G is not easily determined. However, as we already mentioned in Section 1.2, Boggio [63, p. 126] could explicitly calculate the Green function $G_{m,n} := G_{(-\Delta)^m,B}$ for problem (2.61) when Ω is the unit ball in $\mathbb{R}^n$.

Lemma 2.27. *The Green function for the Dirichlet problem (2.61) with $\Omega = B$ is positive and given by*

$$G_{m,n}(x,y) = k_{m,n} |x-y|^{2m-n} \int_{1}^{\left| |x|y - \frac{x}{|x|} \right| / |x-y|} (v^2 - 1)^{m-1} v^{1-n} dv. \tag{2.65}$$

The positive constants $k_{m,n}$ are defined by

$$k_{m,n} = \frac{1}{n e_n 4^{m-1} ((m-1)!)^2}, \qquad e_n = \frac{\pi^{n/2}}{\Gamma(1+n/2)}.$$

Remark 2.28. If $n > 2m$, then by applying the Cayley transform one finds for the half space $\mathbb{R}^n_+ = \{x \in \mathbb{R}^n : x_1 > 0\}$

$$G_{(-\Delta)^m, \mathbb{R}^n_+}(x,y) = k_{m,n} |x-y|^{2m-n} \int_{1}^{|x^*-y|/|x-y|} (v^2 - 1)^{m-1} v^{1-n} dv, \tag{2.66}$$

where $x,y \in \mathbb{R}^n_+$, $x^* = (-x_1, x_2, \ldots, x_n)$. We also emphasise that the assumption $n > 2m$ is required in this half space $\mathbb{R}^n_+$ in order to have uniqueness of the corresponding Green function. When $n \leq 2m$ one may achieve uniqueness in some cases by adding restrictions such as upper bounds for its growth at infinity (see

Remark 6.28 for the case $m = 2$ and $n = 3, 4$) . Alternatively, one may just *impose* that the Green function in the half space is the Cayley transform of its counterpart in the ball and hence given by (2.66).

2.7 The Space $H^2 \cap H_0^1$ and the Sapondžyan-Babuška Paradoxes

In this section, we consider in some detail the space $H^2 \cap H_0^1$ which is in some sense "intermediate" between H^2 and H_0^2. This space is also related to both the homogeneous Navier (2.21) and Steklov (2.22) boundary conditions, see the discussion following (2.34). The norm to be used in this space strongly depends on the smoothness of $\partial \Omega$. It was assumed in Theorem 2.20 that $\partial \Omega \in C^4$. We first show that this assumption may be relaxed in some cases. On the other hand, if it is "too relaxed" then uniqueness, regularity or continuous dependence may fail, leading to some apparent paradoxes. We also point out that the regularity of the boundary plays an important role in the definition of the first Steklov eigenvalue, see Section 3.3.2.

Let us first remark that in the case $m = 2$, Theorem 2.2 reads

Corollary 2.29. *Let $\Omega \subset \mathbb{R}^n$ be a smooth bounded domain. On the space $H_0^2(\Omega)$, the bilinear form*

$$(u, v) \mapsto (u, v)_{H_0^2} := \int_\Omega \Delta u \Delta v \, dx \qquad \text{for all } u, v \in H_0^2(\Omega)$$

defines a scalar product over $H_0^2(\Omega)$ which induces a norm equal to $\|D^2 \cdot \|_{L^2}$ and equivalent to $(\|D^2 \cdot \|_{L^2}^2 + \| \cdot \|_{L^2}^2)^{1/2}$.

We now show the less obvious result that the very same scalar product may also be used in the larger space $\ker \gamma_0 = H^2 \cap H_0^1(\Omega)$ when $\partial \Omega$ is not too bad. For later use, we state this result in general (possibly nonsmooth) domains. The class of domains considered is explained in the following definition taken from [2].

Definition 2.30. We say that a bounded domain $\Omega \subset \mathbb{R}^n$ satisfies an *outer ball condition* if for each $y \in \partial \Omega$ there exists a ball $B \subset \mathbb{R}^n \setminus \Omega$ such that $y \in \partial B$. We say that it satisfies a *uniform outer ball condition* if the radius of the ball B can be taken independently of $y \in \partial \Omega$.

In particular, a convex domain is a Lipschitz domain which satisfies a uniform outer ball condition. We have

Theorem 2.31. *Assume that $\Omega \subset \mathbb{R}^n$ is a Lipschitz bounded domain which satisfies a uniform outer ball condition. Then the space $H^2 \cap H_0^1(\Omega)$ becomes a Hilbert space when endowed with the scalar product*

$$(u, v) \mapsto \int_\Omega \Delta u \Delta v \, dx \qquad \text{for all } u, v \in H^2 \cap H_0^1(\Omega).$$

This scalar product induces a norm equivalent to $\| \cdot \|_{H^2}$.

Proof. Under the assumptions of the theorem Adolfsson [2] proved that there exists a constant $C > 0$ independent of u, such that

$$\|u\|_{H^2} \leq C\|\Delta u\|_{L^2} \qquad \text{for all } u \in H^2 \cap H_0^1(\Omega).$$

For all $u \in H^2 \cap H_0^1(\Omega)$ we also have

$$|D^2 u|^2 = \sum_{i,j=1}^{n} (\partial_{ij} u)^2 \geq \sum_{i=1}^{n} (\partial_{ii} u)^2 \geq \frac{1}{n}|\Delta u|^2 \qquad \text{a.e. in } \Omega. \tag{2.67}$$

This shows that the two norms are equivalent. $\qquad\square$

Remark 2.32. Let $\Omega \subset \mathbb{R}^n$ be a bounded and convex domain with smooth boundary. Consider the set

$$V := \left\{ u \in C^2(\overline{\Omega}); u \geq 0, \frac{\partial u}{\partial v} \geq 0, u\frac{\partial u}{\partial v} = 0 \text{ on } \partial\Omega \right\}.$$

Let W denote the closure of V with respect to the norm $\|.\|_{H^2}$. Then we have

$$\int_{\Omega} |D^2 u|^2 \, dx \leq \int_{\Omega} |\Delta u|^2 \, dx \qquad \text{for all } u \in W,$$

see [199, Theorem 2.1]. This inequality is somehow the converse of (2.67).

The assumptions on $\partial\Omega$ under which Theorem 2.31 holds are related to the so-called Sapondžyan or concave corner paradox. This paradox relies on the fact that, for some nonsmooth domains Ω, the linear Navier problem may have several different solutions, according to the functional space where they are sought.

One way of getting the existence is through a "system solution" which belongs to $H_0^1(\Omega)$ as well as its Laplacian. A second type of solution is obtained using Kondratiev's techniques in the space $H^2 \cap H_0^1(\Omega)$. Note that, since $\|.\|$ defined by $\|u\| := \|\Delta u\|_{L^2}$ is not a norm on $H^2 \cap H_0^1(\Omega)$ when the domain has a reentrant corner, one cannot directly apply the Lax-Milgram theorem. Indeed, Theorem 2.31 may not hold if a uniform outer ball condition fails. The following example appears suitable to illustrate this dichotomy in some detail.

Example 2.33. For $\alpha \in \left(\frac{1}{2}\pi, \pi\right)$ fix the domain

$$\Omega_\alpha = \{(r\cos\varphi, r\sin\varphi) \in \mathbb{R}^2; 0 < r < 1 \text{ and } |\varphi| < \alpha\}.$$

Let $f \in L^2(\Omega_\alpha)$ and consider the homogeneous Navier problem

$$\begin{cases} \Delta^2 u = f & \text{in } \Omega_\alpha, \\ u = 0 & \text{on } \partial\Omega_\alpha, \\ \Delta u = 0 & \text{on } \partial\Omega_\alpha \setminus \{0\}. \end{cases} \tag{2.68}$$

We say that u is a *system solution* to (2.68) if $u, \Delta u \in H_0^1(\Omega_\alpha)$ and

$$\begin{cases} -\Delta u = w \text{ and } -\Delta w = f \text{ in } \Omega_\alpha, \\ u = 0 \quad \text{and} \quad w = 0 \quad \text{on } \partial\Omega_\alpha. \end{cases} \qquad (2.69)$$

By applying twice the Lax-Milgram theorem in $H_0^1(\Omega_\alpha)$, this system solution, as a solution to an iterated Dirichlet Laplace problem on a bounded domain, exists for any $f \in L^2(\Omega_\alpha)$. Using [252] one finds that there also exists a solution in $H^2 \cap H_0^1(\Omega)$ of (2.68), which indeed satisfies $\Delta u = 0$ pointwise on $\partial\Omega \setminus \{0\}$. Since its second derivatives are square summable, let us call this the *energy solution*.

Next we consider a special function. For $\rho = \frac{\pi}{2\alpha}$ the function v_α defined by

$$v_\alpha(r, \varphi) = \left(r^{-\rho} - r^{\rho}\right) \cos(\rho\varphi)$$

satisfies

$$\begin{cases} -\Delta v_\alpha = 0 \text{ in } \Omega_\alpha, \\ v_\alpha = 0 \quad \text{on } \partial\Omega_\alpha \setminus \{0\}. \end{cases}$$

Moreover, one directly checks that $v_\alpha \in L^2(\Omega_\alpha)$ for $\rho \in \left(\frac{1}{2}, 1\right)$. Then there exists a unique solution $b_\alpha \in H_0^1(\Omega_\alpha)$ of

$$\begin{cases} -\Delta b_\alpha = v_\alpha \text{ in } \Omega_\alpha, \\ b_\alpha = 0 \quad \text{on } \partial\Omega_\alpha. \end{cases}$$

One has $\Delta b_\alpha \notin H_0^1(\Omega_\alpha)$ and may check that $b_\alpha \notin H^2(\Omega_\alpha)$. So we have found a nontrivial solution to (2.68) with $f = 0$. This b_α is neither a system solution nor an energy solution. Let u be the system solution. Then the following holds:

1. For all $c \in \mathbb{R}$ we have $u_c := u + c b_\alpha \in H_0^1(\Omega_\alpha)$ and $\Delta u_c \in L^2(\Omega_\alpha)$.
2. For all $c \in \mathbb{R}$, the function u_c satisfies (2.68). Using results in [319] one may show that in fact $u_c \in C^0(\overline{\Omega_\alpha})$ and $\Delta u_c \in C^0_{loc}(\overline{\Omega_\alpha} \setminus \{0\})$ whenever $f \in L^2(\Omega_\alpha)$.
3. One finds $\Delta u_c \in H_0^1(\Omega_\alpha)$ if and only if $c = 0$.
4. For $f \in L^2(\Omega_\alpha)$ let

$$c_\alpha(f) := -\|v_\alpha\|_{L^2}^{-2} \int_{\Omega_\alpha} v_\alpha \mathcal{G}_{-\Delta, \Omega_\alpha} f \, dx.$$

We have $u_c \in H^2 \cap H_0^1(\Omega_\alpha)$ if and only if $c = c_\alpha(f)$.
5. The energy solution to (2.68) is u_c with $c = c_\alpha(f)$. Hence the system solution is different from the energy solution whenever $c_\alpha(f) \neq 0$.

Now let f be positive. A close inspection shows, see [321], that for the system solution the H^2-regularity fails when $c_\alpha(f) \neq 0$ while positivity holds true. On the

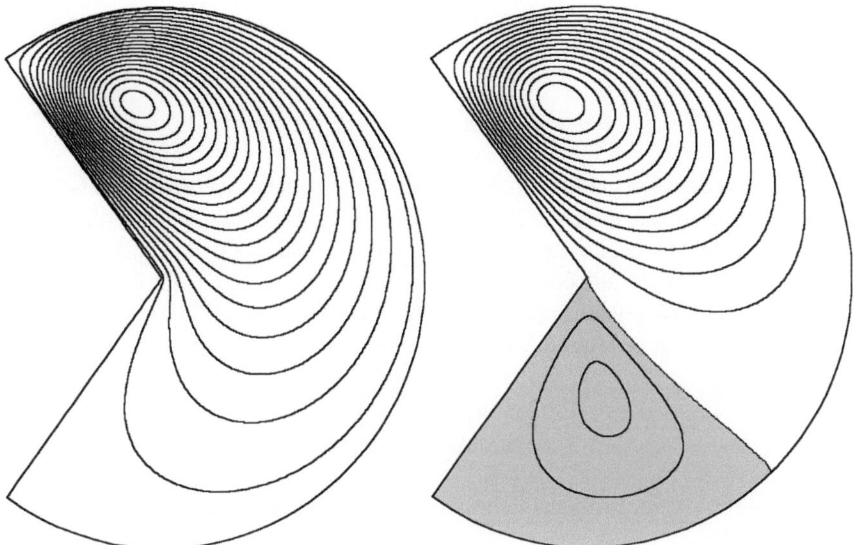

Fig. 2.1 The level lines of u and $u_c = u + cb_\alpha$ with $c = c_\alpha(f)$ for $f \geq 0$ having a small support near the left top of the domain. Grey region $= \{x : u_c(x) < 0\}$; here, a different scale is used for the level lines.

other hand, the energy solution u_c with $c = c_\alpha(f)$ has the appropriate regularity, but positivity fails when $\alpha > \frac{3}{4}\pi$ and

$$\int_{\Omega_\alpha} \left(r^{-\frac{\pi}{\alpha}} - r^{\frac{\pi}{\alpha}} \right) \sin\left(\frac{\pi}{\alpha}\varphi \right) \mathscr{G}_{-\Delta,\Omega_\alpha} f \, dx \neq 0.$$

For $\alpha \in (\frac{1}{2}\pi, \frac{3}{4}\pi)$ there is only numerical evidence of sign-changing energy solutions. See Figure 2.1.

We now discuss in detail another famous paradox due to Babuška, also known as the polygon-circle paradox. The starting point is a planar hinged plate Ω with a load $f \in L^2(\Omega)$. This gives rise to the Steklov problem

$$\begin{cases} \Delta^2 u = f & \text{in } \Omega, \\ u = \Delta u - (1-\sigma)\kappa\frac{\partial u}{\partial v} = 0 & \text{on } \partial\Omega, \end{cases} \tag{2.70}$$

where κ denotes the curvature of the boundary and σ is the Poisson ratio. Problem (2.70) is considered both in the unit disk B and in the sequence (see Figure 1.3) of inscribed regular polygons $(P_m) \subset B$ $(m \geq 3)$ with corners

$$\left\{ \left(\cos\left(\frac{2k}{m}\pi \right), \sin\left(\frac{2k}{m}\pi \right) \right) ; k = 1, \ldots, m \right\}.$$

Since the sides of P_m are flat, the curvature vanishes there and (2.70) becomes

$$\begin{cases} \Delta^2 u = f & \text{in } P_m, \\ u = \Delta u = 0 & \text{on } \partial P_m. \end{cases} \tag{2.71}$$

The so-called Babuška paradox shows that this argument is not correct, that is, (2.71) *is not* the right formulation of (2.70) when $\Omega = P_m$. The "infinite curvature" at the m corners cannot be neglected, one should instead consider Dirac delta-type contributions at each corner. Indeed, the next result states that the sequence of solutions to (2.71) does not converge (as $m \to \infty$) to the unique solution of (2.70) when $\Omega = B$. On the contrary, it converges to the unique solution of the following Navier problem

$$\begin{cases} \Delta^2 u = f & \text{in } B, \\ u = \Delta u = 0 & \text{on } \partial B. \end{cases} \tag{2.72}$$

More precisely, recalling the definition of system solution in (2.69), we have

Proposition 2.34. *Let $P_m \subset B$ with $m \geq 3$ be the interior of the regular polygon with corners $\left\{ \left(\cos\left(\frac{2k}{m}\pi\right), \sin\left(\frac{2k}{m}\pi\right) \right) ; k = 1, \ldots, m \right\}$ and let $f \in L^2(B)$. Then the following holds.*

1. *There exists a unique (weak) system solution u_m of (2.71) so that $u_m, \Delta u_m \in H_0^1(P_m)$.*
2. *There exists a unique minimiser $\tilde{u}_m$ in $\left\{ u \in H^2 \cap H_0^1(P_m) \right\}$ of*

$$J(u) = \int_{P_m} \left(\tfrac{1}{2}(\Delta u)^2 - fu \right) dx.$$

3. *The solution u_m satisfies*

$$u_m \in H^2 \cap C^1(\overline{P_m}) \text{ and } \Delta u_m \in H^2 \cap C^{0,\gamma}(\overline{P_m}),$$

   for $\gamma \in (0,1)$ and hence $\tilde{u}_m = u_m$.
4. *If we extend u_m by 0 on $B \backslash P_m$ and let u_∞ be the solution of (2.72), then*

$$\lim_{m \to \infty} \|u_m - u_\infty\|_{L^\infty(B)} = 0.$$

Note that even if the identity of u_m and $\tilde{u}_m$ may not seem very surprising, Example 2.33 shows that for domains with nonconvex corners these two solutions can be different.

Proof. 1. By the Lax-Milgram theorem, for any bounded domain Ω (in particular, if $\Omega = P_m$) and any $f \in L^2(\Omega)$ the Poisson problem

$$\begin{cases} -\Delta w = f & \text{in } \Omega, \\ w = 0 & \text{on } \partial\Omega, \end{cases} \tag{2.73}$$

has a unique (weak) solution in $H_0^1(\Omega)$. Similarly, one finds a unique solution in $H_0^1(\Omega)$ to $-\Delta u = w$ in Ω with $u = 0$ on $\partial \Omega$. We apply this fact to the case where $\Omega = P_m$.

2. The functional J is convex and coercive on $H^2 \cap H_0^1(P_m)$ in view of Theorem 2.31 since the corner points of P_m all have angles less then π (i.e. P_m satisfies a uniform outer ball condition). As the functional J is weakly lower semi-continuous and the closed unit ball in $H^2 \cap H_0^1(P_m)$ is weakly compact there exists a minimiser. The strict convexity gives uniqueness.

3. Invoking again [2] (or [241] since the P_m are convex), we know that the solution of (2.73) in P_m with source in $L^2(P_m)$ lies in $H^2(P_m)$. Hence $\Delta u_m \in H^2(P_m)$ and, by Theorem 2.6, $\Delta u_m \in C^{0,\gamma}(\overline{P_m})$ for all $\gamma \in (0,1)$. In fact, by Kondratiev [252] one finds that for a convex domain in two dimensions, with all corners having an opening angle less than or equal to $\alpha \in \left(\frac{\pi}{2}, \pi\right)$, the solution of (2.73) for $f \in L^p(\Omega)$ with $p < p_\alpha = \frac{2\alpha}{2\alpha - \pi}$ lies in $W^{2,p}(\Omega)$. Hence for each P_m one finds that $u_m \in W^{2,2+\varepsilon}(P_m)$ for $0 \le \varepsilon < \frac{4}{m-4}$. Theorem 2.6 then implies that $u_m \in C^1(\overline{P_m})$.

4. It is sufficient to prove this result for $f \ge 0$. For $r \in (0,1)$ we compare the solutions w_r of

$$\begin{cases} -\Delta w = f & \text{in } B_r, \\ w = 0 & \text{on } \partial B_r. \end{cases}$$

Extend w_r to $B \backslash B_r$ by 0. Assuming $f \ge 0$ and $f \in L^2(B)$ one finds that for $0 < r_1 < r_2 < 1$ it holds that $w_{r_1} \le w_{r_2}$ and moreover, that

$$\lim_{s \to r} \|w_s - w_r\|_{L^\infty(B)} = 0.$$

Indeed, if $f \in L^2(B)$, then $w_s, w_r \in C^{0,\gamma}(\overline{B})$ and for $s < r$ we find that

$$\begin{cases} -\Delta (w_r - w_s) = 0 & \text{in } B_s, \\ w_r - w_s = w_r & \text{on } \partial B_s. \end{cases}$$

Since $w_r \in C^{0,\gamma}(\overline{B})$, this yields

$$\|w_r\|_{L^\infty(B_r \backslash B_s)} \le C_f |r - s|^\gamma.$$

By the maximum principle

$$\|w_r - w_s\|_{L^\infty(B)} = \|w_r - w_s\|_{L^\infty(B_s)} \le \|w_r\|_{L^\infty(\partial B_s)} \le C_f |r - s|^\gamma.$$

Again using the maximum principle and writing u_r for the solution of (2.72) in B_r (instead of B) we find

$$u_r(x) \le u_1(x) \le u_r(x) + \tilde{C}_f |1 - r|^\gamma. \tag{2.74}$$

Two more applications of the maximum principle result in

$$u_{r_m}(x) \leq u_m(x) \leq u_1(x) \tag{2.75}$$

with $r_m = \cos(\pi/m)$ since $B_{r_m} \subset P_m \subset B$. The last claim follows by combining (2.74) and (2.75). $\qquad\square$

In order to emphasise the role played by smooth/nonsmooth boundaries in this paradox, we prove

Proposition 2.35. *Let $\Omega \subset \mathbb{R}^2$ be a bounded domain with C^2 boundary and let $\sigma \in (-1,1)$. If $f \in L^2(\Omega)$, then there exists a unique minimiser $u_\sigma \in H^2 \cap H_0^1(\Omega)$ of*

$$J(u) = \int_\Omega \left(\tfrac{1}{2}(\Delta u)^2 + (1-\sigma)\left(u_{xy}^2 - u_{xx}u_{yy}\right) - fu \right) dxdy.$$

If $\partial\Omega \in C^4$, then $u_\sigma \in H^4(\Omega)$ and u_σ satisfies

$$\begin{cases} \Delta^2 u = f & \text{in } \Omega, \\ u = \Delta u - (1-\sigma)\,\kappa\,\frac{\partial u}{\partial\nu} = 0 & \text{on } \partial\Omega. \end{cases} \tag{2.76}$$

Proof. Assume first that $0 \leq \sigma < 1$. Since the expression

$$\tfrac{1}{2}(\Delta u)^2 + (1-\sigma)\left(u_{xy}^2 - u_{xx}u_{yy}\right) = \tfrac{\sigma}{2}(u_{xx}+u_{yy})^2 + \tfrac{1-\sigma}{2}\left(u_{xx}^2 + 2u_{xy}^2 + u_{yy}^2\right)$$

is convex in the second derivatives of u, so is J. If $-1 < \sigma < 0$, following [332, Proposition 2.4] we note that

$$\begin{aligned} \langle dJ(v_1) - dJ(v_2), v_1 - v_2 \rangle &= \int_\Omega \left((v_1-v_2)_{xx}^2 + (v_1-v_2)_{yy}^2 + 2\sigma(v_1-v_2)_{xx}(v_1-v_2)_{yy} \right. \\ &\qquad\qquad \left. + 2(1-\sigma)(v_1-v_2)_{xy}^2 \right) dxdy \\ &\geq (1+\sigma)\int_\Omega \left((v_1-v_2)_{xx}^2 + (v_1-v_2)_{yy}^2 + 2(v_1-v_2)_{xy}^2 \right) dxdy \\ &> 0 \end{aligned}$$

for all $v_1, v_2 \in H^2 \cap H_0^1(\Omega)$ ($v_1 \neq v_2$), where we used the simple inequality $2\sigma ab \geq \sigma(a^2 + b^2)$. Hence, also for $-1 < \sigma < 0$, the functional J is convex.

Then existence and uniqueness of a minimiser for J are obtained as in Proposition 2.34. The minimiser u satisfies the weak Euler-Lagrange equation, that is

$$\int_\Omega \left(\Delta u \Delta\varphi + (1-\sigma)\left(2u_{xy}\varphi_{xy} - u_{xx}\varphi_{yy} - u_{yy}\varphi_{xx}\right) - f\varphi \right) dxdy = 0$$

for all $\varphi \in H^2 \cap H_0^1(\Omega)$. Regularity arguments (see Theorem 2.20) show that for $\partial\Omega \in C^4$ and $f \in L^2(\Omega)$ the minimiser lies in $H^4(\Omega)$. The integration by parts in (1.7) and (1.8) show that u_σ satisfies (2.76). $\qquad\square$

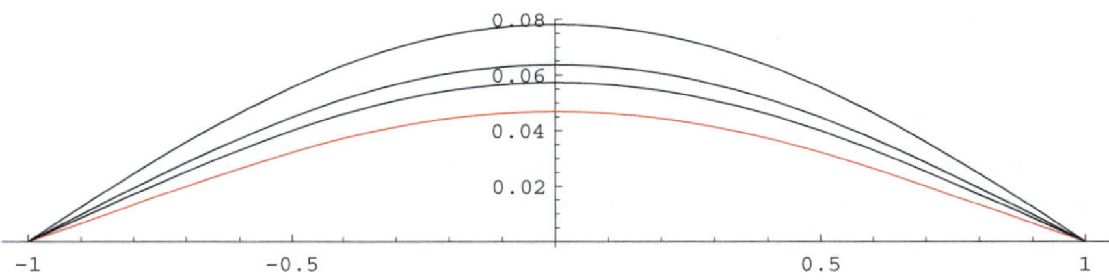

Fig. 2.2 The example of Babuška: $\Omega = B$ and $f = 1$. The solutions u_σ to (1.10) depend on σ; from top to bottom the solutions for $\sigma = 0, .3, .5$ and 1. The solution for $\sigma = 1$ is the limit of u_m from the regular m-polygon with $m \to \infty$.

By combining Propositions 2.34 and 2.35 we may now better explain the Babuška paradox. Assume that $f \in L^2(B)$ and let $\sigma \neq 1$. If u_∞ is as in Proposition 2.34 and u_σ is as in Proposition 2.35, then

$$u_\infty \equiv u_\sigma \text{ in } B \iff \frac{\partial u_\infty}{\partial \nu} \equiv 0 \text{ on } \partial B.$$

But if $0 \neq f \geq 0$, then the maximum principle implies $-\Delta u_\infty > 0$ and $u_\infty > 0$ whereas Hopf's boundary point Lemma even gives $\frac{\partial u_\infty}{\partial \nu} < 0$ on ∂B and hence $u_\infty \neq u_\sigma$. Babuška considered the case where $f = 1$ in B. This simple source allows us to compute all the functions involved. The solution to (2.76) with $f = 1$ on B is the radially symmetric function

$$u_\sigma(x) = \frac{(5+\sigma) - (6+2\sigma)|x|^2 + (1+\sigma)|x|^4}{64(1+\sigma)}.$$

The limit u_∞ in Babuška's example, defined by $u_\infty(x) = \lim_{m\to\infty} u_m(x)$ equals $u_{\sigma=1}(x)$, namely

$$u_\infty(x) = \frac{3}{64} - \frac{1}{16}|x|^2 + \frac{1}{64}|x|^4,$$

see Fig. 2.2 and also [402, p. 499] and [135].

2.8 Bibliographical Notes

Ellipticity and the complementing condition are well explained in [5], see also Section 1 of Chapter 2 in [276]. The polynomial L_m representing the differential operator is taken from [5, Section I.1], see also Section 1.1 of Chapter 2 in [276]. For boundary conditions that do not satisfy assumption (2.15) we refer again to [5]. The complementing condition is sometimes also called Lopatinski-Shapiro condition and may be defined in an equivalent way, see [417, Section 11]. Concerning assumption (2.17), we refer to [417, Theorem 13.1] for a general statement relating ellipticity, the complementing condition, regularity results, Fredholm theory and a priori estimates.

More results about Sobolev spaces may be found in the monographs by Adams [1], Maz'ya [292] and Lions-Magenes [276]. All the embedding theorems in Section 2.2.2 may be derived from [1, Theorems 5.4 and 6.2] whereas for the scalar product in $H^m(\Omega)$ see [276, Théorème 1.1]. Theorem 2.31 is taken from [80] and Theorem 2.7 is taken from [1, Theorem 5.23].

The material from Section 2.4 is taken from Section 9 of Chapter 2 in [276] and from [417]. Theorem 2.13 is a variant of Green's formula, see [417, Theorem 14.8]. A Hilbert triple as in Definition 2.14 is a particular case of a Gelfand triple, see [417, Definition 17.1]. The coercivity condition in (2.41) is given in $H^m(\Omega)$, the framework of our setting; it is taken from [276, Definition 2.9.1]. Agmon's condition which ensures the coercivity of the bilinear form $\Psi(u,v)$ was originally stated in [3]; we also refer to Theorem 9.3 of Chapter 2 in [276] and to [417, Section 18] for equivalent formulations. Theorem 2.16 is a direct consequence of the Lax-Milgram theorem, see [417, Theorem 17.10]. Existence, uniqueness and regularity results for (2.2)-(2.14) with data in the Hilbert spaces H^s with $s \in \mathbb{R}$ (possibly also non integer and negative) are studied in full detail in [276]; in particular, we refer to Remark 7.2 of Chapter 2 in [276] for a statement including all possible cases. Theorem 2.19 is contained in [5, Theorem 9.3] whereas Theorem 2.20 is contained in [5, Theorem 15.2]. Theorem 2.22 and the extension of Corollary 2.23 to all $k \geq 2$ are justified by [5, Theorem 15.3']. Theorem 2.25 follows as a by-product of Theorems 2.6 and 2.20 on one hand and maximum modulus estimates for solutions of higher order elliptic equations on the other hand. This second tool was introduced by Miranda [305, 306] for higher order problems in the plane and subsequently generalised by Agmon [4] in any space dimension. We also refer to [378] for a simple sketch of the proof. Finally, let us mention that partial extensions and counterexamples to Theorem 2.25 in *nonsmooth* domains may be found in works by Pipher-Verchota [338,339], Maz'ya-Rossmann [291], Mayboroda-Maz'ya [285, 286] and references therein. Lemma 2.27 is a fundamental contribution by Boggio [63] and is one of the most frequently used results in this monograph. Results on Green's functions may also be found in the monograph [21]. As for L^p-theory of higher-order elliptic operators and underlying kernel estimates one may also see the survey article [129] by Davies. Theorem 2.31 is a straightforward consequence of results by Adolfsson [2], see also [238,239] for related results. Concerning the Sapondžyan paradox, we refer to the original paper [358] and to more recent results on "multiple" solutions in [321]. Babuška [28] noticed first that by approximating a curvilinear domain by polygons the approximating solutions would not converge to the solution on the curvilinear domain. Engineering approaches to the Babuška or polygon-circle paradox can be found in [242, 315, 355]. A mathematical approach can be found in the work by Maz'ya and Nazarov [293, 294]. These authors dealt with the paradox by a careful asymptotic analysis of the boundary layer and the contribution of the corners in this. Part of their results are based on Γ-convergence results from [423]. More recently, Davini [134] again uses Γ-convergence to find a correct approximation. He focuses on numerical methods that avoid the pitfall of this paradox. Most part of the material in Section 2.7 is taken from [388].

Chapter 3
Eigenvalue Problems

For quite general second order elliptic operators one may use the maximum principle and the Kreĭn-Rutman theorem to show that the eigenfunction corresponding to the first eigenvalue has a fixed sign. It is then a natural question to ask if a similar result holds for higher order Dirichlet problems where a general maximum principle is not available. A partial answer is that a Kreĭn-Rutman type argument can still be used whenever the boundary value problem is positivity preserving. We will also explain in detail an alternative dual cone approach. Both these methods have their own advantages. The Kreĭn-Rutman approach shows under fairly weak assumptions that there exists a real eigenvalue and, somehow as a byproduct, one finds that the eigenvalue and the corresponding eigenfunction are positive. It applies in particular to non-selfadjoint settings. The dual cone decomposition only applies in a selfadjoint framework in a Hilbert space, where the existence of eigenfunctions is well-known. But in this setting it provides a very simple proof for positivity and simplicity of the first eigenfunction. A further quality of this method is that it applies also to some nonlinear situations as we shall see in Chapter 7.

We conclude this first part of the chapter with some further remarks on the connection between the positivity preserving property of the Dirichlet problem and the fixed sign property of the first eigenfunction. In particular, we show that the latter property, as well as the simplicity of the first eigenvalue, may fail.

Then we turn to the minimisation of the first Dirichlet eigenvalue of the biharmonic operator among domains of fixed measure and we show that, in dimensions $n = 2$ and $n = 3$, the ball achieves the minimum. We also consider two further eigenvalue problems, the buckling load of a clamped plate and Steklov eigenvalues. Up to some regularity to be proved, a quite hard open problem, the disk minimises the buckling load among planar domains of given measure. For the Steklov problem we first study in detail the whole spectrum and then we show that an optimal shape of given measure which minimises the first eigenvalue does not exist.

F. Gazzola et al., *Polyharmonic Boundary Value Problems*,
Lecture Notes in Mathematics 1991, DOI 10.1007/978-3-642-12245-3_3,
© Springer-Verlag Berlin Heidelberg 2010

3.1 Dirichlet Eigenvalues

Here we consider the eigenvalue problem

$$\begin{cases} (-\Delta)^m u = \lambda u, \ u \not\equiv 0 & \text{in } \Omega, \\ D^\alpha u|_{\partial\Omega} = 0 & \text{for } |\alpha| \leq m-1 \end{cases} \tag{3.1}$$

on a given bounded domain $\Omega \subset \mathbb{R}^n$ ($n \geq 2$). The first eigenvalue of (3.1) is defined as

$$\Lambda_{m,1}(\Omega) = \min_{u \in H_0^m(\Omega)\setminus\{0\}} \frac{\|u\|_{H_0^m}^2}{\|u\|_{L^2}^2}. \tag{3.2}$$

In this section we discuss several problems related to (3.1) and to its first eigenvalue. We start by showing that the corresponding eigenfunction is of one sign whenever the problem

$$\begin{cases} (-\Delta)^m u = f & \text{in } \Omega, \\ D^\alpha u|_{\partial\Omega} = 0 & \text{for } |\alpha| \leq m-1 \end{cases} \tag{3.3}$$

is positivity preserving, see Definition 3.1 below. This statement can be obtained in two different ways, either with a somehow standard Kreĭn-Rutman type argument or with a decomposition in dual cones which we discuss in detail. Next, we discuss the positivity of the first eigenfunction and its failure in general and we end up with the minimisation of the first eigenvalue among domains of given measure.

3.1.1 A Generalised Kreĭn-Rutman Result

The Kreĭn-Rutman theorem, which can be considered to have its roots in Jentzsch's theorem, appears in many forms with many different and partially overlapping conditions but none of the classical versions are optimal for the solution operator of an elliptic boundary value problem. The main restriction is the necessity of having a positive cone with an open interior, see [258, Theorems 6.2 and 6.3]. As we shall see, this restriction could be removed after a profound result of de Pagter [136].

Consider the linear problem (3.3) and the following notion of positivity preserving.

Definition 3.1. We say that (3.3) has a *positivity preserving property* when the following holds for all u and f satisfying (3.3):

$$f \geq 0 \Rightarrow u \geq 0.$$

In case that a Green function exists, the positivity preserving property holds true if and only if this Green function is nonnegative. We now establish that if (3.3) is positivity preserving then a Kreĭn-Rutman result allows one to verify that the first eigenvalue for (3.1) is simple and corresponds to an eigenfunction of fixed sign.

Let us shortly introduce some terminology.

Definition 3.2. An ordered Banach space $(E, \|\,.\,\|, \geq)$ is called a *Banach lattice* if:

- the least upper bound of two elements in E lies again in E:

$$f, g \in E \text{ implies } f \vee g := \inf\{h \in E; h \geq f \text{ and } h \geq g\} \in E;$$

- the ordering of positive elements is preserved by the norm: setting $|f| = f \vee (-f)$ it holds for all $f, g \in E$ that $|f| \leq |g|$ implies $\|f\| \leq \|g\|$.

A linear subspace $A \subset E$ is called a *lattice ideal* if

$$|f| \leq |g| \text{ and } g \in A \text{ implies } f \in A.$$

We call A invariant under the operator $T : E \to E$ if $T(A) \subset A$.

We can now give a statement which improves the classical Kreĭn-Rutman theorem, see [258].

Theorem 3.3. *Let E be a Banach lattice with $\dim(E) > 1$ and let $T : E \to E$ be a linear operator satisfying:*

- *T is compact;*
- *T is positive: $T(\mathcal{K}) \subset \mathcal{K}$ where $\mathcal{K}$ is the positive cone in E;*
- *T is irreducible: $\{0\}$ and E are the only closed lattice ideals invariant under T.*

Then the spectral radius ρ of T is strictly positive and there exists $v \in \mathcal{K} \setminus \{0\}$ with $Tv = \rho v$. Moreover, the algebraic multiplicity of ρ is one, all other eigenvalues $\tilde{\rho}$ satisfy $|\tilde{\rho}| < \rho$ and no other eigenfunction is positive.

By Lemma 2.27 we know that the Green function in the ball B is positive so that problem (3.3) has a positivity preserving property whenever $\Omega = B$. In fact, Theorem 3.3 applies to any domain Ω where the corresponding Green function G_Ω is strictly positive. In this case, one takes $E = L^2(\Omega)$ or $E = \{v \in C(\overline{\Omega}); v|_{\partial\Omega} = 0\}$ and T as the solution operator for (3.3). Since for each $x \in \Omega$ the Green function $G_\Omega(x,.)$ is strictly positive on $\overline{\Omega}$ except for a set of measure 0 it follows in both settings that T is irreducible.

3.1.2 Decomposition with Respect to Dual Cones

We state and discuss here an abstract result due to Moreau [312] about the decomposition of a Hilbert space H into dual cones; we recall that $\mathcal{K} \subset H$ is a cone if $u \in \mathcal{K}$ and $a \geq 0$ imply that $au \in \mathcal{K}$. In order to exploit the full power of this decomposition, we also establish a generalised Boggio result. This will be used in several different points of this monograph. Finally, we give a first simple application of this decomposition to a capacity problem.

It was Miersemann [302] who first observed that the dual cone decomposition could be quite helpful in the context of fourth order elliptic equations. In the next section we show how this method can be used to prove simplicity and positivity of the first eigenfunction of (3.1) in a ball. Moreover, this decomposition will turn out to be quite useful in a number of semilinear problems considered in this monograph.

Theorem 3.4. *Let H be a Hilbert space with scalar product $(.,.)_H$. Let $\mathcal{K} \subset H$ be a closed convex nonempty cone. Let $\mathcal{K}^*$ be its dual cone, namely*

$$\mathcal{K}^* = \{w \in H;\ (w,v)_H \leq 0 \text{ for all } v \in \mathcal{K}\}.$$

Then for any $u \in H$ there exists a unique $(u_1,u_2) \in \mathcal{K} \times \mathcal{K}^*$ such that

$$u = u_1 + u_2, \qquad (u_1,u_2)_H = 0. \tag{3.4}$$

In particular, $\|u\|_H^2 = \|u_1\|_H^2 + \|u_2\|_H^2$.

Moreover, if we decompose arbitrary $u,v \in H$ according to (3.4), namely $u = u_1 + u_2$ and $v = v_1 + v_2$, then we have that

$$\|u - v\|_H^2 \geq \|u_1 - v_1\|_H^2 + \|u_2 - v_2\|_H^2.$$

In particular, the projection onto $\mathcal{K}$ is Lipschitz continuous with constant 1.

Proof. For a given $u \in H$, we prove separately existence and uniqueness of a decomposition satisfying (3.4).

Existence. Let u_1 be the projection of u onto $\mathcal{K}$ defined by

$$\|u - u_1\| = \min_{v \in \mathcal{K}} \|u - v\|$$

and let $u_2 := u - u_1$. Then for all $t \geq 0$ and $v \in \mathcal{K}$ one has

$$\|u - u_1\|_H^2 \leq \|u - (u_1 + tv)\|_H^2 = \|u - u_1\|_H^2 - 2t(u - u_1, v)_H + t^2\|v\|_H^2$$

so that

$$2t(u_2, v)_H \leq t^2\|v\|_H^2. \tag{3.5}$$

Dividing by $t > 0$ and letting $t \searrow 0$, (3.5) yields $(u_2, v)_H \leq 0$ for all $v \in \mathcal{K}$ so that $u_2 \in \mathcal{K}^*$. Choosing $v = u_1$ also allows for taking $t \in [-1, 0)$ in (3.5), so that dividing by $t < 0$ and letting $t \nearrow 0$ yields $(u_2, u_1)_H \geq 0$ which, combined with the just proved converse inequality, proves that $(u_2, u_1)_H = 0$.

Lipschitz continuity. From the two inequalities $(u_1, v_2)_H \leq 0$ and $(v_1, u_2)_H \leq 0$ and by orthogonality, we obtain

$$
\begin{aligned}
\|u - v\|_H^2 &= (u_1 + u_2 - v_1 - v_2, u_1 + u_2 - v_1 - v_2)_H \\
&= ((u_1 - v_1) + (u_2 - v_2), (u_1 - v_1) + (u_2 - v_2))_H \\
&= \|u_1 - v_1\|_H^2 + \|u_2 - v_2\|_H^2 + 2(u_1 - v_1, u_2 - v_2)_H \\
&= \|u_1 - v_1\|_H^2 + \|u_2 - v_2\|_H^2 - 2(u_1, v_2)_H - 2(v_1, u_2)_H \\
&\geq \|u_1 - v_1\|_H^2 + \|u_2 - v_2\|_H^2
\end{aligned}
$$

and Lipschitz continuity follows.

Uniqueness. It follows from the Lipschitz continuity by taking $u = v$. $\qquad\square$

Remark 3.5. In the context of an abstract Hilbert space it is quite easy to gain an imagination of the projection u_1 of a general element u onto $\mathscr{K}$. However, in the concrete context of function spaces it is difficult to really see how u_1 arises from u and $\mathscr{K}$. Here, a different point of view is helpful: $u_2 := u - u_1 \in H$ is characterised by minimising $\|\,.\,\|_H$ subject to the constraint that $u - u_2 \in \mathscr{K}$. In the framework of the function space $H_0^2(\Omega)$ equipped with the scalar product $(u, v)_{H_0^2(\Omega)} = \int_\Omega \Delta u \, \Delta v$ and the cone $\mathscr{K} \subset H_0^2(\Omega)$ of nonnegative functions this means that u_2 has minimal (quadratic) elastic energy $\int_\Omega (\Delta u_2)^2$ among all H_0^2-functions subject to the constraint that $u - u_2 \geq 0$, i.e. $u_2 \leq u$. This means that one seeks u_2 as the solution of an obstacle problem, see [249]. See Figure 3.1 for an example of a dual cone decomposition in H_0^2. We refer to [68] for further explanations and for some explicit examples of the dual cone decomposition.

Note also that the Lipschitz continuity property stated in Theorem 3.4 strongly depends on the norm considered. To see this, consider the special case $H = H_0^1(-1, 1)$ with $(u, v)_H = \int_{-1}^1 u' v'$ and let $\mathscr{K} = \{v \in H : v \geq 0 \text{ a.e.}\}$. For any $\varepsilon \in (0, 1)$ let

$$
u^\varepsilon(x) := \begin{cases} \dfrac{|x|}{\varepsilon} - 1 & \text{if } |x| \leq \varepsilon, \\[2mm] 0 & \text{otherwise.} \end{cases}
$$

Then

$$
u_2^\varepsilon(x) = |x| - 1, \qquad u_1^\varepsilon(x) = u^\varepsilon(x) - u_2^\varepsilon(x) = \begin{cases} \dfrac{|x|}{\varepsilon} - |x| & \text{if } |x| \leq \varepsilon, \\[2mm] 1 - |x| & \text{otherwise.} \end{cases}
$$

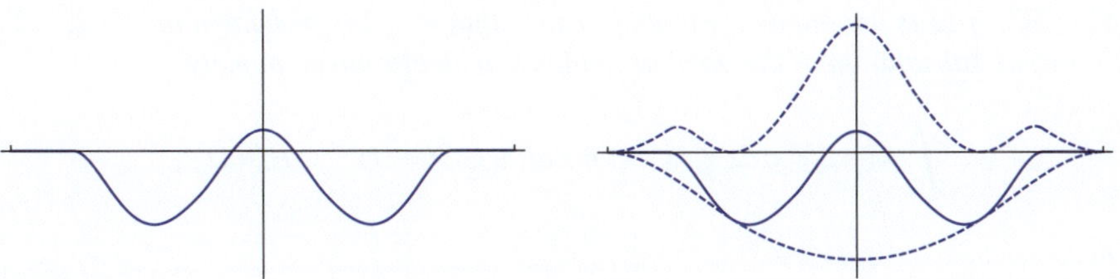

Fig. 3.1 Dual cone decomposition (right) in H_0^2 of the function displayed on the left.

Therefore, $u^\varepsilon \to 0$ in $L^2(-1,1)$ as $\varepsilon \searrow 0$, while $u_1^\varepsilon \to 1 - |x|$ and $u_2^\varepsilon \equiv |x| - 1$. This shows that the decomposition in H_0^1 is not continuous with respect to the L^2-norm.

Let us now explain how we are planning to use the decomposition in Theorem 3.4. We will take H as a Hilbertian functional space (L^2, H^2, $\mathscr{D}^{2,2}\ldots$) and

$$\mathscr{K} = \{u \in H; \ u \geq 0 \text{ a.e.}\}.$$

If $H = L^2(\Omega)$, then $\mathscr{K}^* = -\mathscr{K}$ and Theorem 3.4 simply yields

$$u = u^+ - u^- \qquad \text{for all } u \in L^2(\Omega).$$

If $H = H_0^1(\Omega)$, in order to characterise $\mathscr{K}^*$ we seek $v \in H_0^1(\Omega)$ such that

$$\int_\Omega \nabla u \nabla v \, dx \leq 0 \qquad \text{for all } u \in \mathscr{K}.$$

This means that v is weakly subharmonic (formally, $\int_\Omega u \Delta v \geq 0$) and therefore

$$\mathscr{K}^* = \{v \in H_0^1(\Omega); \ v \text{ is weakly subharmonic}\} \subsetneq -\mathscr{K}.$$

Note that although $\int_\Omega \nabla u^+ \nabla u^- = 0$, the decomposition obtained here is different from $u = u^+ - u^-$.

In higher order Sobolev spaces the decomposition $u = u^+ - u^-$ is no longer admissible because if $u \in H^m$ ($m \geq 2$) then, in general, $u^+, u^- \notin H^m$. In some situations the decomposition into dual cones may substitute the decomposition into positive and negative part. In order to facilitate and strengthen the application of Theorem 3.4 to higher order Sobolev spaces, we generalise Boggio's principle (Lemma 2.27) to weakly subpolyharmonic functions in suitable domains. Let us consider again $\mathscr{K} = \{v \in H_0^m(\Omega); \ v \geq 0 \text{ a.e. in } \Omega\}$ (or $v \in \mathscr{D}^{m,2}(\Omega)$ if Ω is unbounded and $n > 2m$), then

$$\mathscr{K}^* = \left\{w \in H_0^m(\Omega); \ (w,v)_{H_0^m} \leq 0 \text{ for all } v \in \mathscr{K}\right\}.$$

Hence, $\mathscr{K}^* = \{v \in H_0^m(\Omega); \ (-\Delta)^m v \leq 0 \text{ weakly}\}$. In some cases, we have that $\mathscr{K}^* \subset -\mathscr{K}$.

Proposition 3.6. *Assume that either $\Omega = B_R$ (a ball of radius R), or $\Omega = \mathbb{R}^n_+$, or $\Omega = \mathbb{R}^n$; if Ω is unbounded, we also assume that $n > 2m$. Assume that $w \in L^2(\Omega)$ is a weak subsolution of the polyharmonic Dirichlet problem, namely*

$$\int_\Omega w(-\Delta)^m u \, dx \leq 0 \qquad \text{for all } u \in \mathscr{K} \cap H^{2m} \cap H_0^m(\Omega) \ ;$$

then

$$\text{either } w \equiv 0 \text{ or } w < 0 \text{ a.e. in } \Omega. \tag{3.6}$$

In particular, (3.6) holds for all $w \in \mathscr{K}^*$.

Proof. We only prove the result in the case where $\Omega = B$ (the unit ball), the remaining cases being similar. Assuming $n > 2m$, for the half space it suffices to use (2.66) instead of (2.65) whereas for the whole space one takes the fundamental solution of $(-\Delta)^m$.

Take any $\varphi \in \mathcal{K} \cap C_c^\infty(B)$ and let v_φ be the unique (classical) solution of

$$\begin{cases} (-\Delta)^m v_\varphi = \varphi & \text{in } B, \\ D^\alpha v_\varphi = 0 & \text{on } \partial B \qquad \text{for } |\alpha| \le m-1. \end{cases}$$

Then by the classical Boggio's principle (Lemma 2.27) we infer that $v_\varphi \in \mathcal{K}$. Hence, v_φ is a possible test function for all such φ and therefore

$$\int_B w\varphi \, dx = \int_B w(-\Delta)^m v_\varphi \, dx \le 0 \qquad \text{for all } \varphi \in \mathcal{K} \cap C_c^\infty(B).$$

This shows that $w \le 0$ a.e. in B. Assume that $w \not\equiv 0$ a.e. in B and let ϕ denote the characteristic function of the set $\{x \in B; w(x) = 0\}$ so that $\phi \ge 0$, $\phi \not\equiv 0$. Let v_0 be the unique (a.e.) solution of the problem

$$\begin{cases} (-\Delta)^m v_0 = \phi & \text{in } B, \\ D^\alpha v_0 = 0 & \text{on } \partial B \qquad \text{for } |\alpha| \le m-1. \end{cases}$$

Then by Corollary 2.21 and Theorem 2.6, we know that

$$v_0 \in \left(\bigcap_{q \ge 1} W^{2m,q}(B) \right) \subset C^{2m-1}(\overline{B})$$

and again by Boggio's principle we have $v_0 > 0$ in B. One also reads from Boggio's formula (2.65) that $(-\frac{\partial}{\partial \nu})^m v_0 > 0$ on ∂B, see Theorem 5.7 below. We infer that for all $v \in C^{2m}(\overline{B}) \cap H_0^m(B)$ there exist $t_1 \le 0 \le t_0$ such that $v + t_0 v_0 \ge 0$ and $v + t_1 v_0 \le 0$ in B. This, combined with the fact that

$$\int_B w(-\Delta)^m v_0 \, dx = \int_{\{w=0\}} w \, dx = 0,$$

enables us to show that both

$$0 \le \int_B w(-\Delta)^m (v + t_0 v_0) \, dx = \int_B w(-\Delta)^m v \, dx,$$

$$0 \ge \int_B w(-\Delta)^m (v + t_1 v_0) \, dx = \int_B w(-\Delta)^m v \, dx.$$

Hence, we have for all $v \in C^{2m}(\overline{B}) \cap H_0^m(B)$

$$\int_B w(-\Delta)^m v \, dx = 0.$$

We need to show that $C^{2m}(\overline{B}) \cap H_0^m(B)$ is dense in $H^{2m} \cap H_0^m(B)$. For this purpose, take any function $U \in H^{2m} \cap H_0^m(B)$ and put $f := (-\Delta^m)U$. We approximate f in $L^2(B)$ by $C^\infty(\overline{B})$–functions f_k and solve $(-\Delta^m)U_k = f_k$ in B under homogeneous Dirichlet boundary conditions. We then even have $U_k \in C^\infty(\overline{B})$, and by L^2–theory (see Corollary 2.21) it holds that $\|U_k - U\|_{H^{2m}} \to 0$ as $k \to \infty$.

By the previous statement we may now conclude that

$$\int_B w(-\Delta)^m v\, dx = 0 \qquad \text{for all } v \in H^{2m} \cap H_0^m(B).$$

Since $w \in L^2(B)$, we may take as $v \in H^{2m} \cap H_0^m(B)$ the solution of $(-\Delta)^m v = w$ under homogeneous Dirichlet boundary conditions. This finally yields $w \equiv 0$. $\qquad\square$

We conclude this section with a first simple application of Theorem 3.4. We show the positivity of the potential in the second order capacity problem. Given a bounded domain $\Omega \subset \mathbb{R}^n$ $(n > 4)$ we define its second order capacity as

$$\mathrm{cap}(\Omega) = \inf\left\{ \int_{\mathbb{R}^n} |\Delta u|^2\, dx;\ u \in \mathscr{D}^{2,2}(\mathbb{R}^n),\ u \geq 1 \text{ a.e. in } \Omega \right\}.$$

Using Theorem 3.4 we can show that the potential (the minimiser) $\bar{u}$ is nonnegative. Let $H = \mathscr{D}^{2,2}(\mathbb{R}^n)$ and let $\mathscr{K} = \{u \in H : u \geq 0 \text{ a.e. in } \mathbb{R}^n\}$. If $\bar{u}$ is sign-changing, let $\bar{u} = u_1 + u_2$ with $u_1 \in \mathscr{K}$ and $u_2 \in \mathscr{K}^* \setminus \{0\}$ be its decomposition according to Theorem 3.4. Then by Proposition 3.6 we know that $u_2 \leq 0$. Hence, $u_1 \geq 1$ in Ω so that u_1 is an admissible function. Moreover,

$$\int_{\mathbb{R}^n} |\Delta \bar{u}|^2\, dx = \int_{\mathbb{R}^n} |\Delta u_1|^2\, dx + \int_{\mathbb{R}^n} |\Delta u_2|^2\, dx > \int_{\mathbb{R}^n} |\Delta u_1|^2\, dx.$$

This contradicts the minimality of $\bar{u}$ among admissible functions.

3.1.3 *Positivity of the First Eigenfunction*

In this section we study positivity of the first eigenfunction of (3.1) by means of the just explained dual cone decomposition. As already mentioned in Section 3.1.1, whenever we have a positivity preserving solution operator, a Kreĭn-Rutman result yields a positive first eigenfunction with the uniqueness properties stated in Theorem 3.3. In our special self-adjoint situation, the dual cone decomposition yields a direct and simpler proof. Moreover, this strategy can also be exploited for semilinear problems, see e.g. Lemma 7.22 and Theorem 7.58.

Theorem 3.7. *If $\Omega = B \subset \mathbb{R}^n$, then the first eigenvalue*

$$\Lambda_{m,1} = \inf_{H_0^m(B)\setminus\{0\}} \frac{\|u\|_{H_0^m}^2}{\|u\|_{L^2}^2}. \tag{3.7}$$

of (3.1) is simple and the corresponding first eigenfunction u is of one sign.

Proof. Let $H = H_0^m(B)$ and let $\mathscr{K} = \{u \in H : u \geq 0 \text{ a.e. in } B\}$. For contradiction, assume that u changes sign. Then according to Theorem 3.4, we decompose $u = u_1 + u_2$, $u_1 \in \mathscr{K} \setminus \{0\}$, $u_2 \in \mathscr{K}^* \setminus \{0\}$. By Proposition 3.6, we have $u_2 < 0$ a.e. in B. Replacing u with the positive function $u_1 - u_2$ would strictly increase the L^2-norm while by orthogonality we have $\|u_1 + u_2\|_H^2 = \|u_1 - u_2\|_H^2$. The ratio would strictly decrease, a contradiction.

Since a minimiser $u \geq 0$ solves $(-\Delta)^m u = \Lambda_{m,1} u \geq 0$ we have $u > 0$ by Proposition 3.6. By contradiction, assume now that (3.7) admits two linearly independent positive minimisers u and v. Then $w = u + \alpha v$ (for a suitable $\alpha < 0$) is a sign-changing minimiser, contradiction! $\qquad\square$

For $m = 1$ the very same technique used in Theorem 3.7 works in any bounded domain Ω if we wish to show that the first eigenvalue of $-\Delta$ in $H_0^1(\Omega)$ is simple and that the corresponding eigenfunction is of one sign. On the other hand, the L^2-norm remains constant if we replace $u^+ - u^-$ with $u^+ + u^-$. So, for this problem, the decomposition into dual cones works directly, whereas the usual decomposition into positive and negative parts does not prove simplicity of the first eigenvalue nor fixed sign of the first eigenfunction without a further regularity argument.

Theorem 3.7 applies to any bounded domain $\Omega \neq B$ with a positive Green function also for $m \geq 2$. Note that the positivity preserving property (positivity of the Green function) implies fixed sign of the first eigenfunction to (3.1) but the converse implication does not hold in general. One can then wonder whether a positive first eigenfunction can be obtained also for domains which fail to have the positivity preserving property, see Definition 3.1. The answer is delicate and negative in general.

Let us quickly outline what is known for sign-changing first eigenfunctions of

$$\begin{cases} \Delta^2 u = \lambda u, \ u \not\equiv 0 & \text{in } \Omega, \\ u = |Du| = 0 & \text{on } \partial\Omega. \end{cases} \tag{3.8}$$

Basically, only this biharmonic eigenvalue problem on bounded domains has been considered so far. Concerning (3.8) it is proven in [213] that for an appropriately defined family of perturbations starting from the ball the positivity preserving property fails to hold strictly before the first eigenfunction loses its fixed sign. So, the converse implication on ellipses as mentioned above is not true, see also Remark 6.4. It does not seem to be rigorously proven yet that the sign of the first eigenfunction changes on ellipses with a large ratio but there exists numerical evidence.

The first example of a sign-changing first eigenfunction is due to Coffman [107] and deals with squares.

Theorem 3.8. *For $\Omega = (0,1)^2$ problem (3.8) has a sign-changing first eigenfunction.*

Independently of previous results in [111], Kozlov-Kondratiev-Maz'ya [253] proved that domains in any space dimension whose boundaries contain suitable cones also have a sign-changing first eigenfunction for (3.8). Their results cover

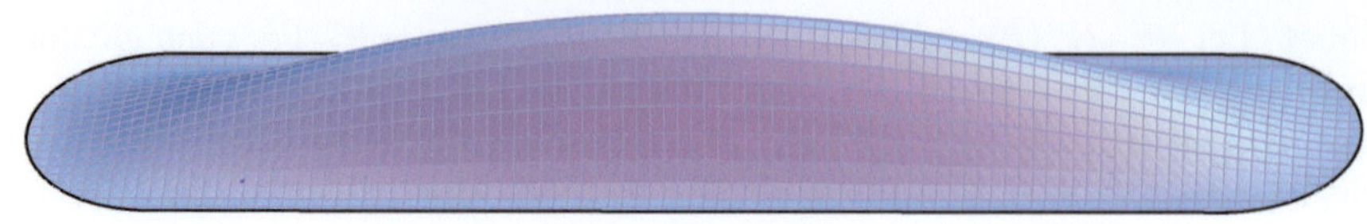

Fig. 3.2 Stadium-like domains seem to have a positive first eigenfunction in the Dirichlet biharmonic case.

a class of elliptic operators of order $2m$ under Dirichlet boundary conditions. Their proof is based on a result which ensures the absence of zeros of infinite order at the vertex of a cone, for nontrivial nonnegative local solutions of the inequality $Au \leq 0$, where A is an elliptic differential operator with real coefficients. Moreover, they constructed a sequence of smooth convex domains that exhaust the cone and since the corresponding first eigenfunctions are proven to converge to the sign-changing first eigenfunction in the cone the same holds eventually for the approximating domains.

A main assumption that often appears is the convexity of Ω. From Theorem 3.8 and the numerical evidence on eccentric ellipses it is clear that this assumption will not be sufficient to ensure positivity of the first eigenfunction. Ellipses suggest that, possibly, a suitable upper bound for the ratio between the radii of the largest inscribed ball in Ω and the largest filling balls of Ω might yield a sufficient condition for a positive eigenfunction. We recall that B is a filling ball for Ω if Ω is the union of translated B. Clearly, for any bounded domain this ratio is always larger than or equal to 1. An interesting family of domains in this sense are elongated disks, the so-called stadiums, where the radius of the largest inscribed ball equals the radius of the largest filling ball. Numerical approximations of the first eigenfunction on such a domain always resulted in functions apparently of fixed sign, see Fig. 3.2.

Domains which are far from being convex are domains with holes. The standard examples are the annuli

$$A_\varepsilon = \{(x,y) \in \mathbb{R}^2; \varepsilon^2 < x^2 + y^2 < 1\} \qquad \text{with } 0 < \varepsilon < 1.$$

For these domains, Coffman-Duffin-Shaffer [109, 110, 155] proved the following somehow surprising statement.

Theorem 3.9. *Let $\Omega = A_\varepsilon$ for some $\varepsilon \in (0,1)$ and consider problem (3.8). There exists $\varepsilon_0 > 0$ such that the following holds.*

1. *If $\varepsilon < \varepsilon_0$, then the first eigenvalue has multiplicity two. There exist two independent eigenfunctions for this first eigenvalue with diametral nodal lines.*
2. *If $\varepsilon = \varepsilon_0$, then the first eigenvalue has multiplicity three. There exists a positive eigenfunction for this eigenvalue and there are two independent eigenfunctions with diametral nodal lines.*
3. *If $\varepsilon > \varepsilon_0$, then the first eigenvalue has multiplicity one and the corresponding eigenfunction is of fixed sign.*

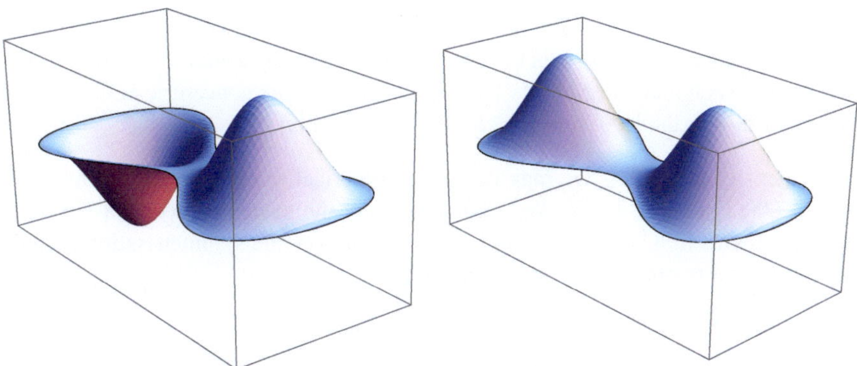

Fig. 3.3 On the left the first eigenfunction for the clamped biharmonic eigenvalue problem on an 8-shaped domain which is sign-changing. On the right the second eigenfunction which is (almost) positive.

It is not surprising that a large hole yields a positive eigenfunction since the domain becomes somehow close to an infinite strip with periodic boundary conditions w.r.t. the unbounded direction, where the first eigenfunction in the appropriate functional space depends only on one variable and is positive.

Even more, numerical experiments indicate that there exist starshaped domains, where the first eigenfunction is anti-symmetric with respect to a nodal line and hence sign-changing. See [76] and Figure 3.3 which shows the first and second eigenfunction for such a domain.

3.1.4 Symmetrisation and Talenti's Comparison Principle

Let $\Lambda_{1,1}(\Omega)$ denote the first Dirichlet eigenvalue for $-\Delta$ in a bounded domain $\Omega \subset \mathbb{R}^n$, see (3.2) with $m = 1$. The celebrated Faber-Krahn [162, 254, 255] inequality states that if one considers the map $\Omega \mapsto \Lambda_{1,1}(\Omega)$ in dependence of domains Ω having all the same measure e_n as the unit ball B, its minimum is achieved precisely for $\Omega = B$ and, moreover, balls of radius 1 are the only minimisers. The crucial tool to prove this statement is *symmetrisation*. We recall here some basic facts about this method.

In 1836, Jacob Steiner noticed that symmetrisation with respect to planes leaves the measure of bounded sets invariant and decreases the measure of their boundary. This is the basic idea for a rigorous proof of the isoperimetric problem. In other words, if Ω^* denotes the ball centered at the origin and having the same measure as a bounded domain Ω, we have $|\Omega^*| = |\Omega|$ and $|\partial\Omega^*| \leq |\partial\Omega|$ with strict inequality if Ω is not a ball. The same principle may be applied to functions.

Definition 3.10. Let $\Omega \subset \mathbb{R}^n$ be a bounded domain and let $u \in C_c^\infty(\Omega)$. The *spherical rearrangement* of u is the unique nonnegative measurable function u^* defined in Ω^* such that its level sets $\{x \in \Omega^*; u^*(x) > t\}$ are concentric balls with the same measure as the level sets $\{x \in \Omega; |u(x)| > t\}$ of $|u|$.

By density arguments we may define the spherical rearrangement of any function in $L^p(\Omega)$ for $p \in [1, \infty)$. We summarise here the basic properties of spherical rearrangements in a statement which makes clear how the symmetrisation method can be applied to obtain the Faber-Krahn result.

Theorem 3.11. *Let $\Omega \subset \mathbb{R}^n$ be a bounded domain.*

1. *If $u \in L^p(\Omega)$ for some $p \in [1, \infty)$ then $u^* \in L^p(\Omega^*)$ and $\|u^*\|_{L^p(\Omega^*)} = \|u\|_{L^p(\Omega)}$.*
2. *If $u \in W_0^{1,p}(\Omega)$ for some $p \in [1, \infty)$ then $u^* \in W_0^{1,p}(\Omega^*)$ and $\|\nabla u^*\|_{L^p(\Omega^*)} \leq \|\nabla u\|_{L^p(\Omega)}$.*
3. *If $u \in L^p(\Omega)$ and $v \in L^{p'}(\Omega)$ for some $p \in (1, \infty)$ with $p' = \frac{p}{p-1}$ its conjugate, then $\|u^* v^*\|_{L^1(\Omega^*)} \geq \|uv\|_{L^1(\Omega)}$.*

Theorem 3.11 has several important applications, for example in the proof of first order Sobolev inequalities and of their sharpness. However, it is unsuitable for higher order derivatives since u^* may not be twice weakly differentiable even if u is very smooth. In their monograph, Pólya-Szegö [344, Section F.5] claim that they can extend the Faber-Krahn result to the Dirichlet biharmonic operator among domains having a first eigenfunction of fixed sign. Not only this assumption does not cover all domains, see Section 3.1.3 above, but also their argument is not correct. They deal with the Laplacian of a symmetrised smooth function and implicitly claim that it belongs to L^2, which is false in general. Incidentally, we point out that this mistake is responsible for the wrong proof in [374], see Section 1.3.3 for the details. This shows that standard symmetrisation methods are not available for higher order problems.

As a possible way out, Cianchi [98] considers larger classes than the Sobolev space, such as the space of functions whose second order distributional derivatives are measures with finite total variation. Alternatively, one can prove an inequality comparing the rearrangement invariant norm of the Hessian matrix of u and a weighted norm in the representation space of $(u^*)'$, see [99]. Unfortunately, none of these tricks works when trying to extend the Faber-Krahn result to the first Dirichlet eigenvalue of the biharmonic operator.

However, as we shall see, in some significant situations a comparison result by Talenti [392] turns out to be extremely useful. For our convenience, we state here an iterated version of this principle which will be used at several different places in the present book.

Theorem 3.12. *Let $\Omega \subset \mathbb{R}^n$ ($n \geq 2$) be a C^m-smooth bounded domain such that $|\Omega| = |B| = e_n$ and let $H_\vartheta^m(\Omega)$ be the space defined in (2.35), namely*

$$H_\vartheta^m(\Omega) := \left\{ v \in H^m(\Omega); \Delta^j v = 0 \text{ on } \partial\Omega \text{ for } j < \frac{m}{2} \right\}.$$

Let $m = 2k$ be an even number, let $f \in L^2(\Omega)$ and let $u \in H^m_\vartheta(\Omega)$ be the unique strong solution to

$$\begin{cases} (-\Delta)^k u = f & \text{in } \Omega, \\ \Delta^j u = 0 & \text{on } \partial\Omega, \end{cases} \qquad j = 0, \dots, k-1. \tag{3.9}$$

Let $f^* \in L^2(B)$ and $u^* \in H^1_0(B)$ denote respectively the spherical rearrangements of f and u (see Definition 3.10) and let $v \in H^m_\vartheta(B)$ be the unique strong solution to

$$\begin{cases} (-\Delta)^k v = f^* & \text{in } B, \\ \Delta^j v = 0 & \text{on } \partial B, \end{cases} \qquad j = 0, \dots, k-1. \tag{3.10}$$

Then $v \geq u^*$ a.e. in B.

Proof. When $k = 1$, Theorem 3.12 is precisely [392, Theorem 1]. For $k \geq 2$ we proceed by finite induction. We may rewrite (3.9) and (3.10) as the following systems:

$$\begin{cases} -\Delta u_1 = f & \text{in } \Omega, \\ u_1 = 0 & \text{on } \partial\Omega, \end{cases} \qquad \begin{cases} -\Delta u_i = u_{i-1} & \text{in } \Omega, \\ u_i = 0 & \text{on } \partial\Omega, \end{cases} \qquad i = 2, \dots, k; \tag{3.11}$$

$$\begin{cases} -\Delta v_1 = f^* & \text{in } B, \\ v_1 = 0 & \text{on } \partial B, \end{cases} \qquad \begin{cases} -\Delta v_i = v_{i-1} & \text{in } B, \\ v_i = 0 & \text{on } \partial B, \end{cases} \qquad i = 2, \dots, k. \tag{3.12}$$

Note that $u_k = u$ and $v_k = v$. By Talenti's principle [392, Theorem 1] applied for $i = 1$, we know that $v_1 \geq u_1^*$ a.e. in B. Assume that the inequality $v_i \geq u_i^*$ a.e. in B has been proved for some $i = 1, \dots, k-1$. By (3.11) and (3.12) we then infer

$$\begin{cases} -\Delta u_{i+1} = u_i \text{ in } \Omega \\ u_{i+1} = 0 \qquad \text{on } \partial\Omega, \end{cases} \qquad \begin{cases} -\Delta v_{i+1} = v_i \geq u_i^* \text{ in } B \\ v_{i+1} = 0 \qquad \text{on } \partial B. \end{cases}$$

By combining the maximum principle for $-\Delta$ in B with a further application of Talenti's principle, we obtain $v_{i+1} \geq u_{i+1}^*$ a.e. in B. This finite induction shows that $v_k \geq u_k^*$ and proves the statement. $\qquad\square$

3.1.5 The Rayleigh Conjecture for the Clamped Plate

We consider here the domain functional given by the first Dirichlet eigenvalue for the biharmonic operator

$$\Omega \mapsto \Lambda_{2,1}(\Omega) = \min_{H^2_0(\Omega)\setminus\{0\}} \frac{\|u\|^2_{H^2_0}}{\|u\|^2_{L^2}}. \tag{3.13}$$

In 1894, Lord Rayleigh [351, p. 382] conjectured that, among planar domains Ω of given area, the disk minimises $\Lambda_{2,1}(\Omega)$. If Ω^* denotes the symmetrised of Ω, namely the ball having the same measure as Ω, Rayleigh's conjecture reads

$$\Lambda_{2,1}(\Omega^*) \leq \Lambda_{2,1}(\Omega). \tag{3.14}$$

After many attempts, see Section 1.3.1, this conjecture was proved one century later by Nadirashvili [316] and immediately extended by Ashbaugh-Benguria [22] to the case of 3-dimensional domains. More precisely, we have

Theorem 3.13. *In dimensions $n = 2$ or $n = 3$ the ball is the unique minimiser of the first eigenvalue of the clamped plate problem (3.13) among bounded domains of given measure. Hence, (3.14) holds whenever $n = 2$ or $n = 3$ with equality only if Ω is a ball.*

Proof. Thanks to the homogeneity of the map $\Omega \mapsto \Lambda_{2,1}(\Omega)$, we may restrict our attention to the case where $|\Omega| = |B| = e_n$. For such a domain, let u denote a first nontrivial eigenfunction so that

$$\frac{\|u\|_{H_0^2}^2}{\|u\|_{L^2}^2} = \Lambda_{2,1}(\Omega).$$

By a bootstrap argument, elliptic regularity theory (see Theorem 2.20) ensures that $u \in C^\infty(\Omega)$. Moreover, the unique continuation principle [337, 346] ensures that u cannot be harmonic (in particular, constant) on a subset of positive measure. In view of Section 3.1.3, both the positive and the negative part of u may be nontrivial so that it makes sense to define

$$\Omega_+ = \{x \in \Omega; u(x) > 0\}\,, \qquad \Omega_- = \{x \in \Omega; u(x) < 0\}.$$

Let $B_a = \Omega_+^*$ and $B_b = \Omega_-^*$ be their spherical symmetrisation, namely the two balls centered at the origin and such that $|\Omega_\pm^*| = |\Omega_\pm|$. Let a and b denote the radii of the balls $B_a = \Omega_+^*$ and $B_b = \Omega_-^*$, then

$$a^n + b^n = 1. \tag{3.15}$$

Let $(\Delta u)_\pm$ denote the positive and negative parts of Δu in Ω; again we point out that they may both be nontrivial. For $s \in [0, e_n]$ let $\sigma(s) := (s/e_n)^{1/n}$ and define the two functions $f, g : [0, e_n] \to \mathbb{R}$ by

$$g(s) := \left((\Delta u)_+\right)^*(\sigma(s)) - \left((\Delta u)_-\right)^*(\sigma(e_n - s)),$$
$$f(s) := -g(e_n - s).$$

Note that at most one of $\left((\Delta u)_+\right)^*(\sigma(s))$ and $\left((\Delta u)_-\right)^*(\sigma(e_n - s))$ can be different from 0 for any s and that

$$\left((\Delta u)_+\right)^*(\sigma(s)) \cdot \left((\Delta u)_-\right)^*(\sigma(e_n - s)) \equiv 0. \tag{3.16}$$

The function g sums the contribution of $\big((\Delta u)_+\big)^*$ starting from the center of B and the contribution of $\big((\Delta u)_-\big)^*$ starting from ∂B. The function f switches these two contributions.

With the change of variable $r = \sigma(s)$, by Theorem 3.11 and the divergence theorem, we obtain

$$\int_0^{e_n} g(s)\,ds = \int_0^{e_n} \Big(\big((\Delta u)_+\big)^*(\sigma(s)) - \big((\Delta u)_-\big)^*(\sigma(e_n - s))\Big)ds$$
$$= \int_0^{e_n} \Big(\big((\Delta u)_+\big)^*(\sigma(s)) - \big((\Delta u)_-\big)^*(\sigma(s))\Big)ds$$
$$= ne_n \int_0^1 r^{n-1}\Big(\big((\Delta u)_+\big)^*(r) - \big((\Delta u)_-\big)^*(r)\Big)dr$$
$$= \int_B \big((\Delta u)_+\big)^* dx - \int_B \big((\Delta u)_-\big)^* dx = \int_\Omega (\Delta u)_+ \, dx - \int_\Omega (\Delta u)_- \, dx$$
$$= \int_\Omega \Delta u \, dx = \int_{\partial\Omega} u_\nu \, d\omega = 0.$$

A similar computation holds for f so that

$$\int_0^{e_n} g(s)\,ds = \int_0^{e_n} f(s)\,ds = 0. \tag{3.17}$$

Let now $v \in H^2 \cap H_0^1(B_a)$ and $w \in H^2 \cap H_0^1(B_b)$ be the solutions of the problems

$$\begin{cases} -\Delta v = f(e_n|x|^n) \text{ in } B_a, \\ v = 0 \qquad\qquad \text{ on } \partial B_a, \end{cases} \qquad \begin{cases} -\Delta w = g(e_n|x|^n) \text{ in } B_b, \\ w = 0 \qquad\qquad \text{ on } \partial B_b. \end{cases}$$

Therefore, from the definition of f and from (3.15) we infer

$$\Delta v(a) + \Delta w(b) = -f(e_n a^n) - g(e_n b^n) = 0.$$

Moreover, by (3.17) and the definition of f we get

$$0 = \int_0^{e_n a^n} f(s)\,ds + \int_{e_n a^n}^{e_n} f(s)\,ds = \int_0^{e_n a^n} f(s)\,ds - \int_0^{e_n b^n} g(s)\,ds$$
$$= ne_n \int_0^a r^{n-1} f(e_n r^n)\,dr - ne_n \int_0^b r^{n-1} g(e_n r^n)\,dr = -\int_{B_a} \Delta v \, dx + \int_{B_b} \Delta w \, dx$$

so that

$$\int_{B_a} \Delta v \, dx = \int_{B_b} \Delta w \, dx. \tag{3.18}$$

In turn, employing the divergence theorem, (3.18) yields

$$a^{n-1} v'(a) = b^{n-1} w'(b). \tag{3.19}$$

In view of (3.16), we remark that

$$\int_{B_b} |\Delta w|^2\, dx = \int_0^{|\Omega_b|} g^2(s)\, ds$$
$$= \int_0^{|\Omega_b|} \left(\left| \left((\Delta u)_+\right)^*(\sigma(s))\right|^2 + \left| \left((\Delta u)_-\right)^*(\sigma(e_n - s))\right|^2 \right) ds.$$

Similarly, we have

$$\int_{B_a} |\Delta v|^2\, dx = \int_0^{|\Omega_a|} \left(\left| \left((\Delta u)_-\right)^*(\sigma(s))\right|^2 + \left| \left((\Delta u)_+\right)^*(\sigma(e_n - s))\right|^2 \right) ds.$$

By adding the last two equations and by Theorem 3.11 we obtain

$$\int_{\Omega} |\Delta u|^2\, dx = \int_{B_a} |\Delta v|^2\, dx + \int_{B_b} |\Delta w|^2\, dx. \tag{3.20}$$

From Talenti [393, (2.7)] we know that

$$u_+^* \le v \quad \text{in } B_a, \qquad u_-^* \le w \quad \text{in } B_b,$$

so that, by Theorem 3.11,

$$\int_{\Omega} u^2\, dx \le \int_{B_a} v^2\, dx + \int_{B_b} w^2\, dx,$$

with strict inequality if $\Omega \ne B$. With this inequality and (3.20) we obtain

$$\Lambda_{2,1}(\Omega) = \frac{\int_{\Omega} |\Delta u|^2\, dx}{\int_{\Omega} u^2\, dx} \ge \frac{\int_{B_a} |\Delta v|^2\, dx + \int_{B_b} |\Delta w|^2\, dx}{\int_{B_a} v^2\, dx + \int_{B_b} w^2\, dx} \tag{3.21}$$

with strict inequality if $\Omega \ne B$. As pointed out by Talenti [393], if $u > 0$ in Ω, then $\Omega_+^* = B_a = B$ and $\Omega_-^* = B_b = \emptyset$ so that (3.21) proves Rayleigh conjecture for domains with first eigenfunction of one sign. Indeed, in this case we have $b = 0$. Therefore, $v_\nu = 0$ on ∂B in view of (3.19).

We define

$$\mu = \mu_{a,b} = \min_{v,w} \frac{\int_{B_a} |\Delta v|^2\, dx + \int_{B_b} |\Delta w|^2\, dx}{\int_{B_a} v^2\, dx + \int_{B_b} w^2\, dx} \tag{3.22}$$

where the minimum is taken over all radially symmetric functions $v \in H^2 \cap H_0^1(B_a)$ and $w \in H^2 \cap H_0^1(B_b)$ satisfying (3.19). Using standard tools of the calculus of variations, it is shown in [22, Appendix 2] that the minimum in (3.22) is attained by a couple of functions satisfying $\Delta^2 v = \mu v$ in B_a and $\Delta^2 w = \mu w$ in B_b, $v(a) = w(b) = 0$, $a^{n-1}v'(a) = b^{n-1}w'(b)$, $\Delta v(a) + \Delta w(b) = 0$; moreover, as shown in formula (3.12) in [316], the functions v and w may be chosen positive and radially decreasing.

By combining (3.21) and (3.22) we obtain $\Lambda_{2,1}(\Omega) \geq \mu_{a,b}$ and, since a and b are unknown,

$$\Lambda_{2,1}(\Omega) \geq \min_{a,b} \mu_{a,b} \tag{3.23}$$

where the minimum is now taken among all couples $(a,b) \in [0,1]^2$ satisfying (3.15). At this point a delicate and technical analysis of fine properties of Bessel functions is needed. A crucial inequality, which only holds for $n = 2,3$, allows to show that $\min_{a,b} \mu_{a,b} = \mu_{1,0} = \Lambda_{2,1}(B)$. This proves the statement when combined with (3.23), see [22, Section 4] for the details. Indeed, recall that if $\Omega \neq B$ then (3.21) is strict. $\qquad\qquad\Box$

We conclude this section by emphasising that a couple of interesting generalisations of Rayleigh's original conjecture are still missing. First, it remains to prove (3.14) in any space dimension $n \geq 2$ and not only for $n = 2,3$. Second, one might wonder whether one could prove that $\Lambda_{m,1}(\Omega^*) \leq \Lambda_{m,1}(\Omega)$ for any $m \geq 2$ and not only for $m = 2$.

3.2 Buckling Load of a Clamped Plate

Similar questions as for the first eigenvalue of the clamped plate (3.2) arise when considering the buckling load of a clamped plate which may be characterised as follows

$$\mu_1(\Omega) = \inf_{H_0^2(\Omega)\setminus\{0\}} \frac{\|\Delta u\|_{L^2}^2}{\|\nabla u\|_{L^2}^2}. \tag{3.24}$$

Here, $\Omega \subset \mathbb{R}^2$ is a bounded planar domain. Minimisers u to (3.24) solve

$$\begin{cases} \Delta^2 u = -\mu_1 \Delta u & \text{in } \Omega, \\ u = u_\nu = 0 & \text{on } \partial\Omega. \end{cases} \tag{3.25}$$

This is the Dirichlet version of the Steklov problem (1.22) considered in Section 1.3.2 which describes the linearised von Kármán equations for an elastic plate. For later use, let us mention that an inequality (which holds true in any space dimension) due to Payne [334] states that for any bounded domain $\Omega \subset \mathbb{R}^2$

$$\mu_1(\Omega) \geq \Lambda_{1,2}(\Omega) \qquad \text{with equality if and only if } \Omega \text{ is a disk,} \tag{3.26}$$

where $\Lambda_{1,2}(\Omega)$ denotes the second Dirichlet eigenvalue for the Laplacian in Ω.

Similarly to (3.14), Pólya-Szegö [344, Note F] conjectured that the disk minimises the buckling load among domains of given measure.

Conjecture 3.14 (Pólya-Szegö). For any bounded domain $\Omega \subset \mathbb{R}^2$

$$\mu_1(\Omega^*) \leq \mu_1(\Omega),$$

where Ω^* denotes the symmetrised of Ω.

A complete proof of this conjecture is not known at the moment. However, we show here two interesting results which give some support to its validity. Consider the following special class of (not necessarily bounded) domains having the same measure as the unit disk:

$$\mathbb{B} = \{\Omega \subset \mathbb{R}^2; \ \Omega \text{ open, connected, simply connected}, \ |\Omega| = \pi\}.$$

The first result, due to Ashbaugh-Bucur [23], states that an optimal domain exists among domains in the class $\mathbb{B}$.

Theorem 3.15. *There exists $\Omega_o \in \mathbb{B}$ such that $\mu_1(\Omega_o) \leq \mu_1(\Omega)$ for any other domain $\Omega \in \mathbb{B}$.*

Proof. Note first that minimising μ_1 in the wider class

$$\mathbb{B}_0 = \{\Omega \subset \mathbb{R}^2; \ \Omega \text{ open, simply connected}, \ |\Omega| = \pi\}$$

is equivalent to minimising μ_1 in $\mathbb{B}$, where we understand that "simply connected" means that "each connected component is simply connected". Indeed, if we find a minimiser in $\mathbb{B}_0$, then it is necessarily connected since otherwise, scaling one of its connected components and noticing that $\Omega \mapsto \mu_1(\Omega)$ is homogeneous of degree -2, would contradict minimality.

So, consider a minimising sequence $(\Omega_m) \subset \mathbb{B}_0$ with (u_m) being the corresponding sequence of normalised eigenfunctions, that is, $\int_{\Omega_m} |\nabla u_m|^2 = 1$. Extending u_m by 0 in $\mathbb{R}^2 \setminus \Omega_m$ we may view (u_m) as a bounded sequence in $H^2(\mathbb{R}^n)$ such that

$$\int_{\mathbb{R}^2} |\Delta u_m|^2 \, dx \leq C_1, \quad \int_{\mathbb{R}^2} |\nabla u_m|^2 \, dx = 1, \quad \int_{\mathbb{R}^2} |u_m|^2 \, dx \leq C_2, \qquad (3.27)$$

for suitable $C_1, C_2 > 0$, the L^2-bound following from Poincaré's inequality in $H_0^1(\Omega_m)$ and the fact that $|\Omega_m| = \pi$ for all m. In particular, (3.27) shows that

$$\inf_{\Omega \in \mathbb{B}_0} \mu_1(\Omega) > 0. \qquad (3.28)$$

Indeed, we have

$$1 = \int_{\mathbb{R}^2} |\nabla u_m|^2 \, dx = -\int_{\mathbb{R}^2} u_m \Delta u_m \, dx \leq C_2^{1/2} \left(\int_{\mathbb{R}^2} |\Delta u_m|^2 \, dx \right)^{1/2},$$

which proves that $\int_{\mathbb{R}^2} |\Delta u_m|^2$ is also bounded away from 0. By (3.27), we may also apply the concentration-compactness principle [277] and deduce that, up to a subsequence, three cases may occur.

1. Vanishing.

$$\lim_{m \to \infty} \sup_{y \in \mathbb{R}^2} \int_{B_R(y)} |\nabla u_m|^2 \, dx = 0 \qquad \text{for all } R > 0.$$

2. Dichotomy. There exists $\alpha \in (0,1)$ such that for all $\varepsilon > 0$ there exist two bounded sequences $(u_m^{(1)}), (u_m^{(2)}) \subset H^2(\mathbb{R}^2)$ such that

$$\lim_{m \to \infty} \mathrm{dist}\left(\mathrm{support}(u_m^{(1)}), \mathrm{support}(u_m^{(2)}) \right) = +\infty,$$

$$\lim_{m \to \infty} \int_{\mathbb{R}^2} |\nabla u_m^{(1)}|^2 \, dx \to \alpha, \qquad \lim_{m \to \infty} \int_{\mathbb{R}^2} |\nabla u_m^{(2)}|^2 \, dx \to 1 - \alpha, \tag{3.29}$$

$$\lim_{m \to \infty} \int_{\mathbb{R}^2} \left| |\nabla u_m|^2 - |\nabla u_m^{(1)}|^2 - |\nabla u_m^{(2)}|^2 \right| dx \le \varepsilon, \tag{3.30}$$

$$\lim_{m \to \infty} \int_{\mathbb{R}^2} \left(|\Delta u_m|^2 - |\Delta u_m^{(1)}|^2 - |\Delta u_m^{(2)}|^2 \right) dx \ge 0. \tag{3.31}$$

3. Compactness. There exists a sequence $(y_m) \subset \mathbb{R}^2$ such that for all $\varepsilon > 0$ there exists $R > 0$ and

$$\int_{B_R(y_m)} |\nabla u_m|^2 \, dx \ge 1 - \varepsilon \qquad \text{for all } m.$$

We first show that vanishing cannot occur. By contradiction, assume that vanishing occurs. Up to a permutation of x_1 and x_2 and up to a subsequence, by (3.27) we have

$$\int_{\mathbb{R}^2} \left(\frac{\partial u_m}{\partial x_1} \right)^2 dx \ge \frac{1}{2}.$$

Moreover, since two integrations by parts yield

$$\int_{\mathbb{R}^2} \frac{\partial^2 u_m}{\partial x_1^2} \frac{\partial^2 u_m}{\partial x_2^2} \, dx = \int_{\mathbb{R}^2} \left(\frac{\partial^2 u_m}{\partial x_1 \partial x_2} \right)^2 dx \ge 0,$$

we remark that

$$\int_{\mathbb{R}^2} |\Delta u_m|^2 \, dx \ge \int_{\mathbb{R}^2} \left| \nabla \frac{\partial u_m}{\partial x_1} \right|^2 dx.$$

Therefore, we infer that

$$\frac{\|\Delta u_m\|_{L^2}^2}{\|\nabla u_m\|_{L^2}^2} \ge \frac{1}{2} \frac{\|\nabla \frac{\partial u_m}{\partial x_1}\|_{L^2}^2}{\|\frac{\partial u_m}{\partial x_1}\|_{L^2}^2}. \tag{3.32}$$

By assumption any translation of $\frac{\partial u_m}{\partial x_1}$ converges weakly to 0 in $L^2(\mathbb{R}^2)$. Moreover, $\frac{\partial u_m}{\partial x_1} \in H_0^1(\Omega_m)$ and $|\Omega_m| = \pi$. Hence, we may apply [81, Lemma 3.3] to get that, up to a subsequence, $\|\nabla \frac{\partial u_m}{\partial x_1}\|_{L^2} \to \infty$. Since the left hand side of (3.32) is supposed to converge to $\inf_{\Omega \in \mathbb{B}_0} \mu_1(\Omega)$, we get a contradiction.

Next, we show that dichotomy cannot occur. By contradiction, assume that dichotomy occurs and fix $\varepsilon > 0$. Then the sequences $(u_m^{(1)})$ and $(u_m^{(2)})$ can be chosen as follows, see [277]. Let B_2 denote the ball of radius 2 centered at the origin and let $\varphi \in C_c^\infty(B_2, [0,1])$ be such that $\varphi \equiv 1$ in B (the unit ball). Then for suitable sequences $(R_m), (\rho_m) \to \infty$, we put

$$u_m^{(1)}(x) := \varphi\left(\frac{x}{R_m}\right) u_m(x), \qquad u_m^{(2)}(x) := \left(1 - \varphi\left(\frac{x}{\rho_m R_m}\right)\right) u_m(x).$$

Note that $\text{support}(u_m^{(1)}) \subset (\overline{\Omega_m} \cap B_{2R_m})$ whereas $\text{support}(u_m^{(2)}) \subset (\overline{\Omega_m} \setminus B_{\rho_m R_m})$. By elementary calculus we know that $\frac{x_1+x_2}{y_1+y_2} \geq \min\{\frac{x_1}{y_1}, \frac{x_2}{y_2}\}$ for all $x_1, x_2, y_1, y_2 > 0$. Hence, by (3.30) and (3.31), up to a switch between $u_m^{(1)}$ and $u_m^{(2)}$, we have

$$\inf_{\Omega \in \mathbb{B}_0} \mu_1(\Omega) = \lim_{m \to \infty} \frac{\|\Delta u_m\|_{L^2(\Omega_m)}^2}{\|\nabla u_m\|_{L^2(\Omega_m)}^2} \geq \limsup_{m \to \infty} \frac{\|\Delta u_m^{(1)}\|_{L^2(\Omega_m \cap B_{2R_m})}^2}{\varepsilon + \|\nabla u_m^{(1)}\|_{L^2(\Omega_m \cap B_{2R_m})}^2}. \tag{3.33}$$

Up to a further subsequence, the above "limsup" becomes a limit.

We now claim that there exists $\delta > 0$ such that for m large enough

$$|\Omega_m \setminus B_{\rho_m R_m}| \geq \delta. \tag{3.34}$$

Indeed, if (3.34) were not true, up to a subsequence we would have $\lim_{m \to \infty} |\Omega_m \setminus B_{\rho_m R_m}| = 0$ implying that $\Lambda_{1,1}(\Omega_m \setminus B_{\rho_m R_m}) \to \infty$. In view of (3.26), this would imply $\mu_1(\Omega_m \setminus B_{\rho_m R_m}) \to \infty$. In turn, since (3.29) states that $\|\nabla u_m^{(2)}\|_{L^2(\mathbb{R}^2)}$ is bounded away from zero, this implies that $\|\Delta u_m^{(2)}\|_{L^2(\mathbb{R}^2)} \to \infty$, contradicting (3.27) and (3.31). A similar argument also shows that $\delta > 0$ in (3.34) may be chosen to be independent of ε.

By (3.34) we know that there exists $\gamma \in (0,1)$, independent of ε, such that

$$\limsup_{m \to \infty} |\Omega_m \cap B_{2R_m}| = \gamma\pi \leq \pi - \delta.$$

Up to a further subsequence, also the above "limsup" becomes a limit. Combined with (3.28), (3.33), and homogeneity of μ_1, this yields

$$\inf_{\Omega \in \mathbb{B}_0} \mu_1(\Omega) \geq \lim_{m \to \infty} \frac{\|\Delta u_m^{(1)}\|_{L^2(\Omega_m \cap B_{2R_m})}^2}{\varepsilon + \|\nabla u_m^{(1)}\|_{L^2(\Omega_m \cap B_{2R_m})}^2}$$

$$\geq \lim_{m \to \infty} \frac{\|\Delta u_m^{(1)}\|_{L^2(\Omega_m \cap B_{2R_m})}^2}{\|\nabla u_m^{(1)}\|_{L^2(\Omega_m \cap B_{2R_m})}^2} \cdot \frac{\|\nabla u_m^{(1)}\|_{L^2(\Omega_m \cap B_{2R_m})}^2}{\varepsilon + \|\nabla u_m^{(1)}\|_{L^2(\Omega_m \cap B_{2R_m})}^2}$$

$$= \frac{\alpha}{\varepsilon + \alpha} \lim_{m \to \infty} \mu_1(\Omega_m \cap B_{2R_m})$$

$$= \frac{\alpha}{\varepsilon + \alpha} \lim_{m \to \infty} \left(\frac{\pi}{|\Omega_m \cap B_{2R_m}|} \right)^2 \mu_1 \left(\pi \frac{\Omega_m \cap B_{2R_m}}{|\Omega_m \cap B_{2R_m}|} \right)$$

$$= \frac{\alpha}{\varepsilon + \alpha} \frac{1}{\gamma^2} \lim_{m \to \infty} \mu_1 \left(\pi \frac{\Omega_m \cap B_{2R_m}}{|\Omega_m \cap B_{2R_m}|} \right)$$

$$\geq \frac{\alpha}{\varepsilon + \alpha} \frac{1}{\gamma^2} \inf_{\Omega \in \mathbb{B}_0} \mu_1(\Omega)$$

since $\pi \frac{\Omega_m \cap B_{2R_m}}{|\Omega_m \cap B_{2R_m}|} \in \mathbb{B}_0$. As $\gamma < 1$, by arbitrariness of ε we get a contradiction which rules out dichotomy.

Since we excluded both vanishing and dichotomy, compactness necessarily occurs. Then by arbitrariness of ε, we infer that there exist $(y_m) \subset \mathbb{R}^2$ and $u \in H^2(\mathbb{R}^2)$ such that

$$u_m(. + y_m) \rightharpoonup u \quad \text{in } H^2(\mathbb{R}^2) \qquad \text{and} \qquad \|\nabla u\|_{L^2(\mathbb{R}^2)} = 1.$$

By combining $u_m \rightharpoonup u$ in $H^1(\mathbb{R}^2)$ and the conservation of norms, we deduce that $u_m \to u$ in the norm topology of $H^1(\mathbb{R}^2)$. In turn, by Poincaré's inequality applied in domains of uniformly bounded measure, this yields $u_m \to u$ in $L^2(\mathbb{R}^2)$. Therefore, it follows that (Ω_m) converges in the Hausdorff topology to some simply connected domain $\widehat{\Omega} \subset \mathbb{R}^2$ such that $\widehat{\Omega} \supset \text{support}(u)$ and

$$|\widehat{\Omega}| \leq \pi. \tag{3.35}$$

From weak convergence in H^2 and strong convergence in H^1, we get

$$\mu_1(\widehat{\Omega}) \leq \frac{\|\Delta u\|^2_{L^2(\widehat{\Omega})}}{\|\nabla u\|^2_{L^2(\widehat{\Omega})}} \leq \liminf_{m \to \infty} \frac{\|\Delta u_m\|^2_{L^2(\Omega_m)}}{\|\nabla u_m\|^2_{L^2(\Omega_m)}} = \inf_{\Omega \in \mathbb{B}_0} \mu_1(\Omega).$$

By (3.35) and homogeneity of μ_1 we infer that all the above inequalities are in fact equalities. So, the minimiser for μ_1 is found. $\qquad\square$

As pointed out in Section 1.3.2, the next step would be to show that the minimiser Ω_o found in Theorem 3.15 has some regularity properties. But already for second order equations this is a very difficult task, see [229]. However, *assuming* smoothness of the boundary, one can show (see [416]) that the optimal domain is indeed the disk.

Theorem 3.16. *If the minimiser Ω_o found in Theorem 3.15 has $C^{2,\gamma}$ boundary, then it is a disk.*

Proof. Let Ω_o be the $C^{2,\gamma}$ minimiser found in Theorem 3.15 and let ϕ denote the corresponding first eigenfunction, namely a solution to (3.25) when $\Omega = \Omega_o$. By performing the shape derivative [229] of the map $\Omega \mapsto \mu_1(\Omega)$ and using the optimality of Ω_o one finds that

$$\Delta \phi \text{ exists and is constant on } \partial \Omega_o. \tag{3.36}$$

We point out that this first step is precisely the part of the proof where smoothness of $\partial\Omega_o$ is needed. Moreover, the connectedness of the boundary $\partial\Omega_o$ is here crucial in order to deduce (3.36).

Since $\phi = 0$ on $\partial\Omega_o$, (3.36) also implies that $\Delta\phi + \mu_1\phi$ is constant on $\partial\Omega_o$. In turn, since $\phi \mapsto \Delta\phi + \mu_1\phi$ is harmonic in Ω_o in view of (3.25), this implies

$$\Delta\phi + \mu_1\phi \quad \text{is constant in } \Omega_o. \tag{3.37}$$

The function ϕ has a critical point in Ω_o which we may assume to be the origin so that $\nabla\phi(0) = 0$.

Next, for $(x,y) \in \Omega_o$ define $w(x,y) := x\phi_y(x,y) - y\phi_x(x,y)$. In polar coordinates (r,θ) this can be written as $w = \phi_\theta$. Therefore, if $w \equiv 0$, then ϕ does not depend on θ so that Ω_o is a disk and we are done. So, assume by contradiction that $w \not\equiv 0$. Since $\phi \in H_0^2(\Omega_o)$, we have $w \in H_0^1(\Omega_o)$ and from (3.37) we deduce that $-\Delta w = \mu_1 w$ in Ω_o. Hence, μ_1 is a Dirichlet eigenvalue for $-\Delta$ in Ω_o and it is the first Dirichlet eigenvalue in each of the nodal zones of w.

Note that $w_x = \phi_y + x\phi_{xy} - y\phi_{xx}$ so that $w_x(0) = 0$, recalling that $\nabla\phi(0) = 0$. Similarly, $w_y(0) = 0$. Hence, both w and ∇w vanish at the origin. This means that the origin 0 is a nodal point of w and a point where a nodal line intersects itself transversally. But then, for topological reasons, this nodal line divides Ω_o into at least three nodal domains and at least one has a measure not exceeding $|\Omega_o|/3$. This would imply the following chain of inequalities

$$\mu_1(\Omega_o) = \Lambda_{1,1}(\text{subdomain of measure } \leq |\Omega_o|/3) \geq \Lambda_{1,1}(\text{ball of measure } |\Omega_o|/3)$$
$$= 3\Lambda_{1,1}(\Omega_o^*) > \Lambda_{1,2}(\Omega_o^*) = \mu_1(\Omega_o^*),$$

which contradicts the minimality of Ω_o. In this chain of inequalities we have used one after the other the monotonicity of $\Lambda_{1,1}$ with respect to domain inclusions, the Faber-Krahn inequality [162,254,255], a scaling argument, an inequality from [336] and (3.26). $\qquad\square$

3.3 Steklov Eigenvalues

Let $\Omega \subset \mathbb{R}^n$ $(n \geq 2)$ be a bounded domain with Lipschitz boundary $\partial\Omega$, let $a \in \mathbb{R}$ and consider the boundary eigenvalue problem

$$\begin{cases} \Delta^2 u = 0 & \text{in } \Omega, \\ u = \Delta u - au_\nu = 0 & \text{on } \partial\Omega. \end{cases} \tag{3.38}$$

We are interested in studying the eigenvalues of (3.38), namely those values of a for which the problem admits nontrivial solutions, the corresponding eigenfunctions. By a solution of (3.38) we mean a function $u \in H^2 \cap H_0^1(\Omega)$ such that

$$\int_{\Omega} \Delta u \Delta v \, dx = a \int_{\partial\Omega} u_{\nu} v_{\nu} \, d\omega \qquad \text{for all } v \in H^2 \cap H_0^1(\Omega). \tag{3.39}$$

By taking $v = u$ in (3.39), it is clear that all the eigenvalues of (3.38) are strictly positive.

3.3.1 The Steklov Spectrum

The least positive eigenvalue of (3.38) may be characterised variationally as

$$\delta_1 = \delta_1(\Omega) := \min \left\{ \frac{\|\Delta u\|_{L^2(\Omega)}^2}{\|u_{\nu}\|_{L^2(\partial\Omega)}^2} \; ; \; u \in [H^2 \cap H_0^1(\Omega)] \setminus H_0^2(\Omega) \right\}. \tag{3.40}$$

We first prove the existence of a function $u \in [H^2 \cap H_0^1(\Omega)] \setminus H_0^2(\Omega)$ which achieves equality in (3.40), provided the domain Ω is smooth (C^2) or satisfies a geometric condition which is fulfilled if Ω has no "reentrant corners" (for instance, if Ω is convex). More precisely, we consider domains satisfying a uniform outer ball condition according to Definition 2.30. Then the following existence result for a minimiser of $\delta_1(\Omega)$ holds.

Theorem 3.17. *Assume that $\Omega \subset \mathbb{R}^n$ is a bounded domain with Lipschitz boundary and satisfying a uniform outer ball condition. Then the minimum in (3.40) is achieved and, up to a multiplicative constant, the minimiser $\bar{u}$ for (3.40) is unique, superharmonic in Ω (in particular, $\bar{u} > 0$ in Ω and $\bar{u}_{\nu} < 0$ on $\partial\Omega$) and it solves (3.38) when $a = \delta_1$. Furthermore, $\bar{u} \in C^{\infty}(\Omega)$ and, up to the boundary, $\bar{u}$ is as smooth as the boundary permits.*

Proof. By Theorem 2.31 we know that $u \mapsto \|\Delta u\|_{L^2}$ is a norm in $H^2(\Omega)$. Let (u_m) be a minimising sequence for $\delta_1(\Omega)$ with $\|\Delta u_m\|_{L^2} = 1$ so that (u_m) is bounded in $H^2(\Omega)$. Up to a subsequence, we may assume that there exists $u \in H^2 \cap H_0^1(\Omega)$ such that $u_m \rightharpoonup u$ in $H^2(\Omega)$. Moreover, since Ω is Lipschitzian and satisfies a uniform outer ball condition, by [322, Chapter 2, Theorem 6.2] we infer that the map

$$H^2 \cap H_0^1(\Omega) \ni u \mapsto \nabla u|_{\partial\Omega} \in \left(L^2(\partial\Omega)\right)^n$$

is well-defined and compact. Hence, we deduce that $(u_m)_{\nu} \to u_{\nu}$ in $L^2(\partial\Omega)$ and that $\delta_1(\Omega) > 0$.

Furthermore, since (u_m) is a minimising sequence, $\|\Delta u_m\|_{L^2} = 1$ holds, and $\|(u_m)_{\nu}\|_{L^2(\partial\Omega)}$ is bounded from below, u_{ν} is not identically zero on $\partial\Omega$ and

$$\|u_{\nu}\|_{L^2(\partial\Omega)}^{-2} = \lim_{m \to \infty} \|(u_m)_{\nu}\|_{L^2(\partial\Omega)}^{-2} = \delta_1(\Omega).$$

Moreover, by weak lower semicontinuity of the norm, we also have

$$\|\Delta u\|_{L^2}^2 \leq \liminf_{m\to\infty} \|\Delta u_m\|_{L^2}^2 = 1$$

and hence $u \in [H^2 \cap H_0^1(\Omega)] \setminus H_0^2(\Omega)$ satisfies

$$\frac{\|\Delta u\|_{L^2(\Omega)}^2}{\|u_v\|_{L^2(\partial\Omega)}^2} \leq \delta_1(\Omega).$$

This proves that u is a minimiser for $\delta_1(\Omega)$.

For all $u \in [H^2 \cap H_0^1(\Omega)] \setminus H_0^2(\Omega)$ put

$$I(u) := \frac{\|\Delta u\|_{L^2(\Omega)}^2}{\|u_v\|_{L^2(\partial\Omega)}^2}.$$

To show that, up to their sign, the minimisers for (3.40) are superharmonic, we observe that for all $u \in [H^2 \cap H_0^1(\Omega)] \setminus H_0^2(\Omega)$ there exists $w \in [H^2 \cap H_0^1(\Omega)] \setminus H_0^2(\Omega)$ such that $-\Delta w \geq 0$ in Ω and $I(w) \leq I(u)$. Indeed, for a given u, let w be the unique solution of

$$\begin{cases} -\Delta w = |\Delta u| & \text{in } \Omega, \\ w = 0 & \text{on } \partial\Omega, \end{cases}$$

so that w is superharmonic. Moreover, both $w \pm u$ are superharmonic in Ω and vanish on $\partial\Omega$. This proves that

$$|u| \leq w \quad \text{in } \Omega, \qquad |u_v| \leq |w_v| \quad \text{on } \partial\Omega.$$

In turn, these inequalities (and $-\Delta w = |\Delta u|$) prove that $I(w) \leq I(u)$. We emphasise that this inequality is strict if Δu changes sign.

Any minimiser $\bar{u}$ for (3.40) solves the Euler equation (3.38) and is a smooth function in view of elliptic theory, see the explanation just after (2.22). In order to conclude the proof we still have to show that the minimiser $\bar{u}$ is unique. By contradiction, let $v \in H^2 \cap H_0^1(\Omega)$ be another positive minimiser and for every $c \in \mathbb{R}$, define $v_c := v + c\bar{u}$. Exploiting the fact that both v and $\bar{u}$ solve (3.38) when $a = \delta_1$, we see that also v_c is a minimiser. But unless v is a multiple of $\bar{u}$, there exists some c such that v_c changes sign in Ω. This leads to a contradiction and completes the proof. $\qquad\qquad\square$

We are now interested in the description of the spectrum of (3.38). To this end, we restrict our attention to smooth domains. As in (2.10), the Hilbert space $H^2 \cap H_0^1(\Omega)$ is endowed with the scalar product

$$(u, v) \mapsto \int_\Omega \Delta u \Delta v \, dx. \qquad\qquad (3.41)$$

Consider the space

$$Z = \left\{v \in C^\infty(\overline{\Omega}) : \Delta^2 u = 0 \text{ in } \Omega, \ u = 0 \text{ on } \partial\Omega\right\} \tag{3.42}$$

and let V denote the completion of Z with respect to the scalar product in (3.41). Then we prove

Theorem 3.18. *Assume that $\Omega \subset \mathbb{R}^n$ $(n \geq 2)$ is a bounded domain with C^2-boundary. Then problem (3.38) admits infinitely many (countable) eigenvalues. The only eigenfunction of one sign is the one corresponding to the first eigenvalue. The set of eigenfunctions forms a complete orthonormal system in V.*

Proof. Let Z be as in (3.42), define on Z the scalar product given by

$$(u,v)_W := \int_{\partial\Omega} u_\nu v_\nu \, d\omega \qquad \text{for all } u, v \in Z$$

and let W denote the completion of Z with respect to this scalar product. We first claim that the (Hilbert) space V is compactly embedded into the (Hilbert) space W. Indeed, by definition of δ_1 we have

$$\|u\|_W = \|u_\nu\|_{L^2(\partial\Omega)} \leq \delta_1^{-1/2} \|\Delta u\|_{L^2(\Omega)} = \delta_1^{-1/2} \|u\|_V \qquad \text{for all } u \in Z. \tag{3.43}$$

Hence any Cauchy sequence in Z with respect to the norm of V is a Cauchy sequence with respect to the norm of W. Since V is the completion of Z with respect to (3.41), it follows that $V \subset W$. The continuity of this inclusion can be obtained by density from (3.43). In order to prove that this embedding is compact, let $u_m \rightharpoonup u$ in V, so that also $u_m \rightharpoonup u$ in $H^2 \cap H_0^1(\Omega)$. Then by the compact trace embedding $H^{1/2}(\partial\Omega) \subset L^2(\partial\Omega)$ we obtain $u_m \to u$ in W. This proves the claim.

Let $I_1 : V \to W$ denote the embedding $V \subset W$ and $I_2 : W \to V'$ the linear continuous operator defined by

$$\langle I_2 u, v \rangle = (u,v)_W \qquad \text{for all } u \in W \text{ and } v \in V.$$

Moreover, let $L : V \to V'$ be the linear operator given by

$$\langle Lu, v \rangle = \int_\Omega \Delta u \Delta v \, dx \qquad \text{for all } u, v \in V.$$

Then by the Lax-Milgram theorem, L is an isomorphism and in view of the compact embedding $V \subset W$, the linear operator $K = L^{-1} I_2 I_1 : V \to V$ is compact. Since for $n \geq 2$, V is an infinite dimensional Hilbert space and K is a compact self-adjoint operator with strictly positive eigenvalues, V admits an orthonormal basis of eigenfunctions of K and the set of the eigenvalues of K can be ordered in a strictly decreasing sequence (μ_i) which converges to zero. Therefore problem (3.39) admits an infinite set of eigenvalues given by $\delta_i = \frac{1}{\mu_i}$ and the eigenfunctions of (3.39) coincide with the eigenfunctions of K.

To complete the proof we need to show that if δ_k is an eigenvalue of (3.38) corresponding to a positive eigenfunction ϕ_k then necessarily $\delta_k = \delta_1$. So, take $\phi_k > 0$ in Ω and $\phi_k = 0$ on $\partial\Omega$; then $(\phi_k)_\nu \le 0$ on $\partial\Omega$ and, in turn, $\Delta\phi_k = \delta_k(\phi_k)_\nu \le 0$ on $\partial\Omega$. Therefore, by $\Delta^2\phi_k = 0$ in Ω and the weak comparison principle, we infer $\Delta\phi_k \le 0$ in Ω. Moreover, since $\phi_k > 0$ in Ω and $\phi_k = 0$ on $\partial\Omega$, the Hopf boundary lemma implies that $(\phi_k)_\nu < 0$ on $\partial\Omega$. Let ϕ_1 be a positive eigenfunction corresponding to the first eigenvalue δ_1, see Theorem 3.17. Then ϕ_1 satisfies $(\phi_1)_\nu < 0$ on $\partial\Omega$ and hence from

$$\delta_k \int_{\partial\Omega} (\phi_k)_\nu (\phi_1)_\nu \, d\omega = \int_\Omega \Delta\phi_k \Delta\phi_1 \, dx = \delta_1 \int_{\partial\Omega} (\phi_k)_\nu (\phi_1)_\nu \, d\omega > 0$$

we obtain $\delta_k = \delta_1$. This completes the proof of Theorem 3.18. $\qquad\square$

The vector space V also has a different interesting characterisation.

Theorem 3.19. *Assume that $\Omega \subset \mathbb{R}^n$ $(n \ge 2)$ is a bounded domain with C^2-boundary. Then the space $H^2 \cap H_0^1(\Omega)$ admits the following orthogonal decomposition with respect to the scalar product (3.41)*

$$H^2 \cap H_0^1(\Omega) = V \oplus H_0^2(\Omega).$$

Moreover, if $v \in H^2 \cap H_0^1(\Omega)$ and if $v = v_1 + v_2$ is the corresponding orthogonal decomposition, then $v_1 \in V$ and $v_2 \in H_0^2(\Omega)$ are weak solutions of

$$\begin{cases} \Delta^2 v_1 = 0 & \text{in } \Omega, \\ v_1 = 0 & \text{on } \partial\Omega, \\ (v_1)_\nu = v_\nu & \text{on } \partial\Omega, \end{cases} \qquad \text{and} \qquad \begin{cases} \Delta^2 v_2 = \Delta^2 v & \text{in } \Omega, \\ v_2 = 0 & \text{on } \partial\Omega, \\ (v_2)_\nu = 0 & \text{on } \partial\Omega. \end{cases} \qquad (3.44)$$

Proof. We start by proving that $Z^\perp = H_0^2(\Omega)$. Let $v \in Z$ and $w \in H^2 \cap H_0^1(\Omega)$. After two integrations by parts we obtain

$$\int_\Omega \Delta v \Delta w \, dx = \int_\Omega \Delta^2 v \, w \, dx + \int_{\partial\Omega} (w_\nu \Delta v - w(\Delta v)_\nu) \, d\omega = \int_{\partial\Omega} w_\nu \Delta v \, d\omega$$

for all $v \in Z$ and $w \in H^2 \cap H_0^1(\Omega)$. This proves that $w_\nu = 0$ on $\partial\Omega$ if and only if $w \in Z^\perp$ and hence $V^\perp = Z^\perp = H_0^2(\Omega)$.

Let $v \in H^2 \cap H_0^1(\Omega)$ and consider the first Dirichlet problem in (3.44), that is

$$\begin{cases} \Delta^2 v_1 = 0 & \text{in } \Omega, \\ v_1 = 0 & \text{on } \partial\Omega, \\ (v_1)_\nu = v_\nu & \text{on } \partial\Omega. \end{cases} \qquad (3.45)$$

Since $v_\nu \in H^{1/2}(\partial\Omega)$, by Lax-Milgram's theorem and [276, Ch. 1, Théorème 8.3], we deduce that (3.45) admits a unique solution $v_1 \in H^2 \cap H_0^1(\Omega)$ such that

$$\|\Delta v_1\|_{L^2(\Omega)} \le C \|v_\nu\|_{H^{1/2}(\partial\Omega)}.$$

This proves that $v_1 \in V$. Let $v_2 = v - v_1$, then $(v_2)_v = 0$ on $\partial\Omega$ and, in turn, $v_2 \in H_0^2(\Omega)$. Moreover, by (3.45) we infer

$$\int_\Omega \Delta v_2 \Delta w \, dx = \int_\Omega \Delta v \Delta w \, dx - \int_\Omega \Delta v_1 \Delta w \, dx = \int_\Omega \Delta v \Delta w \, dx \quad \text{for all } w \in H_0^2(\Omega)$$

which proves that v_2 is a weak solution of the second problem in (3.44). $\qquad\square$

When $\Omega = B$ (the unit ball in $\mathbb{R}^n$, $n \geq 2$) all the eigenvalues of (3.38) can be determined explicitly. To this end, consider the spaces of harmonic homogeneous polynomials

$$\mathscr{P}_k :=$$
$$\{P \in C^\infty(\mathbb{R}^n); \, \Delta P = 0 \text{ in } \mathbb{R}^n, \, P \text{ is a homogeneous polynomial of degree } k-1\}.$$

Also, let μ_k be the dimension of $\mathscr{P}_k$. By [17, p. 450] we know that

$$\mu_k = \frac{(2k+n-4)(k+n-4)!}{(k-1)!(n-2)!}.$$

In particular, we have

$$\mathscr{P}_1 = \text{span}\{1\}, \qquad \mu_1 = 1,$$

$$\mathscr{P}_2 = \text{span}\{x_i; \, i = 1,\ldots,n\}, \qquad \mu_2 = n,$$

$$\mathscr{P}_3 = \text{span}\{x_i x_j; \, x_1^2 - x_h^2; \, i,j = 1,\ldots,n, \, i \neq j, \, h = 2,\ldots,n\}, \quad \mu_3 = \frac{n^2+n-2}{2}.$$

Then we prove

Theorem 3.20. *If $n \geq 2$ and $\Omega = B$, then for all $k = 1,2,3,\ldots$*

1. *the eigenvalues of (3.38) are $\delta_k = n + 2(k-1)$;*
2. *the multiplicity of δ_k equals μ_k;*
3. *for all $\psi_k \in \mathscr{P}_k$, the function $\phi_k(x) := (1 - |x|^2)\psi_k(x)$ is an eigenfunction corresponding to δ_k.*

Proof. Let $u \in C^\infty(\overline{B})$ be an eigenfunction of (3.38) so that $u = 0$ on ∂B. Therefore, we can write

$$u(x) = (1 - |x|^2)\phi(x) \qquad (x \in B) \tag{3.46}$$

for some $\phi \in C^\infty(\overline{B})$. We have $u_{x_i} = -2x_i\phi + (1 - |x|^2)\phi_{x_i}$, and on ∂B,

$$u_v = x \cdot \nabla u = -2\phi. \tag{3.47}$$

Moreover,

$$\Delta u = -2n\phi - 4x \cdot \nabla\phi + (1 - |x|^2)\Delta\phi. \tag{3.48}$$

Hence,

$$\Delta u = -2n\phi - 4\phi_\nu \qquad \text{on } \partial B. \tag{3.49}$$

From (3.48) we get for $i = 1, \ldots, n$,

$$(\Delta u)_{x_i} = -(2n+4)\phi_{x_i} - 4\sum_{j=1}^{n} x_j \phi_{x_j x_i} - 2x_i \Delta\phi + (1 - |x|^2)\Delta\phi_{x_i},$$

and therefore

$$(\Delta u)_{x_i x_i} = -2(n+4)\phi_{x_i x_i} - 4x \cdot \nabla(\phi_{x_i x_i}) - 2\Delta\phi - 4x_i(\Delta\phi)_{x_i} + (1 - |x|^2)\Delta\phi_{x_i x_i}.$$

Summing with respect to i and recalling that u is biharmonic in B, we obtain

$$\begin{aligned}
0 = \Delta^2 u &= -2(n+4)\Delta\phi - 4x \cdot \nabla\Delta\phi - 2n\Delta\phi - 4x \cdot \nabla\Delta\phi + (1 - |x|^2)\Delta^2\phi \\
&= (1 - |x|^2)\Delta^2\phi - 8x \cdot \nabla\Delta\phi - 4(n+2)\Delta\phi.
\end{aligned} \tag{3.50}$$

Writing (3.50) as an equation in $w = \Delta\phi$, we get

$$(1 - |x|^2)\Delta w - 8x \cdot \nabla w - 4(n+2)w = 0 \qquad \text{in } B,$$

so that

$$\begin{aligned}
0 &= -(1 - |x|^2)^4 \Delta w + 8(1 - |x|^2)^3 x \cdot \nabla w + 4(n+2)(1 - |x|^2)^3 w \\
&= -\text{div}\left((1 - |x|^2)^4 \nabla w\right) + 4(n+2)(1 - |x|^2)^3 w.
\end{aligned} \tag{3.51}$$

Multiplying the right hand side of (3.51) by w and integrating by parts over B, we obtain

$$\int_B (1 - |x|^2)^4 |\nabla w|^2 \, dx + 4(n+2) \int_B (1 - |x|^2)^3 w^2 \, dx = \int_{\partial B} (1 - |x|^2)^4 w w_\nu \, d\omega = 0.$$

Hence $\Delta\phi = w = 0$ in B. Now from (3.38), (3.47) and (3.49) we get

$$\phi_\nu = \frac{a - n}{2}\phi \qquad \text{on } \partial B.$$

Therefore, the number a is an eigenvalue of (3.38) with corresponding eigenfunction u if and only if ϕ defined by (3.46) is an eigenfunction of the boundary eigenvalue problem

$$\begin{cases} \Delta\phi = 0 & \text{in } B, \\ \phi_\nu = \gamma\phi & \text{on } \partial B, \end{cases} \tag{3.52}$$

where

$$\gamma = \frac{a - n}{2}. \tag{3.53}$$

We are so led to study the eigenvalues of the second order Steklov problem (3.52). Let us quickly explain how to obtain them. In radial and angular coordinates (r, θ), the equation in (3.52) reads

$$\frac{\partial^2 \phi}{\partial r^2} + \frac{n-1}{r} \frac{\partial \phi}{\partial r} + \frac{1}{r^2} \Delta_\theta \phi = 0,$$

where Δ_θ denotes the Laplace-Beltrami operator on ∂B. From [47, p. 160] we know that $-\Delta_\theta$ admits a sequence of eigenvalues (λ_k) having multiplicity μ_k equal to the number of independent harmonic homogeneous polynomials of degree $k - 1$. Moreover, $\lambda_k = (k-1)(n+k-3)$.

Let us write e_k^j $(j = 1, \ldots, \mu_k)$ for the independent normalised eigenfunctions corresponding to λ_k. Then one seeks functions $\phi = \phi(r, \theta)$ of the kind

$$\phi(r, \theta) = \sum_{k=1}^{\infty} \sum_{j=1}^{\mu_k} \phi_k^j(r) e_k^j(\theta).$$

Hence, by differentiating the series, we obtain

$$\Delta \phi(r, \theta) = \sum_{k=1}^{\infty} \sum_{j=1}^{\mu_k} \left(\frac{d^2}{dr^2} \phi_k^j(r) + \frac{n-1}{r} \frac{d}{dr} \phi_k^j(r) - \frac{\lambda_k}{r^2} \phi_k^j(r) \right) e_k^j(\theta) = 0.$$

Therefore, we are led to solve the equations

$$\frac{d^2}{dr^2} \phi_k^j(r) + \frac{n-1}{r} \frac{d}{dr} \phi_k^j(r) - \frac{\lambda_k}{r^2} \phi_k^j(r) = 0 \qquad k = 1, 2 \ldots \qquad j = 1, \ldots, \mu_k. \quad (3.54)$$

With the change of variables $r = e^t$ $(t \leq 0)$, equation (3.54) becomes a linear constant coefficients equation. It has two linearly independent solutions, but one is singular. Hence, up to multiples, the only regular solution of (3.54) is given by $\phi_k^j(r) = r^{k-1}$ because

$$\frac{2 - n + \sqrt{(n-2)^2 + 4\lambda_k}}{2} = k - 1.$$

Since the boundary condition in (3.52) reads $\frac{d}{dr} \phi_k^j(1) = \gamma \phi_k^j(1)$ we immediately infer that $\gamma = \bar{k} - 1$ for some $\bar{k}$. In turn, (3.53) tells us that

$$\delta_{\bar{k}} = n + 2(\bar{k} - 1).$$

The proof of Theorem 3.20 is so complete.   □

Remark 3.21. Theorems 3.18 and 3.20 become false if $n = 1$ since the problem

$$u^{iv} = 0 \text{ in } (-1, 1), \quad u(\pm 1) = u''(-1) + au'(-1) = u''(1) - au'(1) = 0, \quad (3.55)$$

admits only *two* eigenvalues, $\delta_1 = 1$ and $\delta_2 = 3$, each one of multiplicity 1. The reason of this striking difference is that the "boundary space" of (3.55) has precisely dimension 2, one for each endpoint of the interval $(-1, 1)$. This result is consistent with Theorem 3.20 since $\mu_1 = \mu_2 = 1$ and $\mu_3 = 0$ whenever $n = 1$.

By combining Theorems 3.18 and 3.20 we obtain

Corollary 3.22. *Assume that $n \geq 2$ and that $\Omega = B$. Assume moreover that for all $k \in \mathbb{N}^+$ the set $\{\psi_k^j : j = 1, \ldots, \mu_k\}$ is a basis of $\mathscr{P}_k$ chosen in such a way that the corresponding functions ϕ_k^j are orthonormal with respect to the scalar product (3.41). Then for any $u \in V$ there exists a sequence $(\alpha_k^j) \subset \ell^2$ ($k \in \mathbb{N}^+$; $j = 1, \ldots, \mu_k$) such that*

$$u(x) = (1 - |x|^2) \sum_{k=1}^{\infty} \sum_{j=1}^{\mu_k} \alpha_k^j \psi_k^j(x) \qquad \text{for a.e. } x \in B.$$

3.3.2 Minimisation of the First Eigenvalue

In this section we take advantage of Theorem 3.17 and we study several aspects of the first Steklov eigenvalue δ_1.

We first give an alternative characterisation of $\delta_1(\Omega)$. Let

$$C_H^2\left(\overline{\Omega}\right) := \left\{v \in C^2\left(\overline{\Omega}\right) ; \Delta v = 0 \text{ in } \Omega\right\}$$

and consider the norm defined by $\|v\|_H := \|v\|_{L^2(\partial\Omega)}$ for all $v \in C_H^2\left(\overline{\Omega}\right)$. Then define

$$\mathbf{H} := \text{the completion of } C_H^2\left(\overline{\Omega}\right) \text{ with respect to the norm } \|\cdot\|_H.$$

Since Ω is assumed to have a Lipschitz boundary, we infer by [239] that $\mathbf{H} \subset H^{1/2}(\Omega) \subset L^2(\Omega)$. Therefore, the quantity

$$\sigma_1(\Omega) := \inf_{h \in \mathbf{H} \setminus \{0\}} \frac{\|h\|_{L^2(\partial\Omega)}^2}{\|h\|_{L^2(\Omega)}^2}$$

is well-defined. Our purpose is now to relate σ_1 with δ_1, see (3.40). To this end, we make use of a suitable version of Fichera's principle of duality [170]. However, in its original version, this principle requires smoothness of the boundary $\partial\Omega$. Since we aim to deal with most general domains, we need to drop this assumption. We consider Lipschitz domains satisfying a uniform outer ball condition, see Definition 2.30. Then regularity results by Jerison-Kenig [238, 239] enable us to prove the following result.

Theorem 3.23. *If $\Omega \subset \mathbb{R}^n$ is a bounded domain with Lipschitz boundary, then $\sigma_1(\Omega)$ admits a minimiser $h \in \mathbf{H} \backslash \{0\}$. If Ω also satisfies a uniform outer ball condition then the minimiser is positive, unique up to a constant multiplier and $\sigma_1(\Omega) = \delta_1(\Omega)$.*

Proof. In the first part of this proof, we just assume that Ω is a domain with Lipschitz boundary. Let $(h_m) \subset \mathbf{H} \backslash \{0\}$ be a minimising sequence for $\sigma_1(\Omega)$ with $\|h_m\|_H = \|h_m\|_{L^2(\partial\Omega)} = 1$. Up to a subsequence, we may assume that there exists $h \in \mathbf{H}$ such that $h_m \rightharpoonup h$ in $\mathbf{H}$. By regularity estimates [238, 239], we infer that there exists a constant $C > 0$ such that

$$\|h\|_{H^{1/2}(\Omega)} \leq C \|h\|_{L^2(\partial\Omega)} \qquad \text{for all } h \in \mathbf{H}$$

so that $\sigma_1(\Omega) > 0$ and the sequence (h_m) is bounded in $H^{1/2}(\Omega)$, $h_m \rightharpoonup h$ in $H^{1/2}(\Omega)$ up to a subsequence and, by compact embedding, we also have $h_m \to h$ in $L^2(\Omega)$. Therefore, since (h_m) is a minimising sequence, $\|h_m\|_{L^2(\partial\Omega)} = 1$ and $\|h_m\|_{L^2(\Omega)}$ is bounded it follows that $h \in \mathbf{H} \backslash \{0\}$ and

$$\|h\|_{L^2(\Omega)}^{-2} = \lim_{m \to \infty} \|h_m\|_{L^2(\Omega)}^{-2} = \sigma_1(\Omega).$$

Moreover, by weak lower semicontinuity of $\|.\|_H$ we also have

$$\|h\|_{L^2(\partial\Omega)}^2 = \|h\|_H^2 \leq \liminf_{m \to \infty} \|h_m\|_H^2 = 1$$

and hence $h \in \mathbf{H} \backslash \{0\}$ satisfies

$$\frac{\|h\|_{L^2(\partial\Omega)}^2}{\|h\|_{L^2(\Omega)}^2} \leq \sigma_1(\Omega).$$

This proves that h is a minimiser for $\sigma_1(\Omega)$.

In the rest of the proof, we assume furthermore that Ω satisfies a uniform outer ball condition. Under this condition, we have the existence of a minimiser for $\delta_1(\Omega)$ by Theorem 3.17. We say that σ is a *harmonic boundary eigenvalue* if there exists $g \in \mathbf{H}$ such that

$$\sigma \int_\Omega gv \, dx = \int_{\partial\Omega} gv \, d\omega \qquad \text{for all } v \in \mathbf{H}.$$

Clearly, σ_1 is the least harmonic boundary eigenvalue. We prove that $\sigma_1 = \delta_1$ by showing two inequalities.

Proof of $\sigma_1 \geq \delta_1$. Let h be a minimiser for σ_1, then

$$\sigma_1 \int_\Omega hv \, dx = \int_{\partial\Omega} hv \, d\omega \qquad \text{for all } v \in \mathbf{H}. \tag{3.56}$$

Let $u \in [H^2 \cap H_0^1(\Omega)] \setminus H_0^2(\Omega)$ be the unique solution to

$$\begin{cases} \Delta u = h & \text{in } \Omega, \\ u = 0 & \text{on } \partial\Omega. \end{cases}$$

Integrating by parts we have

$$\int_\Omega h v \, dx = \int_\Omega v \Delta u \, dx = \int_{\partial\Omega} v u_\nu \, d\omega \quad \text{for all } v \in \mathbf{H} \cap C^2(\overline{\Omega}).$$

By a density argument, the latter follows for all $v \in \mathbf{H}$. Inserting this into (3.56) gives

$$\sigma_1 \int_{\partial\Omega} v u_\nu \, d\omega = \int_{\partial\Omega} v \Delta u \, d\omega \qquad \text{for all } v \in \mathbf{H}.$$

This yields $\Delta u = \sigma_1 u_\nu$ on $\partial\Omega$. Therefore,

$$\sigma_1 = \frac{\|h\|_{L^2(\partial\Omega)}^2}{\|h\|_{L^2(\Omega)}^2} = \frac{\|\Delta u\|_{L^2(\partial\Omega)}^2}{\|\Delta u\|_{L^2(\Omega)}^2} = \sigma_1^2 \frac{\|u_\nu\|_{L^2(\partial\Omega)}^2}{\|\Delta u\|_{L^2(\Omega)}^2}.$$

In turn, this implies that

$$\sigma_1 = \frac{\|\Delta u\|_{L^2(\Omega)}^2}{\|u_\nu\|_{L^2(\partial\Omega)}^2} \geq \min \left\{ \frac{\|\Delta v\|_{L^2(\Omega)}^2}{\|v_\nu\|_{L^2(\partial\Omega)}^2} \; ; \; v \in [H^2 \cap H_0^1(\Omega)] \setminus H_0^2(\Omega) \right\} = \delta_1.$$

Proof of $\sigma_1 \leq \delta_1$. Let u be a minimiser for δ_1 in (3.40), then $\Delta u = \delta_1 u_\nu$ on $\partial\Omega$ so that $\Delta u \in H^{1/2}(\partial\Omega) \subset L^2(\partial\Omega)$ and

$$\int_{\partial\Omega} v \Delta u \, d\omega = \delta_1 \int_{\partial\Omega} v u_\nu \, d\omega \qquad \text{for all } v \in \mathbf{H}. \tag{3.57}$$

Let $h := \Delta u$ so that $h \in L^2(\Omega) \cap L^2(\partial\Omega)$. Moreover, $\Delta h = \Delta^2 u = 0$ in a distributional sense and hence $h \in \mathbf{H}$. Two integrations by parts and a density argument yield

$$\int_\Omega h v \, dx = \int_{\partial\Omega} v u_\nu \, d\omega \qquad \text{for all } v \in \mathbf{H}.$$

Replacing this into (3.57) gives

$$\int_{\partial\Omega} h v \, d\omega = \delta_1 \int_\Omega h v \, dx \qquad \text{for all } v \in \mathbf{H}.$$

This proves that h is an eigenfunction with corresponding harmonic boundary eigenvalue δ_1. Since σ_1 is the least harmonic boundary eigenvalue, we obtain $\delta_1 \geq \sigma_1$.

Then $\sigma_1 = \delta_1$ and there is a one-to-one correspondence between minimisers of $\sigma_1(\Omega)$ and $\delta_1(\Omega)$. Hence, uniqueness of a minimiser for $\sigma_1(\Omega)$ up to a constant multiplier follows from Theorem 3.17. $\qquad\square$

We now show that an optimal shape for δ_1 under volume or perimeter constraint does not exist in any space dimension $n \geq 2$.

Theorem 3.24. *Let $D_\varepsilon = \{x \in \mathbb{R}^2; \varepsilon < |x| < 1\}$ and let $\Omega_\varepsilon \subset \mathbb{R}^n \, (n \geq 2)$ be such that*

$$\Omega_\varepsilon = D_\varepsilon \times (0,1)^{n-2} \, ;$$

in particular, if $n = 2$ we have $\Omega_\varepsilon = D_\varepsilon$. Then

$$\lim_{\varepsilon \searrow 0} \delta_1(\Omega_\varepsilon) = 0.$$

Proof. We assume first that $n = 2$. For any $\varepsilon \in (0,1)$ let $w_\varepsilon \in H^2 \cap H_0^1(D_\varepsilon)$ be defined by

$$w_\varepsilon(x) = \frac{1 - |x|^2}{4} - \frac{1 - \varepsilon^2}{4 \log \varepsilon} \log |x| \qquad \text{for all } x \in D_\varepsilon. \tag{3.58}$$

Then we have

$$\Delta w_\varepsilon = -1 \qquad \text{in } \Omega_\varepsilon$$

and

$$|\nabla w_\varepsilon(x)|^2 = \left(\frac{|x|}{2} + \frac{1 - \varepsilon^2}{4 \log \varepsilon} \frac{1}{|x|} \right)^2 \qquad \text{for all } x \in \overline{\Omega}_\varepsilon$$

so that

$$\int_{\Omega_\varepsilon} |\Delta w_\varepsilon|^2 \, dx = \pi \left(1 - \varepsilon^2 \right)$$

and

$$\int_{\partial \Omega_\varepsilon} (w_\varepsilon)_\nu^2 \, d\omega = 2\pi \left(\frac{1}{2} + \frac{1 - \varepsilon^2}{4 \log \varepsilon} \right)^2 + 2\pi\varepsilon \left(\frac{\varepsilon}{2} + \frac{1 - \varepsilon^2}{4\varepsilon \log \varepsilon} \right)^2$$

$$= \frac{\pi}{8} \frac{1}{\varepsilon \log^2 \varepsilon} + o\left(\frac{1}{\varepsilon \log^2 \varepsilon} \right) \to +\infty \qquad \text{as } \varepsilon \searrow 0. \tag{3.59}$$

It follows immediately that

$$\lim_{\varepsilon \searrow 0} \delta_1(\Omega_\varepsilon) \leq \lim_{\varepsilon \searrow 0} \frac{\displaystyle\int_{\Omega_\varepsilon} |\Delta w_\varepsilon|^2 \, dx}{\displaystyle\int_{\partial \Omega_\varepsilon} (w_\varepsilon)_\nu^2 \, d\omega} = 0.$$

This completes the proof of the theorem for $n = 2$.

We now consider the case $n \geq 3$. Let

$$u_\varepsilon(x) = \left(\prod_{i=3}^{n} x_i(1-x_i) \right) w_\varepsilon(x_1,x_2) \qquad \text{for all } x \in \Omega_\varepsilon$$

where w_ε is as in (3.58); note that u_ε vanishes on $\partial\Omega_\varepsilon$ and $u_\varepsilon \in H^2 \cap H_0^1(\Omega_\varepsilon)$. Then we have

$$\Delta u_\varepsilon = -\prod_{i=3}^{n} x_i(1-x_i) - 2w_\varepsilon(x_1,x_2) \sum_{j=3}^{n} \prod_{\substack{i=3 \\ i \neq j}}^{n} x_i(1-x_i)$$

(with the convention that $\prod_{i \in \emptyset} \beta_i = 1$) and

$$\int_{\Omega_\varepsilon} |\Delta u_\varepsilon|^2\, dx \leq 2 \int_{\Omega_\varepsilon} \prod_{i=3}^{n} x_i^2(1-x_i)^2\, dx + 8 \int_{\Omega_\varepsilon} w_\varepsilon^2(x_1,x_2) \sum_{j=3}^{n} \prod_{\substack{i=3 \\ i \neq j}}^{n} x_i^2(1-x_i)^2\ dx.$$

Hence, since $|w_\varepsilon(x)| < \frac{1}{2}$ for all $x \in D_\varepsilon$, there exists $C > 0$ such that

$$\int_{\Omega_\varepsilon} |\Delta u_\varepsilon|^2\, dx \leq C \qquad \text{for all } \varepsilon \in (0,1). \tag{3.60}$$

On the other hand, we have

$$|\nabla u_\varepsilon|^2 = \prod_{i=3}^{n} x_i^2(1-x_i)^2 \left(\left(\frac{\partial w_\varepsilon}{\partial x_1} \right)^2 + \left(\frac{\partial w_\varepsilon}{\partial x_2} \right)^2 \right)$$

$$+ \sum_{j=3}^{n} \left((1-2x_j)^2 w_\varepsilon^2(x_1,x_2) \prod_{\substack{i=3 \\ i \neq j}}^{n} x_i^2(1-x_i)^2 \right)$$

and since w_ε vanishes on ∂D_ε we obtain

$$\int_{\partial\Omega_\varepsilon} (u_\varepsilon)_\nu^2\, d\omega = \int_{\partial\Omega_\varepsilon} |\nabla u_\varepsilon|^2\, d\omega \geq \int_{\partial D_\varepsilon \times (0,1)^{n-2}} |\nabla u_\varepsilon|^2\, d\omega$$

$$\geq \int_{\partial D_\varepsilon} (w_\varepsilon)_\nu^2\, d\omega \cdot \prod_{i=3}^{n} \int_0^1 x_i^2(1-x_i)^2\, dx_i \to +\infty$$

as $\varepsilon \searrow 0$ in view of (3.59). Therefore, by (3.60) we obtain

$$\lim_{\varepsilon \searrow 0} \delta_1(\Omega_\varepsilon) \leq \lim_{\varepsilon \searrow 0} \frac{\displaystyle\int_{\Omega_\varepsilon} |\Delta u_\varepsilon|^2\, dx}{\displaystyle\int_{\partial\Omega_\varepsilon} (u_\varepsilon)_\nu^2\, d\omega} = 0$$

which proves the theorem also when $n \geq 3$. $\qquad\qquad\square$

Theorem 3.24 has several important consequences. First, it shows that $\delta_1(\Omega)$ has no optimal shape under the constraint that Ω is contained in a fixed ball.

Corollary 3.25. *Let* $B_R = \{x \in \mathbb{R}^n; \, |x| < R\}$. *Then for any* $R > 0$

$$\inf_{\Omega \subset B_R} \delta_1(\Omega) = 0$$

where the infimum is taken over all domains $\Omega \subset B_R$ *such that* $\partial\Omega \in C^\infty$ *if* $n = 2$ *and* $\partial\Omega$ *is Lipschitzian if* $n \geq 3$.

A second consequence of Theorem 3.24 is that it disproves the conjecture by Kuttler [259] which states that the disk has the smallest δ_1 among all planar regions having the same perimeter. Let us also mention that, although the ball has no isoperimetric property, it is a stationary domain for the map $\Omega \mapsto \delta_1(\Omega)$ in the class of C^4 domains under smooth perturbations which preserve measure, see [80] for the details.

Theorem 3.24 also shows that the map $\Omega \mapsto \delta_1(\Omega)$ is *not* monotonically decreasing with respect to domain inclusion.

Finally, Theorem 3.24 raises several natural questions. Why do we consider an annulus in the plane and the region between two cylinders in space dimensions $n \geq 3$? What happens if we consider an annulus in any space dimension? The quite surprising answer is given in

Theorem 3.26. *Let* $n \geq 3$ *and let* $\Omega^\varepsilon = \{x \in \mathbb{R}^n; \, \varepsilon < |x| < 1\}$.

1. *If* $n = 3$ *then*

$$\lim_{\varepsilon \searrow 0} \delta_1(\Omega^\varepsilon) = 2.$$

2. *If* $n \geq 4$ *then*

$$\lim_{\varepsilon \searrow 0} \delta_1(\Omega^\varepsilon) = n.$$

For the proof of Theorem 3.26 we refer to [80]. Theorems 3.24 and 3.26 highlight a striking difference between dimension $n = 2$, dimension $n = 3$ and dimensions $n \geq 4$. Since the set Ω^ε is smooth, by Theorem 3.23 it follows that $\delta_1(\Omega^\varepsilon) = \sigma_1(\Omega^\varepsilon)$. Moreover, since the proof of Theorem 3.26 in [80] uses radial harmonic functions $h = h(r)$ $(r = |x|)$, we may rewrite the ratio defining $\sigma_1(\Omega^\varepsilon)$ as

$$\frac{\int_{\partial\Omega^\varepsilon} h^2 \, d\omega}{\int_{\Omega^\varepsilon} h^2 \, dx} = \frac{h(1)^2 + \varepsilon^{n-1} h(\varepsilon)^2}{\int_\varepsilon^1 h(r)^2 r^{n-1} \, dr}.$$

In this setting, we can treat the space dimension n as a real number. Then we have

Theorem 3.27. *Let* $\varepsilon \in (0,1)$, *let* $K_\varepsilon = \{h \in C^2([\varepsilon,1]); h''(r) + \frac{n-1}{r}h'(r) = 0, r \in [\varepsilon,1]\}$ *and, for all* $n \in [1,\infty)$, *let*

$$\gamma_\varepsilon(n) = \inf_{h \in K_\varepsilon \setminus \{0\}} \frac{h(1)^2 + \varepsilon^{n-1}h(\varepsilon)^2}{\displaystyle\int_\varepsilon^1 h(r)^2 r^{n-1} dr}.$$

Then

$$\lim_{\varepsilon \to 0} \gamma_\varepsilon(n) = \begin{cases} 2 & \text{if } n = 1, \\ 0 & \text{if } 1 < n < 3, \\ 2 & \text{if } n = 3, \\ n & \text{if } n > 3. \end{cases}$$

Theorem 3.27 is proved in [80] and shows that dimensions $n = 1$ and $n = 3$ are "discontinuous" dimensions for the behaviour of γ_ε. The reason of this discontinuity is not clear to us.

Finally, we point out that Steklov boundary conditions, producing a boundary integral in the denominator of the Rayleigh quotient, require a *strong* geometric convergence (namely a very fine topology) in order to preserve the perimeter. However, contrary to the Babuška paradox (see Section 1.4.2), we notice that we do have stability of the first eigenvalue on the sequence of regular polygons converging to the disk.

Theorem 3.28. *Let* $n = 2$ *and let* (P_k) *be a sequence of regular polygons with k edges circumscribed to the unit disk D centered at the origin. Then*

$$\lim_{k \to \infty} \delta_1(P_k) = \delta_1(D) = 2.$$

The proof of Theorem 3.28 is lengthy and delicate. This is why we refer again to [80].

3.4 Bibliographical Notes

An interesting survey on spectral properties of higher-order elliptic operators is also provided by Davies [129].

For the original version of the Kreĭn-Rutman theorem, which generalises Jentzsch's [237] theorem, we refer to [258, Theorem 6.2 and 6.3]. Theorem 3.3 is taken from the appendix of [55] and it follows by combining the variant of the Kreĭn-Rutman result in [360, Theorem 6.6, p. 337] with a result by de Pagter [136].

Theorem 3.4 is due to Moreau [312]. A first application of this decomposition is given in the paper by Miersemann [302] for the positivity in a buckling eigenvalue problem. Proposition 3.6 is the generalisation of [19, Lemma 16] from $m = 2$ to the case $m \geq 2$.

Theorem 3.7 is a straightforward consequence of Kreĭn-Rutman's theorem and Lemma 2.27 but the elementary proof suggested here is taken from [181]. The rest of Section 3.1.3 is taken from the survey paper [386].

Concerning Theorem 3.8, numerical results in 1972 and 1980 already predicted that the first eigenfunction on a square changes sign, see [34, 221]. Subsequently, in 1982 Coffman [107] gave an analytic proof of Theorem 3.8. More recently, in 1996, the numerical results on the square have been revisited by Wieners [413] who proved that the sign-changing of the numerically approximated first eigenfunction is rigorous, that is, the sign changing effect is too large to be explained by numerical errors.

Theorem 3.9 is due to Coffman-Duffin-Shaffer. The eigenvalue problem for domains with holes was first studied by Duffin-Shaffer [155]. Subsequently, with Coffman [110] they could show that for the annuli with a small hole, the first eigenfunction changes sign. They used an explicit formula and explicit values of the Bessel functions involved and obtained even a critical number for the ratio of the inner and outer radius. The proof has been simplified in [109].

For further results on sign-changing first eigenfunctions to (3.8), for numerical experiments, and for conjectures on simple domains (such as ellipses, elongated disks, dumb-bells, and limaçons) we refer again to [386].

For some first properties of spherical rearrangements, we refer to [344]. A complete proof of Theorem 3.11 can be found in [10] while its essential Item 2 goes back to Sperner [379] and Talenti [391]. Kawohl [244] discusses the question whether equality in Item 2 of Theorem 3.11 implies symmetry; he shows that the answer is affirmative for analytic functions while it is negative in general. A more general condition ensuring symmetry was subsequently obtained by Brothers-Ziemer [74], see also [100] and references therein for further results on this topic. Theorem 3.12 is an iteration of Talenti's principle [392].

For a fairly complete story of Rayleigh's conjecture [351], we refer to Section 1.3.1. Although it was Nadirashvili [316] who proved first the Rayleigh conjecture in dimension $n = 2$, the proof of Theorem 3.13 follows closely the one by Ashbaugh-Benguria [22] which is more general since it also holds for $n = 3$. It uses some results by Talenti [393].

Theorem 3.15 is due to Ashbaugh-Bucur [23]. Minimisation of the buckling load can be also performed in different classes of domains. For instance, one could argue in the class of convex domains like in [245, Proposition 4.5]. For further classes of domains, such as open sets, quasi-open sets or multiply connected sets, we refer again to [23]. On the occasion of an Oberwolfach meeting in 1995, Willms gave a talk with the proof of Theorem 3.16 according to joint work with Weinberger [416] but the proof was never written by them. With their permission, Kawohl [245, Proposition 4.4] wrote down the proof of the talk by Willms and this is where we have taken it, see also [23]. For more results on buckling eigenvalues, mainly under Dirichlet boundary conditions, we refer to [48, 49, 172, 229, 246, 302, 377] and references therein.

Elliptic problems with parameters in the boundary conditions are called *Steklov problems* from their first appearance in [380]. For the biharmonic operator, these conditions were first considered by Kuttler-Sigillito [261] and Payne [335] who studied the isoperimetric properties of the first eigenvalue δ_1, see also subsequent work by Smith [374, 375] and Kuttler [259, 260]. We also refer to the monograph by Kuttler-Sigillito [262] for some numerical experiments and for a survey of results known at that time. Finally, we refer to Section 1.3.3 for the complete story about the minimisation of δ_1.

Theorem 3.17 is taken from [80] although it was already known in the smooth case $\partial \Omega \in C^2$, see [42]. The characterisation of the first Steklov eigenvalue in the ball and Remark 3.21 are taken from [42]. Subsequently, the whole spectrum of the biharmonic Steklov problem was studied by Ferrero-Gazzola-Weth [165] from where Theorems 3.18, 3.19, and 3.20 are taken. Theorem 3.23 is a generalisation to nonsmooth domains of a particular application of Fichera's principle of duality [170]; in this final form it is proved in [80], see also [165, 170] for previous work in the case $\partial \Omega \in C^2$. All the other statements in Section 3.3.2 are taken from Bucur-Ferrero-Gazzola [80].

Chapter 4
Kernel Estimates

In Chapters 5 and 6 we discuss positivity and almost positivity for higher order boundary value problems. The goal of the present chapter is to provide the required estimates, which are also interesting in themselves. In order to avoid a too technical exposition, in many cases the discussion is restricted to fourth order problems. However, whenever it does not require too many additional distinctions, the general case of $2m$-th order operators is also covered.

4.1 Consequences of Boggio's Formula

Throughout this chapter we will exploit the following notations.

Notation 4.1 Let $f, g \geq 0$ be functions defined on the same set D.

- We write $f \preceq g$ if there exists $c > 0$ such that $f(x) \leq cg(x)$ for all $x \in D$.
- We write $f \simeq g$ if both $f \preceq g$ and $g \preceq f$.

Notation 4.2 For a smooth bounded domain Ω, we define the distance function to the boundary

$$d(x) := \operatorname{dist}(x, \partial\Omega) = \min_{y \in \partial\Omega} |x - y|, \quad x \in \Omega. \tag{4.1}$$

Many estimates will be for coordinates inside the unit ball B in $\mathbb{R}^n$ and for this special domain the following expression will be used repeatedly.

Notation 4.3 For $x, y \in \overline{B}$ we write

$$[XY] := \sqrt{|x|^2 |y|^2 - 2x \cdot y + 1} = \left| |x|y - \frac{x}{|x|} \right| = \left| |y|x - \frac{y}{|y|} \right|. \tag{4.2}$$

As (4.2) shows, $[XY]$ is the distance from $|y|x$ to the projection of y on the unit sphere, which is larger than $|x - y|$. Indeed

$$[XY]^2 - |x - y|^2 = (1 - |x|^2)(1 - |y|^2) > 0 \text{ for } x, y \in B. \tag{4.3}$$

F. Gazzola et al., *Polyharmonic Boundary Value Problems*,
Lecture Notes in Mathematics 1991, DOI 10.1007/978-3-642-12245-3_4,
© Springer-Verlag Berlin Heidelberg 2010

Since $1 - |x| = d(x)$ for $x \in \overline{B}$ it even shows that

$$|x-y|^2 + d(x)d(y) \le [XY]^2 \le |x-y|^2 + 4d(x)d(y). \tag{4.4}$$

We focus on the polyharmonic analogue of the clamped plate boundary value problem

$$\begin{cases} (-\Delta)^m u = f & \text{in } \Omega, \\ D^\alpha u|_{\partial\Omega} = 0 & \text{for } |\alpha| \le m-1. \end{cases} \tag{4.5}$$

Here $\Omega \subset \mathbb{R}^n$ is a bounded smooth domain, f a datum in a suitable functional space and u denotes the unknown solution.

In bounded smooth domains, a unique Green function $G_{(-\Delta)^m,\Omega}$ for problem (4.5) exists and the representation formula

$$u(x) = \int_\Omega G_{(-\Delta)^m,\Omega}(x,y)f(y)\,dy, \quad x \in \Omega, \tag{4.6}$$

holds true, see (2.64). Having a positivity preserving property in Ω is equivalent to $G_{(-\Delta)^m,\Omega} \ge 0$. *Almost positivity* will mean that the negative part of $G_{(-\Delta)^m,\Omega}$ is small in a sense to be specified when compared with its positive part. The main goal of Chapters 5 and 6 is to identify domains and also further differential operators enjoying almost positivity or even a positivity preserving property. To this end, we provide in the present chapter fine estimates for the Green function and the other kernels involved in the solution of higher order boundary value problems.

With $[XY]$ as in (4.2), the Green function from Lemma 2.27 by Boggio for the Dirichlet problem (4.5) with $\Omega = B$, the unit ball, is given by

$$G_{m,n}(x,y) = k_{m,n}|x-y|^{2m-n} \int_1^{[XY]/|x-y|} (v^2-1)^{m-1}v^{1-n}\,dv . \tag{4.7}$$

In Section 4.2 we give the following characterisation of $G_{m,n}$, which will be much more convenient than Boggio's original formula in discussing positivity issues:

$$G_{m,n}(x,y) \simeq \begin{cases} |x-y|^{2m-n} \min\left\{1, \dfrac{d(x)^m d(y)^m}{|x-y|^{2m}}\right\} & \text{if } n > 2m, \\[2ex] \log\left(1 + \dfrac{d(x)^m d(y)^m}{|x-y|^{2m}}\right) & \text{if } n = 2m, \\[2ex] d(x)^{m-\frac{n}{2}} d(y)^{m-\frac{n}{2}} \min\left\{1, \dfrac{d(x)^{\frac{n}{2}} d(y)^{\frac{n}{2}}}{|x-y|^n}\right\} & \text{if } n < 2m. \end{cases}$$

A more detailed discussion of the boundary terms will be given below. We further deduce related estimates for the derivatives $|D_x^\alpha G_{m,n}(x,y)|$. All these are used to prove so-called 3-G-type theorems in Section 4.2.2, which will help us to develop a perturbation theory of positivity.

It is an obvious question whether in general domains $\Omega \subset \mathbb{R}^n$, where one does not have positivity preserving, estimates for $|G_{(-\Delta)^m,\Omega}|$ as above are available. This question is addressed in Section 4.5. In order to avoid too many technicalities, we confine ourselves here to the biharmonic case. The following estimate is proven in any bounded domain $\Omega \subset \mathbb{R}^n$ with $\partial\Omega \in C^{4,\gamma}$.

$$
|G_{\Delta^2,\Omega}(x,y)| \preceq
\begin{cases}
|x-y|^{4-n} \min\left\{1, \dfrac{d(x)^2 d(y)^2}{|x-y|^4}\right\} & \text{if } n > 4, \\[2ex]
\log\left(1 + \dfrac{d(x)^2 d(y)^2}{|x-y|^4}\right) & \text{if } n = 4, \\[2ex]
d(x)^{2-\frac{n}{2}} d(y)^{2-\frac{n}{2}} \min\left\{1, \dfrac{d(x)^{\frac{n}{2}} d(y)^{\frac{n}{2}}}{|x-y|^n}\right\} & \text{if } n < 4.
\end{cases}
\tag{4.8}
$$

These estimates, also being quite interesting in themselves, will prove to be basic for the positivity and almost positivity results in Chapter 6.

Finally, kernel estimates and 3-G-type results are collected in Section 4.3 to prepare the discussion of positivity in the Steklov problem which will be given in Section 5.4.

4.2 Kernel Estimates in the Ball

4.2.1 Direct Green Function Estimates

Let $G_{m,n} : \overline{B} \times \overline{B} \to \mathbb{R} \cup \{\infty\}$ denote the Green function for $(-\Delta)^m$ under homogeneous Dirichlet boundary conditions, see (4.7), and let

$$
\mathscr{G}_{m,n} : L^p(B) \to W^{2m,p} \cap W_0^{m,p}(B), \quad (\mathscr{G}_{m,n}f)(x) = \int_B G_{m,n}(x,y) f(y)\, dy, \tag{4.9}
$$

be the corresponding Green operator. In order to base a perturbation theory of positivity on this formula, we first condense the key information on the behaviour of $G_{m,n}$ and its derivatives in more convenient expressions, which also allow for a more direct interpretation of its behaviour, see Theorems 4.6 and 4.7 below.

The first lemma characterises the crucial distinction between the cases "x and y are closer to the boundary ∂B than to each other" and vice versa.

Lemma 4.4. *Let $x,y \in \overline{B}$. If $|x-y| \geq \frac{1}{2}[XY]$, then*

$$
d(x)d(y) \leq 3|x-y|^2, \tag{4.10}
$$

$$
\max\{d(x), d(y)\} \leq 3|x-y|. \tag{4.11}
$$

If $|x-y| \leq \frac{1}{2}[XY]$, then

$$\tfrac{3}{4}|x-y|^2 \leq \tfrac{3}{16}[XY]^2 \leq d(x)d(y), \tag{4.12}$$

$$\tfrac{1}{4}d(x) \leq d(y) \leq 4d(x), \tag{4.13}$$

$$|x-y| \leq 3\min\{d(x),d(y)\}, \tag{4.14}$$

$$[XY] \leq 5\min\{d(x),d(y)\}. \tag{4.15}$$

Moreover, for all $x,y \in \overline{B}$ we have

$$d(x) \leq [XY], \qquad d(y) \leq [XY], \tag{4.16}$$

$$[XY] \simeq d(x)+d(y)+|x-y|. \tag{4.17}$$

Proof. Let $|x-y| \geq \frac{1}{2}[XY]$. Then one has

$$d(x)d(y) \leq \left(1-|x|^2\right)\left(1-|y|^2\right) = [XY]^2 - |x-y|^2 \leq 3|x-y|^2,$$

hence (4.10). The estimate (4.11) follows from

$$d(x)^2 \leq d(x)\,(d(y)+|x-y|) \leq 3|x-y|^2 + |x-y|d(x) \leq 4|x-y|^2 + \tfrac{1}{4}d(x)^2$$
$$\Rightarrow \quad d(x)^2 \leq \tfrac{16}{3}|x-y|^2,$$

and a corresponding estimate for y.

Now, let $|x-y| \leq \frac{1}{2}[XY]$. Then it follows

$$d(x)d(y) \geq \tfrac{1}{4}\left(1-|x|^2\right)\left(1-|y|^2\right) = \tfrac{1}{4}\left([XY]^2 - |x-y|^2\right) \geq \tfrac{3}{16}[XY]^2 \geq \tfrac{3}{4}|x-y|^2,$$

hence (4.12). Inequalities (4.13) can be deduced from

$$d(y) \leq d(x)+|x-y| \leq d(x)+\left(\tfrac{4}{3}d(x)d(y)\right)^{1/2} \leq \left(1+\tfrac{2}{3}\right)d(x)+\tfrac{1}{2}d(y)$$
$$\Rightarrow \quad d(y) \leq \tfrac{10}{3}d(x),$$

and the analogous computation with x and y interchanged; (4.14) and (4.15) are now obvious.

Finally, for all $x,y \in \overline{B}$ we have

$$[XY]^2 = \left||x|y-\frac{x}{|x|}\right|^2 \geq 1-2|x||y|+|x|^2|y|^2 = (1-|x||y|)^2 \geq \begin{cases} (1-|x|)^2 = d(x)^2 \\ (1-|y|)^2 = d(y)^2 \end{cases}$$

thereby proving (4.16). For (4.17), formulae (4.3) and (4.16) show "$\succeq$". On the other hand, $[XY]^2 - |x-y|^2 = \left(1-|x|^2\right)\left(1-|y|^2\right) \leq 4d(x)d(y) \leq 2d(x)^2 + 2d(y)^2$ showing also "$\preceq$". $\square$

In the ball, the following lemma is a direct consequence of the preceding one. However, since the result is needed also in general domains we prove it in this framework.

Lemma 4.5. *Let $\Omega \subset \mathbb{R}^n$ be a bounded domain and let $p,q \geq 0$ be fixed. For $(x,y) \in \overline{\Omega}^2$ we have:*

$$\min\left\{1, \frac{d(y)}{|x-y|}\right\} \simeq \min\left\{1, \frac{d(y)}{d(x)}, \frac{d(y)}{|x-y|}\right\}, \tag{4.18}$$

$$\min\left\{1, \frac{d(x)d(y)}{|x-y|^2}\right\} \simeq \min\left\{\frac{d(y)}{d(x)}, \frac{d(x)}{d(y)}, \frac{d(x)d(y)}{|x-y|^2}\right\}, \tag{4.19}$$

$$\min\left\{1, \frac{d(x)^p d(y)^q}{|x-y|^{p+q}}\right\} \simeq \min\left\{1, \frac{d(x)^p}{|x-y|^p}, \frac{d(y)^q}{|x-y|^q}, \frac{d(x)^p d(y)^q}{|x-y|^{p+q}}\right\}, \tag{4.20}$$

$$\min\left\{1, \frac{d(x)^p d(y)^q}{|x-y|^{p+q}}\right\} \simeq \min\left\{1, \frac{d(x)}{|x-y|}\right\}^p \min\left\{1, \frac{d(y)}{|x-y|}\right\}^q, \tag{4.21}$$

and assuming moreover that $p+q > 0$, we also have

$$\log\left(1 + \frac{d(x)^p d(y)^q}{|x-y|^{p+q}}\right) \simeq \log\left(2 + \frac{d(y)}{|x-y|}\right) \min\left\{1, \frac{d(x)^p d(y)^q}{|x-y|^{p+q}}\right\}. \tag{4.22}$$

Proof. Case $d(x) \geq 2|x-y|$ or $d(y) \geq 2|x-y|$.
If $d(x) \geq 2|x-y|$ we also have

$$d(y) \geq d(x) - |x-y| \geq d(x) - \tfrac{1}{2}d(x) = \tfrac{1}{2}d(x) \geq |x-y|,$$
$$d(y) \leq d(x) + |x-y| \leq \tfrac{3}{2}d(x).$$

If, on the other hand, $d(y) \geq 2|x-y|$ one concludes similarly that

$$|x-y| \leq \tfrac{1}{2}d(y) \leq d(x) \leq \tfrac{3}{2}d(y).$$

Hence, in what follows we may use that

$$|x-y| \leq d(x) \text{ and } |x-y| \leq d(x) \text{ and } d(x) \simeq d(y). \tag{4.23}$$

This shows that in (4.18) - (4.21) we have that the left hand sides as well as the right hand sides all satisfy $\simeq 1$. As for (4.22), we have in this case thanks to $p+q > 0$

$$\log\left(1 + \frac{d(x)^p d(y)^q}{|x-y|^{p+q}}\right) \simeq \log\left(2 + \frac{d(y)}{|x-y|}\right) \simeq \log\left(2 + \frac{d(y)}{|x-y|}\right) \min\left\{1, \frac{d(x)^p d(y)^q}{|x-y|^{p+q}}\right\}.$$

Case $d(x) < 2|x-y|$ and $d(y) < 2|x-y|$.
As for (4.18), inequality "$\succeq$" is obvious, while "$\preceq$" follows from

$$\min\left\{1, \frac{d(y)}{|x-y|}\right\} \simeq \frac{d(y)}{|x-y|} \leq 2\frac{d(y)}{d(x)}.$$

For (4.19) one uses $\min\{t, \frac{1}{t}\} \leq 1$ (for all $t > 0$) to prove "$\succeq$". For "$\preceq$"one may observe that

$$\frac{1}{|x-y|^2} \leq \frac{4}{d(x)^2} \quad \text{and} \quad \frac{1}{|x-y|^2} \leq \frac{4}{d(y)^2}.$$

In the case considered claims (4.20) and (4.21) are obvious. Finally, through

$$\log\left(1 + \frac{d(x)^p d(y)^q}{|x-y|^{p+q}}\right) \simeq \frac{d(x)^p d(y)^q}{|x-y|^{p+q}} \simeq \min\left\{1, \frac{d(x)^p d(y)^q}{|x-y|^{p+q}}\right\}$$
$$\simeq \log\left(2 + \frac{d(y)}{|x-y|}\right) \min\left\{1, \frac{d(x)^p d(y)^q}{|x-y|^{p+q}}\right\}$$

we find (4.22). $\square$

We are now ready to establish the basic Green function estimates. In what follows the estimates of $G_{m,n}$ *from below* will be crucial.

Theorem 4.6. *(Two-sided estimates of the Green function) In $\overline{B} \times \overline{B}$ we have*

$$G_{m,n}(x,y) \simeq \begin{cases} |x-y|^{2m-n} \min\left\{1, \dfrac{d(x)^m d(y)^m}{|x-y|^{2m}}\right\} & \text{if } n > 2m; \\[3mm] \log\left(1 + \dfrac{d(x)^m d(y)^m}{|x-y|^{2m}}\right) & \text{if } n = 2m; \\[3mm] d(x)^{m-\frac{n}{2}} d(y)^{m-\frac{n}{2}} \min\left\{1, \dfrac{d(x)^{\frac{n}{2}} d(y)^{\frac{n}{2}}}{|x-y|^n}\right\} & \text{if } n < 2m. \end{cases} \qquad (4.24)$$

Proof. According to Lemma 4.4 it is essential to distinguish the two cases "$|x-y| \geq \frac{1}{2}[XY]$" and "$|x-y| \leq \frac{1}{2}[XY]$".
1st case: $|x-y| \leq \frac{1}{2}[XY]$. Here (4.10) applies and we have to show

$$G_{m,n}(x,y) \simeq \begin{cases} |x-y|^{2m-n} & \text{if } n > 2m, \\[3mm] \log\left(1 + \dfrac{d(x)^m d(y)^m}{|x-y|^{2m}}\right) & \text{if } n = 2m, \\[3mm] d(x)^{m-\frac{n}{2}} d(y)^{m-\frac{n}{2}} & \text{if } n < 2m. \end{cases} \qquad (4.25)$$

It is not too hard to see that

$$a \in [2, \infty) \Rightarrow \int_1^a (v^2 - 1)^{m-1} v^{1-n}\, dv \simeq \int_1^a v^{2m-n-1}\, dv$$

holds true. According to our assumption we may conclude in this case from formula (4.7) for the Green function

$$G_{m,n}(x,y) \simeq |x-y|^{2m-n} \int_1^{[XY]/|x-y|} (v^2-1)^{m-1} v^{1-n}\, dv$$

$$\simeq |x-y|^{2m-n} \int_1^{[XY]/|x-y|} v^{2m-n-1}\, dv$$

$$\simeq \begin{cases} |x-y|^{2m-n} & \text{if } n > 2m, \\[2mm] \log\left(\dfrac{[XY]}{|x-y|}\right) & \text{if } n = 2m, \\[2mm] [XY]^{2m-n} - |x-y|^{2m-n} \simeq [XY]^{2m-n} & \text{if } n < 2m. \end{cases} \qquad (4.26)$$

If $n > 2m$, statement (4.25) is already proved. In order to proceed also in small dimensions $n \le 2m$ we combine (4.15) and (4.16). We obtain in this case

$$[XY] \simeq d(x) \simeq d(y).$$

Hence, (4.25) is now obvious also for $n < 2m$. If $n = 2m$, we observe further that

$$a \in [2,\infty) \;\Rightarrow\; \log a \simeq \log(1+a^m). \qquad (4.27)$$

The discussion of the case $|x-y| \le \frac{1}{2}[XY]$ is now complete.

2nd case: $|x-y| \ge \frac{1}{2}[XY]$.
In this case we have $\frac{d(x)}{|x-y|} \le 3$, $\frac{d(y)}{|x-y|} \le 3$. Hence, independently of whether $n > 2m$, $n = 2m$ or $n < 2m$, we have to show

$$G_{m,n}(x,y) \simeq |x-y|^{-n} d(x)^m d(y)^m. \qquad (4.28)$$

When using formula (4.7) for $G_{m,n}$, we note that the upper integration bound $[XY]/|x-y|$ is in $[1,2]$. On this interval one has $v^{-n} \simeq 1$ and may conclude

$$G_{m,n}(x,y) \simeq |x-y|^{2m-n} \int_1^{[XY]/|x-y|} (v^2-1)^{m-1} v\, dv$$

$$\simeq |x-y|^{2m-n} \left(\frac{[XY]^2}{|x-y|^2} - 1\right)^m = |x-y|^{-n}\left([XY]^2 - |x-y|^2\right)^m$$

$$= |x-y|^{-n}\left((1-|x|^2)(1-|y|^2)\right)^m \simeq |x-y|^{-n} d(x)^m d(y)^m.$$

The proof of (4.28), and hence of Theorem 4.6, is complete. $\qquad\qquad\square$

In the spirit of Theorem 4.6 we also have estimates for the derivatives.

Theorem 4.7. *(Estimates of the derivatives of the Green function)*
Let $\alpha \in \mathbb{N}^n$ be a multiindex. Then in $\overline{B} \times \overline{B}$ we have

$$|D_x^\alpha G_{m,n}(x,y)| \preceq (*)$$

with $(*)$ *as follows:*

1. if $|\alpha| \geq 2m - n$ *and* n *odd, or if* $|\alpha| > 2m - n$ *and* n *even*

$$(*) = \begin{cases} |x-y|^{2m-n-|\alpha|} \min\left\{1, \dfrac{d(x)^{m-|\alpha|}d(y)^m}{|x-y|^{2m-|\alpha|}}\right\} & \text{for } |\alpha| < m, \\[3ex] |x-y|^{2m-n-|\alpha|} \min\left\{1, \dfrac{d(y)^m}{|x-y|^m}\right\} & \text{for } |\alpha| \geq m; \end{cases}$$

2. if $|\alpha| = 2m - n$ *and* n *even*

$$(*) = \begin{cases} \log\left(2 + \dfrac{d(y)}{|x-y|}\right)\min\left\{1, \dfrac{d(x)^{m-|\alpha|}d(y)^m}{|x-y|^{2m-|\alpha|}}\right\} & \text{for } |\alpha| < m, \\[3ex] \log\left(2 + \dfrac{d(y)}{|x-y|}\right)\min\left\{1, \dfrac{d(y)^m}{|x-y|^m}\right\} & \text{for } |\alpha| \geq m; \end{cases}$$

3. if $|\alpha| \leq 2m - n$ *and* n *odd, or if* $|\alpha| < 2m - n$ *and* n *even*

$$(*) = \begin{cases} d(x)^{m-\frac{n}{2}-|\alpha|}d(y)^{m-\frac{n}{2}} \min\left\{1, \dfrac{d(x)^{\frac{n}{2}}d(y)^{\frac{n}{2}}}{|x-y|^n}\right\} & \text{for } |\alpha| < m - \frac{n}{2}, \\[3ex] d(y)^{2m-n-|\alpha|} \min\left\{1, \dfrac{d(x)^{m-|\alpha|}d(y)^{n-m+|\alpha|}}{|x-y|^n}\right\} & \text{for } m - \frac{n}{2} \leq |\alpha| < m, \\[3ex] d(y)^{2m-n-|\alpha|} \min\left\{1, \dfrac{d(y)^{n-m+|\alpha|}}{|x-y|^{n-m+|\alpha|}}\right\} & \text{for } |\alpha| \geq m. \end{cases}$$

Proof. 1. We claim that on $\{(x,y) \in \overline{B} \times \overline{B} : |x-y| \geq \frac{1}{2}[XY]\}$ it holds true that

$$|D_x^\alpha G_{m,n}(x,y)| \preceq |x-y|^{2m-n-|\alpha|} \left(\frac{d(x)}{|x-y|}\right)^{\max\{m-|\alpha|,0\}} \left(\frac{d(y)}{|x-y|}\right)^m. \tag{4.29}$$

To this end we use the transformation $s = 1 - \frac{1}{v^2}$ in formula (4.7) in order to show the boundary behaviour of the Green function more clearly. We have

$$G_{m,n}(x,y) = \frac{k_{m,n}}{2} |x-y|^{2m-n} f_{m,n}(A_{x,y}), \tag{4.30}$$

where

$$f_{m,n}(t) := \int_0^t s^{m-1}(1-s)^{\frac{n}{2}-m-1}\,ds, \tag{4.31}$$

$$A_{x,y} := \frac{[XY]^2 - |x-y|^2}{[XY]^2} = \frac{(1-|x|^2)(1-|y|^2)}{[XY]^2} \simeq \frac{d(x)d(y)}{[XY]^2}. \tag{4.32}$$

According to the assumption we have

$$A_{x,y} \le \frac{3}{4}. \tag{4.33}$$

Here, i.e for $t \in [0, \frac{3}{4}]$, we know

$$\left| f_{m,n}^{(j)}(t) \right| \preceq t^{\max\{m-j,0\}}. \tag{4.34}$$

Since $d(x) \le [XY]$, by (4.16), for every multiindex $\beta \in \mathbb{N}^n$ one has

$$\left| D_x^{\beta} A_{x,y} \right| \preceq d(y)[XY]^{-1-|\beta|}. \tag{4.35}$$

Application of a general product and chain rule yields

$$|D_x^{\alpha} G_{m,n}(x,y)| \preceq \sum_{\beta \le \alpha} \left| D_x^{\alpha-\beta} |x-y|^{2m-n} \right| \cdot \left| D_x^{\beta} f_{m,n}(A_{x,y}) \right|$$

$$\preceq |x-y|^{2m-n-|\alpha|} \cdot \left| f_{m,n}(A_{x,y}) \right|$$

$$+ \sum_{\substack{\beta \le \alpha \\ \beta \ne 0}} |x-y|^{2m-n-|\alpha|+|\beta|} \cdot \sum_{j=1}^{|\beta|} \left\{ \left| f_{m,n}^{(j)}(A_{x,y}) \right| \cdot \sum_{\substack{\sum_{i=1}^{j} \beta^{(i)} = \beta \\ |\beta^{(i)}| \ge 1}} \prod_{i=1}^{j} \left| D_x^{\beta^{(i)}} A_{x,y} \right| \right\}$$

$$\preceq |x-y|^{2m-n-|\alpha|} \frac{d(x)^m d(y)^m}{[XY]^{2m}}$$

$$+ \sum_{\substack{\beta \le \alpha \\ \beta \ne 0}} |x-y|^{2m-n-|\alpha|+|\beta|} \cdot \sum_{j=1}^{|\beta|} \left\{ \left(\frac{d(x)d(y)}{[XY]^2} \right)^{\max\{m-j,0\}} \cdot \frac{d(y)^j}{[XY]^{j+|\beta|}} \right\}$$

$$\text{by (4.32), (4.34), (4.35)}$$

$$\preceq \sum_{\beta \leq \alpha} |x-y|^{2m-n-|\alpha|} \left(\frac{d(x)}{[XY]}\right)^{\max\{m-|\beta|,0\}} \left(\frac{d(y)}{[XY]}\right)^m \left(\frac{|x-y|}{[XY]}\right)^{|\beta|}$$

by (4.16)

$$\preceq |x-y|^{2m-n-|\alpha|} \left(\frac{d(x)}{[XY]}\right)^{\max\{m-|\alpha|,0\}} \left(\frac{d(y)}{[XY]}\right)^m$$

by (4.3) and (4.16).

Thanks to inequality (4.3) the estimate (4.29) follows.

2. We claim that in $\left\{(x,y) \in \overline{B} \times \overline{B} : |x-y| \leq \frac{1}{2}[XY]\right\}$ one has

$$|D_x^\alpha G_{m,n}(x,y)| \preceq \begin{cases} |x-y|^{2m-n-|\alpha|} & \text{if } |\alpha| > 2m-n, \\[2mm] \log\left(\dfrac{[XY]}{|x-y|}\right) & \text{if } |\alpha| = 2m-n \text{ and } n \text{ even}, \\[2mm] 1 & \text{if } |\alpha| = 2m-n \text{ and } n \text{ odd}, \\[2mm] [XY]^{2m-n-|\alpha|} & \text{if } |\alpha| < 2m-n. \end{cases} \tag{4.36}$$

In contrast with the proof of (4.29) here we do not have to discuss the behaviour of the Green function close to the boundary but "close to the singularity $x = y$". For this reason it is suitable to expand formula (4.7) first and then to carry out the integration explicitly. The integrand contains a term like $\frac{1}{v}$ if and only if n even and $n \leq 2m$. It follows for suitable numbers $c_j = c_j(m,n) \in \mathbb{R}$, $j = 0,\dots,m$:

$$G_{m,n}(x,y) = \begin{cases} c_m|x-y|^{2m-n} + \displaystyle\sum_{j=0}^{m-1} c_j[XY]^{2m-n-2j}|x-y|^{2j} \\[2mm] \qquad\qquad \text{if } n > 2m \text{ or } n \text{ odd}, \\[4mm] c_m|x-y|^{2m-n}\log\dfrac{[XY]}{|x-y|} + \displaystyle\sum_{j=0}^{m-1} c_j[XY]^{2m-n-2j}|x-y|^{2j} \\[2mm] \qquad\qquad \text{if } n \leq 2m \text{ and } n \text{ even}. \end{cases} \tag{4.37}$$

When differentiating we take into account that $|x-y|^{2j}$ is a polynomial of degree $2j$, whose derivatives of order $> 2j$ vanish identically. Moreover, taking advantage of $|x-y| \leq [XY]$, $|D_x^\alpha [XY]^k| \preceq [XY]^{k-|\alpha|}$ and $|D_x^\alpha |x-y|^k| \preceq |x-y|^{k-|\alpha|}$:

$$|D_x^\alpha G_{m,n}(x,y)| \preceq \begin{cases} |x-y|^{2m-n-|\alpha|} + [XY]^{2m-n-|\alpha|} \\[2mm] \qquad\qquad \text{if } n > 2m - |\alpha| \text{ or } n \text{ odd}, \\[4mm] |x-y|^{2m-n-|\alpha|}\left(1 + \log\dfrac{[XY]}{|x-y|}\right) + [XY]^{2m-n-|\alpha|} \\[2mm] \qquad\qquad \text{if } n \leq 2m - |\alpha| \text{ and } n \text{ even}. \end{cases} \tag{4.38}$$

This already proves (4.36) except in the case where n is even and $n < 2m - |\alpha|$. In this case, we use $a \in [1, \infty) \Rightarrow 0 \leq \log a \leq a$ and conclude from (4.38):

$$|D_x^\alpha G_{m,n}(x,y)| \preceq |x-y|^{2m-n-|\alpha|-1}[XY] + [XY]^{2m-n-|\alpha|} \preceq [XY]^{2m-n-|\alpha|}.$$

Therefore, (4.36) holds in any case.

3. We conclude the proof of the theorem by using (4.29) and (4.36). Let $x, y \in \overline{B}$ be arbitrary. According to Lemma 4.4 two cases have to be distinguished.

1st case: $|x-y| \leq \frac{1}{2}[XY]$.

Here $d(x) \simeq d(y)$, and using Lemma 4.4 we obtain for $p, q \geq 0$:

$$\min\left\{1, \left(\frac{d(x)}{|x-y|}\right)^p \left(\frac{d(y)}{|x-y|}\right)^q\right\} \simeq 1.$$

We have to show that

$$|D_x^\alpha G_{m,n}(x,y)| \preceq \begin{cases} |x-y|^{2m-n-|\alpha|} & \text{if } |\alpha| > 2m - n, \\ \log\left(2 + \dfrac{d(y)}{|x-y|}\right) & \text{if } |\alpha| = 2m - n \text{ and } n \text{ even}, \\ 1 & \text{if } |\alpha| = 2m - n \text{ and } n \text{ odd}, \\ d(y)^{2m-n-|\alpha|} & \text{if } |\alpha| < 2m - n. \end{cases}$$

This estimate follows from (4.36), since $d(x) \simeq d(y) \simeq [XY]$ according to (4.15) and (4.16). For the logarithmic term one should observe further (4.27). Making use of $[XY]/|x-y| \geq 2$ and $[XY] \leq 5d(y)$ we obtain

$$\log\frac{[XY]}{|x-y|} \preceq \log\left(1 + \frac{1}{5}\frac{[XY]}{|x-y|}\right) \leq \log\left(2 + \frac{d(y)}{|x-y|}\right).$$

2nd case: $|x-y| \geq \frac{1}{2}[XY]$.

According to Lemma 4.4 we have for $p, q \geq 0$:

$$\log\left(2 + \frac{d(y)}{|x-y|}\right) \simeq 1,$$

$$\min\left\{1, \left(\frac{d(x)}{|x-y|}\right)^p \left(\frac{d(y)}{|x-y|}\right)^q\right\} \simeq \left(\frac{d(x)}{|x-y|}\right)^p \left(\frac{d(y)}{|x-y|}\right)^q.$$

The estimates for $(*)$ as in the statement follow immediately from (4.29). $\qquad\square$

The Green function for the Laplacian ($m = 1, n > 2$) satisfies the estimates above in arbitrary bounded $C^{2,\gamma}$-smooth domains, see e.g. [412]. This result is proved with the help of general maximum principles and Harnack's inequality. For higher order equations we proceed just in the opposite way, namely, we deduce the above estimates from Boggio's explicit formula and, in turn, use them to prove some comparison principles.

In general the following estimate is weaker than Item 3 of Theorem 4.7 but still appropriate and more convenient for our purposes.

Corollary 4.8. *For $|\alpha| \leq 2m - n$ and n odd, or, $|\alpha| < 2m - n$ and n even we have*

$$|D_x^\alpha G_{m,n}(x,y)| \preceq d(x)^{m-\frac{n}{2}-|\alpha|} d(y)^{m-\frac{n}{2}} \min\left\{1, \frac{d(x)^{\frac{n}{2}} d(y)^{\frac{n}{2}}}{|x-y|^n}\right\}.$$

4.2.2 A 3-G-Type Theorem

In Chapter 5 we develop a perturbation theory of positivity for Boggio's prototype situation of the polyharmonic operator in the ball. This will be achieved by means of Neumann series and estimates of iterated Green operators. The latter are consequences of the following 3-G-type result, which provides an estimate for a term of three Green functions.

Theorem 4.9 (3-G-theorem). *Let $\alpha \in \mathbb{N}^n$ be a multiindex. Then on $B \times B \times B$ we have*

$$\frac{G_{m,n}(x,z)\left|D_z^\alpha G_{m,n}(z,y)\right|}{G_{m,n}(x,y)}$$

$$\preceq \begin{cases} |x-z|^{2m-n-|\alpha|} + |y-z|^{2m-n-|\alpha|} & \text{if } |\alpha| > 2m-n, \\[2mm] \log\left(\dfrac{3}{|x-z|}\right) + \log\left(\dfrac{3}{|y-z|}\right) & \text{if } |\alpha| = 2m-n \text{ and } n \text{ even}, \\[2mm] 1 & \text{if } |\alpha| = 2m-n \text{ and } n \text{ odd}, \\[2mm] 1 & \text{if } |\alpha| < 2m-n. \end{cases} \tag{4.39}$$

The proof is crucially based on the Green function estimates in Theorems 4.6 and 4.7 and a number of technical inequalities and equivalencies which we are going to prove first.

Lemma 4.10. *For $s, t > 0$ it holds that*

$$\frac{\log(1+t)}{\log(1+s)} \leq 1 + \frac{t}{s}. \tag{4.40}$$

Proof. For $s > 0$ and $\alpha \geq 1$ concavity of the logarithm yields

$$\log(1+s) = \log\left(\frac{1}{\alpha}(1+\alpha s) + (1 - \frac{1}{\alpha} \cdot 1)\right) \geq \frac{1}{\alpha}\log(1+\alpha s),$$

i.e. $\log(1+\alpha s)/\log(1+s) \leq \alpha$. For $0 < \alpha \leq 1$ it is obvious that $\log(1+\alpha s) \leq \log(1+s)$. Combining these estimates we have for $s, \alpha > 0$

$$\frac{\log(1+\alpha s)}{\log(1+s)} \leq 1 + \alpha.$$

The claim (4.40) follows by taking $\alpha = \frac{t}{s}$. $\qquad \Box$

Boggio's formula is the reason that we can prove the 3-G-theorem 4.9 only in balls. The following lemmas, however, hold true in any bounded domain.

Lemma 4.11. *Let $\Omega \subset \mathbb{R}^n$ be a bounded domain. Assume that $p, q, r \geq 0$, $r \leq p + q$. Further let $s \in \mathbb{R}$ be such that $\frac{r}{2} - p \leq s \leq q - \frac{r}{2}$. Then, on $\overline{\Omega} \times \overline{\Omega}$, we have*

$$\min\left\{1, \left(\frac{d(x)}{|x-y|}\right)^p \left(\frac{d(y)}{|x-y|}\right)^q\right\} \asymp \left(\frac{d(y)}{d(x)}\right)^s \min\left\{1, \frac{d(x)d(y)}{|x-y|^2}\right\}^{\frac{r}{2}}. \qquad (4.41)$$

Proof. We make the same distinction as in the proof of Lemma 4.5.
Case $d(x) \geq 2|x-y|$ or $d(y) \geq 2|x-y|$. According to (4.23) we have that

$$|x-y| \leq d(x) \text{ and } |x-y| \leq d(x) \text{ and } d(x) \simeq d(y)$$

which directly yields (4.41).
Case $d(x) < 2|x-y|$ and $d(y) < 2|x-y|$. Under this assumption we obtain

$$\min\left\{1, \left(\frac{d(x)}{|x-y|}\right)^p \left(\frac{d(y)}{|x-y|}\right)^q\right\} \simeq \left(\frac{d(x)}{|x-y|}\right)^p \left(\frac{d(y)}{|x-y|}\right)^q$$

$$= \left(\frac{d(y)}{d(x)}\right)^s \left(\frac{d(x)d(y)}{|x-y|^2}\right)^{\frac{r}{2}} \left(\frac{d(x)}{|x-y|}\right)^{p+s-\frac{r}{2}} \left(\frac{d(y)}{|x-y|}\right)^{q-s-\frac{r}{2}}$$

and, since $p + s - \frac{r}{2}$ and $q - s - \frac{r}{2}$ are nonnegative, the estimate in (4.41). $\qquad \Box$

Lemma 4.12. *Let $\Omega \subset \mathbb{R}^n$ be a bounded domain. On $\Omega \times \Omega \times \Omega$, it holds true that*

$$Q(x,y,z) := \frac{\min\left\{1, \frac{d(x)d(z)}{|x-z|^2}\right\} \min\left\{1, \frac{d(z)d(y)}{|z-y|^2}\right\}}{\min\left\{1, \frac{d(x)d(y)}{|x-y|^2}\right\}} \asymp 1, \qquad (4.42)$$

$$R(x,y,z) := \frac{\min\left\{1, \frac{d(x)d(z)}{|x-z|^2}\right\} \min\left\{1, \frac{d(y)}{|z-y|}\right\}}{\min\left\{1, \frac{d(x)d(y)}{|x-y|^2}\right\}} \asymp 1 + \frac{|y-z|}{|x-z|}, \qquad (4.43)$$

$$S(x,y,z) := \frac{\log\left(1 + \frac{d(x)d(z)}{|x-z|^2}\right) \min\left\{1, \frac{d(y)}{|z-y|}\right\}}{\log\left(1 + \frac{d(x)d(y)}{|x-y|^2}\right)} \asymp 1 + \frac{|y-z|}{|x-z|}, \qquad (4.44)$$

$$T(x,y,z) := \frac{|x-y|}{|x-z||z-y|} \leq \frac{1}{|x-z|} + \frac{1}{|z-y|}. \qquad (4.45)$$

Proof. Estimate (4.45) is an immediate consequence of the triangle inequality. To prove the remaining estimates we distinguish several cases as in Lemmas 4.5 and 4.11.

Case $d(x) \geq 2|x-y|$ or $d(y) \geq 2|x-y|$. Again, we refer to (4.23):

$$|x-y| \leq d(x) \text{ and } |x-y| \leq d(x) \text{ and } d(x) \simeq d(y).$$

This shows that the denominators of Q and R are bounded from below. Estimating the numerators by 1 from above proves (4.42) and (4.43). In order to estimate S, we make also use of (4.22) and Lemma 4.10.

$$S(x,y,z) \precsim \frac{\log\left(2 + \frac{d(x)}{|x-z|}\right)}{\log\left(2 + \frac{d(x)}{|x-y|}\right)} \cdot 1 \precsim 1 + \frac{1 + \frac{d(x)}{|x-z|}}{1 + \frac{d(x)}{|x-y|}}$$

$$\leq 2 + \frac{\frac{d(x)}{|x-z|}}{1 + \frac{d(x)}{|x-y|}} \leq 2 + \frac{|x-y|}{|x-z|} \leq 3 + \frac{|y-z|}{|x-z|}.$$

Case $d(x) < 2|x-y|$ and $d(y) < 2|x-y|$. Under this assumption we have

$$\min\left\{1, \frac{d(x)d(y)}{|x-y|^2}\right\} \simeq \log\left(1 + \frac{d(x)d(y)}{|x-y|^2}\right) \simeq \frac{d(x)d(y)}{|x-y|^2}.$$

A further distinction with respect to z seems inevitable.

Assume first that $|x-z| \geq \frac{1}{2}|x-y|$. Then (4.18), (4.19), and $\log(1+x) \leq x$ yield

$$\left.\begin{array}{r} Q(x,y,z) \\ R(x,y,z) \\ S(x,y,z) \end{array}\right\} \precsim \frac{|x-y|^2}{d(x)d(y)} \cdot \frac{d(x)d(z)}{|x-z|^2} \cdot \frac{d(y)}{d(z)} \precsim 1.$$

Assume now that $|x-z| < \frac{1}{2}|x-y|$. Then $|y-z| \geq |y-x| - |x-z| \geq \frac{1}{2}|x-y|$. We obtain by applying Lemma 4.5

$$Q(x,y,z) \precsim \frac{|x-y|^2}{d(x)d(y)} \cdot \frac{d(x)}{d(z)} \cdot \frac{d(z)d(y)}{|y-z|^2} \precsim 1,$$

$$R(x,y,z) \precsim \frac{|x-y|^2}{d(x)d(y)} \cdot \frac{d(x)}{|x-z|} \cdot \frac{d(y)}{|y-z|} \precsim \frac{|x-y|}{|x-z|} \precsim 1 + \frac{|y-z|}{|x-z|},$$

$$S(x,y,z) \precsim \frac{|x-y|^2}{d(x)d(y)} \cdot \log\left(2 + \frac{d(x)}{|x-z|}\right) \cdot \min\left\{1, \frac{d(x)}{|x-z|}\right\} \cdot \frac{d(y)}{|y-z|}$$

$$\precsim \frac{|x-y|}{|x-z|} \precsim 1 + \frac{|y-z|}{|x-z|}.$$

□

Proof of the 3-G-theorem 4.9. According to Theorems 4.6 and 4.7 several cases have to be distinguished.

The case: $n > 2m$.

$$\frac{G_{m,n}(x,z)\left|D_z^\alpha G_{m,n}(z,y)\right|}{G_{m,n}(x,y)}$$

$$\preceq \frac{|x-y|^{n-2m}\min\left\{1,\frac{d(x)^m d(z)^m}{|x-z|^{2m}}\right\}\min\left\{1,\left(\frac{d(z)}{|z-y|}\right)^{\max\{m-|\alpha|,0\}}\left(\frac{d(y)}{|z-y|}\right)^m\right\}}{|x-z|^{n-2m}|z-y|^{n+|\alpha|-2m}\min\left\{1,\frac{d(x)^m d(y)^m}{|x-y|^{2m}}\right\}}$$

$$\preceq \frac{1}{|y-z|^{|\alpha|}}\left(T(x,y,z)\right)^{n-2m}\left(Q(x,y,z)\right)^{\max\{m-|\alpha|,0\}}\left(R(x,y,z)\right)^{\min\{|\alpha|,m\}} =: (\star_0)$$

thanks to (4.21). We continue by Lemma 4.12 to find

$$(\star_0) \preceq \frac{1}{|y-z|^{|\alpha|}}\left(\frac{1}{|x-z|}+\frac{1}{|y-z|}\right)^{n-2m}\left(1+\frac{|y-z|}{|x-z|}\right)^{\min\{|\alpha|,m\}}$$

$$\preceq |x-z|^{2m-n}|y-z|^{-|\alpha|}+|y-z|^{2m-n-|\alpha|}$$

$$\quad + |x-z|^{2m-n-\min\{|\alpha|,m\}}|y-z|^{-|\alpha|+\min\{|\alpha|,m\}}$$

$$\quad + |x-z|^{-\min\{|\alpha|,m\}}|y-z|^{2m-n-|\alpha|+\min\{|\alpha|,m\}}$$

$$\preceq |x-z|^{2m-n-|\alpha|}+|y-z|^{2m-n-|\alpha|}.$$

The case: $n = 2m$ and $\alpha = 0$.

$$\frac{G_{m,n}(x,z)G_{m,n}(z,y)}{G_{m,n}(x,y)}$$

$$\preceq \frac{\log\left(2+\frac{d(x)}{|x-z|}\right)\log\left(2+\frac{d(y)}{|y-z|}\right)\min\left\{1,\frac{d(x)^m d(z)^m}{|x-z|^{2m}}\right\}\min\left\{1,\frac{d(z)^m d(y)^m}{|z-y|^{2m}}\right\}}{\max\left\{\log\left(2+\frac{d(x)}{|x-y|}\right),\log\left(2+\frac{d(y)}{|x-y|}\right)\right\}\min\left\{1,\frac{d(x)^m d(y)^m}{|x-y|^{2m}}\right\}}$$

$$\text{by virtue of (4.22)}$$

$$\preceq \frac{\log\left(2+\frac{d(x)}{|x-z|}\right)\log\left(2+\frac{d(y)}{|y-z|}\right)}{\max\left\{\log\left(2+\frac{d(x)}{|x-y|}\right),\log\left(2+\frac{d(y)}{|x-y|}\right)\right\}}\left(Q(x,y,z)\right)^m =: (\star_1).$$

If $|x-z| \geq \frac{1}{2}|x-y|$, then $\log\left(2+\frac{d(x)}{|x-z|}\right) \preceq \log\left(2+\frac{d(x)}{|x-y|}\right)$. If, on the other hand, $|x-z| < \frac{1}{2}|x-y|$, then the reverse inequality $|y-z| \geq |x-y|-|x-z| \geq \frac{1}{2}|x-y|$ follows and hence $\log\left(2+\frac{d(y)}{|y-z|}\right) \preceq \log\left(2+\frac{d(y)}{|x-y|}\right)$. Combining this estimate with Lemma 4.12, (4.42) yields

$$(\star_1) \preceq \log\left(2+\frac{d(x)}{|x-z|}\right)+\log\left(2+\frac{d(y)}{|y-z|}\right) \preceq \log\left(\frac{3}{|x-z|}\right)+\log\left(\frac{3}{|y-z|}\right).$$

The case: $n = 2m$ and $|\alpha| > 0$.

$$\frac{G_{m,n}(x,z)\left|D_z^\alpha G_{m,n}(z,y)\right|}{G_{m,n}(x,y)}$$

$$\preceq \frac{\log\left(1 + \frac{d(x)d(z)}{|x-z|^2}\right)\min\left\{1, \frac{d(x)^{m-1}d(z)^{m-1}}{|x-z|^{2m-2}}\right\}\min\left\{1, \frac{d(z)^{\max\{m-|\alpha|,0\}}d(y)^m}{|z-y|^{m+\max\{m-|\alpha|,0\}}}\right\}}{\log\left(1 + \frac{d(x)d(y)}{|x-y|^2}\right)|y-z|^{|\alpha|}\min\left\{1, \frac{d(x)^{m-1}d(y)^{m-1}}{|x-y|^{2m-2}}\right\}}$$

$$\preceq |y-z|^{-|\alpha|}\, S(x,y,z)\,(Q(x,y,z))^{\max\{m-|\alpha|,0\}}\,(R(x,y,z))^{\min\{|\alpha|,m\}-1}$$

by (4.21)

$$\preceq |y-z|^{-|\alpha|}\left(1 + \frac{|y-z|}{|x-z|}\right)^{\min\{|\alpha|,m\}} \qquad\qquad \text{by Lemma 4.12}$$

$$\preceq |x-z|^{-|\alpha|} + |y-z|^{-|\alpha|}.$$

The case: $n < 2m$ and $|\alpha| < 2m - n$,
or: $n < 2m$ and $|\alpha| \le 2m - n$ and n odd.
Here we use Corollary 4.8. Together with Theorem 4.6 we obtain

$$\frac{G_{m,n}(x,z)\left|D_z^\alpha G_{m,n}(z,y)\right|}{G_{m,n}(x,y)}$$

$$\preceq \frac{d(x)^{m-\frac{n}{2}}d(z)^{2m-n-|\alpha|}d(y)^{m-\frac{n}{2}}\min\left\{1, \frac{d(x)^{\frac{n}{2}}d(z)^{\frac{n}{2}}}{|x-z|^n}\right\}\min\left\{1, \frac{d(z)^{\frac{n}{2}}d(y)^{\frac{n}{2}}}{|z-y|^n}\right\}}{d(x)^{m-\frac{n}{2}}d(y)^{m-\frac{n}{2}}\min\left\{1, \frac{d(x)^{\frac{n}{2}}d(y)^{\frac{n}{2}}}{|x-y|^n}\right\}}$$

$$\preceq d(z)^{2m-n-|\alpha|}\,(Q(x,y,z))^{\frac{n}{2}} \preceq 1 \qquad\qquad \text{due to (4.42).}$$

The case: $n < 2m$ and $|\alpha| = 2m - n$ and n even.
We employ Lemma 4.11 with $p = \max\{m - |\alpha|, 0\}$, $q = m$, $s = m - \frac{n}{2}$ and $r = n$.
In the present case, due to $|\alpha| = 2m - n$, one has: $p + q = \max\{n - m, 0\} + m = \max\{n, m\} \ge n = r$; $q - \frac{r}{2} = m - \frac{n}{2} = s = \frac{n}{2} - (n - m) \ge \frac{r}{2} - p$.

$$\frac{G_{m,n}(x,z)\left|D_z^\alpha G_{m,n}(z,y)\right|}{G_{m,n}(x,y)}$$

$$\preceq \frac{d(x)^{m-\frac{n}{2}}d(z)^{m-\frac{n}{2}}\min\left\{1, \frac{d(x)^{\frac{n}{2}}d(z)^{\frac{n}{2}}}{|x-z|^n}\right\}}{d(x)^{m-\frac{n}{2}}d(y)^{m-\frac{n}{2}}\min\left\{1, \frac{d(x)^{\frac{n}{2}}d(y)^{\frac{n}{2}}}{|x-y|^n}\right\}}$$

$$\times \log\left(2 + \frac{d(y)}{|z-y|}\right)\min\left\{1, \left(\frac{d(z)}{|z-y|}\right)^{\max\{m-|\alpha|,0\}}\left(\frac{d(y)}{|z-y|}\right)^m\right\}$$

$$\preceq \frac{\log\left(2+\frac{d(y)}{|z-y|}\right)d(z)^{m-\frac{n}{2}}\min\left\{1,\frac{d(x)^{\frac{n}{2}}d(z)^{\frac{n}{2}}}{|x-z|^n}\right\}\left(\frac{d(y)}{d(z)}\right)^{m-\frac{n}{2}}\min\left\{1,\frac{d(z)^{\frac{n}{2}}d(y)^{\frac{n}{2}}}{|z-y|^n}\right\}}{d(y)^{m-\frac{n}{2}}\min\left\{1,\frac{d(x)^{\frac{n}{2}}d(y)^{\frac{n}{2}}}{|x-y|^n}\right\}}$$

$$\preceq \log\left(2+\frac{d(y)}{|z-y|}\right)(Q(x,y,z))^{\frac{n}{2}} \preceq \log\left(\frac{3}{|y-z|}\right) \qquad \text{by virtue of (4.42).}$$

The case: $n < 2m$ and $|\alpha| > 2m - n$.

$$\frac{G_{m,n}(x,z)\,|D_z^\alpha G_{m,n}(z,y)|}{G_{m,n}(x,y)} \preceq \frac{d(x)^{m-\frac{n}{2}}d(z)^{m-\frac{n}{2}}\min\left\{1,\frac{d(x)^{\frac{n}{2}}d(z)^{\frac{n}{2}}}{|x-z|^n}\right\}}{d(x)^{m-\frac{n}{2}}d(y)^{m-\frac{n}{2}}\min\left\{1,\frac{d(x)^{\frac{n}{2}}d(y)^{\frac{n}{2}}}{|x-y|^n}\right\}}$$

$$\times |z-y|^{2m-n-|\alpha|}\min\left\{1,\left(\frac{d(z)}{|z-y|}\right)^{\max\{m-|\alpha|,0\}}\right.$$

$$\left.\times\left(\frac{d(y)}{|z-y|}\right)^m\right\}$$

$$= |y-z|^{2m-n-|\alpha|}\frac{d(z)^{m-\frac{n}{2}}\min\left\{1,\frac{d(x)^{\frac{n}{2}}d(z)^{\frac{n}{2}}}{|x-z|^n}\right\}}{d(y)^{m-\frac{n}{2}}\min\left\{1,\frac{d(x)^{\frac{n}{2}}d(y)^{\frac{n}{2}}}{|x-y|^n}\right\}}$$

$$\times \min\left\{1,\left(\frac{d(z)}{|z-y|}\right)^{\max\{m-|\alpha|,0\}}\left(\frac{d(y)}{|z-y|}\right)^m\right\} =: (\star_2).$$

In order to proceed, we have to distinguish further cases.

In addition, we assume first that $|\alpha| \le 2m - \frac{n}{2}$.
We apply (4.21) of Lemma 4.5 to the "dangerous" term in $(\star_2)$. Here one has to observe that $|\alpha| + n - 2m > 0$ as well as $3m - n - |\alpha| \ge 3m - n - 2m + \frac{n}{2} = m - \frac{n}{2} > 0$. In a second step we make use of Lemma 4.11 with $p = \max\{m-|\alpha|,0\} \ge 0$, $q = 3m - n - |\alpha| \ge 0$, $r = 4m - n - 2|\alpha| \ge 0$ and $s = m - \frac{n}{2}$. Obviously $p + q - r = \max\{|\alpha| - m,0\} \ge 0$, $q - \frac{r}{2} = s$, $\frac{r}{2} - p = m - \frac{n}{2} - \max\{|\alpha| - m,0\} \le s$.

$$\min\left\{1,\left(\frac{d(z)}{|z-y|}\right)^{\max\{m-|\alpha|,0\}}\left(\frac{d(y)}{|z-y|}\right)^m\right\}$$

$$\simeq \min\left\{1,\frac{d(y)}{|y-z|}\right\}^{|\alpha|+n-2m}\min\left\{1,\left(\frac{d(z)}{|z-y|}\right)^{\max\{m-|\alpha|,0\}}\left(\frac{d(y)}{|z-y|}\right)^{3m-n-|\alpha|}\right\}$$

$$\preceq \min\left\{1,\frac{d(y)}{|y-z|}\right\}^{|\alpha|+n-2m}\left(\frac{d(y)}{d(z)}\right)^{m-\frac{n}{2}}\min\left\{1,\frac{d(z)d(y)}{|z-y|^2}\right\}^{2m-\frac{n}{2}-|\alpha|}.$$

With the aid of this estimate and of Lemma 4.12 we obtain further

$$(\star_2) \preceq |y-z|^{2m-n-|\alpha|} \, (Q(x,y,z))^{2m-\frac{n}{2}-|\alpha|} \, (R(x,y,z))^{|\alpha|+n-2m}$$

$$\preceq |y-z|^{2m-n-|\alpha|} \left(1+\frac{|y-z|}{|x-z|}\right)^{|\alpha|+n-2m} \preceq |y-z|^{2m-n-|\alpha|} + |x-z|^{2m-n-|\alpha|}.$$

Now we assume that additionally $|\alpha| > 2m - \frac{n}{2}$ holds true.

Here one has to deal with the "dangerous" term in $(\star_2)$ in a different manner. Obviously, one has that $|\alpha| > m + \left(m - \frac{n}{2}\right) > m$. We apply repeatedly Lemma 4.5, observing that $\frac{n}{2} < m$.

$$\min\left\{1, \left(\frac{d(z)}{|z-y|}\right)^{\max\{m-|\alpha|,0\}}\right\} \left(\frac{d(y)}{|z-y|}\right)^m \simeq \min\left\{1, \left(\frac{d(y)}{|z-y|}\right)^m\right\}$$

$$\simeq \min\left\{1, \frac{d(y)}{|y-z|}\right\}^{\frac{n}{2}} \min\left\{1, \frac{d(y)}{|z-y|}\right\}^{m-\frac{n}{2}} \preceq \min\left\{1, \frac{d(y)}{|y-z|}\right\}^{\frac{n}{2}} \left(\frac{d(y)}{d(z)}\right)^{m-\frac{n}{2}}.$$

By means of this estimate and of Lemma 4.12 we further conclude that

$$(\star_2) \preceq |y-z|^{2m-n-|\alpha|} \, (R(x,y,z))^{\frac{n}{2}} \preceq |y-z|^{2m-n-|\alpha|} \left(1+\frac{|y-z|}{|x-z|}\right)^{\frac{n}{2}}$$

$$\preceq |y-z|^{2m-n-|\alpha|} + |y-z|^{2m-\frac{n}{2}-|\alpha|} \, |x-z|^{-\frac{n}{2}}$$

$$\preceq |y-z|^{2m-n-|\alpha|} + |x-z|^{2m-n-|\alpha|}.$$

To apply Young's inequality in the last step, one has to exploit the assumption $2m - \frac{n}{2} - |\alpha| < 0$ of this case. $\qquad\square$

4.3 Estimates for the Steklov Problem

In the previous section we considered the higher order operator $(-\Delta)^m$ under Dirichlet boundary conditions starting from the explicit formula of Boggio and hence we necessarily had to restrict ourselves to the ball as domain. Under different boundary conditions the boundary value problem may be rewritten as a second order system. The present section prepares for such a situation so that general bounded smooth domains are allowed. So we consider the second order Green operator $\mathcal{G}$ and the Poisson operator $\mathcal{K}$ on a general domain Ω, that is, $w = \mathcal{G}f + \mathcal{K}g$ formally solves

$$\begin{cases} -\Delta w = f & \text{in } \Omega, \\ w = g & \text{on } \partial\Omega. \end{cases}$$

For bounded C^2-domains the operators $\mathscr{G}$ and $\mathscr{K}$ can be represented by integral kernels G and K, namely

$$(\mathscr{G}f)(x) = \int_\Omega G(x,y)f(y)\,dy \quad \text{and} \quad (\mathscr{K}g)(x) = \int_{\partial\Omega} K(x,y)g(y)\,d\omega_y. \qquad (4.46)$$

Moreover, it holds that

$$K(x,y) = \frac{-\partial}{\partial v_y}G(x,y) \qquad \text{for all } (x,y) \in \Omega \times \partial\Omega. \qquad (4.47)$$

According to (4.6), the Green function G in (4.46) should be written as $G_{-\Delta,\Omega}$. However, since this function is frequently used in this section, we drop the subscripts for a simpler notation.

In this section we prove some estimates for the kernels G and K in general bounded domains Ω such that $\partial\Omega \in C^2$. These estimates will be intensively used in Section 5.4 in order to prove positivity properties for the biharmonic Steklov problem.

Based on several estimates due to Zhao [421, 422] (see also [118, 385]), Grunau-Sweers [214] were able to show:

Proposition 4.13. *Let $\Omega \subset \mathbb{R}^n$ be a bounded domain with $\partial\Omega \in C^2$. Then the following uniform estimates hold for $(x,y) \in \Omega \times \Omega$:*

$$\textit{for } n > 4: \quad \int_\Omega G(x,z)G(z,y)dz \simeq |x-y|^{4-n} \min\left\{1, \frac{d(x)d(y)}{|x-y|^2}\right\}, \qquad (4.48)$$

$$\textit{for } n = 4: \quad \int_\Omega G(x,z)G(z,y)dz \simeq \log\left(1 + \frac{d(x)d(y)}{|x-y|^2}\right), \qquad (4.49)$$

$$\textit{for } n = 3: \quad \int_\Omega G(x,z)G(z,y)dz \simeq \sqrt{d(x)d(y)}\min\left\{1, \frac{\sqrt{d(x)d(y)}}{|x-y|}\right\}, \qquad (4.50)$$

$$\textit{for } n = 2: \quad \int_\Omega G(x,z)G(z,y)dz \simeq d(x)d(y)\log\left(2 + \frac{1}{|x-y|^2 + d(x)d(y)}\right). \quad (4.51)$$

We will exploit these estimates combined with the following "geometric" result:

Lemma 4.14. *Let $\Omega \subset \mathbb{R}^n$ $(n \geq 2)$ be a bounded domain with $\partial\Omega \in C^2$. For $x \in \Omega$ let $x^* \in \partial\Omega$ be any point such that $d(x) = |x - x^*|$.*

- *Then there exists $r_\Omega > 0$ such that for $x \in \Omega$ with $d(x) \leq r_\Omega$ there is a unique $x^* \in \partial\Omega$.*
- *Then the following uniform estimates hold:*

$$\textit{for } (x,y) \in \Omega \times \Omega: \quad |x-y| \preceq d(x)+d(y)+|x^*-y^*|, \qquad (4.52)$$

$$\text{for } (x,y) \in \Omega \times \Omega: \quad \frac{d(x)}{d(x)+d(y)+|x^*-y^*|} \preceq \min\left\{1, \frac{d(x)}{|x-y|}\right\}, \quad (4.53)$$

$$\text{for } (x,z) \in \Omega \times \partial\Omega: \quad |x-z| \simeq d(x)+|x^*-z|. \quad (4.54)$$

And for $(x,y,z) \in \Omega \times \Omega \times \partial\Omega$:

$$\text{if } d(y) \le d(x) \text{ and } |x^*-y^*| \le d(x)+d(y), \text{ then } |x-z| \simeq d(x)+|y^*-z|. \quad (4.55)$$

Proof. Since $\partial\Omega \in C^2$, there exists $r_1 > 0$ such that Ω can be filled with balls of radius r_1. Set $r_\Omega = \frac{1}{2}r_1$. For $x \in \Omega$ with $d(x) \le r_\Omega$ there is a unique $x^* \in \partial\Omega$.

Estimate (4.52) is just the triangle inequality. Estimate (4.54) follows from the three inequalities

$$|x-z| \le |x-x^*|+|x^*-z| = d(x)+|x^*-z|,$$
$$d(x) \le |x-z| \quad \text{and} \quad |x^*-z| \le |x^*-x|+|x-z| \le 2|x-z|.$$

In order to prove (4.55), we first remark that under the assumptions made we have $d(x) \ge \frac{1}{2}|x^*-y^*|$. This yields the two inequalities

$$d(x)+|x^*-z| \le d(x)+|x^*-y^*|+|y^*-z| \le 3d(x)+|y^*-z| \le 3(d(x)+|y^*-z|),$$
$$d(x)+|y^*-z| \le d(x)+|x^*-y^*|+|x^*-z| \le 3d(x)+|x^*-z| \le 3(d(x)+|x^*-z|).$$

In turn these inequalities read as $d(x)+|x^*-z| \simeq d(x)+|y^*-z|$. This, combined with (4.54), proves (4.55).

To prove (4.53), we distinguish two cases. If $|x-y| \le \frac{1}{2}\max(d(x),d(y))$, then $\frac{1}{2}d(x) \le d(y) \le 2d(x)$ and $|x-y| \preceq d(x) \simeq d(y)$. It follows that

$$\frac{d(x)}{d(x)+d(y)+|x^*-y^*|} \preceq 1 \simeq \min\left\{1, \frac{d(x)}{|x-y|}\right\}$$

and a similar estimate with x and y interchanged. If $|x-y| \ge \frac{1}{2}\max(d(x),d(y))$, we use (4.52) to find that

$$\frac{d(x)}{d(x)+d(y)+|x^*-y^*|} \preceq \frac{d(x)}{|x-y|} \simeq \min\left\{1, \frac{d(x)}{|x-y|}\right\}$$

and a similar estimate with x and y interchanged. $\qquad\square$

We are now ready to prove the estimates which are needed for the study of the Steklov problem.

Lemma 4.15. *Let* $\Omega \subset \mathbb{R}^n$ $(n \ge 2)$ *be a bounded domain with* $\partial\Omega \in C^2$. *Then the following uniform estimates hold for* $(x,z) \in \Omega \times \partial\Omega$:

$$\int_\Omega G(x,\xi)K(\xi,z)d\xi \simeq \begin{cases} d(x)\,|x-z|^{2-n} & \text{for } n \ge 3, \\ d(x)\,\log\left(2+\frac{1}{|x-z|^2}\right) & \text{for } n = 2. \end{cases}$$

Proof. Let

$$H(x,z) := \int_{\Omega} G(x,\xi)G(\xi,z)d\xi \qquad \text{for all } (x,z) \in \Omega \times \partial\Omega.$$

In view of (4.47), and since $H(x,z) = 0$ for $z \in \partial\Omega$, we have

$$\int_{\Omega} G(x,\xi)K(\xi,z)d\xi = \frac{-\partial}{\partial v_z}H(x,z) = \lim_{t\to 0} \frac{H(x,z-tv_z)}{t}. \qquad (4.56)$$

Note also that if r_Ω is as in Lemma 4.14, then $d(z-tv_z) = t$ for all $z \in \partial\Omega$ and $t \le r_\Omega$. Hence, by (4.48) we obtain for $n > 4$

$$\lim_{t\to 0} \frac{H(x,z-tv_z)}{t} \simeq \lim_{t\to 0} \frac{|x-z+tv_z|^{4-n}\,\min\left\{1,\frac{td(x)}{|x-z+tv_z|^2}\right\}}{t} = d(x)\,|x-z|^{2-n}.$$

For $n = 4$ we use (4.49) to obtain

$$\lim_{t\to 0} \frac{H(x,z-tv_z)}{t} \simeq \lim_{t\to 0} \frac{\log\left(1+\frac{td(x)}{|x-z+tv_z|^2}\right)}{t} \simeq d(x)\,|x-z|^{-2}.$$

For $n = 3$ we use (4.50) to obtain

$$\lim_{t\to 0} \frac{H(x,z-tv_z)}{t} \simeq \lim_{t\to 0} \frac{\sqrt{td(x)}\,\min\left\{1,\frac{\sqrt{td(x)}}{|x-z+tv_z|}\right\}}{t} = d(x)\,|x-z|^{-1}.$$

And finally for $n = 2$ we use (4.51) to obtain

$$\lim_{t\to 0} \frac{H(x,z-tv_z)}{t} \simeq \lim_{t\to 0} \frac{td(x)\log\left(2+\frac{1}{|x-z+tv_z|^2+td(x)}\right)}{t} = d(x)\,\log\left(2+\frac{1}{|x-z|^2}\right).$$

By (4.56) the statement is so proved for any $n \ge 2$. $\qquad\square$

Lemma 4.16. *Let $\Omega \subset \mathbb{R}^n$ ($n \ge 2$) be a bounded domain with $\partial\Omega \in C^2$. Then the following uniform estimates hold for $(x,y) \in \Omega \times \Omega$:*

$$\int_{\Omega}\int_{\partial\Omega}\int_{\Omega} G(x,\xi)K(\xi,z)\frac{-\partial}{\partial v_z}G(z,w)G(w,y)d\xi\,d\omega_z dw$$

$$\preceq \begin{cases} d(x)d(y)\,(d(x)+d(y)+|x^*-y^*|)^{2-n} & \text{for } n \ge 3, \\ d(x)d(y)\,\log\left(2+\frac{1}{d(x)+d(y)+|x^*-y^*|}\right) & \text{for } n = 2, \end{cases} \qquad (4.57)$$

respectively for $(x,y) \in \Omega \times \partial\Omega$:

$$\int_\Omega \int_{\partial\Omega} \int_\Omega G(x,\xi)K(\xi,z)\frac{-\partial}{\partial v_z}G(z,w)K(w,y)d\xi\, d\omega_z dw$$

$$\preceq \begin{cases} d(x)\,|x-y|^{2-n} & \text{for } n \geq 3, \\ d(x)\,\log\left(2 + \frac{1}{|x-y|)}\right) & \text{for } n = 2. \end{cases} \tag{4.58}$$

Proof. Setting

$$R(x,y) := \int_\Omega \int_{\partial\Omega} \int_\Omega G(x,\xi)K(\xi,z)\frac{-\partial}{\partial v_z}G(z,w)G(w,y)d\xi\, d\omega_z dw,$$

and using (4.47) and the estimates from Lemma 4.15, the following holds:

$$R(x,y) \preceq d(x)d(y)\int_{\partial\Omega}|x-z|^{2-n}|z-y|^{2-n}d\omega_z \qquad \text{for } n \geq 3,$$

$$R(x,y) \preceq d(x)d(y)\int_{\partial\Omega}\log\left(2+\frac{1}{|x-z|^2}\right)\log\left(2+\frac{1}{|y-z|^2}\right)d\omega_z \text{ for } n = 2.$$

Let r_Ω be as in Lemma 4.14. We distinguish three cases, according to the positions of $x,y \in \Omega$.

Case 1: $\max(d(x),d(y)) \geq r_\Omega$.
By symmetry we may assume that $d(y) \geq r_\Omega$ and find for $n \geq 3$ that

$$\int_{\partial\Omega}|x-z|^{2-n}|z-y|^{2-n}d\omega_z \preceq \int_{\partial\Omega}|x-z|^{2-n}d\omega_z \preceq \int_0^1 \frac{r^{n-2}}{(d(x)+r)^{n-2}}dr \preceq 1,$$

and for $n = 2$

$$\int_{\partial\Omega}\log\left(2+\frac{1}{|x-z|^2}\right)\log\left(2+\frac{1}{|z-y|^2}\right)d\omega_z \preceq \int_0^1 \log\left(2+\frac{1}{(d(x)+r)^2}\right)dr \preceq 1,$$

which imply (4.57) since $d(y) \geq r_\Omega$.

Case 2: $\max(d(x),d(y)) < r_\Omega$ and $|x^* - y^*| \geq d(x)+d(y)$.
In this case, in view of Lemma 4.14, we have that (4.54) holds for both x and y. So, for $n \geq 3$ we have

$$\int_{\partial\Omega}|x-z|^{2-n}|z-y|^{2-n}d\omega_z \preceq \int_{\partial\Omega}\frac{1}{(d(x)+|x^*-z|)^{n-2}}\frac{1}{(d(y)+|y^*-z|)^{n-2}}d\omega_z.$$

We split this integral as $I_x + I_y$ where I_x is over $\partial\Omega_x = \{z \in \partial\Omega; |x^*-z| \leq |y^*-z|\}$ and I_y over $\partial\Omega_y = \partial\Omega\setminus\partial\Omega_x$. Over $\partial\Omega_x$ we have

$$|x^*-z| + |x^*-y^*| \leq |x^*-z| + |x^*-z| + |y^*-z| \leq 3|y^*-z|.$$

Hence we find

$$I_x \preceq \int_{\partial\Omega_x} \frac{1}{(d(x)+|x^*-z|)^{n-2}} \frac{1}{(d(y)+|x^*-z|+|x^*-y^*|)^{n-2}} d\omega_z$$

$$\preceq \frac{1}{|x^*-y^*|^{n-2}} \int_0^1 \frac{r^{n-2}}{(d(x)+r)^{n-2}} dr \preceq |x^*-y^*|^{2-n}$$

$$\preceq (d(x)+d(y)+|x^*-y^*|)^{2-n}$$

where we used $|x^*-y^*| \geq d(x)+d(y)$ in the last estimate.

Similarly, for $n=2$ we find

$$I_x \preceq \int_{\partial\Omega_x} \log\left(2+\frac{1}{d(x)+|x^*-z|}\right) \log\left(2+\frac{1}{d(y)+|x^*-z|+|x^*-y^*|}\right) d\omega_z$$

$$\preceq \log\left(2+\frac{1}{d(y)+|x^*-y^*|}\right) \int_0^1 \log\left(2+\frac{1}{d(x)+r}\right) dr$$

$$\preceq \log\left(2+\frac{1}{d(y)+|x^*-y^*|}\right) \preceq \log\left(2+\frac{1}{d(x)+d(y)+|x^*-y^*|}\right).$$

Analogous estimates hold for I_y. All together these estimates prove (4.57) in Case 2.

Case 3: $\max(d(x),d(y)) < r_\Omega$ and $|x^*-y^*| \leq d(x)+d(y)$.
By symmetry, we may assume that $d(y) \leq d(x)$. Then we may use both (4.54) and (4.55). So, for $n \geq 3$ we find

$$\int_{\partial\Omega} |x-z|^{2-n} |z-y|^{2-n} d\omega_z$$

$$\preceq \int_{\partial\Omega} \frac{1}{(d(x)+|y^*-z|)^{n-2}} \frac{1}{(d(y)+|y^*-z|)^{n-2}} d\omega_z$$

$$\preceq \int_0^1 \frac{r^{n-2}}{(d(x)+r)^{n-2}} \frac{1}{(d(y)+r)^{n-2}} dr$$

$$\preceq \frac{1}{d(x)^{n-2}} \preceq (d(x)+d(y)+|x^*-y^*|)^{2-n},$$

and for $n=2$

$$\int_{\partial\Omega} \log\left(2+\frac{1}{|x-z|}\right) \log\left(2+\frac{1}{|y-z|}\right) d\omega_z$$

$$\preceq \int_{\partial\Omega} \log\left(2+\frac{1}{d(x)+|y^*-z|}\right) \log\left(2+\frac{1}{d(y)+|y^*-z|}\right) d\omega_z$$

$$\preceq \int_0^1 \log\left(2+\frac{1}{d(x)+r}\right) \log\left(2+\frac{1}{d(y)+r}\right) dr$$

$$\preceq \log\left(2+\frac{1}{d(x)}\right) \preceq \log\left(2+\frac{1}{d(x)+d(y)+|x^*-y^*|}\right).$$

This proves (4.57) in Case 3.

For the estimates in (4.58) one divides the estimates in (4.57) by $d(y)$, takes the limit for $d(y) \to 0$, and uses (4.54), namely that $d(x) + |x^* - y| \simeq |x - y|$ for $y \in \partial\Omega$. $\qquad\qquad\square$

4.4 General Properties of the Green Functions

In this section we collect some smoothness properties of biharmonic functions and derive some preliminary pointwise estimates. That is, we first give a more precise statement concerning the smoothness of the Green functions simultaneously with respect to *both* variables. Next we will show some pointwise estimates for the Green function that follow almost directly from its construction through the fundamental solution.

4.4.1 Regularity of the Biharmonic Green Function

For brevity we here write $G = G_{\Delta^2,\Omega}$ for the Green function in the domain Ω, see (4.6).

Proposition 4.17. *Let Ω be a bounded $C^{4,\gamma}$-smooth domain. Let G be the Green function for the biharmonic Dirichlet problem. Then*

$$G \in C^{4,\gamma}\left(\overline{\Omega} \times \overline{\Omega} \setminus \{(x,x) : x \in \overline{\Omega}\}\right).$$

Proof. Suppose $\alpha \in \mathbb{N}^n$ with $i = |\alpha| \leq 3$ and let $p \in (n, n+1)$. In particular it holds that $4 - \frac{n}{p} > i$. Let $\varphi \in C_c^\infty(\Omega)$ and consider $\psi \in C^{4,\gamma}(\overline{\Omega})$ such that $\Delta^2 \psi = \varphi$ in Ω and $\psi = \psi_\nu = 0$ on $\partial\Omega$. It follows from Theorem 2.20 and Sobolev's embedding theorem 2.6 that

$$\|\psi\|_{C^{i,\mu}(\overline{\Omega})} \leq C\|\psi\|_{W^{4,p}(\Omega)} \leq C\|\varphi\|_{L^p(\Omega)}$$

for all $\mu \in (0,1)$ with $i + \mu \leq 4 - \frac{n}{p}$. Since $\psi(x) = \int_\Omega G(x,y)\varphi(y)\,dy$, we get that

$$\left| \int_\Omega \left(D_x^\alpha G(x,y) - D_x^\alpha G(x',y)\right) \varphi(y)\,dy \right| \leq C_2 \|\varphi\|_{L^p(\Omega)} |x - x'|^\mu \quad \text{for all } x, x' \in \Omega.$$

By duality, we then obtain $y \mapsto D_x^\alpha G(x,y) \in L^q(\Omega)$ for all $q \in (\frac{n+1}{n}, \frac{n}{n-1})$ and moreover, for all $\mu \leq 4 - i - n + \frac{n}{q}$ with $\mu \in (0,1)$, that

$$\|D_x^\alpha G(x, \cdot) - D_x^\alpha G(x', \cdot)\|_q \leq C(q) |x - x'|^\mu \quad \text{for all } x, x' \in \Omega.$$

Since the functions $y \mapsto G(x,y)$ are biharmonic in $\Omega \setminus \{x\}$, so is $y \mapsto D_x^\alpha G(x,y)$. Fix x and consider $y \mapsto D_x^\alpha G(x,y)$. Since $D_x^\alpha G(x,\,.\,) = \frac{\partial}{\partial \nu} D_x^\alpha G(x,\,.\,) = 0$ on $\partial \Omega$, regularity theory, as one may find in Theorem 2.19, gives that $D_x^\alpha G(x,\,.\,) \in C^{4,\gamma}$ $(\overline{\Omega} \setminus \{x\})$. Moreover, for all $\delta > 0$ there exists $C(\delta) > 0$ such that

$$\left\| D_x^\alpha G(x,\,.\,) - D_x^\alpha G(x',\,.\,) \right\|_{C^{4,\gamma}(\overline{\Omega} \setminus (B_\delta(x) \cup B_\delta(x')))} \le C(\delta) \left| x - x' \right|^\mu \quad \text{for all } x, x' \in \Omega.$$

This is valid for $|\alpha| \le 3$. Using the symmetry of the Green function, we have a similar result for $|\alpha| = 4$ with respect to the $C^{3,\gamma}(\overline{\Omega} \setminus (B_\delta(x) \cup B_\delta(x')))$-norm. So, all derivatives of order 4 are covered and we find that $G \in C^{4,\gamma}(\overline{\Omega} \times \overline{\Omega} \setminus \{(x,x) : x \in \overline{\Omega}\})$. $\qquad \square$

4.4.2 Preliminary Estimates for the Green Function

We start with a relatively straightforward application of the Schauder theory to the construction of Green's functions.

Lemma 4.18. *Let $\Omega \subset \mathbb{R}^n$ be a bounded $C^{4,\gamma}$-smooth domain and let $d(\,.\,)$ be as in (4.1). Then for the biharmonic Green function $G_{\Delta^2,\Omega}$ the following estimates hold true:*

$$\left| G_{\Delta^2,\Omega}(x,y) \right| \le C(\Omega) \cdot \begin{cases} |x-y|^{4-n} + \max\{d(x), d(y)\}^{4-n} & \text{if } n > 4, \\ \log\left(1 + |x-y|^{-1} + \max\{d(x), d(y)\}^{-1}\right) & \text{if } n = 4, \\ 1 & \text{if } n = 2,3. \end{cases}$$

$$(4.59)$$

For $n = 2,3,4$ also the following gradient estimates hold true:

$$\left| \nabla_x G_{\Delta^2,\Omega}(x,y) \right| \le C(\Omega) \cdot \begin{cases} |x-y|^{-1} + \max\{d(x), d(y)\}^{-1} & \text{if } n = 4, \\ 1 & \text{if } n = 2,3. \end{cases} \qquad (4.60)$$

By symmetry (4.60) also holds for $\left| \nabla_y G_{\Delta^2,\Omega}(x,y) \right|$. The dependence of the constants $C(\Omega)$ on Ω is explicit via the $C^{4,\gamma}$-properties of $\partial \Omega$.

Proof. For brevity we write $G(x,y) = G_{\Delta^2,\Omega}(x,y)$. We recall a fundamental solution for Δ^2 on $\mathbb{R}^n$:

$$F_n(x) = \begin{cases} c_n |x|^{4-n} & \text{if } n \notin \{2,4\}, \\ -2c_4 \log|x| & \text{if } n = 4, \\ 2c_2 |x|^2 \log|x| & \text{if } n = 2, \end{cases} \qquad (4.61)$$

where c_n is defined through $e_n = |B|$ and

$$c_n = \begin{cases} \dfrac{1}{2(n-4)(n-2)ne_n} & \text{if } n \notin \{2,4\}, \\[2ex] \dfrac{1}{8ne_n} & \text{if } n \in \{2,4\}. \end{cases}$$

The Green function is given by $G(x,y) = F_n(|x-y|) + h(x,y)$, where $h(x,.)$ is a solution of the following Dirichlet problem:

$$\begin{cases} \Delta_y^2 h(x,y) = 0 & \text{in } \Omega, \\ h(x,y) = -F_n(|x-y|) & \text{for } y \in \partial\Omega, \\ \dfrac{\partial}{\partial v_y} h(x,y) = -\dfrac{\partial}{\partial v_y} F_n(|x-y|) & \text{for } y \in \partial\Omega. \end{cases} \tag{4.62}$$

We first discuss extensively the generic case $n > 4$. At the end we comment on the changes and additional arguments which have to be made for $n \in \{2,3,4\}$.

Case $n > 4$. In (4.62), the $C^{1,\gamma}$-norm of the datum for $h(x,.)|_{\partial\Omega}$ and the $C^{0,\gamma}$-norm of the datum for $\frac{\partial}{\partial v_y} h(x,.)|_{\partial\Omega}$ are bounded by $C(\partial\Omega)d(x)^{3-n-\gamma}$. The dependence of the constant $C(\partial\Omega)$ on $\partial\Omega$ is constructive and explicit via its curvature properties and their derivatives. According to $C^{1,\gamma}$-estimates for boundary value problems in variational form like (4.62) we see with the help of Theorem 2.19 that

$$\|h(x,.)\|_{C^{1,\gamma}(\overline{\Omega})} \le C(\partial\Omega)d(x)^{3-n-\gamma}. \tag{4.63}$$

One should observe that the differential operators are uniformly coercive, so that no $h(x,.)$-term needs to appear on the right-hand-side of (4.63).

As long as $d(y) \le d(x)$, (4.63) shows $h(x,y) \le C(\partial\Omega)d(x)^{4-n}$ and hence that

$$|G(x,y)| \le C(\partial\Omega)\left(|x-y|^{4-n} + d(x)^{4-n}\right). \tag{4.64}$$

For $d(y) > d(x)$, we conclude from (4.64) by exploiting the symmetry of the Green function:

$$|G(x,y)| = |G(y,x)| \le C(\partial\Omega)\left(|x-y|^{4-n} + d(y)^{4-n}\right). \tag{4.65}$$

Combining (4.64) and (4.65) yields (4.59) for $n > 4$.

Case $n = 4$. As above we find that

$$\|h(x,.)\|_{C^{1,\gamma}(\overline{\Omega})} \le C(\partial\Omega)d(x)^{-1-\gamma}. \tag{4.66}$$

As long as $d(y) \le d(x)$, (4.66) shows that

$$|\nabla_y G(x,y)| \le C(\partial\Omega)\left(|x-y|^{-1} + d(x)^{-1}\right). \tag{4.67}$$

In order to exploit the symmetry of $G(x,y)$ we need a similar estimate also for $|\nabla_x G(x,y)|$. To this end one has to differentiate (4.62) with respect to x being considered here as a parameter and obtains as before that for $d(y) \le d(x)$

$$|\nabla_x G(x,y)| \le C(\partial\Omega)\left(|x-y|^{-1}+d(x)^{-1}\right). \tag{4.68}$$

By symmetry $G(x,y) = G(y,x)$, and (4.68) shows that for $d(x) \le d(y)$ one has

$$\left|\nabla_y G(x,y)\right| \le C(\partial\Omega)\left(|x-y|^{-1}+d(y)^{-1}\right) \tag{4.69}$$

while (4.67) yields

$$|\nabla_x G(x,y)| \le C(\partial\Omega)\left(|x-y|^{-1}+d(y)^{-1}\right). \tag{4.70}$$

Combining (4.67)-(4.70) proves (4.60) and hence (4.59) in the case $n=4$.

    Case $n=3$. As in the previous cases one comes up with

$$\|h(x,.)\|_{C^{1,\gamma}(\overline{\Omega})} \le C(\partial\Omega)d(x)^{-\gamma}.$$

Proceeding as for $n=4$ yields (4.60) and hence (4.59) also in the case $n=3$.

    Case $n=2$. Here, one directly finds that

$$\|h(x,.)\|_{C^{1,\gamma}(\overline{\Omega})} \le C(\partial\Omega)$$

and the claims (4.59), (4.60) immediately follow. $\square$

4.5 Uniform Green Functions Estimates in $C^{4,\gamma}$-Families of Domains

Later on we will need convergence properties of Green functions defined on a converging family of domains. For the sake of simplicity we restrict ourselves also in this section to biharmonic operators. Moreover, in order to avoid too many technicalities, we restrict ourselves to special families of bounded domains that may be parametrised with the help of global coordinate charts over the closure of a fixed bounded smooth domain.

To be more precise: we will consider the family of the biharmonic Green functions $G_k = G_{\Delta^2,\Omega_k}$ and $G = G_{\Delta^2,\Omega}$ in Ω_k and Ω respectively, where $(\Omega_k)_{k\in\mathbb{N}}$ is a family of domains converging to a bounded domain $\Omega \subset \mathbb{R}^n$ in the following sense.

Definition 4.19. We say that the sequence $(\Omega_k)_{k\in\mathbb{N}}$ is a $C^{4,\gamma}$-perturbation of the bounded $C^{4,\gamma}$-smooth domain Ω, if there exists a neighbourhood U of $\overline{\Omega}$ and for each $k \in \mathbb{N}$ a $C^{4,\gamma}$-diffeomorphism $\Psi_k : U \to \Psi_k(U)$ with $\Psi_k(\overline{\Omega}) = \overline{\Omega_k}$ such that one has

$$\lim_{k\to\infty} \|Id - \Psi_k\|_{C^{4,\gamma}(\overline{U})} = 0.$$

The remaining section is divided in a part without and a part with boundary terms and we finish with some results on the convergence of these Green functions.

4.5.1 Uniform Global Estimates Without Boundary Terms

As for the diffeomorphisms Ψ_k we refer to Definition 4.19.

Theorem 4.20. *Assume that $(\Omega_k)_{k\in\mathbb{N}}$ is a $C^{4,\gamma}$-perturbation of the bounded $C^{4,\gamma}$-smooth domain $\Omega \subset \mathbb{R}^n$ and let $G_k = G_{\Delta^2,\Omega_k}$ be the biharmonic Green function in Ω_k under Dirichlet boundary conditions. Then there exists a constant $C = C((\Omega_k)_{k\in\mathbb{N}})$, which is independent of k, such that for all $k \in \mathbb{N}$ and $\alpha, \beta \in \mathbb{N}^n$ with $|\alpha| + |\beta| \le 4$:*

- *If $|\alpha| + |\beta| + n > 4$:*

$$\left| D_x^\alpha D_y^\beta G_k(x,y) \right| \le C \, |x-y|^{4-n-|\alpha|-|\beta|} \text{ for all } x, y \in \Omega_k. \tag{4.71}$$

- *If $|\alpha| + |\beta| + n = 4$ and n is even*

$$\left| D_x^\alpha D_y^\beta G_k(x,y) \right| \le C \, \log\left(1 + |x-y|^{-1}\right) \text{ for all } x, y \in \Omega_k. \tag{4.72}$$

- *If $|\alpha| + |\beta| + n = 4$ and n is odd, or if $|\alpha| + |\beta| + n < 4$*

$$\left| D_x^\alpha D_y^\beta G_k(x,y) \right| \le C \text{ for all } x, y \in \Omega_k. \tag{4.73}$$

This kind of estimates was given by Krasovskiĭ [256, 257] in a very general framework. Here we provide an independent proof, which is simpler but nevertheless still quite involved. We shall proceed in several steps, where the proof of Proposition 4.22 is the most important part.

Lemma 4.21. *Assume that $n \ge 4$. Let $(\Omega_k)_{k\in\mathbb{N}}$ be a $C^{4,\gamma}$-perturbation of the bounded $C^{4,\gamma}$-smooth domain Ω. Let G_k denote the Green functions for Δ^2 in Ω_k under Dirichlet boundary conditions and $d(\,.\,)$ the distance function to the boundary $\partial\Omega_k$. For any $q \in \left(\frac{n}{n-3}, \frac{n}{n-4}\right)$ there exists $C(q) > 0$ such that for all k and all $x \in \Omega_k$ we have*

$$\|G_k(x, \cdot)\|_{L^q(\Omega_k)} \le C(q)\, d(x)^{4-n+\frac{n}{q}}. \tag{4.74}$$

The constant $C(q)$ can be chosen uniformly for the family $(\Omega_k)_{k\in\mathbb{N}}$.

Proof. We proceed with the help of a duality argument. Let $\psi \in C_c^\infty(\Omega_k)$ and let $\varphi \in C^{4,\gamma}(\overline{\Omega_k})$ be a solution of

$$\begin{cases} \Delta^2\varphi = \psi & \text{in } \Omega_k, \\ \varphi = \varphi_\nu = 0 & \text{on } \partial\Omega_k. \end{cases}$$

Let $q \in \left(\frac{n}{n-3}, \frac{n}{n-4}\right)$ and let $q' = \frac{q}{q-1}$ be the dual exponent, so that in particular $\frac{n}{4} < q' < \frac{n}{3}$. It follows from Theorem 2.20 (in particular, Corollary 2.21) that there exists $C_3 > 0$ independent of φ, ψ and k such that

$$\|\varphi\|_{W^{4,q'}(\Omega_k)} \le C_3 \|\psi\|_{L^{q'}(\Omega_k)}.$$

The embedding $W^{4,q'}(\Omega_k) \subset C^{0,\mu}(\overline{\Omega_k})$ (see Theorem 2.6) with $\mu = 4 - \frac{n}{q'} = 4 - n + \frac{n}{q}$ being continuous uniformly in k shows that there exists $C_4 > 0$ independent of φ and k such that $\|\varphi\|_{C^{0,\mu}(\overline{\Omega_k})} \leq C_4 \|\varphi\|_{W^{4,q'}(\Omega_k)}$. Let $x \in \Omega_k$ and $x' \in \partial\Omega_k$. We then get that

$$|\varphi(x)| = |\varphi(x) - \varphi(x')| \leq \|\varphi\|_{C^{0,\mu}(\overline{\Omega_k})} |x - x'|^\mu \leq C_3 C_4 \|\psi\|_{L^{q'}(\Omega_k)} |x - x'|^\mu.$$

Moreover, it follows from Green's representation formula that

$$\varphi(x) = \int_{\Omega_k} G_k(x,y)\psi(y)\,dy \text{ for all } x \in \Omega_k.$$

Therefore, taking the infimum with respect to $x' \in \partial\Omega_k$, we have that

$$\left| \int_{\Omega_k} G_k(x,y)\psi(y)\,dy \right| \leq C_3 C_4 \|\psi\|_{L^{q'}(\Omega_k)} d(x)^\mu$$

for all $\psi \in C_c^\infty(\Omega_k)$. Inequality (4.74) then follows. $\hfill\square$

4.5.1.1 Zero and First Derivative Estimates

Proposition 4.22. *Let $(\Omega_k)_{k\in\mathbb{N}}$ be a $C^{4,\gamma}$-perturbation of the bounded $C^{4,\gamma}$-smooth domain Ω. Let G_k be as in Lemma 4.21. Then there exists a constant $C_1 > 0$ such that for all k and all $x,y \in \Omega_k$ with $x \neq y$ one has that*

$$|G_k(x,y)| \leq C_1 \cdot \begin{cases} |x-y|^{4-n} & \text{if } n > 4, \\ \log\left(1 + |x-y|^{-1}\right) & \text{if } n = 4, \\ 1 & \text{if } n = 2,3. \end{cases} \tag{4.75}$$

Moreover, for $n = 2,3,4$ and for all $k \in \mathbb{N}$ and $x,y \in \Omega_k$ with $x \neq y$

$$|\nabla_x G_k(x,y)| \leq C_1 \cdot \begin{cases} |x-y|^{-1} & \text{if } n = 4, \\ 1 & \text{if } n = 2,3. \end{cases} \tag{4.76}$$

By symmetry the last estimate also holds for $\left|\nabla_y G_k(x,y)\right|$.

Proof. If $n = 2,3$, the statement of Lemma 4.18 is already strong enough and nothing remains to be proved.

We start with the case $n > 4$. We use an argument by contradiction and assume that there exist two sequences $(x_k)_{k\in\mathbb{N}}$, $(y_k)_{k\in\mathbb{N}}$ with $x_k, y_k \in \Omega_{\ell_k}$ for a suitable sequence $(\ell_k) \subset \mathbb{N}$ such that $x_k \neq y_k$ for all $k \in \mathbb{N}$ and such that

$$\lim_{k\to+\infty} |x_k - y_k|^{n-4} \left|G_{\ell_k}(x_k,y_k)\right| = +\infty. \tag{4.77}$$

It is enough to consider $\ell_k = k$; other situations may be reduced to this by relabeling or are even more special. After possibly passing to a subsequence, it follows from (4.59) that there exists $x_\infty \in \partial\Omega$ such that

$$\lim_{k\to+\infty} x_k = x_\infty \quad \text{and} \quad \lim_{k\to+\infty} \frac{d(x_k)}{|x_k - y_k|} = 0. \tag{4.78}$$

We remark that the constant in (4.59) can be chosen uniformly for the family $(\Omega_k)_{k\in\mathbb{N}}$.

Next we claim that if (4.77) holds, then

$$\lim_{k\to+\infty} |x_k - y_k| = 0. \tag{4.79}$$

Assume by contradiction that $|x_k - y_k|$ does not converge to 0. After extracting a subsequence we may then assume that there exists $\delta > 0$ such that for all k we have $x_k \in B_\delta(x_\infty)$ and $y_k \in \Omega_k \setminus \overline{B_{3\delta}(x_\infty)}$. We consider q as in Lemma 4.21. In particular we know that $\|G_k(x, \cdot)\|_{L^q(\Omega_k)} \le C$ uniformly in k. By applying local elliptic estimates (see Theorem 2.20) combined with Sobolev embeddings in $\Omega_k \setminus \overline{B_{2\delta}(x_\infty)}$ we find that

$$\|G_k(x_k, \cdot)\|_{L^\infty(\Omega_k \setminus \overline{B_{3\delta}(x_\infty)})} \le C(q, \delta)$$

uniformly in k. In particular, we would have

$$|G_k(x_k, y_k)| \le C(q, \delta) \quad \text{and} \quad |x_k - y_k|^{n-4} |G_k(x_k, y_k)| \le C(q, \delta)$$

independent of k. This contradicts the hypothesis (4.77) and proves the claim in (4.79).

Let $\Phi : U \to \mathbb{R}^n$, $\Phi(0) = x_\infty$ be a fixed coordinate chart for $\overline{\Omega}$ around ∞. We put $\Phi_k := \Psi_k \circ \Phi$ and have that

$$\Phi_k(U \cap \{x_1 < 0\}) = \Phi_k(U) \cap \Omega_k \quad \text{and} \quad \Phi_k(U \cap \{x_1 = 0\}) = \Phi_k(U) \cap \partial\Omega_k.$$

Let $x_k = \Phi_k(x'_k)$ and $y_k = \Phi_k(y'_k)$. Therefore, (4.78) rewrites as

$$\lim_{k\to+\infty} x'_k = 0 \quad \text{and} \quad \lim_{k\to+\infty} \frac{x'_{k,1}}{|x'_k - y'_k|} = 0. \tag{4.80}$$

We define for R and k large enough

$$\tilde{G}_k(z) = |x'_k - y'_k|^{n-4} G_k\left(\Phi_k(x'_k), \Phi_k\left(x'_k + |x'_k - y'_k|(z - \rho_k e_1)\right)\right)$$

in $B_R(0) \cap \{x_1 < 0\}$, where $\rho_k := \dfrac{x'_{k,1}}{|x'_k - y'_k|}$ and e_1 is the first unit vector. The biharmonic equation $\Delta^2 G_k(x, \cdot) = 0$, complemented with Dirichlet boundary conditions, is rewritten as

$$\Delta^2_{g_k} \tilde{G}_k = 0 \text{ in } (B_R(0) \cap \{z_1 < 0\}) \setminus \{\rho_k e_1\}, \qquad \tilde{G}_k = \partial_1 \tilde{G}_k = 0 \text{ on } \{z_1 = 0\}.$$

Here, $g_k(z) = \Phi_k^*(\mathscr{E})(x_k' + |x_k' - y_k'|(z - \rho_k \mathbf{e}_1))$, $\mathscr{E} = (\delta_{ij})$ the Euclidean metric, and Δ_{g_k} denotes the Laplace-Beltrami operator with respect to this scaled and translated pull back of the Euclidean metric under Φ_k. Then for some $q \in \left(\frac{n}{n-3}, \frac{n}{n-4}\right)$ and $\tau > 0$ being chosen suitably small, it follows from elliptic estimates (see Theorem 2.20) and Sobolev embeddings that there exists $C(R, \tau, q) > 0$ such that

$$|\tilde{G}_k(z)| \le C(R, q, \tau) \left\| \tilde{G}_k \right\|_{L^q(B_R(0) \setminus B_\tau(0))} \tag{4.81}$$

for all $z \in B_{R/2}(0) \setminus B_{2\tau}(0)$, $z_1 \le 0$. In order to estimate the L^q-norm on the right-hand side we use (4.74) and obtain that

$$\int_{B_R(0) \cap \{\zeta_1 < 0\}} |\tilde{G}_k(\zeta)|^q \, d\zeta \le C |x_k' - y_k'|^{q(n-4)-n} \int_{\Omega_k} |G_k(x_k, y)|^q \, dy$$

$$\le C |x_k' - y_k'|^{q(n-4)-n} d(x_k)^{(4-n)q+n}$$

$$\le C \left(\frac{d(x_k)}{|x_k' - y_k'|} \right)^{n-q(n-4)}.$$

Therefore, with (4.78), we get that $\lim_{k \to +\infty} \left\| \tilde{G}_k \right\|_{L^q(B_R(0) \setminus B_\tau(0))} = 0$, and (4.81) yields

$$\lim_{k \to +\infty} \tilde{G}_k = 0 \text{ in } C^0((B_{R/2}(0) \setminus B_{2\tau}(0)) \cap \{z_1 \le 0\}).$$

In particular, since $\lim_{k \to +\infty} \rho_k = 0$, we have that

$$\lim_{k \to +\infty} \tilde{G}_k \left(\frac{y_k' - x_k'}{|x_k' - y_k'|} + \rho_k \mathbf{e}_1 \right) = 0.$$

This limit rewrites as

$$\lim_{k \to +\infty} |x_k - y_k|^{n-4} |G_k(x_k, y_k)| = 0,$$

contradicting (4.77). This completes the proof of Proposition 4.22 for the case $n > 4$.

Now let us consider the case $n = 4$. Here it is enough to prove (4.76) for ∇_y, exploiting the symmetry of the Green function. We argue by contradiction and, as in the proof for $n > 4$, we may assume that there exist two sequences $(x_k)_{k \in \mathbb{N}}, (y_k)_{k \in \mathbb{N}}$ with $x_k, y_k \in \Omega_k$ such that $x_k \ne y_k$ and

$$\lim_{k \to +\infty} |x_k - y_k| \, |\nabla_y G_k(x_k, y_k)| = +\infty. \tag{4.82}$$

After possibly passing to a subsequence it follows from (4.60) that there exists $x_\infty \in \partial\Omega$ such that

$$\lim_{k \to +\infty} x_k = x_\infty \text{ and } \lim_{k \to +\infty} \frac{d(x_k)}{|x_k - y_k|} = 0. \tag{4.83}$$

Lemma 4.21 may be applied with some $q > 4$. The analogue of (4.79) is proved in exactly the same way as above. Like above we now put for R and k large enough

$$\tilde{G}_k(z) = G_k\left(\Phi_k(x_k'), \Phi_k(x_k' + |x_k' - y_k'|(z - \rho_k \mathbf{e}_1))\right)$$

in $B_R(0) \cap \{z_1 < 0\}$, where $x_k = \Phi_k(x_k')$, $y_k = \Phi_k(y_k')$, $\rho_k := \frac{x_{k,1}'}{|x_k' - y_k'|}$. As above we find for $\tau > 0$ small enough that there exists $C(R, \tau, q) > 0$ such that

$$\left|\nabla \tilde{G}_k(z)\right| \leq C(R, q, \tau) \left\|\tilde{G}_k\right\|_{L^q(B_R(0) \setminus B_\tau(0))}$$

for all $z \in B_{R/2}(0) \setminus B_{2\tau}(0)$, $z_1 \leq 0$. Using (4.74) we obtain that

$$\int_{B_R(0) \cap \{\zeta_1 < 0\}} |\tilde{G}_k(\zeta)|^q \, d\zeta \leq C |x_k' - y_k'|^{-4} \int_{\Omega_k} |G_k(x_k, y)|^q \, dy$$
$$\leq C \left(\frac{d(x_k)}{|x_k' - y_k'|}\right)^4.$$

In the same way as in the generic case $n > 4$ this yields first that

$$\lim_{k \to +\infty} \nabla \tilde{G}_k = 0 \text{ in } C^0((B_{R/2}(0) \setminus B_{2\tau}(0)) \cap \{z_1 \leq 0\})$$

and back in the original coordinates

$$\lim_{k \to \infty} |x_k - y_k| \left|\nabla_y G_k(x_k, y_k)\right| = 0.$$

So, we achieve a contradiction also if $n = 4$. This proves (4.76). By integrating (4.76) we get (4.75). The proof of Proposition 4.22 is complete. $\square$

4.5.1.2   First and Higher Derivatives for $n \geq 3$

Proposition 4.23. *Suppose that $n \geq 3$ and let $(\Omega_k)_{k \in \mathbb{N}}$ be a $C^{4,\gamma}$-perturbation of the bounded $C^{4,\gamma}$-smooth domain Ω. Let G_k be as in Lemma 4.21. Then there exists a constant $C > 0$ such that for all k, all $\alpha, \beta \in \mathbb{N}^n$ with $1 \leq |\alpha| + |\beta| < 4$, and all $x, y \in \Omega_k$ with $x \neq y$ one has that*

$$\left|D_x^\alpha D_y^\beta G_k(x, y)\right| \leq C \cdot |x - y|^{4-n-|\alpha|-|\beta|}. \tag{4.84}$$

Proof. For $|\alpha| + |\beta| = 1$ with $n = 3, 4$ the result is found in (4.76).

To obtain estimates for (higher) derivatives we will use the following local estimate, see (2.19) for biharmonic functions. This local estimate is fundamental. Moreover, it also holds near the boundary part where homogeneous Dirichlet boundary conditions are satisfied. For any two concentric balls $B_R \subset B_{2R}$ and $|\alpha| \leq 4$ we have

$$\|D^\alpha v\|_{L^\infty(B_R \cap \Omega_k)} \leq \frac{C}{R^{|\alpha|}} \|v\|_{L^\infty(B_{2R} \cap \Omega_k)}. \tag{4.85}$$

The constant is uniform in k and R. The behaviour with respect to (small) R is obtained by means of scaling.

Case $n > 4$. Keeping $x \in \Omega_k$ fixed, for any $y \in \Omega_k \setminus \{x\}$ we choose $R = |x-y|/4$ and apply (4.85) and (4.75) of Proposition 4.22 in $B_R(y) \subset B_{2R}(y)$ to $G_k(x, .)$. This proves (4.84) for $|\alpha| = 0$. By symmetry the same estimate holds for $|\alpha| > 0$ and $|\beta| = 0$. Since also $D_x^\alpha G_k(x, .)$ is biharmonic with homogeneous Dirichlet boundary conditions we may repeat the argument to find estimates for mixed derivatives.

Case $n = 3,4$. The result follows from a similar argument as above but now starting with the first order estimate in (4.76). $\qquad\square$

4.5.1.3 Second and Higher Derivatives for $n = 2$

Lemma 4.24. *Let $n = 2$ and $\delta > 0$. Then there exists a constant $C = C(\delta, (\Omega_k)_{k \in \mathbb{N}})$ such that for $\alpha, \beta \in \mathbb{N}^2$ with $|a| + |\beta| = 2$*

$$x, y \in \Omega_k, \quad \max\{d(x), d(y)\} \geq \delta \Rightarrow \left|D_x^\alpha D_y^\beta G_k(x,y)\right| \leq C \log\left(1 + |x-y|^{-1}\right).$$

Proof. The Green function can be written as $G(x,y) = F_{n,2}(|x-y|) + h(x,y)$ with $h(x, .)$ the solution of (4.62). For $d(x) > \delta$ one finds as a direct consequence of Schauder estimates that $\|h(x, .)\|_{C^m(\overline{\Omega})} < C(\delta, m)$ for any $m \in \mathbb{N}$ and uniformly for all x with $d(x) > \delta$. Hence, for $|\beta| = 2$ one obtains

$$\left|D_y^\beta G(x,y)\right| \leq C_1 \left|D_y^\beta F_{n,2}(x,y)\right| + C(\delta),$$

which shows the estimate in Lemma 4.24 for $\alpha = 0$. For $|\beta| < 2$ and hence $|\alpha| > 0$ one considers the function $D_x^\alpha h(x, .)$ and proceeds similar as before. So one has found the estimates in Lemma 4.24 for $d(x) > \delta$. Since the Green function is symmetric one may interchange the role of x and y and a similar result holds when $d(y) > \delta$. $\qquad\square$

Proposition 4.25. *Let $n = 2$. There exists a constant $C = C((\Omega_k)_{k \in \mathbb{N}})$ such that for $\alpha, \beta \in \mathbb{N}^2$ with $|a| + |\beta| \geq 3$*

$$\left|D_x^\alpha D_y^\beta G_k(x,y)\right| \leq C |x-y|^{2-|\alpha|-|\beta|}.$$

The proposition requires a somehow technical proof which will be performed in several steps. However, combining the lemma and the proposition obviously gives a proof of the remaining cases of (4.72)-(4.73) and the proof of Theorem 4.20 will then be complete for $n = 2$.

As a starting point we prove an L^q-estimate for second derivatives of the Green functions.

Lemma 4.26. *Let $n = 2$. For any $q > 2$, there exists a constant $C = C(q, (\Omega_k)_{k \in \mathbb{N}})$ such that*

$$\left\| \nabla_y^2 G_k(x, \cdot) \right\|_{L^q(\Omega_k)} \leq C\, d(x)^{2/q}; \tag{4.86}$$

$$\left\| \nabla_x \nabla_y G_k(x, \cdot) \right\|_{L^q(\Omega_k)} \leq C\, d(x)^{2/q}. \tag{4.87}$$

Proof. We argue along the lines of the proof of Lemma 4.21 to which we refer for more detailed arguments. We prove first (4.86). For $\psi \in L^{q'}(\Omega_k)$, $q' = \frac{q}{q-1} \in (1,2)$ let $\varphi \in W^{2,q'}(\Omega_k)$ be the solution of

$$\begin{cases} \Delta^2 \varphi = \nabla^2 \psi & \text{in } \Omega_k, \\ \varphi = \varphi_\nu = 0 & \text{on } \partial\Omega_k. \end{cases}$$

For biharmonic equations in integral form $L^{q'}$-estimates (see Theorem 2.22) yield

$$\|\varphi\|_{W^{2,q'}} \leq C \|\psi\|_{L^{q'}}.$$

Since $q' \in (1,2)$ we have that $2 - 2/q' \in (0,1)$ and employing also Sobolev's embedding theorem gives

$$|\varphi(x)| \leq C \|\psi\|_{L^{q'}}\, d(x)^{2-2/q'}. \tag{4.88}$$

We observe the following representation formula, homogeneous Dirichlet boundary data of the Green functions and integrate by parts:

$$\varphi(x) = \int_{\Omega_k} G_k(x,y) \nabla_y^2 \psi(y)\, dy = \int_{\Omega_k} \nabla_y^2 G_k(x,y) \psi(y)\, dy.$$

Together with (4.88) and $2 - 2/q' = 2/q$ this shows (4.86).

In order to prove (4.87) we solve

$$\begin{cases} \Delta^2 \varphi = \nabla \psi & \text{in } \Omega_k, \\ \varphi = \varphi_\nu = 0 & \text{on } \partial\Omega_k. \end{cases}$$

and get

$$\|\varphi\|_{W^{3,q'}} \leq C \|\psi\|_{L^{q'}}.$$

We proceed similarly as above and find

$$|\nabla\varphi(x)| \leq C \|\psi\|_{L^{q'}}\, d(x)^{2-2/q'}$$

as well as

$$\nabla\varphi(x) = - \int_{\Omega_k} \nabla_x \nabla_y G_k(x,y) \psi(y)\, dy$$

and so, finally, (4.87). $\qquad\square$

Proof of Proposition 4.25. We first prove the statement for $D_y^\beta G_k(x,y)$ with $|\beta| = 3$. We assume by contradiction that, after suitably relabeling, there exist sequences (x_k), (y_k) with $x_k, y_k \in \Omega_k$ and $x_k \neq y_k$, such that

$$\lim_{k \to \infty} |x_k - y_k| \, D_y^\beta G_k(x_k, y_k) = \infty. \tag{4.89}$$

As in Proposition 4.22, local elliptic estimates show that

$$\lim_{k \to \infty} |x_k - y_k| = 0.$$

Hence, we may assume that there exists $x_\infty \in \overline{\Omega}$ with $x_\infty = \lim_{k \to \infty} x_k = \lim_{k \to \infty} y_k$. This shows that local elliptic estimates around x_k and y_k may be rescaled and hold with uniform constants.

First case: $d(x_k) < 2|x_k - y_k|$. Here we work in $B_{4|x_k-y_k|}(x_k) \setminus B_{|x_k-y_k|/2}(x_k)$, which certainly hit the boundaries $\partial \Omega_k$ where we have homogeneous Dirichlet boundary data for $G_k(x_k, \cdot)$. These allow to apply local rescaled elliptic estimates and a localised Poincaré inequality to show that

$$|D_y^\beta G_k(x_k, y_k)| \leq C|x_k - y_k|^{-3-2/q} \|G_k(x_k, \cdot)\|_{L^q(\Omega \cap (B_{4|x_k-y_k|}(x_k) \setminus B_{|x_k-y_k|/2}(x_k)))}$$

$$\leq C|x_k - y_k|^{-1-2/q} \|\nabla_y^2 G_k(x_k, \cdot)\|_{L^q(\Omega \cap (B_{4|x_k-y_k|}(x_k) \setminus B_{|x_k-y_k|/2}(x_k)))}$$

$$\leq C|x_k - y_k|^{-1-2/q} d(x_k)^{2/q} \leq C|x_k - y_k|^{-1},$$

where $q > 2$ is some arbitrarily chosen number. This inequality contradicts the assumption (4.89).

Second case: $d(x_k) \geq 2|x_k - y_k|$. We change our point of view and consider now y_k as parameter and the boundary value problem for the regular part of $\nabla_y^3 G_k(\cdot, y_k)$. Arguing as in Lemma 4.18 and integrating local Schauder estimates yields

$$|D_y^\beta G_k(x_k, y_k)| \leq C \left(\frac{1}{|x_k - y_k|} + \frac{d(x_k)^{1+\gamma}}{d(y_k)^{2+\gamma}} \right). \tag{4.90}$$

By assumption we have $d(x_k) \geq 2|x_k - y_k|$, which implies that

$$d(x_k) \leq |x_k - y_k| + d(y_k) \leq \frac{1}{2} d(x_k) + d(y_k),$$

$$\Rightarrow d(x_k) \leq 2d(y_k).$$

Inserting this into (4.90) gives

$$|D_y^\beta G_k(x_k, y_k)| \leq C \left(\frac{1}{|x_k - y_k|} + \frac{1}{d(x_k)} \right) \leq C \frac{1}{|x_k - y_k|},$$

again a contradiction to the assumption (4.89).

We comment now on how to prove the statement for $\nabla_x \nabla_y^2 G_k(x, \cdot)$. The remaining cases then follow by exploiting the symmetry of the Green functions. We assume by contradiction that – after a suitable relabeling – there exist sequences (x_k), (y_k), with $x_k, y_k \in \Omega_k$ and $x_k \neq y_k$, such that

$$\lim_{k \to \infty} |x_k - y_k| \cdot \nabla_x \nabla_y^2 G_k(x_k, y_k) = \infty. \tag{4.91}$$

As for (4.79), local elliptic estimates show that

$$\lim_{k \to \infty} |x_k - y_k| = 0.$$

Hence, we may assume that there exists $x_\infty \in \overline{\Omega}$ with $x_\infty = \lim_{k \to \infty} x_k = \lim_{k \to \infty} y_k$. This shows that local elliptic estimates around x_k and y_k may be rescaled and hold with uniform constants.

First case: $d(x_k) < 2|x_k - y_k|$. As above we work in $B_{4|x_k - y_k|}(x_k) \setminus B_{|x_k - y_k|/2}(x_k)$ and find that

$$
\begin{aligned}
|\nabla_x \nabla_y^2 G_k(x_k, y_k)| &\leq C|x_k - y_k|^{-2-2/q} \|\nabla_x G_k(x_k, \cdot)\|_{L^q(\Omega \cap (B_{4|x_k - y_k|}(x_k) \setminus B_{|x_k - y_k|/2}(x_k)))} \\
&\leq C|x_k - y_k|^{-1-2/q} \|\nabla_x \nabla_y G_k(x_k, \cdot)\|_{L^q(\Omega \cap (B_{4|x_k - y_k|}(x_k) \setminus B_{|x_k - y_k|/2}(x_k)))} \\
&\leq C|x_k - y_k|^{-1-2/q} d(x_k)^{2/q} \leq C|x_k - y_k|^{-1},
\end{aligned}
$$

where $q > 2$ is some arbitrarily chosen number. This inequality contradicts the assumption (4.91).

Second case: $d(x_k) \geq 2|x_k - y_k|$. Again we change our point of view and consider now y_k as parameter and the boundary value problem for the regular part of $\nabla_y^2 G_k(\cdot, y_k)$. Arguing as in Lemma 4.18, integrating local Schauder estimates yield

$$|\nabla_x \nabla_y^2 G_k(x_k, y_k)| \leq C \left(\frac{1}{|x_k - y_k|} + \frac{d(x_k)^\gamma}{d(y_k)^{1+\gamma}} \right). \tag{4.92}$$

As above we may insert $d(x_k) \leq 2d(y_k)$ into (4.92) and obtain

$$|\nabla_x \nabla_y^2 G_k(x_k, y_k)| \leq C \left(\frac{1}{|x_k - y_k|} + \frac{1}{d(x_k)} \right) \leq C \frac{1}{|x_k - y_k|},$$

again a contradiction to the assumption (4.91).

Once the estimates for $|\alpha| + |\beta| = 3$ have been derived, we may proceed as in the proof of Proposition 4.23 to obtain the estimates for $|\alpha| + |\beta| = 4$. $\qquad \square$

Proposition 4.27. *Let $n = 2$. There exists a constant $C = C((\Omega_k)_{k \in \mathbb{N}})$ such that for $\alpha, \beta \in \mathbb{N}^2$ with $|a| + |\beta| = 2$*

$$\left| D_x^\alpha D_y^\beta G_k(x, y) \right| \leq C \log \left(1 + |x - y|^{-1} \right).$$

Proof. For x or y away from the boundary the result is found in Lemma 4.24. For $\delta > 0$ small enough take $y_0 \in \Omega_k$ with $d(y_0) > 2\delta$ and assume both $d(x) < \delta$ and $d(y) < \delta$. Let $r_0 > 0$ be small enough such that $\Omega_k \setminus B_{2r_0}(z)$ is still connected for each $z \in \Omega_k$. Using the estimate from Proposition 4.23 and integrating along a path γ from y_0 to y that avoids $B_r(x)$ with $r = \min\{r_0, |x-y|\}$, one finds

$$\left| D_x^\alpha D_y^\beta G_k(x,y) \right| \leq \left| D_x^\alpha D_y^\beta G_k(x,y_0) \right| + \int_\gamma \left| D_x^\alpha \nabla_y D_y^\beta G_k(x,\gamma(s)) \right| d\gamma(s)$$

$$\leq C\left(1 + \int_{|x-y|}^M r^{-1} dr \right) \leq C' \log\left(1 + |x-y|^{-1}\right),$$

which shows the claim. $\qquad\square$

4.5.1.4 The Proof of the Uniform Estimates

Proof of Theorem 4.20. For $n \geq 3$ and $|\alpha| + |\beta| + n > 4$ the estimate in (4.71) follows from Propositions 4.22 and 4.23. For $n = 2$ the estimate (4.71) follows from Proposition 4.25. The estimate in (4.72) is stated in Proposition 4.22 for $n = 4$ and in Proposition 4.27 for $n = 2$. The estimate in (4.73) is contained in Proposition 4.22. $\qquad\square$

4.5.2 Uniform Global Estimates Including Boundary Terms

Theorem 4.28. *We assume that* $(\Omega_k)_{k \in \mathbb{N}}$ *is a* $C^{4,\gamma}$*-perturbation of the bounded* $C^{4,\gamma}$*-smooth domain* $\Omega \subset \mathbb{R}^n$*. Let* $G_k = G_{\Delta^2, \Omega_k}$ *denote the biharmonic Green function in* Ω_k *under Dirichlet boundary conditions. Then there exists a constant* $C = C((\Omega_k)_{k \in \mathbb{N}})$*, independent of k, such that for all $k \in \mathbb{N}$ it holds that*

$$\left| D_x^\alpha D_y^\beta G_k(x,y) \right| \leq C\,(*), \tag{4.93}$$

where $(*)$ *is as in Table 4.1. In this table the following abbreviations are used:*

$$d(x) = d(x, \partial\Omega_k) \quad \text{and} \quad d(y) = d(y, \partial\Omega_k),$$

$$W_x = \min\left\{1, \frac{d(x)}{|x-y|}\right\} \quad \text{and} \quad W_y = \min\left\{1, \frac{d(y)}{|x-y|}\right\}.$$

Proof. The ingredients of this proof are the estimates in Theorem 4.20, the construction of appropriate curves connecting x with a boundary point x^* that avoid singular points, and integral estimates along these curves. The $C^{4,\gamma}$-smoothness only comes in through the constant that appear in Theorem 4.20. So we may suppress the dependence on k in the present proof.

Table 4.1 Expressions to be inserted into $(*)$ of (4.93)

$$\boxed{\begin{array}{c} |\alpha| = |\beta| = 2: \\ |x-y|^{-n} \end{array}}$$

$$\boxed{\begin{array}{c} |\alpha| = 1 \text{ and } |\beta| = 2: \\ |x-y|^{1-n} W_x \end{array}} \qquad \boxed{\begin{array}{c} |\alpha| = 2 \text{ and } |\beta| = 1: \\ |x-y|^{1-n} W_y \end{array}}$$

$$\boxed{\begin{array}{c} |\alpha| = 0 \text{ and } |\beta| = 2: \\ |x-y|^{2-n} W_x^2 \quad \text{for } n \geq 3 \\ \log\left(1 + \frac{d(x)^2}{|x-y|^2}\right) \text{ for } n = 2 \end{array}} \quad \boxed{\begin{array}{c} |\alpha| = |\beta| = 1: \\ |x-y|^{2-n} W_x W_y \quad \text{for } n \geq 3 \\ \log\left(1 + \frac{d(x)d(y)}{|x-y|^2}\right) \text{ for } n = 2 \end{array}} \quad \boxed{\begin{array}{c} |\alpha| = 2 \text{ and } |\beta| = 0: \\ |x-y|^{2-n} W_y^2 \quad \text{for } n \geq 3 \\ \log\left(1 + \frac{d(y)^2}{|x-y|^2}\right) \text{ for } n = 2 \end{array}}$$

$$\boxed{\begin{array}{c} |\alpha| = 0 \text{ and } |\beta| = 1: \\ |x-y|^{3-n} W_x^2 W_y \text{ for } n \geq 4 \\ W_x^2 W_y \qquad \text{for } n = 3 \\ d(x) W_x W_y \qquad \text{for } n = 2 \end{array}} \qquad \boxed{\begin{array}{c} |\alpha| = 1 \text{ and } |\beta| = 0: \\ |x-y|^{3-n} W_x W_y^2 \text{ for } n \geq 4 \\ W_x W_y^2 \qquad \text{for } n = 3 \\ d(y) W_x W_y \qquad \text{for } n = 2 \end{array}}$$

$$\boxed{\begin{array}{c} |\alpha| = |\beta| = 0: \\ |x-y|^{4-n} W_x^2 W_y^2 \qquad \text{for } n \geq 5 \\ \log\left(1 + \frac{d(x)^2 d(y)^2}{|x-y|^4}\right) \qquad \text{for } n = 4 \\ d(x)^{1/2} d(y)^{1/2} W_x^{3/2} W_y^{3/2} \text{ for } n = 3 \\ d(x) d(y) W_x W_y \qquad \text{for } n = 2 \end{array}}$$

Claim 1: Let $x, y \in \Omega$. There exists a piecewise smooth curve Γ_x connecting x with the boundary $\partial\Omega$ such that $d(\Gamma_x, y) \geq \frac{1}{2}|x-y|$ and if we parametrise Γ_x by arclength, it holds that:

$$\tfrac{2}{3}s \leq |\Gamma_x(s) - x| \leq s, \tag{4.94}$$

$$|\Gamma_x(s) - y| \geq \tfrac{1}{8}|x-y| + \tfrac{1}{8}|\Gamma_x(s) - x|. \tag{4.95}$$

Let x^* be such that $d(x) = |x-x^*|$. If the interval $[x,x^*]$ does not intersect $B_{\frac{1}{2}|x-y|}(y)$, then we take $\Gamma_x = [x,x^*]$. If the set $[x,x^*] \cap B_{\frac{1}{2}|x-y|}(y)$ is nonempty while $B_{\frac{1}{2}|x-y|}(y) \cap \partial\Omega$ is empty, one modifies Γ_x by replacing $[x,x^*] \cap B_{\frac{1}{2}|x-y|}(y)$ by a shortest path on $\partial B_{\frac{1}{2}|x-y|}(y)$ that connects the two points of $[x,x^*] \cap \partial B_{\frac{1}{2}|x-y|}(y)$. If both $[x,x^*] \cap B_{\frac{1}{2}|x-y|}(y)$ and $B_{\frac{1}{2}|x-y|}(y) \cap \partial\Omega$ are nonempty, the part of $[x,x^*] \cap B_{\frac{1}{2}|x-y|}(y)$ is replaced by the shortest path on $\partial B_{\frac{1}{2}|x-y|}(y)$ that connects with the boundary, see Figure 4.1.

Geometric arguments show that $s \leq \frac{1}{3}(\pi+1)|\Gamma_x(s) - x|$. Using $\frac{1}{3}(\pi+1) < \frac{3}{2}$, (4.94) follows. Moreover, writing $z = \Gamma_x(s)$, we have $|z-y| \geq \frac{1}{2}|x-y|$. So, if

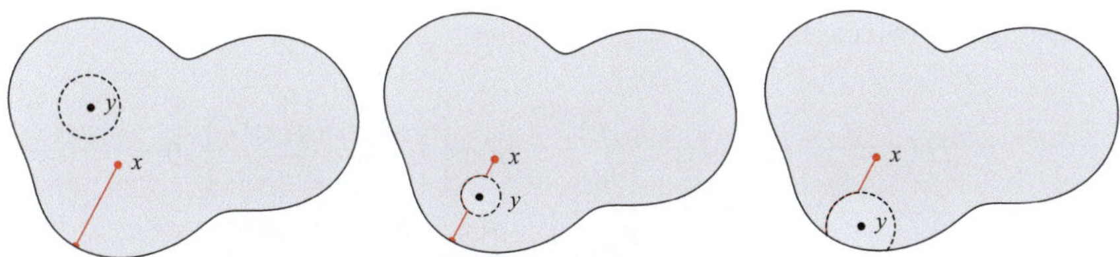

Fig. 4.1 Curves connecting x with the boundary by a path of length less than $\frac{3}{2}d(x)$ that avoid the singularity in y by staying outside of $B_{\frac{1}{2}|x-y|}(y)$.

$|z-x| \leq 2|x-y|$, then $|z-x| \leq 4|z-y|$. If $|z-x| \geq 2|x-y|$, then $|z-y| \geq |z-x| - |x-y| \geq \frac{1}{2}|z-x|$. Combining we obtain (4.95).

Claim 2: Let $k \geq 2$ and $v_1, v_2 \geq 0$. If $H(x,y) = 0$ for all $x \in \partial\Omega$ and $y \in \Omega$ and if for some $C \in \mathbb{R}^+$

$$|\nabla_x H(x,y)| \leq C\, |x-y|^{-k} \min\left\{1, \frac{d(x)}{|x-y|}\right\}^{v_1} \min\left\{1, \frac{d(y)}{|x-y|}\right\}^{v_2} \text{ for } x,y \in \Omega,$$

then there is $\tilde{C} \in \mathbb{R}^+$ such that

$$|H(x,y)| \leq \tilde{C}\, |x-y|^{1-k} \min\left\{1, \frac{d(x)}{|x-y|}\right\}^{v_1+1} \min\left\{1, \frac{d(y)}{|x-y|}\right\}^{v_2} \text{ for } x,y \in \Omega.$$

Let $s \mapsto x(s)$ parametrise Γ_x as above by arclength connecting $x^* \in \partial\Omega$ with x. Then

$$H(x,y) = H(x^*,y) + \int_{\Gamma_x} \nabla_x H(x(s),y) \cdot \tau(s)\, ds \tag{4.96}$$

and using Lemma 4.5

$$|H(x,y)| \leq \int_{\Gamma_x} |\nabla_x H(x(s),y)|\, ds$$

$$\leq \int_{\Gamma_x} C|x(s)-y|^{-k} \min\left\{1, \frac{d(x(s))^{v_1} d(y)^{v_2}}{|x(s)-y|^{v_1+v_2}}\right\}\, ds.$$

It follows from (4.95) that

$$|H(x,y)| \leq c_1 \int_0^{\frac{3}{2}d(x)} (|x-y|+s)^{-k} \min\left\{1, \frac{d(x)^{v_1} d(y)^{v_2}}{(|x-y|+s)^{v_1+v_2}}\right\}\, ds$$

$$= c_1 |x-y|^{1-k} \int_0^{\frac{3}{2}\frac{d(x)}{|x-y|}} (1+t)^{-k} \min\left\{1, \frac{d(x)^{v_1} d(y)^{v_2}}{(1+t)^{v_1+v_2} |x-y|^{v_1+v_2}}\right\}\, dt.$$

We distinguish two cases. If $d(x) \leq |x-y|$, then

$$|H(x,y)| \leq c_1 |x-y|^{1-k} \int_0^{\frac{3}{2}\frac{d(x)}{|x-y|}} \min\left\{1, \frac{d(x)^{v_1} d(y)^{v_2}}{|x-y|^{v_1+v_2}}\right\} dt$$

$$\leq c_2 |x-y|^{1-k} \min\left\{1, \frac{d(x)^{v_1} d(y)^{v_2}}{|x-y|^{v_1+v_2}}\right\} \frac{d(x)}{|x-y|}$$

$$\leq c_3 |x-y|^{1-k} \min\left\{1, \frac{d(x)^{v_1+1} d(y)^{v_2}}{|x-y|^{v_1+v_2+1}}\right\}. \tag{4.97}$$

If $d(x) \geq |x-y|$, then

$$|H(x,y)| \leq c_1 |x-y|^{1-k} \int_0^{\frac{3}{2}\frac{d(x)}{|x-y|}} (1+t)^{-k} dt \cdot \min\left\{1, \frac{d(y)^{v_2}}{|x-y|^{v_2}}\right\}$$

$$\leq c_2 |x-y|^{1-k} \min\left\{1, \frac{d(y)^{v_2}}{|x-y|^{v_2}}\right\}$$

$$\leq c_3 |x-y|^{1-k} \min\left\{1, \frac{d(x)^{v_1+1} d(y)^{v_2}}{|x-y|^{v_1+v_2+1}}\right\}.$$

Claim 3: Let $v_1, v_2 \geq 0$. If $H(x,y) = 0$ for $x \in \partial\Omega$ and for some $C \in \mathbb{R}^+$

$$|\nabla_x H(x,y)| \leq C |x-y|^{-1} \min\left\{1, \frac{d(x)}{|x-y|}\right\}^{v_1} \min\left\{1, \frac{d(y)}{|x-y|}\right\}^{v_2} \text{ for } x,y \in \Omega,$$

then there is $\tilde{C} \in \mathbb{R}^+$ such that for all $x,y \in \Omega$

$$|H(x,y)| \leq \tilde{C} \log\left(2 + \frac{d(x)}{|x-y|}\right) \min\left\{1, \frac{d(x)}{|x-y|}\right\}^{v_1+1} \min\left\{1, \frac{d(y)}{|x-y|}\right\}^{v_2}.$$

The steps of Claim 2 remain valid until

$$|H(x,y)| \leq c_1 \int_0^{\frac{3}{2}\frac{d(x)}{|x-y|}} (1+t)^{-1} \min\left\{1, \frac{d(x)^{v_1} d(y)^{v_2}}{(1+t)^{v_1+v_2} |x-y|^{v_1+v_2}}\right\} dt,$$

and inclusive (4.97). For $d(x) \leq |x-y|$ the claim follows. If $d(x) \geq |x-y|$, then

$$\int_0^{\frac{3}{2}\frac{d(x)}{|x-y|}} (1+t)^{-1} \min\left\{1, \frac{d(x)^{v_1} d(y)^{v_2}}{(1+t)^{v_1+v_2} |x-y|^{v_1+v_2}}\right\} dt$$

$$\leq \log\left(2 + \frac{d(x)}{|x-y|}\right) \min\left\{1, \frac{d(x)^{v_1} d(y)^{v_2}}{|x-y|^{v_1+v_2}}\right\}$$

$$\leq \log\left(2 + \frac{d(x)}{|x-y|}\right) \min\left\{1, \frac{d(x)^{v_1+1} d(y)^{v_2}}{|x-y|^{v_1+v_2+1}}\right\}.$$

Claim 4: Let $k \geq 2$ and $v_1, v_2, \alpha_1, \alpha_2 \geq 0$. If $H(x,y) = 0$ for $x \in \partial\Omega$ and for some $C \in \mathbb{R}^+$

$$|\nabla_x H(x,y)| \leq C\, d(x)^{\alpha_1} d(y)^{\alpha_2} \min\left\{1, \frac{d(x)}{|x-y|}\right\}^{v_1} \min\left\{1, \frac{d(y)}{|x-y|}\right\}^{v_2} \textit{ for } x,y \in \Omega,$$

then there is $\tilde{C} \in \mathbb{R}^+$ such that

$$|H(x,y)| \leq \tilde{C}\, d(x)^{\alpha_1+1} d(y)^{\alpha_2} \min\left\{1, \frac{d(x)}{|x-y|}\right\}^{v_1} \min\left\{1, \frac{d(y)}{|x-y|}\right\}^{v_2} \textit{ for } x,y \in \Omega.$$

This is a direct consequence of (4.96), (4.94) and (4.95).

Claim 5: Let $k \geq 2$ and $v_1, v_2, \alpha_1, \alpha_2 \geq 0$. If $H(x,y) = 0$ for $x \in \partial\Omega$ and if there exists $C \in \mathbb{R}^+$ such that

$$|\nabla_x H(x,y)| \leq C \log\left(2 + \frac{d(x)d(y)}{|x-y|^2}\right) \min\left\{1, \frac{d(x)}{|x-y|}\right\}^{v_1} \min\left\{1, \frac{d(y)}{|x-y|}\right\}^{v_2}$$

for $x,y \in \Omega$, then there is $\tilde{C} \in \mathbb{R}^+$ such that

$$|H(x,y)| \leq \tilde{C}\, d(x) \min\left\{1, \frac{d(x)}{|x-y|}\right\}^{v_1} \min\left\{1, \frac{d(y)}{|x-y|}\right\}^{v_2} \textit{ for } x,y \in \Omega.$$

We first observe that

$$\log\left(2 + \frac{d(x)d(y)}{|x-y|^2}\right) \min\left\{1, \frac{d(x)}{|x-y|}\right\}^{v_1} \min\left\{1, \frac{d(y)}{|x-y|}\right\}^{v_2}$$
$$\simeq \log\left(2 + \frac{d(x)}{|x-y|}\right) \min\left\{1, \frac{d(x)}{|x-y|}\right\}^{v_1} \min\left\{1, \frac{d(y)}{|x-y|}\right\}^{v_2}.$$

If $\frac{d(x)}{|x-y|} \leq 1$, then $\log\left(2 + \frac{d(x)}{|x-y|}\right)$ is bounded and the result is again a direct consequence of (4.96), (4.94) and (4.95). If $\frac{d(x)}{|x-y|} \geq 1$, then for $z \in \Gamma_x$

$$\frac{d(z)}{|x-z|} \leq \frac{8d(z)}{|x-y|+s}$$

and one finds

$$|\nabla_x H(x(s),y)| \leq c_1 \log\left(2 + \frac{d(x)}{|x-y|+s}\right) \cdot \min\left\{1, \frac{d(y)^{v_2}}{|x-y|^{v_2}}\right\}.$$

Hence

$$|H(x,y)| \le c_1 \int_0^{\frac{3}{2}d(x)} \log\left(2 + \frac{d(x)}{|x-y|+s}\right) ds \cdot \min\left\{1, \frac{d(y)^{\nu_2}}{|x-y|^{\nu_2}}\right\}$$
$$\le c_2 d(x) \min\left\{1, \frac{d(y)^{\nu_2}}{|x-y|^{\nu_2}}\right\}.$$

In order to complete the proof of Theorem 4.28 one starts from the estimates of Theorem 4.20. We find, using the Claims 2 to 5 and working our way down, the estimates as in Table 4.1 except for $n = 3$ with $|\alpha| + |\beta| \le 1$. Suppose $\alpha = 0$ and $|\beta| = 1$. Then

$$\left|\nabla_x D_y^\beta G(x,y)\right| \le C |x-y|^{-1} \min\left\{1, \frac{d(x)}{|x-y|}\right\} \min\left\{1, \frac{d(y)}{|x-y|}\right\}$$

implies

$$\left|D_y^\beta G(x,y)\right| \le C \log\left(1 + \frac{d(x)^2 d(y)}{|x-y|^3}\right).$$

Together with (4.76) we obtain

$$\left|D_y^\beta G(x,y)\right| \le C \min\left\{1, \frac{d(x)}{|x-y|}\right\}^2 \min\left\{1, \frac{d(y)}{|x-y|}\right\}.$$

For the zeroth order in case $n = 3$ one finds through

$$\left|\nabla_y G(x,y)\right| \le C \min\left\{1, \frac{d(x)}{|x-y|}\right\}^2 \min\left\{1, \frac{d(y)}{|x-y|}\right\}$$

that

$$|G(x,y)| \le C d(y) \min\left\{1, \frac{d(x)}{|x-y|}\right\}^2 \min\left\{1, \frac{d(y)}{|x-y|}\right\}$$

and through the similar estimate for $|\nabla_x G(x,y)|$ that

$$|G(x,y)| \le C M(x,y) \min\left\{1, \frac{d(x)d(y)}{|x-y|^2}\right\}$$

with

$$M(x,y) = \min\left\{d(y) \min\left\{1, \frac{d(x)}{|x-y|}\right\}, d(x) \min\left\{1, \frac{d(y)}{|x-y|}\right\}\right\}.$$

Since $M(x,y) \le \sqrt{d(x)d(y)} \min\left\{1, \frac{d(x)d(y)}{|x-y|^2}\right\}^{3/2}$ the proof is complete. $\qquad\square$

In a similar way one may derive estimates for the Poisson kernels. Consider

$$\begin{cases} \Delta^2 u = f & \text{in } \Omega, \\ u|_{\partial\Omega} = \psi, & -\frac{\partial u}{\partial \nu}|_{\partial\Omega} = \varphi. \end{cases} \tag{4.98}$$

If $G = G_{\Delta^2,\Omega}$ is the Green function for this boundary value problem, then the solution of (4.98) is written as

$$u(x) = \int_\Omega G(x,y)\, f(y)\, dy + \int_{\partial\Omega} K(x,y)\, \psi(y)\, d\omega_y + \int_{\partial\Omega} L(x,y)\, \varphi(y)\, d\omega_y,$$

with $K, L : \overline{\Omega} \times \partial\Omega \to \mathbb{R}$ defined by

$$K(x,y) = \frac{\partial}{\partial \nu_y} \Delta_y G(x,y),$$

$$L(x,y) = \Delta_y G(x,y).$$

Theorem 4.29. *Let $(\Omega_k)_{k\in\mathbb{N}}$ be as in Theorem 4.28 and let K_{Ω_k} and L_{Ω_k} be the corresponding Poisson kernels. Then there exists $C = C\left((\Omega_k)_{k\in\mathbb{N}}\right)$ such that for all $(x,y) \in \Omega \times \partial\Omega$:*

$$\left| K_{\Omega_k}(x,y) \right| \le C \frac{d(x)^2}{|x-y|^{n+1}} \quad \text{and} \quad \left| L_{\Omega_k}(x,y) \right| \le C \frac{d(x)^2}{|x-y|^n}.$$

For $n = 2$ one obtains $\left| L_{\Omega_k}(x,y) \right| \le C$.

4.5.3 Convergence of the Green Function in Domain Approximations

Proposition 4.30. *Let $x_k \in \Omega_k$ and assume that $\lim_{k\to\infty} x_k = x_\infty \in \Omega$. Then we have*

$$G_k(x_k, .) \to G(x_\infty, .) \text{ in } C^4_{loc}(\Omega \setminus \{x_\infty\}),$$
$$G_k(x_k, .) \to G(x_\infty, .) \text{ in } L^1(\mathbb{R}^n),$$
$$G_k(x_k, .) \circ \Psi_k \to G(x_\infty, .) \text{ in } C^4_{loc}(\overline{\Omega} \setminus \{x_\infty\}).$$

If $n = 3$ we have in addition that

$$G_k(.,.) \to G(.,.) \text{ in } C^0_{loc}(\Omega \times \Omega).$$

Proof. According to Theorem 4.20 we know that

$$|G_k(x,y)| \le C \begin{cases} |x-y|^{4-n} & \text{if } n > 4, \\ \log\left(1 + |x-y|^{-1}\right) & \text{if } n = 4, \\ 1 & \text{if } n = 3; \end{cases} \tag{4.99}$$

uniformly in k. This shows that in particular

$$\|G_k(x, \cdot)\|_{L^1(\Omega_k)} \le C \text{ uniformly in } x \text{ and } k.$$

Moreover, since $x_k \to x_\infty$, we may assume that all x_k are in a small neighbourhood around x_∞. Let $\Omega_0 \subset\subset \Omega$ be arbitrary; local Schauder estimates (see Theorem 2.19) show that $(G_k(x_k, \cdot))_{k \in \mathbb{N}}$ is locally bounded in $C^4_{loc}\left(\overline{\Omega_0} \setminus \{x_\infty\}\right)$. Hence, after selecting a suitable subsequence we see that for each such $\Omega_0 \subset\subset \Omega$ one has $G_k(x_k, \cdot) \to \varphi$ in $C^4_{loc}\left(\overline{\Omega_0} \setminus \{x_\infty\}\right)$ and $G_k(x_k, \cdot) \circ \Psi_k \to \varphi$ in $C^4_{loc}(\overline{\Omega} \setminus \{x_\infty\})$ with a suitable $\varphi \in C^{4,\gamma}(\overline{\Omega} \setminus \{x_\infty\})$. Thanks to this compactness and the fact that in any case the limit is the uniquely determined Green function, we have convergence on the whole sequence towards $G(x_\infty, \cdot)$.

Finally, since we have pointwise convergence, (4.99) allows for applying Vitali's convergence theorem to show that

$$G_k(x_k, \cdot) \to G(x_\infty, \cdot) \text{ in } L^1(\mathbb{R}^n).$$

The statement concerning $C^0_{loc}(\Omega \times \Omega)$-convergence in $n = 3$ is a consequence of $|\nabla G_k(\cdot, \cdot)| \le C$, see (4.60). $\square$

In order to have enough smoothness to conclude also for the last case in Theorem 6.30 we also need a convergence result simultaneous in both variables.

Proposition 4.31. *We have that*

$$G_k(\cdot, \cdot) \circ (\Psi_k \times \Psi_k) \to G(\cdot, \cdot) \text{ in } C^4_{loc}\left(\overline{\Omega} \times \overline{\Omega} \setminus \{(x,x) ; x \in \overline{\Omega}\}\right).$$

Proof. We combine the ideas of the proofs of Propositions 4.30 and 4.17. One should observe that Theorem 4.20 guarantees uniform L^1-bounds for G_k as in the proof of Proposition 4.30. $\square$

4.6 Weighted Estimates for the Dirichlet Problem

As a side result the estimates in the previous section for the homogeneous biharmonic Dirichlet problem allow weighted L^p-L^q estimates for elliptic boundary value problems under homogeneous boundary conditions with $d(\cdot)^\theta$ as weight function. Here we will restrict ourselves to the biharmonic case. A more general version of this theorem is found in [118].

If u and f are such that

$$\begin{cases} \Delta^2 u = f & \text{in } \Omega, \\ u = u_\nu = 0 & \text{on } \partial\Omega, \end{cases} \tag{4.100}$$

then we have

Theorem 4.32. *Let Ω be a bounded $C^{4,\gamma}$-smooth domain and let $u \in C^4\left(\overline{\Omega}\right)$, $f \in C\left(\overline{\Omega}\right)$ be as in (4.100). Then the following hold:*

1. For $n < 4$ there exists $C = C(\Omega)$ such that for all $\theta \in [0,1]$:

$$\left\| d\,(.)^{\theta n - 2}\, u \right\|_{L^\infty(\Omega)} \leq C \left\| d(.)^{2-(1-\theta)n} f \right\|_{L^1(\Omega)}. \tag{4.101}$$

2. For all $n \geq 2$ if $p,q \in [1,\infty]$ are such that $0 \leq \frac{1}{p} - \frac{1}{q} < \alpha \leq \min\left\{1, \frac{4}{n}\right\}$, then there exists $C = C(\Omega, \alpha)$ such that for all $\theta \in [0,1]$:

$$\left\| d\,(.)^{\theta n \alpha - 2}\, u \right\|_{L^q(\Omega)} \leq C \left\| d(.)^{2-(1-\theta)n\alpha} f \right\|_{L^p(\Omega)}. \tag{4.102}$$

Remark 4.33. Notice that the shift in the exponent of $d\,(.)$ in (4.102) is $4 - n\alpha$ with $\alpha > 0$. If $p = q$ this shift can be arbitrarily close to 4 but will not reach 4.

Before proving this theorem we recall an estimate involving the Riesz potential

$$(K_\gamma * f)\,(x) := \int_\Omega |x - y|^{-\gamma} f(y) dy.$$

We prove a classical convolution estimate which can e.g. be found in [232, Corollary 4.5.2].

Lemma 4.34. *Let $\Omega \subset \mathbb{R}^n$ be bounded, $\gamma \in (0,n)$ and $1 \leq p,q \leq \infty$. If $\frac{\gamma}{n} < \frac{1}{r} :=$ $\min\left\{1, 1 + \frac{1}{q} - \frac{1}{p}\right\}$, then there exists $C = C(\text{diam}(\Omega), n - \gamma r) \in \mathbb{R}^+$ such that for all $f \in L^p(\Omega)$:*

$$\|K_\gamma * f\|_{L^q(\Omega)} \leq C \|f\|_{L^p(\Omega)}. \tag{4.103}$$

Proof. We let $p' \in [1,\infty]$ denote the conjugate of $p \in [1,\infty]$: $\frac{1}{p} + \frac{1}{p'} = 1$ etc. Set

$$C_{\Omega,s} = \max_{x \in \Omega} \left\| |x - .|^{-s} \right\|_{L^1(\Omega)}$$

and notice that $C_{\Omega,s}$ is bounded for $s < n$.

If $q = 1$, then $r = 1$ and a change in the order of integration gives

$$\|K_\gamma * f\|_{L^1(\Omega)} \leq \int_\Omega \left(\int_\Omega |x - y|^{-\gamma} dx \right) |f(y)| dy$$
$$\leq C_{\Omega,\gamma r} \|f\|_{L^1(\Omega)} \leq C \|f\|_{L^p(\Omega)}. \tag{4.104}$$

If $p \geq q$, then $r = 1$ and

$$\|K_\gamma * f\|_{L^q(\Omega)}^q = \int_\Omega \left| \int_\Omega |x-y|^{-\gamma} f(y) dy \right|^q dx$$

$$\leq \int_\Omega \left(\int_\Omega |x-y|^{-\gamma} dy \right)^{\frac{q}{q'}} \left(\int_\Omega |x-y|^{-\gamma} |f(y)|^q dx \right) dy$$

$$\leq C_{\Omega,\gamma r}^{q/q'} \|K_\gamma * |f|^q\|_{L^1(\Omega)}$$

and one continues with (4.104).

One finds for $1 < p < q < \infty$, since $\frac{1}{r'} + \frac{1}{p'} + \frac{1}{q} = 1$, that

$$\left| (K_\gamma * f)(x) \right|$$

$$\leq \left(\int_\Omega |f(y)|^p dy \right)^{\frac{1}{r'}} \left(\int_\Omega |x-y|^{-r\gamma} dy \right)^{\frac{1}{p'}} \left(\int_\Omega |x-y|^{-r\gamma} |f(y)|^p dy \right)^{\frac{1}{q}}$$

$$\tag{4.105}$$

and, by changing the order of integration and using $pq/r' + p = q$, also

$$\|K_\gamma * f\|_{L^q(\Omega)}^q \leq \|f\|_{L^p(\Omega)}^{pq/r'} C_{\Omega,\gamma r}^{q/p'} \int_\Omega \left(\int_\Omega |x-y|^{-r\gamma} dx \right) |f(y)|^p dy$$

$$\leq C_{\Omega,\gamma r}^{1+q/p'} \|f\|_{L^p(\Omega)}^q. \tag{4.106}$$

For $q = \infty$ and has $r = p'$ and the proof reduces to

$$\left| (K_\gamma * f)(x) \right| \leq \left\| |x - \cdot|^{-\gamma} \right\|_{L^{p'}(\Omega)} \|f\|_{L^p(\Omega)} \leq C_{\Omega,\gamma r}^{1/r} \|f\|_{L^p(\Omega)}.$$

For $p = 1$ one replaces (4.105) by

$$\left| (K_\gamma * f)(x) \right| \leq \left(\int_\Omega |f(y)| dy \right)^{\frac{1}{q'}} \left(\int_\Omega |x-y|^{-q\gamma} |f(y)| dy \right)^{\frac{1}{q}}$$

and continues similar as in (4.106) by changing the order of integration. $\square$

Proof of Theorem 4.32. Again we use the notation (4.1). The estimate that we will use repeatedly is a consequence of Lemma 4.5:

$$\min \left\{ 1, \frac{d(x) d(y)}{|x-y|^2} \right\} \leq C \left(\frac{d(x) d(y)}{|x-y|^2} \right)^{1-s} \left(\frac{d(y)}{d(x)} \right)^{s(2\theta-1)} \quad \text{for all } s, \theta \in [0,1].$$

$$\tag{4.107}$$

To prove Item 1 in Theorem 4.32 we apply this estimate for $s = 1$ to find for $n < 4$

$$|G(x,y)| \leq Cd(x)^{2-\frac{1}{2}n}d(y)^{2-\frac{1}{2}n}\left(\frac{d(y)}{d(x)}\right)^{\frac{1}{2}n(2\theta-1)} = Cd(x)^{2-n\theta}d(y)^{2-n(1-\theta)}$$

and a straightforward integration shows (4.101).

For Item 2, we need Lemma 4.34. If $n > 4$ we use from Theorem 4.28 the estimate for the Green function itself, and (4.107) for $s = \frac{1}{4}n\alpha$, which is allowed since $0 \leq \frac{1}{4}n\alpha \leq 1$,

$$|G(x,y)| \leq C|x-y|^{4-n}\min\left\{1, \frac{d(x)d(y)}{|x-y|^2}\right\}^2$$

$$\leq C|x-y|^{4-n}\left(\frac{d(x)d(y)}{|x-y|^2}\right)^{2\left(1-\frac{1}{4}n\alpha\right)}\left(\frac{d(y)}{d(x)}\right)^{\frac{1}{2}n\alpha(2\theta-1)}$$

$$= C|x-y|^{n(\alpha-1)}d(x)^{2-\theta n\alpha}d(y)^{2-(1-\theta)n\alpha}.$$

For $n = 4$ it holds that

$$|G(x,y)| \leq C\log\left(1 + \frac{d(x)^2 d(y)^2}{|x-y|^2}\right) \leq C|x-y|^{-\varepsilon}\min\left\{1, \frac{d(x)d(y)}{|x-y|^2}\right\}^2$$

and we may continue as before if we choose ε small enough such that $\alpha - \frac{1}{4}\varepsilon > \frac{1}{p} - \frac{1}{q}$.

If $n = 3$, one has for $\alpha, \theta \in [0,1]$

$$|G(x,y)| \leq Cd(x)^{\frac{1}{2}}d(y)^{\frac{1}{2}}\min\left\{1, \frac{d(x)d(y)}{|x-y|^2}\right\}^{\frac{3}{2}}$$

$$\leq Cd(x)^{\frac{1}{2}}d(y)^{\frac{1}{2}}\left(\frac{d(x)d(y)}{|x-y|^2}\right)^{\frac{3}{2}(1-\alpha)}\left(\frac{d(y)}{d(x)}\right)^{\frac{3}{2}\alpha(2\theta-1)}$$

$$= C|x-y|^{-3(1-\alpha)}d(x)^{2-3\alpha\theta}d(y)^{2-3\alpha(1-\theta)}.$$

Similarly, if $n = 2$ one finds for $\alpha, \theta \in [0,1]$

$$|G(x,y)| \leq C|x-y|^{-2(1-\alpha)}d(x)^{2-2\alpha\theta}d(y)^{2-2\alpha(1-\theta)}.$$

We have found for all $n \neq 4$ (and with a minor change for $n = 4$) that

$$\left|d(x)^{\theta n\alpha-2}u(x)\right| \leq C\left|\int_{\Omega}|x-y|^{-n(1-\alpha)}d(y)^{2-(1-\theta)n\alpha}f(y)dy\right|.$$

By Lemma 4.34 we have

$$\left\| d(.)^{\theta n\alpha - 2} u \right\|_{L^q(\Omega)} \leq C \left\| d(.)^{2-(1-\theta)n\alpha} f \right\|_{L^p(\Omega)}$$

whenever $\alpha > \frac{1}{p} - \frac{1}{q}$ and with $\alpha \leq \frac{1}{4}n$ for $n \geq 4$ and $\alpha \leq 1$ for $n < 4$. $\qquad\square$

4.7 Bibliographical Notes

Characterisations of Green's functions like in Section 4.2.1 are well-known by now in the context of second order problems on arbitrary smooth domains. Estimates from above go back to works of Grüter-Widman [216, 412]. The two-sided sharp estimates for the Green function of the Dirichlet-Laplacian are due to Zhao in [420–422]. Hueber-Sieveking [234] and Cranston-Fabes-Zhao [114] proved bounds for the Green function for general second order operators based on Harnack inequalities. For the importance of so-called 3-G-theorems in the potential theory of Schrödinger operators and the link with stochastic processes we refer to [97]. A first place where the optimal two-sided estimates are listed is [385].

For the higher order problems considered here the situation is quite different, namely, no general maximum principles and in particular no general Harnack inequalities are available. In the polyharmonic situation the starting point is Boggio's explicit formula for the Green's function in balls [63] from 1905, see (2.65). His formula led to optimal two-sided estimates for the polyharmonic Green function in case of a ball and inspired the estimates for the absolute value of the Green function in general domains. The subsequent estimates and 3-G-theorems were developed by Grunau-Sweers [211]. For further classical material on polyharmonic operators we refer to the book of Nicolesco [324].

As for the biharmonic Steklov boundary value problem in Section 4.3, we follow Gazzola-Sweers [191]. Proposition 4.13 is taken from [214, Lemmas 3.1 and 3.2] and is based on previous estimates by Zhao [421, 422], see also [118, 385]. Some of the results in Section 4.3 can be obtained under the assumption that $\partial\Omega \in C^{1,1}$, see [191].

Estimates of Green's functions for general higher order elliptic operators are due to Krasovskiĭ [256, 257]. However, due the general situation considered there, high regularity was imposed on the boundary. Since we restrict ourselves to biharmonic Green's functions and for the reader's convenience, we give a more elementary derivation of such estimates which only need to refer to Agmon-Douglis-Nirenberg [5], i.e. to Section 2.5 of the present book. The actual estimates are based on Dall'Acqua-Sweers and Grunau-Robert [118, 208]. For generalisations of Green function estimates to nonsmooth domains see also Mayboroda-Maz'ya [287].

Chapter 5
Positivity and Lower Order Perturbations

As already mentioned in Section 1.2, in general one does not have positivity preserving for higher order Dirichlet problems. Nevertheless, in Chapter 6 we shall identify some families of domains where the biharmonic — or more generally the polyharmonic — Dirichlet problem enjoys a positivity preserving property. Moreover, there we shall prove "almost positivity" for the biharmonic Dirichlet problem in any bounded smooth domain $\Omega \subset \mathbb{R}^n$.

As an intermediate step, taking advantage of the kernel estimates proved in Chapter 4, we study lower order perturbations of the prototype $((-\Delta)^m, B \subset \mathbb{R}^n)$ where B is again the unit ball. In Theorem 5.1 we prove positivity for Dirichlet problems

$$\begin{cases} Lu := (-\Delta)^m u + \displaystyle\sum_{|\beta| \leq 2m-1} a_\beta(x) D^\beta u = f \text{ in } B, \\ D^\alpha u|_{\partial B} = 0 \qquad\qquad\qquad\qquad\quad \text{for } |\alpha| \leq m-1, \end{cases} \tag{5.1}$$

with "small" coefficients a_β. Its proof is based on Green's function estimates, estimates for iterated Green's functions via the 3-G-theorem 4.9, and Neumann series. With the help of Riemann's theorem on conformal mappings and a reduction to normal form, this result will be used to prove the more general Theorem 6.3 where this approach permits to consider also highest-order perturbations in two dimensions.

If we remove in (5.1) the smallness assumptions on the coefficients a_β, we are still able to prove a *local* maximum principle for differential inequalities, which is true also in arbitrary domains Ω, see Theorem 5.19.

In the same spirit we study in Section 5.4 positivity preserving for the Steklov boundary value problem

$$\begin{cases} \Delta^2 u = f & \text{in } \Omega, \\ u = \Delta u - au_\nu = 0 & \text{on } \partial\Omega, \end{cases}$$

with data $a \in C^0(\partial\Omega)$ and suitable $f \geq 0$. It turns out that when a is below the corresponding positive first Steklov eigenvalue (see (3.40)) and above a negative critical parameter, one has positivity preserving, i.e. $f \geq 0 \Rightarrow u \geq 0$. This critical parameter

F. Gazzola et al., *Polyharmonic Boundary Value Problems*,
Lecture Notes in Mathematics 1991, DOI 10.1007/978-3-642-12245-3_5,

may also be $-\infty$. This issue is somehow related to positivity in the corresponding Dirichlet problem. As an application we will see that in a *convex* planar domain the hinged plate described by (1.10) satisfies the positivity preserving property, namely upwards pushing yields upwards bending.

It is another interesting question to ask which is the role of nontrivial Dirichlet boundary data with regard to the positivity of the solution. We look first at the inhomogeneous problem for the clamped plate equation:

$$\begin{cases} \Delta^2 u = 0 & \text{in } \Omega, \\[2ex] u|_{\partial\Omega} = \psi, & -\dfrac{\partial u}{\partial v}\Big|_{\partial\Omega} = \varphi. \end{cases} \tag{5.2}$$

One could think that, at least in the unit ball B, nonnegative data $\psi \geq 0$, $\varphi \geq 0$ yield a nonnegative solution $u \geq 0$. Actually, for B this is true with respect to φ in any dimension and with respect to ψ if $n \leq 4$. But for $n \geq 5$, the corresponding integral kernel changes sign! This issue will be discussed in detail in Section 5.2. A perturbation theory of positivity, allowing also for highest order perturbations, can be developed also with respect to φ in the special case $\psi = 0$. This can be generalised to equations of arbitrary order, see Theorems 5.6 and 5.7 and Remark 6.8. With respect to ψ we can prove only a rather restricted perturbation result, see Theorem 5.15.

5.1 A Positivity Result for Dirichlet Problems in the Ball

In order to avoid unnecessarily strong assumptions on the coefficients, a reasonable framework for the positivity results is L^p-theory. For existence and regularity we refer to Chapter 2. The next statement should be compared with Corollary 5.5 below for the necessity of the smallness assumptions on the coefficients.

Theorem 5.1. *There exists $\varepsilon_0 = \varepsilon_0(m,n) > 0$ such that if the $a_\beta \in C^0(\overline{B})$ satisfy the smallness condition $\|a_\beta\|_{C^0(\overline{B})} \leq \varepsilon_0$, $|\beta| \leq 2m - 1$, then for every $f \in L^p(B)$, $1 < p < \infty$, there exists a solution $u \in W^{2m,p} \cap W_0^{m,p}(B)$ of the Dirichlet problem (5.1). Moreover, if $f \gneq 0$ then the solution is strictly positive,*

$$u > 0 \text{ in } B.$$

To explain the strategy of the proof, we rewrite the boundary value problem (5.1) as

$$\begin{cases} ((-\Delta)^m + \mathscr{A})u = f & \text{in } B, \\[2ex] D^\alpha u|_{\partial B} = 0 & \text{for } |\alpha| \leq m - 1, \end{cases} \tag{5.3}$$

where we put

$$\mathscr{A}u := \sum_{|\beta|\leq 2m-1} a_\beta(.)D^\beta u(.), \qquad a_\beta \in C^0(\overline{B}).$$

We recall from Section 4.2.1 the definition of the Green operator $\mathscr{G}_{m,n}$ for the boundary value problem (5.3) with $\mathscr{A}=0$. In order to prove Theorem 5.1, we write the solution of (5.3) in the form

$$u = (\mathscr{I} + \mathscr{G}_{m,n}\mathscr{A})^{-1}\mathscr{G}_{m,n}f,$$

where $\mathscr{I}$ is the identity operator, and we shall estimate

$$(\mathscr{I} + \mathscr{G}_{m,n}\mathscr{A})^{-1}\mathscr{G}_{m,n} \geq \frac{1}{C}\mathscr{G}_{m,n}.$$

For this purpose we introduce the following notation.

Definition 5.2. For two operators $\mathscr{S}, \mathscr{T} : L^p(B) \to L^p(B)$ we write

$$\mathscr{S} \geq \mathscr{T},$$

if for all $f \in L^p(B)$ satisfying $f \geq 0$, one has that $\mathscr{S}f \geq \mathscr{T}f$.

Lemma 5.3. *Let* $1 < p < \infty$. *Then* $\mathscr{G}_{m,n}\mathscr{A} : W^{2m,p} \cap W_0^{m,p}(B) \to W^{2m,p} \cap W_0^{m,p}(B)$ *is a bounded linear operator. Furthermore, there exists* $\varepsilon_1 = \varepsilon_1(m,n) > 0$ *such that the following holds true.*

Assume that $\|a_\beta\|_{C^0(\overline{B})} \leq \varepsilon_1$ *for all* $|\beta| \leq 2m-1$. *Then* $\mathscr{I} + \mathscr{G}_{m,n}\mathscr{A} : W^{2m,p} \cap W_0^{m,p}(B) \to W^{2m,p} \cap W_0^{m,p}(B)$ *is boundedly invertible. For each* $f \in L^p(B)$ *the boundary value problem (5.3) has precisely one solution. This solution is given by*

$$u = (\mathscr{I} + \mathscr{G}_{m,n}\mathscr{A})^{-1}\mathscr{G}_{m,n}f. \tag{5.4}$$

The Green operator for the boundary value problem (5.3)

$$(\mathscr{I} + \mathscr{G}_{m,n}\mathscr{A})^{-1}\mathscr{G}_{m,n} : L^p(B) \to W^{2m,p} \cap W_0^{m,p}(B) \subset L^p(B)$$

is compact.

Proof. This is an immediate consequence of Corollary 2.21: $\mathscr{G}_{m,n} : L^p(B) \to W^{2m,p} \cap W_0^{m,p}(B)$ is a bounded linear operator. $\qquad\square$

The Green operator for (5.3) is studied with the help of a Neumann series. Theorem 5.1 then follows from the next result.

Theorem 5.4. *Assume that* $1 < p < \infty$. *There exists* $\varepsilon_0 = \varepsilon_0(m,n) > 0$ *such that if* $\|a_\beta\|_{C^0(\overline{B})} \leq \varepsilon_0$ *for all* $|\beta| \leq 2m-1$, *then the Green operator*

$$\mathscr{G}_{m,n,\mathscr{A}} := (\mathscr{I} + \mathscr{G}_{m,n}\mathscr{A})^{-1}\mathscr{G}_{m,n} : L^p(B) \to W^{2m,p} \cap W_0^{m,p}(B)$$

for the boundary value problem (5.3) exists. The corresponding Green function $G_{m,n,\mathscr{A}} : \overline{B} \times \overline{B} \to \mathbb{R} \cup \{\infty\}$ *is defined according to:*

$$(\mathscr{G}_{m,n,\mathscr{A}} f)(x) = \int_B G_{m,n,\mathscr{A}}(x,y) f(y)\, dy$$

and satisfies the following estimate with a constant $C = C(m,n) > 0$:

$$\frac{1}{C}\,\mathscr{G}_{m,n} \leq \mathscr{G}_{m,n,\mathscr{A}} \leq C\,\mathscr{G}_{m,n}. \tag{5.5}$$

On $\overline{B} \times \overline{B}$, *this reads:*

$$\frac{1}{C}\, G_{m,n}(x,y) \leq G_{m,n,\mathscr{A}}(x,y) \leq C\, G_{m,n}(x,y). \tag{5.6}$$

Proof. Let $\varepsilon := \max_{|\beta| \leq 2m-1} \|a_\beta\|_{C^0(\overline{B})}$. If $\varepsilon \leq \varepsilon_1$, then according to Lemma 5.3, $\mathscr{G}_{m,n,\mathscr{A}}$ exists and enjoys the properties listed there. If we choose moreover ε small enough such that $\|\mathscr{G}_{m,n}\mathscr{A}\| < 1$ in the sense of bounded linear operators in $W^{2m,p} \cap W_0^{m,p}(B)$, we may use a Neumann series to see that for all $f \in L^p(B)$

$$\mathscr{G}_{m,n,\mathscr{A}} f = (\mathscr{I} + \mathscr{G}_{m,n}\mathscr{A})^{-1} \mathscr{G}_{m,n} f = \sum_{i=0}^{\infty} (-1)^i (\mathscr{G}_{m,n}\mathscr{A})^i \mathscr{G}_{m,n} f.$$

Here, with the aid of the Fubini-Tonelli theorem and analogously to [198, Lemma 4.1], we conclude:

$$\begin{aligned}
\mathscr{G}^{(i)} f &:= (-1)^i (\mathscr{G}_{m,n}\mathscr{A})^i \mathscr{G}_{m,n} f \\
&= (-1)^i \int_B G_{m,n}(\,.\,,z_1) \mathscr{A}_{z_1} \int_B G_{m,n}(z_1,z_2) \\
&\quad\quad \times \mathscr{A}_{z_2} \ldots \mathscr{A}_{z_i} \int_B G_{m,n}(z_i,y) f(y)\, dy\, dz_i \ldots dz_1 \\
&= \int_B \Big\{ (-1)^i \int_B \cdots \int_B G_{m,n}(\,.\,,z_1)(\mathscr{A}_{z_1} G_{m,n}(z_1,z_2)) \ldots \\
&\quad\quad \ldots (\mathscr{A}_{z_i} G_{m,n}(z_i,y))\, d(z_1,\ldots,z_i) \Big\} f(y)\, dy \\
&=: \int_B G^{(i)}(\,.\,,y) f(y)\, dy.
\end{aligned}$$

We use the following version of the 3-G-theorem 4.9:

$$\int_B \frac{G_{m,n}(x,z)\,|\mathscr{A}_z G_{m,n}(z,y)|}{G_{m,n}(x,y)}\, dz \leq \varepsilon M < \infty$$

where $M = M(m,n) > 0$ is independent of ε and obtain:

$$
\begin{aligned}
\left| G^{(i)}(x,y) \right| &= \left| \int_B \cdots \int_B \frac{G_{m,n}(x,z_1)\left(\mathscr{A}_{z_1} G_{m,n}(z_1,z_2)\right)}{G_{m,n}(x,z_2)} \right. \\
&\quad \times \frac{G_{m,n}(x,z_2)\left(\mathscr{A}_{z_2} G_{m,n}(z_2,z_3)\right)}{G_{m,n}(x,z_3)} \cdots \\
&\quad \left. \cdots \frac{G_{m,n}(x,z_i)\left(\mathscr{A}_{z_i} G_{m,n}(z_i,y)\right)}{G_{m,n}(x,y)} G_{m,n}(x,y)\, d(z_1,\ldots,z_i) \right| \\
&\le G_{m,n}(x,y) \prod_{j=1}^{i} \sup_{\xi,\eta\in B} \int_B \frac{G_{m,n}(\xi,z_j)\left|\mathscr{A}_{z_j} G_{m,n}(z_j,\eta)\right|}{G_{m,n}(\xi,\eta)}\, dz_j \\
&\le (\varepsilon M)^i\, G_{m,n}(x,y).
\end{aligned}
$$

(5.7)

For $\varepsilon M < 1$, thanks to $\sum_{i=0}^{\infty}(\varepsilon M)^i = (1-\varepsilon M)^{-1} < \infty$, we have absolute local uniform convergence in $x \neq y$ of

$$
G_{m,n,\mathscr{A}}(x,y) := \sum_{i=0}^{\infty} G^{(i)}(x,y). \tag{5.8}
$$

It follows that

$$
\left| G_{m,n,\mathscr{A}}(x,y) \right| \le \frac{1}{1-\varepsilon M} G_{m,n}(x,y). \tag{5.9}
$$

On the other hand, Lebesgue's theorem yields

$$
\begin{aligned}
\left(\mathscr{G}_{m,n,\mathscr{A}} f\right)(x) &= \sum_{i=0}^{\infty}\left(\mathscr{G}^{(i)} f\right)(x) = \sum_{i=0}^{\infty} \int_B G^{(i)}(x,y) f(y)\, dy \\
&= \int_B \left(\sum_{i=0}^{\infty} G^{(i)}(x,y) f(y)\right) dy = \int_B G_{m,n,\mathscr{A}}(x,y) f(y)\, dy.
\end{aligned}
$$

Finally, thanks to (5.7) we have

$$
\begin{aligned}
G_{m,n,\mathscr{A}}(x,y) &= G_{m,n}(x,y) + \sum_{i=1}^{\infty} G^{(i)}(x,y) \\
&\ge G_{m,n}(x,y) - \left(\sum_{i=1}^{\infty}(\varepsilon M)^i\right) G_{m,n}(x,y) = \frac{1-2\varepsilon M}{1-\varepsilon M} G_{m,n}(x,y).
\end{aligned}
$$

Choosing $\varepsilon_0 \le 1/(4M)$ yields the crucial part of the estimate (5.6) from below for the Green function of the perturbed boundary value problem (5.3), provided $\varepsilon \in [0,\varepsilon_0]$. $\qquad\square$

If we confine ourselves to perturbations of order zero, we may show the necessity of the smallness conditions in Theorem 5.1. Consider the problem

$$
\begin{cases}
(-\Delta)^m u + a(x)u = f & \text{in } \Omega, \\
D^\alpha u|_{\partial\Omega} = 0 & \text{for } |\alpha| \le m-1,
\end{cases}
\tag{5.10}
$$

where $\Omega \subset \mathbb{R}^n$ is a bounded $C^{2m,\gamma}$-smooth domain. For coefficients $a \in C^0(\overline{\Omega})$, where we have uniqueness and hence existence in (5.10), let $\mathscr{G}_{m,\Omega,a}$ be the corresponding Green operator. In case of a constant coefficient a, $\mathscr{G}_{m,\Omega,a}$ is the resolvent operator. Recalling the meaning of the symbols

$$\phi > 0, \quad \phi \gneqq 0, \quad \phi \geqq 0,$$

in the Notations-Section, the positivity properties of Dirichlet problem (5.10) may be summarised as follows.

Corollary 5.5. *Let $m > 1$ and let $\Omega \subset \mathbb{R}^n$ be a bounded smooth domain such that the (Dirichlet-) Green function for $(-\Delta)^m$ is positive in $\Omega \times \Omega$. Let $\Lambda_{m,1}$ denote the first Dirichlet eigenvalue of $(-\Delta)^m$ in Ω.*

Then there exists a critical number $a_c \in [0,\infty)$ such that for $a \in C^0(\overline{\Omega})$ we have:

1. *If $a > a_c$ on $\overline{\Omega}$, then $\mathscr{G}_{m,\Omega,a}$ does not preserve positivity:*

$$\text{there exists } f \gneqq 0: \quad \mathscr{G}_{m,\Omega,a}f \ngeqq 0. \tag{5.11}$$

   On the other hand we have:

$$\text{for all } f \gneqq 0: \mathscr{G}_{m,\Omega,a}f \nleqq 0, \tag{5.12}$$
$$\text{there exists } f \gneqq 0: \mathscr{G}_{m,\Omega,a}f \geq 0. \tag{5.13}$$

2. *If $-\Lambda_{m,1} < a \leq a_c$ (or $-\Lambda_{m,1} < a < a_c$, respectively), then $\mathscr{G}_{m,\Omega,a}$ is positivity preserving (or strongly positivity preserving, respectively), that is,*

$$\text{for all } f \gneqq 0: \quad \mathscr{G}_{m,\Omega,a}f \geq 0 \ (\text{or } \mathscr{G}_{m,\Omega,a}f > 0 \text{ in } \Omega, \text{ respectively}). \tag{5.14}$$

3. *If $a = -\Lambda_{m,1}$ and $f \gneqq 0$, then (5.10) has no solution.*
4. *If $a < -\Lambda_{m,1}$, then (5.10) kills positivity, that is, if $f \gneqq 0$ and u is a solution to (5.10), then $u \ngeqq 0$ in Ω.*

For a proof, the strategy of which is related to – but simpler than – the arguments in Section 5.4, we refer to [211, Section 6] and [213, Lemma 1].

Case 1 does not occur in second order equations. This different behaviour may be responsible for the difficulties in classical solvability of semilinear boundary value problems of higher order, see e.g. [210, 396, 405, 406]. If $m > 1$, we have that for a large enough the resolvent is always sign changing, see e.g. Coffman-Grover [111]. As was noted e.g. by Bernis [51], this is equivalent to instantaneous change of sign for the corresponding parabolic heat kernel, see also related contributions to local eventual positivity by Ferrero-Gazzola-Grunau [164, 183, 184].

5.2 The Role of Positive Boundary Data

This section is devoted to the role of nonhomogeneous boundary data with regard to the sign of the solution. As already mentioned in the introduction this problem is rather subtle. In general we cannot expect that fixed sign of any particular Dirichlet datum leads to fixed sign of the solution. It seems that a perturbation theory of positivity (analogous to that above with regard to the right-hand side) exists in general only for the Dirichlet datum of highest order. As a dual result we obtain also a Hopf-type boundary lemma in the ball for perturbed polyharmonic Dirichlet problems.

With respect to lower order data, only much more restricted results can be achieved, see Theorem 5.15 below.

5.2.1 The Highest Order Dirichlet Datum

Here, we consider the following boundary value problem:

$$
\begin{cases}
((-\Delta)^m + \mathscr{A})\,u = f & \text{in } B, \\
D^\alpha u|_{\partial B} = 0 & \text{for } |\alpha| \leq m-2, \\[4pt]
\begin{cases}
-\dfrac{\partial}{\partial \nu}\Delta^{(m/2)-1}u|_{\partial B} = \varphi & \text{if } m \text{ is even,} \\[8pt]
\Delta^{(m-1)/2}u|_{\partial B} = \varphi & \text{if } m \text{ is odd.}
\end{cases}
\end{cases}
\tag{5.15}
$$

Here $f \in C^0(\overline{B})$, $\varphi \in C^0(\partial B)$ and

$$
\mathscr{A} = \sum_{|\beta| \leq 2m-1} a_\beta(.)D^\beta, \qquad a_\beta \in C^{|\beta|}(\overline{B}),
\tag{5.16}
$$

is a sufficiently small lower order perturbation. For existence of solutions $u \in W_{\text{loc}}^{2m,p}(B) \cap C^{m-1}(\overline{B})$, $p > 1$, we refer to the local L^p-theory in Theorem 2.20 and the lines following it and to the Agmon-Miranda maximum estimates of Theorem 2.25. The latter already require the strong regularity assumptions on the coefficients a_β. These have to be imposed whenever the adjoint operator $(-\Delta)^m + \mathscr{A}^*$ is involved.

Theorem 5.6. *There exists $\varepsilon_0 = \varepsilon_0(m,n) > 0$ such that the following holds true.*
If for all $|\beta| \leq 2m-1$ the smallness condition $\|a_\beta\|_{C^{|\beta|}(\overline{B})} \leq \varepsilon_0$ is fulfilled, then for every $f \in C^0(\overline{B})$ and $\varphi \in C^0(\partial B)$ there exists a solution $u \in W_{\text{loc}}^{2m,p}(B) \cap C^{m-1}(\overline{B})$, $1 < p < \infty$, to the Dirichlet problem (5.15). Moreover, $f \geq 0$ and $\varphi \geq 0$, with $f \not\equiv 0$ or $\varphi \not\equiv 0$, implies that $u > 0$.

If $m = 1$, we recover a special form of the strong maximum principle for second order elliptic equations. The next result, in some sense dual to the previous one, may be viewed as a higher order analogue to the Hopf boundary lemma.

Theorem 5.7. *Assume* $a_\beta \in C^0(\overline{B})$, $|\beta| \leq 2m-1$. *There exists* $\varepsilon_0 = \varepsilon_0(m,n) > 0$ *such that the following holds:*

If $\|a_\beta\|_{C^0(\overline{B})} \leq \varepsilon_0$, $|\beta| \leq 2m-1$, *then for every* $f \in C^0(\overline{B})$ *the Dirichlet problem* *(5.1) has a solution* $u \in W^{2m,p}(B) \cap C^{2m-1}(\overline{B})$, $p > 1$ *arbitrary. Moreover* $0 \not\equiv f \geq 0$ *implies* $u > 0$ *in* B *and for every* $x \in \partial B$

$$\begin{cases} \Delta^{(m/2)} u(x) > 0 & \text{if } m \text{ even,} \\ -\dfrac{\partial}{\partial \nu} \Delta^{(m-1)/2} u(x) > 0 & \text{if } m \text{ odd.} \end{cases} \tag{5.17}$$

The common key point in the proof of both theorems is the observation that the corresponding Green function $G_{m,n,\mathscr{A}}$ vanishes (in both variables) on ∂B precisely of order m, see Theorem 4.7 and estimate (5.6). In the proof of Theorem 5.6 we observe further that, for $x \in B$, $y \in \partial B$, the Poisson kernel for φ is given by

$$\begin{cases} \Delta_y^{m/2} G_{m,n,\mathscr{A}}(x,y) & \text{if } m \text{ even,} \\ \left(-\dfrac{\partial}{\partial \nu_y} \Delta_y^{(m-1)/2} \right) G_{m,n,\mathscr{A}}(x,y) & \text{if } m \text{ odd.} \end{cases}$$

In order to prove the theorems we need a precise characterisation of the growth properties near ∂B of the Green function $G_{m,n,\mathscr{A}}$ for the boundary value problem (5.15). These estimates were proved in a more general setting but under more restrictive assumptions on the coefficients by Krasovskiĭ [256, 257], see also Theorem 4.20. In the present special situation we provide an elementary proof which combines Theorems 4.6 and 5.4. Moreover, we need to verify suitable smoothness.

Lemma 5.8. *We assume that* $a_\beta \in C^0(\overline{B})$. *Then there exists* $\varepsilon_1 = \varepsilon_1(m,n) > 0$ *such that the following holds true.*

If $\|a_\beta\|_{C^0(\overline{B})} \leq \varepsilon_1$ *for all* $|\beta| \leq 2m-1$, *then the Green function* $G_{m,n,\mathscr{A}}(.,.)$ *for the boundary value problem (5.15) exists. For each* $y \in B$, $G_{m,n,\mathscr{A}}(.,y) \in C^{2m-1}(\overline{B} \setminus \{y\})$. *Furthermore, there exist constants* $C = C(m,n)$ *such that for* $|\beta| \leq 2m-1$, $x,y \in \overline{B}$

$$\begin{cases} G_{m,n,\mathscr{A}}(.,y) \in C^{|\beta|}(\overline{B}) & \text{if } 0 \leq |\beta| < 2m - n, \\[2mm] \left| D_x^\beta G_{m,n,\mathscr{A}}(x,y) \right| \leq C & \text{if } 0 \leq |\beta| < 2m - n, \\[2mm] \left| D_x^\beta G_{m,n,\mathscr{A}}(x,y) \right| \leq C \log \left(\dfrac{3}{|x-y|} \right) & \text{if } |\beta| = 2m - n \text{ and } n \text{ even,} \\[2mm] \left| D_x^\beta G_{m,n,\mathscr{A}}(x,y) \right| \leq C & \text{if } |\beta| = 2m - n \text{ and } n \text{ odd,} \\[2mm] \left| D_x^\beta G_{m,n,\mathscr{A}}(x,y) \right| \leq C |x-y|^{2m-n-|\beta|} & \text{if } 2m - n < |\beta| < 2m. \end{cases} \tag{5.18}$$

Moreover, one has $D_x^\beta G_{m,n,\mathscr{A}} \in C^0\left(\overline{B} \times \overline{B} \setminus \{(x,y) : x = y\} \right)$.

Proof. We come back to Theorem 5.4, making use of the notations and formulae in its proof. The following holds true, provided ε_1 is chosen sufficiently small.

The Green function $G_{m,n,\mathscr{A}}$ exists, and one has

$$G_{m,n,\mathscr{A}}(x,y) = \sum_{i=0}^{\infty} G^{(i)}(x,y),$$

where

$$G^{(0)}(x,y) = G_{m,n}(x,y),$$
$$G^{(i)}(x,y) = (-1)^i \int_B \cdots \int_B G_{m,n}(x,z_1)\,(\mathscr{A}_{z_1} G_{m,n}(z_1,z_2))$$
$$\times \ldots (\mathscr{A}_{z_i} G_{m,n}(z_i,y))\,d(z_1,\ldots,z_i).$$

In particular, we have $G^{(i)}(\,.\,,y) \in C^{2m-1}(\overline{B}\setminus\{y\})$, $G^{(i)}(\,.\,,y) \in C^{|\beta|}(\overline{B})$ for $0 \le |\beta| < 2m - n$. If $|\beta| \le 2m - 1$ and $i \ge 1$, one has with constants $C_j = C_j(m,n)$ which are independent of i

$$\left| D_x^\beta G^{(i)}(x,y) \right|$$
$$\le \int_B \cdots \int_B \left| D_x^\beta G_{m,n}(x,z_1) \right| \left| \mathscr{A}_{z_1} G_{m,n}(z_1,z_2) \right| \ldots \left| \mathscr{A}_{z_i} G_{m,n}(z_i,y) \right| d(z_1,\ldots,z_i)$$
$$\le \varepsilon_1^i\, C_1^{i+1} \int_B \cdots \int_B \Gamma(|x-z_1|)\,|z_1-z_2|^{1-n} \ldots |z_i-y|^{1-n} d(z_1,\ldots,z_i).$$

Here, in view of Theorem 4.7, we define

$$\Gamma(\rho) := \begin{cases} 1 & \text{if } 0 \le |\beta| < 2m - n, \\ \log\left(\dfrac{3}{\rho}\right) & \text{if } |\beta| = 2m - n \text{ and } n \text{ even}, \\ 1 & \text{if } |\beta| = 2m - n \text{ and } n \text{ odd}, \\ \rho^{2m-n-|\beta|} & \text{if } |\beta| > 2m - n. \end{cases}$$

Applying repeatedly $\int_B |\xi - z|^{1-n} |z - \eta|^{1-n} dz \le C_2 |\xi - \eta|^{1-n}$, we conclude:

$$\left| D_x^\beta G^{(i)}(x,y) \right| \le \varepsilon_1^i\, C_1^{i+1}\, C_2^{i-1} \int_B \Gamma(|x-z_1|)\,|z_1-y|^{1-n}\,dz_1$$
$$\le \varepsilon_1^i\, C_1^{i+1}\, C_2^{i-1} \begin{cases} C_2 & \text{if } |\beta| \le 2m - n, \\ C_2 |x-y|^{2m-n-|\beta|} & \text{if } 2m - n < |\beta| < 2m. \end{cases}$$

For sufficiently small $\varepsilon_1 > 0$ we achieve absolute uniform convergence of the series $\sum_{i=0}^{\infty} D_x^\beta G^{(i)}(\,.\,,y)$ in $\overline{B}$ if $|\beta| \le 2m - n$, and in $\overline{B}\setminus B_\delta(y)$ otherwise, where $\delta > 0$ is arbitrary. Taking the properties of $G^{(0)} = G_{m,n}$ into account we obtain the estimates for $D_x^\beta G_{m,n,\mathscr{A}}$ as well as the stated smoothness. $\qquad\square$

Lemma 5.9. *We assume that $a_\beta \in C^{|\beta|}(\overline{B})$. Then there exists $\varepsilon_2 = \varepsilon_2(m,n) > 0$ such that the following holds true.*

If $\|a_\beta\|_{C^{|\beta|}(\overline{B})} \le \varepsilon_2$ for all $|\beta| \le 2m - 1$, the Green function $G_{m,n,\mathscr{A}}(.,.)$ for the boundary value problem (5.15) exists. Moreover, for each $x \in B$ we have $G_{m,n,\mathscr{A}}(x,.) \in C^{2m-1}(\overline{B} \setminus \{x\})$. Furthermore, for $|\beta| \le 2m-1$ one has with constants $C = C(m,n)$ being independent of x,y:

$$
\begin{cases}
G_{m,n,\mathscr{A}}(x,.) \in C^{|\beta|}(\overline{B}) & \text{if } 0 \le |\beta| < 2m - n, \\[2mm]
\left| D_y^\beta G_{m,n,\mathscr{A}}(x,y) \right| \le C & \text{if } 0 \le |\beta| < 2m - n, \\[2mm]
\left| D_y^\beta G_{m,n,\mathscr{A}}(x,y) \right| \le C \log\left(\dfrac{3}{|x-y|} \right) & \text{if } |\beta| = 2m - n \text{ and } n \text{ even}, \\[2mm]
\left| D_y^\beta G_{m,n,\mathscr{A}}(x,y) \right| \le C & \text{if } |\beta| = 2m - n \text{ and } n \text{ odd}, \\[2mm]
\left| D_y^\beta G_{m,n,\mathscr{A}}(x,y) \right| \le C |x - y|^{2m-n-|\beta|} & \text{if } 2m - n < |\beta| < 2m.
\end{cases}
\tag{5.19}
$$

Moreover, $D_y^\beta G_{m,n,\mathscr{A}}$ is continuous outside the diagonal of $\overline{B} \times \overline{B}$.

Proof. Thanks to the strong differentiability assumptions on the coefficients a_β we may consider the adjoint boundary value problem

$$
\begin{cases}
(-\Delta)^m u + \mathscr{A}^* u = f & \text{in } B, \\[2mm]
D^\alpha u|_{\partial B} = 0 & \text{for } |\alpha| \le m - 1,
\end{cases}
$$

where $(\mathscr{A}^* u)(x) = \sum_{|\beta| \le 2m-1} (-1)^{|\beta|} D^\beta \left(a_\beta(x) u(x) \right)$. If ε_2 is small enough, the corresponding Green function $G_{m,n,\mathscr{A}^*}$ exists and satisfies $G_{m,n,\mathscr{A}}(x,y) = G_{m,n,\mathscr{A}^*}(y,x)$. This observation allows us to apply Lemma 5.8 and the claim follows. $\qquad\square$

Proof of Theorem 5.6. Let $\varepsilon_0 > 0$ be sufficiently small so that Theorem 5.4 and Lemma 5.9 are applicable.

The required smoothness of the Green function $G_{m,n,\mathscr{A}}$ has just been proved in Lemma 5.9. For solutions of the boundary value problem (5.15) we have the following representation formula:

$$
u(x) =
\begin{cases}
\displaystyle \int_B G_{m,n,\mathscr{A}}(x,y) f(y)\, dy + \int_{\partial B} \Delta_y^{m/2} G_{m,n,\mathscr{A}}(x,y) \varphi(y)\, d\omega(y) \\[4mm]
\hspace{8cm} \text{if } m \text{ even}, \\[4mm]
\displaystyle \int_B G_{m,n,\mathscr{A}}(x,y) f(y)\, dy + \int_{\partial B} \left(-\frac{\partial}{\partial \nu_y} \Delta_y^{(m-1)/2} \right) G_{m,n,\mathscr{A}}(x,y) \varphi(y)\, d\omega(y) \\[4mm]
\hspace{8cm} \text{if } m \text{ odd}.
\end{cases}
$$

We keep arbitrary $x \in B$ fixed and consider y "close" to ∂B. Then an application of Theorems 5.4 and 4.6 yields

$$
G_{m,n,\mathscr{A}}(x,y) \succeq G_{m,n}(x,y) \succeq |x-y|^{-n} d(x)^m d(y)^m \succeq d(y)^m.
$$

It follows for each $x \in B$ that

$$
\begin{cases}
\Delta_y^{m/2} G_{m,n,\mathscr{A}}(x, \cdot)|_{\partial B} > 0 & \text{for even } m, \\
-\dfrac{\partial}{\partial \nu_y} \Delta_y^{(m-1)/2} G_{m,n,\mathscr{A}}(x, \cdot)|_{\partial B} > 0 & \text{for odd } m.
\end{cases}
$$

Together with the positivity of $G_{m,n,\mathscr{A}}$, the claim of Theorem 5.6 is now obvious. $\square$

Proof of Theorem 5.7. This proof is "dual" to the previous one. Let $\varepsilon_0 > 0$ be sufficiently small. Differentiating the representation formula

$$
u(x) = \int_B G_{m,n,\mathscr{A}}(x,y) f(y)\, dy
$$

gives for $x \in \partial B$:

$$
\begin{cases}
\Delta^{m/2} u(x) = \displaystyle\int_B \left(\Delta_x^{m/2} G_{m,n,\mathscr{A}}(x,y) \right) f(y)\, dy & m \text{ even}, \\
-\dfrac{\partial}{\partial \nu} \Delta^{(m-1)/2} u(x) = \displaystyle\int_B \left(-\dfrac{\partial}{\partial \nu_x} \Delta_x^{(m-1)/2} G_{m,n,\mathscr{A}}(x,y) \right) f(y)\, dy & m \text{ odd}.
\end{cases}
$$

Keeping an arbitrary $y \in B$ fixed, we see that for $\tilde{x}$ "close" to ∂B

$$
G_{m,n,\mathscr{A}}(\tilde{x},y) \succeq G_{m,n}(\tilde{x},y) \succeq |\tilde{x} - y|^{-n} d(\tilde{x})^m d(y)^m \succeq d(\tilde{x})^m
$$

and consequently for $x \in \partial B$, $y \in B$

$$
\begin{cases}
\Delta_x^{m/2} G_{m,n,\mathscr{A}}(x,y) > 0 & \text{for even } m, \\
-\dfrac{\partial}{\partial \nu_x} \Delta_x^{(m-1)/2} G_{m,n,\mathscr{A}}(x,y) > 0 & \text{for odd } m.
\end{cases}
$$

Now it is immediate that $\Delta^{m/2} u(x) > 0$ or $-\frac{\partial}{\partial \nu} \Delta^{(m-1)/2} u(x) > 0$, according to whether m is even or odd. $\square$

5.2.2 Also Nonzero Lower Order Boundary Terms

Now we turn to investigating further conditions on $\varphi \geq 0$ and $\psi \geq 0$ such that the solution u of the Dirichlet problem

$$
\begin{cases}
\Delta^2 u = 0 & \text{in } B, \\
u|_{\partial B} = \psi, \quad \left(-\dfrac{\partial u}{\partial \nu} \right)\Big|_{\partial B} = \varphi,
\end{cases}
\tag{5.20}
$$

is positive, i.e. $u \geq 0$. We recall that we have the following explicit formula for the solution u of (5.20), see [324, p.34]:

$$u(x) = \int_{\partial B} K_{2,n}(x,y)\,\psi(y)\,d\omega(y) + \int_{\partial B} L_{2,n}(x,y)\,\varphi(y)\,d\omega(y), \quad x \in B, \qquad (5.21)$$

where

$$K_{2,n}(x,y) = \frac{1}{2ne_n} \frac{(1-|x|^2)^2}{|x-y|^{n+2}} \left(2 + (n-4)x\cdot y - (n-2)|x|^2\right), \qquad (5.22)$$

$$L_{2,n}(x,y) = \frac{1}{2ne_n} \frac{(1-|x|^2)^2}{|x-y|^n}, \qquad (5.23)$$

with $x \in B$, $y \in \partial B$, and $ne_n = |\partial B|$. Evidently $L_{2,n} > 0$ for any n, while $K_{2,n} > 0$ only for $n \leq 4$ and $K_{2,n}$ changes sign for $n \geq 5$.

We will show that the Dirichlet problem (5.20) may be reformulated in such a way that we have a positivity result with respect to both boundary data in any dimension. Moreover for $n \leq 3$ and in particular for $n = 2$ the above mentioned result may be sharpened so that if $\psi(x_0) > 0$ for some $x_0 \in \partial B$. Also negative values for φ near x_0 are admissible.

Finally, we switch to polyharmonic Dirichlet problems of arbitrary order $2m$. We will admit some "small" lower order perturbations of the differential operator. Positivity with respect to the Dirichlet data of order $(m-1)$ and $(m-2)$ will be shown in any dimension n, provided the other boundary data are prescribed homogeneously and the positivity assumption is posed in a suitable way.

5.2.2.1 The Appropriate Positivity Assumption for the Clamped Plate Equation

In order to find the adequate positivity assumption on the boundary data in the Dirichlet problem (5.20), one may observe that adding a suitable multiple of $L_{2,n}$ to $K_{2,n}$ yields a positive kernel.

Lemma 5.10. *Let $s \in \mathbb{R}$, $s \geq \frac{1}{2}(n-4)$. Then for*

$$\hat{K}_{2,n,s}(x,y) := K_{2,n}(x,y) + sL_{2,n}(x,y), \quad x \in B, \quad y \in \partial B, \qquad (5.24)$$

we have

$$\hat{K}_{2,n,s}(x,y) > 0.$$

Proof. We observe that for $x \in B$, $y \in \partial B$ (i.e. $|y| = 1$) we have

$$K_{2,n}(x,y) = \frac{1}{2ne_n} \frac{(1-|x|^2)^2}{|x-y|^{n+2}} \left(\frac{n}{2}(1-|x|^2) - \frac{1}{2}(n-4)|x-y|^2 \right)$$

$$= \frac{1}{4e_n} \frac{(1-|x|^2)^3}{|x-y|^{n+2}} - \frac{1}{2}(n-4)L_{2,n}(x,y), \quad ne_n = |\partial B|.$$

$\square$

Proposition 5.11. *Let* $\varphi \in C^0(\partial B)$, $\psi \in C^1(\partial B)$ *and* $s \geq \frac{1}{2}(n-4)$. *If we assume that*

$$\psi(x) \geq 0 \quad and \quad \varphi(x) \geq s\psi(x) \quad for\ x \in \partial B,$$

then the uniquely determined solution $u \in C^4(B) \cap C^1(\overline{B})$ *of the Dirichlet problem (5.20) is positive:*

$$u \geq 0 \quad in\ B.$$

Proof. From (5.21) and (5.24) we obtain:

$$u(x) = \int_{\partial B} K_{2,n}(x,y)\psi(y)\,d\omega(y) + \int_{\partial B} L_{2,n}(x,y)\varphi(y)\,d\omega(y)$$

$$= \int_{\partial B} \hat{K}_{2,n,s}(x,y)\psi(y)\,d\omega(y) + \int_{\partial B} L_{2,n}(x,y)\Big(\varphi(y) - s\,\psi(y) \Big)\,d\omega(y).$$

$\square$

One may observe that for $n = 1,2,3$ also negative values for s are admissible. On $B_R(0)$ the condition on s is $s \geq \frac{1}{2R}(n-4)$.

We are interested in whether this positivity result remains under perturbations of the prototype problem (5.20). Since in higher order Dirichlet problems quite similar phenomena can be observed, we develop the perturbation theory for the biharmonic Dirichlet problem (5.20) as a special case of the perturbation theory for the polyharmonic Dirichlet problem (5.25) below.

5.2.2.2 Higher Order Equations. Perturbations

In this section we assume $m \geq 2$.

First we consider the polyharmonic prototype Dirichlet problem:

$$\begin{cases} (-\Delta)^m u = 0 & \text{in } B, \\[2mm] \left(-\dfrac{\partial}{\partial \nu} \right)^j u = 0 & \text{on } \partial B \quad \text{for } j = 0,\ldots,m-3, \\[2mm] \left(-\dfrac{\partial}{\partial \nu} \right)^{m-2} u = \psi & \text{on } \partial B, \\[2mm] \left(-\dfrac{\partial}{\partial \nu} \right)^{m-1} u = \varphi & \text{on } \partial B. \end{cases} \tag{5.25}$$

No uniform positivity result can be expected with respect to the boundary data of order $0,\ldots,m-3$, as we will explain below in Example 5.14. So, these data are prescribed homogeneously. Such behaviour is in contrast with the radially symmetric case $u = u(|x|)$, where $(-\Delta)^m u \geq 0$ in B, $(-1)^j u^{(j)}(1) \geq 0$ $(j = 0,\ldots,m-1)$ implies that $u \geq 0$ in B, see Soranzo [376, Proposition 1, Remark 9].

After some elementary calculations we find from Boggio's formula (2.65) (see also [158]) that for $\varphi \in C^0(\partial B)$, $\psi \in C^1(\partial B)$ the solution $u \in C^{2m}(B) \cap C^{m-1}(\overline{B})$ to the Dirichlet problem (5.25) is given by

$$u(x) = \int_{\partial B} K_{m,n}(x,y)\psi(y)\,d\omega(y) + \int_{\partial B} L_{m,n}(x,y)\varphi(y)\,d\omega(y), \quad x \in B. \quad (5.26)$$

Here, the Poisson kernels are defined by

$$K_{m,n}(x,y) = \frac{1}{2^m(m-2)!\,n e_n}\frac{(1-|x|^2)^m}{|x-y|^{n+2}}\left(n(1-|x|^2) - (n-2-m)|x-y|^2\right), \quad (5.27)$$

$$L_{m,n}(x,y) = \frac{1}{2^{m-1}(m-1)!\,n e_n}\frac{(1-|x|^2)^m}{|x-y|^n}, \quad (5.28)$$

with $x \in B$, $y \in \partial B$. The following result generalises Lemma 5.10.

Lemma 5.12. *Let $s \in \mathbb{R}$ satisfy $s \geq \frac{1}{2}(n-2-m)(m-1)$. Then for*

$$\hat{K}_{m,n,s}(x,y) := K_{m,n}(x,y) + sL_{m,n}(x,y), \quad x \in B, \quad y \in \partial B, \quad (5.29)$$

we have

$$\hat{K}_{m,n,s}(x,y) > 0.$$

Proof.

$$\hat{K}_{m,n,s}(x,y) = \frac{1}{2^m(m-2)!\,n e_n}\frac{(1-|x|^2)^m}{|x-y|^{n+2}}$$
$$\times \left(n(1-|x|^2) + \left(\frac{2s}{m-1} - (n-2-m)\right)|x-y|^2\right). \quad (5.30)$$

$\square$

Proposition 5.13. *Let $\varphi \in C^0(\partial B)$, $\psi \in C^1(\partial B)$ and $s \geq \frac{1}{2}(n-2-m)(m-1)$. If*

$$\psi(x) \geq 0 \quad \text{and} \quad \varphi(x) \geq s\psi(x) \quad \text{for } x \in \partial B,$$

then the uniquely determined solution $u \in C^{2m}(B) \cap C^{m-1}(\overline{B})$ of the Dirichlet problem (5.25) is positive:

$$u \geq 0 \quad \text{in } B.$$

Example 5.14. In the triharmonic Dirichlet problem

$$\begin{cases} (-\Delta)^3 u = 0 & \text{in } B, \\[2mm] u = \chi & \text{on } \partial B, \\[2mm] \left(-\dfrac{\partial}{\partial v}\right) u = \psi & \text{on } \partial B, \\[2mm] \left(\dfrac{\partial}{\partial v}\right)^2 u = \varphi & \text{on } \partial B, \end{cases}$$

the solution is given by

$$u(x) = \int_{\partial B} H_{3,n}(x,y)\chi(y)\,d\omega(y) + \int_{\partial B} K_{3,n}(x,y)\psi(y)\,d\omega(y)$$
$$+ \int_{\partial B} L_{3,n}(x,y)\varphi(y)\,d\omega(y), \qquad x \in B.$$

The kernels $K_{3,n}$ and $L_{3,n}$ are defined above and

$$H_{3,n}(x,y) = \frac{1}{16ne_n}\frac{(1-|x|^2)^3}{|x-y|^{n+4}}\left(n(n+2)(1-|x|^2)^2 + (n-4)(n-8)|x-y|^4 \right.$$
$$\left. -2n(n-7)(1-|x|^2)|x-y|^2 - 4n|x-y|^2\right)$$

with $x \in B$, $y \in \partial B$. For any n, $x \to y$, x "very close" to the boundary, $H_{3,n}$ takes on also negative values. By adding multiples of $L_{3,n}$ and $K_{3,n}$, only the terms $|x-y|^4$ and $(1-|x|^2)|x-y|^2$ in the curved brackets could be effected. In any case the most dangerous term $-4n|x-y|^2$ remains.

Theorem 5.15. *Let $s > \frac{1}{2}(n-2-m)(m-1)$. Then there exists $\varepsilon_0 = \varepsilon_0(m,n,s) > 0$ such that the following holds.*

If $\|a_\beta\|_{C^{|\beta|}(\overline{B})} \le \varepsilon_0$ for $|\beta| \le 2m-2$, then for every $\varphi \in C^0(\partial B)$ and $\psi \in C^1(\partial B)$ with

$$\left. \begin{array}{l} \psi \ge 0 \\[4mm] \varphi \ge s\psi \end{array} \right\} \text{ on } \partial B, \ \psi \not\equiv 0 \text{ or } \varphi \not\equiv 0,$$

the Dirichlet problem

$$\begin{cases} (-\Delta)^m u + \displaystyle\sum_{|\beta|\le 2m-2} a_\beta(x)D^\beta u = 0 & \text{in } B, \\[4mm] \left(-\dfrac{\partial}{\partial v}\right)^j u = 0 & \text{on } \partial B \text{ for } j = 0,\dots,m-3, \\[4mm] \left(-\dfrac{\partial}{\partial v}\right)^{m-2} u = \psi & \text{on } \partial B, \\[4mm] \left(-\dfrac{\partial}{\partial v}\right)^{m-1} u = \varphi & \text{on } \partial B, \end{cases} \qquad (5.31)$$

has a solution $u \in W_{\text{loc}}^{2m,p}(B) \cap C^{m-1}(\overline{B})$ ($p > 1$ arbitrary) which is strictly positive: $u > 0$ in B.

In order to prove this result we first need to describe the essential properties of the integral kernels $\hat{K}_{m,n,s}$ and $L_{m,n}$.

Lemma 5.16. *1. Let $s \geq \frac{1}{2}(n-2-m)(m-1)$. On $B \times \partial B$ (i.e. for $x \in B$, $y \in \partial B$) we have*

$$\hat{K}_{m,n,s}(x,y) \begin{cases} \preceq |x-y|^{-n-1}d(x)^m, \\ \succeq |x-y|^{-n-2}d(x)^{m+1}, \end{cases} \tag{5.32}$$

$$L_{m,n}(x,y) \simeq |x-y|^{-n}d(x)^m. \tag{5.33}$$

2. If we assume additionally that $s > \frac{1}{2}(n-2-m)(m-1)$, then we have on $B \times \partial B$:

$$\hat{K}_{m,n,s}(x,y) \begin{cases} \preceq |x-y|^{-n-1}d(x)^m, \\ \succeq |x-y|^{-n}d(x)^m. \end{cases} \tag{5.34}$$

Proof. The claim follows from $1 - |x|^2 \simeq d(x)$, $d(x) \leq |x-y|$ and the expression in (5.30). $\qquad\square$

Remark 5.17. 1. The estimation constants in (5.34) depend strongly on s.
2. If $s = \frac{1}{2}(n-2-m)(m-1)$ then we have $\hat{K}_{m,n,s}(x,y) \simeq |x-y|^{-n-2}d(x)^{m+1}$, i.e.
   for $x \to \partial B \setminus \{y\}$ we have a zero of order $(m+1)$. We would have expected,
   and actually need in order to prove perturbation results, a zero of order m. Consequently in what follows we have to assume $s > \frac{1}{2}(n-2-m)(m-1)$. The
   estimate (5.34) is more appropriate. But as $\hat{K}_{m,n,s}(x,y) \not\simeq |x-y|^{-n-1}d(x)^m$ our
   perturbation result Theorem 5.15 below is less general than the corresponding
   results in Theorems 5.1, 6.3 and 6.29. In particular, domain perturbations are not
   considered.

For our purposes the following "3-G-type" estimates are essential. We recall that $G_{m,n} = G_{(-\Delta)^m,B}$ denotes the Dirichlet Green function for $(-\Delta)^m$ in the unit ball $B \subset \mathbb{R}^n$.

Lemma 5.18. *Let $s > \frac{1}{2}(n-2-m)(m-1)$, $\beta \in \mathbb{N}^n$. Then on $B \times \partial B \times B$ (i.e. for $x \in B$, $y \in \partial B$, $z \in B$) we have the following:*

$$\frac{\left| D_z^\beta G_{m,n}(x,z) \right| \hat{K}_{m,n,s}(z,y)}{\hat{K}_{m,n,s}(x,y)}$$

$$\preceq \begin{cases} 1 & \text{if } |\beta| < 2m-n, \\ |x-z|^{2m-1-n-|\beta|} + |y-z|^{2m-1-n-|\beta|} & \text{if } |\beta| \geq 2m-n; \end{cases} \tag{5.35}$$

$$\frac{\left| D_z^\beta G_{m,n}(x,z) \right| L_{m,n}(z,y)}{L_{m,n}(x,y)}$$

$$\preceq \begin{cases} 1 & \text{if } |\beta| < 2m - n, \\[2ex] 1 & \text{if } |\beta| = 2m - n \text{ and } n \text{ odd}, \\[2ex] \log\left(\dfrac{3}{|x-z|}\right) & \text{if } |\beta| = 2m - n \text{ and } n \text{ even}, \\[2ex] |x-z|^{2m-n-|\beta|} + |y-z|^{2m-n-|\beta|} & \text{if } |\beta| > 2m - n. \end{cases} \qquad (5.36)$$

The proof is quite similar to that of Theorem 4.9 and is based on the Green's functions estimates of Theorems 4.6 and 4.7, Corollary 4.8 and the boundary kernel estimates of Lemma 5.16. For this reason we skip the proof here and refer to [212, Lemma 3.4].

Proof of Theorem 5.15. For existence and regularity we refer to Theorem 2.25. First, we assume additionally that $\psi \in C^{m+2,\gamma}(\partial B)$, $\varphi \in C^{m+1,\gamma}(\partial B)$. We write $\hat{\varphi}_s = \varphi - s\psi$ and let $p > 1$ be arbitrary. The operator

$$\mathscr{L}_{m,n}\hat{\varphi}_s(x) := \int_{\partial B} L_{m,n}(x,y)\hat{\varphi}_s(y)\,d\omega(y)$$

satisfies $\mathscr{L}_{m,n} : C^{m+1,\gamma}(\partial B) \to C^{2m,\gamma}(\overline{B}) \subset W^{2m,p}(B)$, the operator

$$\hat{\mathscr{K}}_{m,n,s}\psi(x) := \int_{\partial B} \hat{K}_{m,n,s}(x,y)\psi(y)\,d\omega(y)$$

satisfies $\hat{\mathscr{K}}_{m,n,s} : C^{m+2,\gamma}(\partial B) \to C^{2m,\gamma}(\overline{B}) \subset W^{2m,p}(B)$, while the Green operator

$$\mathscr{G}_{m,n}f(x) := \int_B G_{m,n}(x,y)f(y)\,dy$$

satisfies $\mathscr{G}_{m,n} : L^p(B) \to W^{2m,p} \cap W_0^{m,p}(B)$, see Theorem 2.19 and Corollary 2.21. We write $\mathscr{A} := \sum_{|\beta| \le 2m-2} a_\beta(.)D^\beta$. The solution of (5.31) is given by

$$u = -\mathscr{G}_{m,n}\mathscr{A}u + \hat{\mathscr{K}}_{m,n,s}\psi + \mathscr{L}_{m,n}\hat{\varphi}_s \quad \text{or} \quad (\mathscr{I} + \mathscr{G}_{m,n}\mathscr{A})u = \hat{\mathscr{K}}_{m,n,s}\psi + \mathscr{L}_{m,n}\hat{\varphi}_s.$$

Here, $\mathscr{I} + \mathscr{G}_{m,n}\mathscr{A}$ is a bounded linear operator in $W^{2m,p}(B)$ which for sufficiently small ε_0 is invertible. Hence

$$u = (\mathscr{I} + \mathscr{G}_{m,n}\mathscr{A})^{-1}\hat{\mathscr{K}}_{m,n,s}\psi + (\mathscr{I} + \mathscr{G}_{m,n}\mathscr{A})^{-1}\mathscr{L}_{m,n}\hat{\varphi}_s$$

$$= \hat{\mathscr{K}}_{m,n,s}\psi + \sum_{i=1}^\infty (-\mathscr{G}_{m,n}\mathscr{A})^i \hat{\mathscr{K}}_{m,n,s}\psi + \mathscr{L}_{m,n}\hat{\varphi}_s + \sum_{i=1}^\infty (-\mathscr{G}_{m,n}\mathscr{A})^i \mathscr{L}_{m,n}\hat{\varphi}_s.$$

We only show how to deal with the first Neumann series containing $\hat{\mathscr{K}}_{m,n,s}$, the second series containing $\mathscr{L}_{m,n}$ can be treated in the same way with some obvious simplifications. For $i \geq 1$ we integrate by parts. As $\mathscr{A}$ is of order $\leq 2m-2$ and $\hat{\mathscr{K}}_{m,n,s}\psi$ vanishes on ∂B of order $m-2$ no additional boundary integrals arise. By means of the Fubini-Tonelli theorem we obtain for $x \in B$

$$(-\mathscr{G}_{m,n}\mathscr{A})^i \, \hat{\mathscr{K}}_{m,n,s}\psi(x) = (-1)^i \int_{z_1 \in B} G_{m,n}(x,z_1)$$

$$\times \mathscr{A}_{z_1} \int_{z_2 \in B} G_{m,n}(z_1,z_2) \times \ldots \mathscr{A}_{z_{i-1}} \int_{z_i \in B} G_{m,n}(z_{i-1},z_i)$$

$$\times \mathscr{A}_{z_i} \int_{y \in \partial B} \hat{K}_{m,n,s}(z_i,y)\psi(y)\,d\omega(y)dz_i \ldots dz_1$$

$$= (-1)^i \int_{z_1 \in B} \left(\mathscr{A}_{z_1}^* G_{m,n}(x,z_1)\right) \int_{z_2 \in B} \left(\mathscr{A}_{z_2}^* G_{m,n}(z_1,z_2)\right)$$

$$\times \ldots \int_{z_i \in B} \left(\mathscr{A}_{z_i}^* G_{m,n}(z_{i-1},z_i)\right)$$

$$\times \int_{y \in \partial B} \hat{K}_{m,n,s}(z_i,y)\psi(y)d\omega(y)dz_i \ldots dz_1$$

$$= (-1)^i \int_B \cdots \int_B \int_{\partial B} \left(\mathscr{A}_{z_1}^* G_{m,n}(x,z_1)\right) \left(\mathscr{A}_{z_2}^* G_{m,n}(z_1,z_2)\right)$$

$$\times \ldots \left(\mathscr{A}_{z_i}^* G_{m,n}(z_{i-1},z_i)\right) \hat{K}_{m,n,s}(z_i,y)\psi(y)d\omega(y)d(z_1,\ldots,z_i).$$

Here $\mathscr{A}_{\cdot}^* = \sum_{|\beta| \leq 2m-2}(-1)^{|\beta|}D^\beta(a_\beta \cdot)$ is the (formally) adjoint operator of the perturbation $\mathscr{A}$. The estimates (5.35) and (5.36) in Lemma 5.18 are integrable with respect to $z \in B$ uniformly in $x \in B$, $y \in \partial B$ if $|\beta| \leq 2m-2$. They yield

$$\left|(-\mathscr{G}_{m,n}\mathscr{A})^i \, \hat{\mathscr{K}}_{m,n,s}\psi(x)\right| \leq \int_{\partial B}\int_B \cdots \int_B \hat{K}_{m,n,s}(x,y)\frac{\left|\mathscr{A}_{z_1}^* G_{m,n}(x,z_1)\right| \hat{K}_{m,n,s}(z_1,y)}{\hat{K}_{m,n,s}(x,y)}$$

$$\times \frac{\left|\mathscr{A}_{z_2}^* G_{m,n}(z_1,z_2)\right| \hat{K}_{m,n,s}(z_2,y)}{\hat{K}_{m,n,s}(z_1,y)}$$

$$\times \ldots \frac{\left|\mathscr{A}_{z_i}^* G_{m,n}(z_{i-1},z_i)\right| \hat{K}_{m,n,s}(z_i,y)}{\hat{K}_{m,n,s}(z_{i-1},y)}$$

$$\times \psi(y)\,d(z_1,\ldots,z_i)d\omega(y)$$

$$\leq (C_0\varepsilon_0)^i \int_{\partial B} \hat{K}_{m,n,s}(x,y)\psi(y)\,d\omega(y)$$

$$= (C_0\varepsilon_0)^i \left(\hat{\mathscr{K}}_{m,n,s}\psi\right)(x).$$

Analogously we have:

$$\left|(-\mathscr{G}_{m,n}\mathscr{A})^i \, \mathscr{L}_{m,n}\hat{\varphi}_s(x)\right| \leq \left(\hat{C}_0\varepsilon_0\right)^i (\mathscr{L}_{m,n}\hat{\varphi}_s)(x).$$

The constants $C_0 = C_0(m,n,s)$, $\hat{C}_0 = \hat{C}_0(m,n)$ do not depend on i.

If $\varepsilon_0 = \varepsilon_0(m,n,s) > 0$ is chosen sufficiently small, we come up with

$$u \geq \frac{1}{C}\hat{\mathscr{K}}_{m,n,s}\psi + \frac{1}{C}\mathscr{L}_{m,n}\hat{\varphi}_s.$$ (5.37)

The general case $\varphi \in C^0(\partial B)$, $\psi \in C^1(\partial B)$ follows from (5.37) with the help of an approximation, the maximum modulus estimates of Theorem 2.25 and local L^p-estimates, see Theorem 2.20. $\qquad\square$

5.3 Local Maximum Principles for Higher Order Differential Inequalities

The comparison results of Section 5.1 together with the observations of Section 5.2 on the Poisson boundary kernels will yield local maximum principles for differential inequalities, which are valid for a large class of operators. Here lower order perturbations are no longer subject to smallness restrictions.

We consider $C^{2m,\gamma}$-smooth domains $\Omega \subset \mathbb{R}^n$ and differential operators L like

$$Lu := \left(-\sum_{i,j=1}^{n} \tilde{a}_{ij}\frac{\partial^2}{\partial x_i \partial x_j}\right)^m u + \sum_{|\beta|\leq 2m-1} a_\beta(.)D^\beta u,$$ (5.38)

with constant highest order coefficients $\tilde{a}_{ij} = \tilde{a}_{ji}$ obeying the ellipticity condition

$$\lambda|\xi|^2 \leq \sum_{i,j=1}^{n} \tilde{a}_{ij}\xi_i\xi_j \leq \Lambda|\xi|^2 \qquad \text{for all } \xi \in \mathbb{R}^n.$$

The ellipticity constants are subject to the condition $0 < \lambda \leq \Lambda$. The lower order coefficients are merely assumed to be smooth,

$$a_\beta \in C^{|\beta|,\gamma}(\overline{\Omega}).$$

Under these assumptions, we have:

Theorem 5.19. *Assume that $q \geq 1$, $q > \frac{n}{2m}$ and that $K \subset \Omega$ is a compact subset. Then there exists a constant*

$$C = C\left(n,m,\lambda,\Lambda,q, \max_{|\beta|\leq 2m-1}\|a_\beta\|_{C^{|\beta|}(\overline{\Omega})}, \mathrm{dist}(K,\partial\Omega)\right)$$

such that, for every $f \in C^0(\overline{\Omega})$ and every subsolution $u \in C^{2m}(\overline{\Omega})$ of the differential inequality

$$Lu \leq f,$$

the following local maximum estimate holds true:

$$\sup_K u \leq C\left(\|f^+\|_{L^q} + \|u\|_{W^{m-1,1}}\right).$$ (5.39)

Proof. With the help of a linear transformation we may achieve $\tilde{a}_{ij} = \delta_{ij}$. So, in what follows we consider the principal part $(-\Delta)^m$.

We want to apply Theorem 5.4 and Lemma 5.9. Let $\varepsilon_0 = \varepsilon_0(m,n) > 0$ be such that both results hold true in the unit ball B for all differential operators $\tilde{L} = (-\Delta)^m + \sum_{|\beta| \leq 2m-1} \tilde{a}_\beta D^\beta$ with $\max_{|\beta| \leq 2m-1} \|\tilde{a}_\beta\|_{C^{|\beta|}(\overline{B})} \leq \varepsilon_0$. For the differential operator $L = (-\Delta)^m + \sum_{|\beta| \leq 2m-1} a_\beta D^\beta$ defined in Ω we want to achieve the required smallness by means of scaling.

Let $x_0 \in K$ be arbitrary, after translation we may assume $x_0 = 0$. We put

$$M := \max_{|\beta| \leq 2m-1} \|a_\beta\|_{C^{|\beta|}(\overline{\Omega})},$$

$$\rho_0 := \min\left\{ 1, \frac{1}{2}\operatorname{dist}(K, \partial\Omega), \frac{\varepsilon_0}{M} \right\}. \tag{5.40}$$

For $\rho \in (0, \rho_0]$ we introduce the following scaled functions $\overline{B} \to \mathbb{R}$:

$$u_\rho(x) := u(\rho x), \qquad f_\rho(x) := \rho^{2m} f(\rho x), \qquad a_{\beta,\rho}(x) := \rho^{2m-|\beta|} a_\beta(\rho x).$$

For these functions we have on $\overline{B}$ the following differential inequality

$$L_\rho u_\rho(x) := (-\Delta)^m u_\rho(x) + \sum_{|\beta| \leq 2m-1} a_{\beta,\rho}(x) D^\beta u_\rho(x) \leq f_\rho(x). \tag{5.41}$$

Here, thanks to our choice (5.40) of ρ_0, on $\overline{B}$ the coefficients $a_{\beta,\rho}$, $|\beta| \leq 2m-1$, are subject of the following smallness condition

$$\|a_{\beta,\rho}\|_{C^{|\beta|}(\overline{B})} = \sum_{|\mu| \leq |\beta|} \max_{x \in \overline{B}} |D^\mu a_{\beta,\rho}(x)| = \sum_{|\mu| \leq |\beta|} \max_{x \in \overline{B}} \left(\rho^{2m-|\beta|+|\mu|} |(D^\mu a_\beta)(\rho x)| \right)$$

$$\leq \rho_0 \|a_\beta\|_{C^{|\beta|}(\overline{\Omega})} \leq \rho_0 M \leq \varepsilon_0.$$

Let $G_{L_\rho,B}$ be the Green function for L_ρ in B. Theorem 5.4 and Lemma 5.9 show that there exist constants $C = C(m,n,\varepsilon_0(m,n)) = C(m,n)$, independent of $\rho \in (0,\rho_0]$, such that we have:

$$\begin{cases} G_{L_\rho,B}(x,y) > 0 & \text{in } B \times B, \\[2mm] G_{L_\rho,B}(x,y) \leq C|x-y|^{2m-n} & \text{in } B \times B \text{ if } n > 2m, \\[2mm] G_{L_\rho,B}(x,y) \leq C\log\left(\dfrac{3}{|x-y|}\right) & \text{in } B \times B \text{ if } n = 2m, \\[2mm] G_{L_\rho,B}(x,y) \leq C & \text{in } B \times B \text{ if } n < 2m, \\[2mm] \left| D_y^\beta G_{L_\rho,B}(0,y) \right| \leq C & \text{for } |\beta| \leq 2m-1, \; y \in \partial B. \end{cases} \tag{5.42}$$

To estimate $u(0) = u_\rho(0)$ we use the representation formula for u_ρ. Beside the Dirichlet data $D^\beta u_\rho$, $|\beta| \leq m-1$ and terms of the kind $D_y^\beta G_{L_\rho,B}(0,y)$, $m \leq |\beta| \leq 2m-1$, the boundary integrals contain factors $a_{\beta,\rho}$ and their derivatives up to order $\leq \max\{0, |\beta| - m - 1\}$. Making use of (5.42), we obtain independently of $\rho \in (0, \rho_0]$:

$$
\begin{aligned}
u(0) = u_\rho(0) &\leq \int_B G_{L_\rho,B}(0,y) f_\rho^+(y)\,dy \\
&\quad + C(m,n,M) \sum_{|\beta| \leq m-1} \int_{\partial B} \left| D^\beta u_\rho(y) \right| d\omega(y) \\
&\leq C(m,n,q)\|f_\rho^+\|_{L^q(B)} + C(m,n,M) \sum_{|\beta| \leq m-1} \rho^{|\beta|} \int_{\partial B} \left| \left(D^\beta u \right)(\rho y) \right| d\omega(y) \\
&\leq C(m,n,q)\rho^{2m-(n/q)}\|f^+\|_{L^q(B_\rho)} \\
&\quad + C(m,n,M) \sum_{|\beta| \leq m-1} \rho^{|\beta|-n+1} \int_{|y|=\rho} \left| D^\beta u(y) \right| d\omega(y).
\end{aligned}
$$

Integration with respect to $\rho \in [\tfrac{1}{2}\rho_0, \rho_0]$ yields

$$
u(0) \leq C \left(\|f^+\|_{L^q(\Omega)} + \|u\|_{W^{m-1,1}(\Omega)} \right)
$$

with a constant $C = C(m,n,q,M,\rho_0)$. Here $C = O(\rho_0^{-n})$ for $\rho_0 \searrow 0$. $\qquad\square$

Remark 5.20. This local maximum principle may also be applied to nonlinear problems which are not subject to the standard (controllable) growth conditions as in [405, 406], see [210]. For instance, one finds "almost" classical solutions $u \in C^{2m,\gamma}(\Omega) \cap H_0^m(\Omega)$ to the Dirichlet problem for $Lu = e^u$ where L is as in (5.38).

5.4 Steklov Boundary Conditions

Let Ω be a bounded domain of $\mathbb{R}^n$ ($n \geq 2$) with $\partial\Omega \in C^2$ and consider the boundary value problem

$$
\begin{cases}
\Delta^2 u = f & \text{in } \Omega, \\
u = \Delta u - a \frac{\partial u}{\partial v} = 0 & \text{on } \partial\Omega,
\end{cases}
\tag{5.43}
$$

where $a \in C^0(\partial\Omega)$, $f \in L^2(\Omega)$ and v is the outside normal (we will also use $u_v = \frac{\partial u}{\partial v}$). In this section we study the positivity preserving property for (5.43), namely under which conditions on Ω and on the boundary coefficient a the assumption $f \geq 0$ implies that the solution u exists and is positive. Let us first make precise what is meant by a solution of (5.43).

Definition 5.21. For $f \in L^2(\Omega)$ we say that u is a *weak solution* of (5.43) if $u \in H^2 \cap H_0^1(\Omega)$ and

$$\int_\Omega \Delta u \Delta v \, dx - \int_{\partial\Omega} a \, u_\nu v_\nu \, d\omega = \int_\Omega f v \, dx \qquad \text{for all } v \in H^2 \cap H_0^1(\Omega). \quad (5.44)$$

Note that weak solutions are well-defined for $a \in C^0(\partial\Omega)$. For $u \in H^4(\Omega)$ one may integrate by parts to find indeed that a weak solution of (5.44) satisfies the boundary value problem in (5.43). This means that the second boundary condition in (5.43) is hidden in the choice of the space $H^2 \cap H_0^1(\Omega)$ of admissible testing functions. For regularity results related to problem (5.43) we refer to Corollary 2.23.

In the next section we state the positivity preserving properties for (5.43) and we give the first part of their proof. The second part of their proof is more delicate and requires a Schauder setting and a different notion of solution. This is the reason why it is postponed to Section 5.4.3. In turn, the Schauder setting takes advantage of the positivity properties of the operators involved in the solution of (5.43). These properties are proven in Section 5.4.2 with a strong use of the kernel estimates of Section 4.3.

5.4.1 Positivity Preserving

The first statement describes existence, uniqueness and positivity of a weak solution to (5.43). A crucial role is played by a "weighted first eigenvalue". Fix a nontrivial positive weight function $b \in C^0(\partial\Omega)$ and set

$$J_b(u) = \frac{\int_\Omega |\Delta u|^2 \, dx}{\int_{\partial\Omega} b \, u_\nu^2 \, d\omega} \qquad \text{for} \qquad \int_{\partial\Omega} b \, u_\nu^2 \, d\omega \neq 0 \qquad (5.45)$$

and $J_b(u) = \infty$ otherwise. For every $u \in H^2 \cap H_0^1(\Omega)$ the functional in (5.45) is strictly positive, possibly ∞. Since the linear map $H^2(\Omega) \to L^2(\partial\Omega)$ defined by $u \mapsto u_\nu|_{\partial\Omega}$ is compact, there exists a minimiser for the problem

$$\delta_{1,b} = \delta_{1,b}(\Omega) := \inf_{u \in H^2 \cap H_0^1(\Omega)} J_b(u). \qquad (5.46)$$

Hence $\delta_{1,b} > 0$ and it may be viewed as a kind of first Steklov eigenvalue with respect to the weight function b and any minimiser as a corresponding eigenfunction. This definition should be compared with (3.40) in Section 3.3.1.

The next statement summarises the positivity preserving results for (5.43), see the Notations-Section for the interpretation of the symbols.

Theorem 5.22. *Let $\Omega \subset \mathbb{R}^n$ ($n \geq 2$) be a bounded domain with $\partial\Omega \in C^2$ and let $0 \lneqq b \in C^0(\partial\Omega)$. Let the eigenvalue $\delta_{1,b}$ be as defined in (5.46). Then there* *exists a number $\delta_{c,b} := \delta_{c,b}(\Omega) \in [-\infty, 0)$ such that the following holds for any function $a \in C^0(\partial\Omega)$.*

1. *If $a \geq \delta_{1,b}b$ and if $0 \lneqq f \in L^2(\Omega)$, then (5.43) has no nontrivial positive weak solution.*
2. *If $a = \delta_{1,b}b$, then there exists a positive eigenfunction, that is, problem (5.43) with $f = 0$ admits a weak solution $u_{1,b}$ that satisfies $u_{1,b} > 0$ and $-\Delta u_{1,b} > 0$ in Ω, $\frac{\partial}{\partial v} u_{1,b} < 0$ on $\partial\Omega$. This eigenfunction $u_{1,b}$ is unique, up to a constant multiplier.*
3. *If $a \lneqq \delta_{1,b}b$, then for any $f \in L^2(\Omega)$ problem (5.43) admits a unique weak solution u.*

   a. *If $\delta_{c,b}b \leq a \lneqq \delta_{1,b}b$ and if $0 \lneqq f \in L^2(\Omega)$, then $u \gneqq 0$.*
   b. *If $\delta_{c,b}b < a \lneqq \delta_{1,b}b$ and if $0 \lneqq f \in L^2(\Omega)$, then for some $c_f > 0$ it holds that $u \geq c_f d$ with d as in (4.1). Furthermore, if $a(x_0) < 0$ for some $x_0 \in \partial\Omega$, then $-\Delta u \ngeqq 0$ in Ω, whereas if $a \geq 0$, then $0 \lneqq f$ implies $-\Delta u \geq 0$ in Ω.*
   c. *If $a < \delta_{c,b}b$, then there are $0 \lneqq f \in L^2(\Omega)$ such that the corresponding solution u of (5.43) is not positive: $0 \nleq u$.*

Proof. We first prove Item 2, then Item 1 and we end with Item 3.

Proof of Item 2. Let $u_1 := u_{1,b} \in H^2 \cap H_0^1(\Omega)$ be a minimiser for (5.46) and let $\tilde{u}_1$ be the unique solution in $H^2 \cap H_0^1(\Omega)$ of $-\Delta\tilde{u}_1 = |\Delta u_1|$. Then by the maximum principle we infer that $|u_1| \leq \tilde{u}_1$ in Ω and $\left|\frac{\partial}{\partial v}u_1\right| \leq \left|\frac{\partial}{\partial v}\tilde{u}_1\right|$ on $\partial\Omega$. If Δu_1 changes sign, then these inequalities are strict and imply $J_b(u_1) > J_b(\tilde{u}_1)$. Hence, Δu_1 is of fixed sign, say $-\Delta u_1 \geq 0$, so that the maximum principle implies $\frac{\partial}{\partial v}u_1 < 0$ on $\partial\Omega$ and $u_1 \geq cd$ in Ω, where d is as in (4.1). Similarly, if u_1 and u_2 are two minimisers which are not multiples of each other, then there is a linear combination which is a sign changing minimiser and one proceeds as above to find a contradiction. This proves Item 2.

Proof of Item 1. Let us suppose by contradiction that $a \geq \delta_{1,b}$, that $f \gneqq 0$ and that u is a nontrivial positive solution to (5.43). Hence $u_v \leq 0$ on $\partial\Omega$. Let $u_{1,b}$ be a minimiser for (5.46) as obtained above. By taking $v = u_{1,b}$ in (5.44) one finds

$$0 < \int_\Omega f u_{1,b}\, dx = \int_\Omega \Delta u \Delta u_{1,b}\, dx - \int_{\partial\Omega} a\, u_v\, (u_{1,b})_v\, d\omega$$

$$\leq \int_\Omega \Delta u \Delta u_{1,b}\, dx - \int_{\partial\Omega} \delta_{1,b}\, b\, u_v\, (u_{1,b})_v\, d\omega = 0,$$

a contradiction. The last equality follows by the fact that $u_{1,b}$ minimises (5.46). This proves Item 1.

Proof of Item 3. On the space $H^2 \cap H_0^1(\Omega)$ we define the energy functional

$$I(u) := \tfrac{1}{2}\int_\Omega |\Delta u|^2\, dx - \tfrac{1}{2}\int_{\partial\Omega} a u_v^2\, d\omega - \int_\Omega fu\, dx \qquad u \in H^2 \cap H_0^1(\Omega).$$

Critical points of I are weak solutions of (5.43) in the sense of Definition 5.21. We will show that for $a \lneqq \delta_{1,b}b$ the functional I has a unique critical point.

If $a < \delta_{1,b}b$, one sets

$$\varepsilon := \frac{\min\{\delta_{1,b}b(x) - a(x); x \in \partial\Omega\}}{\max\{\delta_{1,b}b(x); x \in \partial\Omega\}} > 0, \qquad (5.47)$$

and finds that $a \le (1 - \varepsilon)\delta_{1,b}b$. By the definition of $\delta_{1,b}$ we have for all $u \in H^2 \cap H_0^1(\Omega)$

$$\int_\Omega |\Delta u|^2 \, dx - \int_{\partial\Omega} a \, u_\nu^2 \, d\omega$$

$$\ge \varepsilon \int_\Omega |\Delta u|^2 \, dx + (1 - \varepsilon)\left(\int_\Omega |\Delta u|^2 \, dx - \int_{\partial\Omega} \delta_{1,b}b \, u_\nu^2 \, d\omega\right)$$

$$\ge \varepsilon \int_\Omega |\Delta u|^2 \, dx, \qquad (5.48)$$

so that the functional I is coercive. Since it is also strictly convex the functional I admits a unique critical point which is its global minimum over $H^2 \cap H_0^1(\Omega)$.

In order to deal with the case that $a^+ \lneqq \delta_{1,b}b$, but $a^+(x) = \delta_{1,b}b(x)$ for some $x \in \partial\Omega$, we set

$$\tilde{b} := \frac{1}{2}\left(b + \delta_{1,b}^{-1}a^+\right).$$

Let u_1 be a minimiser of $J_{\tilde{b}}$ and u_2 of J_b. For the definition see (5.45). Then, since $0 \lneqq \tilde{b} \lneqq b$ and $\left(\frac{\partial}{\partial\nu}u_1\right)^2 > 0$ on $\partial\Omega$, we find $\delta_{1,\tilde{b}} = J_{\tilde{b}}(u_1) > J_b(u_1) \ge J_b(u_2) = \delta_{1,b}$. Instead of (5.47) we set

$$\varepsilon := 1 - \delta_{1,b}/\delta_{1,\tilde{b}} > 0,$$

find for $x \in \partial\Omega$ that $a \le a^+ = \delta_{1,b}(2\tilde{b} - b) \le \delta_{1,b}\tilde{b} = (1 - \varepsilon)\delta_{1,\tilde{b}}\tilde{b}$ and proceed by replacing all b in (5.48) with $\tilde{b}$.

If $a^+ = \delta_{1,b}b$ and $a^- \gneqq 0$, then one may not proceed directly as before. However, instead of the functional in (5.45), one may use

$$J_b^{a^-}(u) = \left(\int_\Omega |\Delta u|^2 \, dx + \int_{\partial\Omega} a^- \, u_\nu^2 \, d\omega\right)\left(\int_{\partial\Omega} b \, u_\nu^2 \, d\omega\right)^{-1}.$$

Then, defining $\delta_{1,b}^{a^-}$ for $J_b^{a^-}$ as in (5.46), this minimum is assumed, say by $u_{1,b}^{a^-}$. Since

$$\delta_{1,b}^{a^-} = J_b^{a^-}(u_{1,b}^{a^-}) \ge J_b(u_{1,b}^{a^-}) \ge J_b(u_{1,b}) = \delta_{1,b},$$

with the last inequality strict if $u_{1,b}^{a^-} \ne c\, u_{1,b}$ and with the first inequality strict if $u_{1,b}^{a^-} = c\, u_{1,b}$ since $(u_{1,b})_\nu^2 > 0$, we find $\delta_{1,b}^{a^-} > \delta_{1,b}$. So,

$$\int_\Omega |\Delta u|^2 \, dx + \int_{\partial\Omega} a^- \, u_\nu^2 \, d\omega \ge \delta_{1,b}^{a^-} \int_{\partial\Omega} b \, u_\nu^2 \, d\omega \quad \text{for all } u \in H^2 \cap H_0^1(\Omega)$$

and by setting

$$\varepsilon := 1 - \delta_{1,b}/\delta_{1,b}^{a^-} > 0$$

we find the result that replaces (5.48). Indeed

$$\int_{\Omega} |\Delta u|^2 \, dx - \int_{\partial\Omega} a \, u_\nu^2 \, d\omega = \int_{\Omega} |\Delta u|^2 \, dx + \int_{\partial\Omega} a^- \, u_\nu^2 \, d\omega - \int_{\partial\Omega} \delta_{1,b} b \, u_\nu^2 \, d\omega$$

$$\geq \varepsilon \int_{\Omega} |\Delta u|^2 \, dx + (1-\varepsilon) \left(\int_{\Omega} |\Delta u|^2 \, dx \right.$$

$$\left. + \int_{\partial\Omega} a^- \, u_\nu^2 \, d\omega - \int_{\partial\Omega} \delta_{1,b}^{a^-} b \, u_\nu^2 \, d\omega \right)$$

$$\geq \varepsilon \int_{\Omega} |\Delta u|^2 \, dx.$$

Hence, I is coercive and strictly convex and we conclude as for (5.48). The existence and uniqueness is so proved.

Assume now that there exists $x_0 \in \partial\Omega$ such that $a(x_0) < 0$. If the weak solution u were superharmonic, then by Hopf's boundary lemma we would have $u_\nu(x_0) < 0$. Using the second boundary condition in (5.43), we would then obtain $\Delta u(x_0) > 0$, a contradiction.

If $a \geq 0$ and $f \gneqq 0$, let $\tilde{u}$ be the unique solution in $H^2 \cap H_0^1(\Omega)$ of $-\Delta \tilde{u} = |\Delta u|$ in Ω. Since $\tilde{u} > u$ or $\tilde{u} = u$ in Ω, and $|\tilde{u}_\nu| \geq |u_\nu|$ on $\partial\Omega$, one finds for $f \gneqq 0$ that

$$I(\tilde{u}) - I(u) = -\frac{1}{2} \int_{\partial\Omega} a \left(\tilde{u}_\nu^2 - u_\nu^2 \right) d\omega - \int_{\Omega} f \, (\tilde{u} - u) \, dx \leq 0.$$

Equality occurs only when $\tilde{u} = u$. Since I is strictly convex there is at most one critical point which is a minimum. So $u = \tilde{u} > 0$ and $-\Delta u = -\Delta \tilde{u} = |\Delta u| \geq 0$. This completes the proof of existence and uniqueness whenever $a \lneqq \delta_{1,b} b$. The proof of the remaining statements (a), (b), (c) in Item 3 is more lengthy and delicate and we give it in Section 5.4.3, see Theorem 5.37. $\square$

Note that in Theorem 5.22 it may happen that $b(x) = 0$ on some part $\Gamma_1 \subset \partial\Omega$ and $b(x) > 0$ on the remaining part $\Gamma_0 = \partial\Omega \setminus \Gamma_1$. If moreover $\delta_{c,b} = -\infty$, then the limit problem for which the positivity preserving property holds (that is, $a = \delta_{c,b} b$) becomes a mixed Dirichlet-Navier problem with boundary conditions

$$u = 0 \text{ on } \partial\Omega, \qquad u_\nu = 0 \text{ on } \Gamma_0, \qquad \Delta u = 0 \text{ on } \Gamma_1.$$

As a first consequence of Theorem 5.22 we have the positivity preserving property for the hinged plate model in planar convex domains. As we have seen in Section 1.1.2, the physical bounds for the Poisson ratio are given by

$$-1 < \sigma < 1. \tag{5.49}$$

Under this constraint, the following result holds.

Corollary 5.23. *Let $\Omega \subset \mathbb{R}^2$ be a bounded convex domain with C^2-boundary and assume (5.49). For any $f \in L^2(\Omega)$ there exists a unique minimiser $u \in H^2 \cap H_0^1(\Omega)$ of the elastic energy functional (1.11) that is, of*

$$J(u) = \int_\Omega \left(\frac{|\Delta u|^2}{2} - fu \right) dx - \frac{1-\sigma}{2} \int_{\partial\Omega} \kappa u_\nu^2 \, d\omega,$$

where κ denotes the curvature of $\partial\Omega$. The minimiser u is the unique weak solution to

$$\begin{cases} \Delta^2 u = f & \text{in } \Omega, \\ u = \Delta u - (1-\sigma)\kappa u_\nu = 0 & \text{on } \partial\Omega. \end{cases}$$

Moreover, $f \gneqq 0$ implies that there exists $c_f > 0$ such that $u(x) \geq c_f d(x)$ and u is superharmonic in Ω.

Proof. We first show that the energy functional J coincides with the form given in (1.5). This can be done on a dense subset of smooth functions. Since $u|_{\partial\Omega} = 0$, one has $u_x = u_\nu \nu_1$ and $u_y = u_\nu \nu_2$ and may conclude that

$$2\int_\Omega \left(u_{xy}^2 - u_{xx}u_{yy} \right) dxdy$$

$$= \int_{\partial\Omega} \left(u_{xy}u_y \nu_1 + u_{xy}u_x \nu_2 - u_{xx}u_y \nu_2 - u_{yy}u_x \nu_1 \right) d\omega$$

$$= \int_{\partial\Omega} u_\nu \left(2u_{xy}\nu_1\nu_2 - u_{xx}\nu_2^2 - u_{yy}\nu_1^2 \right) d\omega = -\int_{\partial\Omega} \kappa u_\nu^2 \, d\omega,$$

where in the last step we used (1.8). Hence, existence and uniqueness of a minimiser u follow from Proposition 2.35.

Since $\partial\Omega \in C^2$ and Ω is convex we have $0 \lneqq \kappa \in C^0(\partial\Omega)$. In Proposition 2.35, it is also shown that J is strictly convex so that $(1-\sigma)\kappa \lneqq \delta_{1,\kappa}\kappa$. Hence, if $f \gneqq 0$ it follows first from statement 3.(a) in Theorem 5.22 that $u \gneqq 0$ in Ω and so that $u_\nu|_{\partial\Omega} \leq 0$. In view of the boundary value problem solved by u we obtain $-\Delta u \gneqq 0$ in Ω. This superharmonicity finally yields the other properties stated for u. $\square$

More generally, if we take $b = 1$ in Theorem 5.22, we obtain the following

Corollary 5.24. *Let $\Omega \subset \mathbb{R}^n$ ($n \geq 2$) be a bounded domain with $\partial\Omega \in C^2$ and let*

$$\delta_1 := \delta_1(\Omega) := \inf \left\{ \frac{\int_\Omega |\Delta u|^2 \, dx}{\int_{\partial\Omega} u_\nu^2 \, d\omega} ; u \in H^2 \cap H_0^1(\Omega) \setminus H_0^2(\Omega) \right\} \in (0, \infty) \quad (5.50)$$

be the first Steklov eigenvalue. Then there exists a number $\delta_c := \delta_c(\Omega) \in [-\infty, 0)$ such that the following holds for any function $a \in C^0(\partial\Omega)$.

1. If $a \geq \delta_1$ and if $0 \lneqq f \in L^2(\Omega)$, then (5.43) has no nontrivial positive weak solution.

2. *If $a = \delta_1$, then there exists a positive eigenfunction, that is, problem (5.43) admits a nontrivial weak solution u_1 with $u_1 > 0$ in Ω for $f = 0$. Moreover, the function u_1 is, up to multiples, the unique solution of (5.43) with $f = 0$ and $a = \delta_1$.*
3. *If $a \lneqq \delta_1$, then for any $f \in L^2(\Omega)$ problem (5.43) admits a unique weak solution u.*

   a. *If $\delta_c \le a \lneqq \delta_1$, then $0 \lneqq f \in L^2(\Omega)$ implies $u \gneqq 0$ in Ω.*
   b. *If $\delta_c < a \lneqq \delta_1$, then $0 \lneqq f \in L^2(\Omega)$ implies $u \ge c_f d > 0$ in Ω for some $c_f > 0$.*
   c. *If $a < \delta_c$, then there are $0 \lneqq f \in L^2(\Omega)$ with $0 \not\le u$.*

The result described in Corollary 5.24 quite closely resembles the structure for the resolvent of the biharmonic operator under Navier boundary conditions – see McKenna-Walter [298] and Kawohl-Sweers [247] – or for the biharmonic operator under Dirichlet boundary conditions in case the domain is a ball – see Corollary 5.5 where $(-a)$ plays the same role as a here. For all these problems the scheme is as follows.

$\exists f > 0$ with $u \not\ge 0$	$\forall f > 0 : \exists u$ and $u \ge 0$	$\forall f > 0$ if $\exists u$ then $u \not\ge 0$
δ_c	$0 \qquad\qquad \delta_1$	$a \longrightarrow$

Under Dirichlet boundary conditions such that the corresponding Green function is positive, Corollary 5.5 tells us that the set of constant coefficients $a \in \mathbb{R}$ for which $\Delta^2 u \ge au$ implies $u \ge 0$ is an interval $(a_c, \Lambda_{2,1})$ with $a_c \in (-\infty, 0]$. By combining Theorem 5.22 with Lemma 5.35 below, we immediately see that a similar result holds for (5.43).

Theorem 5.25. *Let $\Omega \subset \mathbb{R}^n$ ($n \ge 2$) be a bounded domain with $\partial\Omega \in C^2$ and let $a_i \in C^0(\partial\Omega)$ with $i = 1, 2$. Suppose that $a_1 \le 0 \le a_2$ are such that both for $a = a_1$ and $a = a_2$ we have the following: for all $f \in L^2(\Omega)$ there exists a weak solution $u = u_i$ ($i = 1, 2$) for (5.43), and moreover*

$$f \gneqq 0 \text{ implies } u \gneqq 0. \tag{5.51}$$

Then for any $a \in C^0(\partial\Omega)$ satisfying $a_1 \le a \le a_2$ and for each $f \in L^2(\Omega)$, a unique weak solution of (5.43) exists and (5.51) holds true.

However, a crucial difference with the Dirichlet boundary value problem for $\Delta^2 u \ge au$ is that $a_c \in (-\infty, 0]$ while for problem (5.43) it might happen that $\delta_c(\Omega) = -\infty$ although for general domains one cannot expect to have the positivity preserving property for any negative a. This is stated in the next results which show that the limit situation where $\delta_c(\Omega) = -\infty$ is closely related to the positivity preserving property for the biharmonic Dirichlet problem

$$\begin{cases} \Delta^2 u = f & \text{in } \Omega, \\ u = u_\nu = 0 & \text{on } \partial\Omega. \end{cases} \tag{5.52}$$

To this end, let us recall once more that the positivity preserving property does not hold in general domains $\Omega \subset \mathbb{R}^n$ for (5.52), see Section 1.2. It is clear that (5.43) with $|a| = +\infty$ corresponds to (5.52). However, if $a \to +\infty$ then a crosses the spectrum of $-\Delta$ under Steklov boundary conditions, see Theorem 3.18, whereas the next statement justifies the feeling that (5.52) only corresponds to the limit case $a = -\infty$.

Theorem 5.26. *Let $\Omega \subset \mathbb{R}^n$ ($n \geq 2$) be a bounded domain with $\partial\Omega \in C^2$. If for every $m \in \mathbb{N}^+$ and $0 \lneqq f \in L^2(\Omega)$ the weak solution of (5.43) with $a = -m$ is nontrivial and positive, then for every $0 \lneqq f \in L^2(\Omega)$ the solution $u \in H_0^2(\Omega)$ of (5.52) satisfies $u \gneqq 0$.*

Proof. Let us first recall the two boundary value problems addressed in the statement, namely

$$\begin{cases} \Delta^2 u = f & \text{in } \Omega, \\ u = \left(\Delta + m\frac{\partial}{\partial v}\right)u = 0 \text{ on } \partial\Omega, \end{cases} \quad \text{and} \quad \begin{cases} \Delta^2 u = f & \text{in } \Omega, \\ u = u_v = 0 \text{ on } \partial\Omega, \end{cases} \tag{5.53}$$

For all $m > 0$ let $u_m \in H^2 \cap H_0^1(\Omega)$ be the unique weak solution of the problem on the left in (5.53). Then according to (5.44) we have

$$\int_\Omega \Delta u_m \Delta\phi \, dx + m \int_{\partial\Omega} \frac{\partial u_m}{\partial v} \frac{\partial\phi}{\partial v} \, d\omega = \int_\Omega f\phi \, dx \text{ for all } \phi \in H^2 \cap H_0^1(\Omega). \tag{5.54}$$

Taking $\phi = u_m$ in (5.54) and using Hölder and Poincaré inequalities, gives for all $m > 0$

$$\|\Delta u_m\|_{L^2}^2 \leq \|\Delta u_m\|_{L^2}^2 + m \int_{\partial\Omega} \left|\frac{\partial u_m}{\partial v}\right|^2 d\omega = \int_\Omega f u_m \, dx \leq c\|f\|_{L^2}\|\Delta u_m\|_{L^2}. \tag{5.55}$$

Inequality (5.55) shows that the sequence (u_m) is bounded in $H^2(\Omega)$ so that, up to a subsequence, we have

$$u_m \rightharpoonup \bar{u} \quad \text{in } H^2(\Omega) \quad \text{as } m \to \infty \tag{5.56}$$

for some $\bar{u} \in H^2 \cap H_0^1(\Omega)$. Once boundedness is established, if we let $m \to \infty$ then (5.55) also tells us that

$$\frac{\partial u_m}{\partial v} \to 0 \quad \text{in } L^2(\partial\Omega) \quad \text{as } m \to \infty.$$

Therefore, $\bar{u} \in H_0^2(\Omega)$. Now take any function $\phi \in H_0^2(\Omega)$ in (5.54) and let $m \to \infty$. By (5.56) we obtain

$$\int_\Omega \Delta\bar{u}\Delta\phi \, dx = \int_\Omega f\phi \, dx \quad \text{for all } \phi \in H_0^2(\Omega).$$

Hence, $\bar{u}$ is the unique solution of the corresponding Dirichlet problem (5.52). Since (5.56) also implies that, up to a subsequence, $u_m(x) \to \bar{u}(x)$ for a.e. $x \in \Omega$, one finds that $\bar{u} \not\geq 0$. $\qquad\square$

Theorem 5.26 states that there exists some link between the Steklov and the Dirichlet problems. This link is confirmed by the special case when Ω is the unit ball. In this case, from Theorem 3.20 we know that the first Steklov eigenvalue as defined in (5.50) satisfies $\delta_1 = n$ and the following holds.

Theorem 5.27. *Let $\Omega = B$, the unit ball in $\mathbb{R}^n$ ($n \geq 2$). Then, for all $0 \lneqq f \in L^2(B)$ and all $a \in C^0(\partial B)$ such that $a \lneqq n$, there exists $c > 0$ such that the weak solution u of (5.43) satisfies $u(x) \geq cd(x)$ in B.*

The constant c depends both on f and a, $c = c_{f,a}$. For a fixed $0 \lneqq f \in L^2(B)$ we expect that $c = c_{f,a} \to 0$ as $a \to -\infty$.

Also the proof of Theorem 5.27 requires a Schauder setting and an approximation procedure. For this reason it is postponed to the end of Section 5.4.3.

5.4.2 Positivity of the Operators Involved in the Steklov Problem

We consider the second order Green operator $\mathscr{G}$ and the Poisson operator $\mathscr{K}$, that is, $w = \mathscr{G}f + \mathscr{K}g$ formally solves

$$\begin{cases} -\Delta w = f & \text{in } \Omega, \\ w = g & \text{on } \partial\Omega. \end{cases}$$

For C^2-domains, the operators $\mathscr{G}$ and $\mathscr{K}$ can be represented by integral kernels G and K, see (4.46) in Section 4.3. Let $(\mathscr{P}w)(x) := -v \cdot \nabla w(x) = -w_v(x)$ for $x \in \partial\Omega$. In this section we use the kernel estimates obtained in Section 4.3 in order to prove some positivity properties of these operators. First, we fix the appropriate setting so that $\mathscr{G}$, $\mathscr{K}$ and $\mathscr{P}$ are well-defined operators.

Notation 5.28 Let d denote the distance to $\partial\Omega$ as defined in (4.1). Set

$$C_d(\overline{\Omega}) = \{u \in C^0(\overline{\Omega}); \text{ there exists } w \in C^0(\overline{\Omega}) \text{ such that } u = dw\}$$

with norm

$$\|u\|_{C_d(\overline{\Omega})} = \sup\left\{\frac{|u(x)|}{d(x)}; x \in \Omega\right\}.$$

Set also $C_0(\overline{\Omega}) = \{u \in C^0(\overline{\Omega}); u = 0 \text{ on } \partial\Omega\}$ so that $C_d(\overline{\Omega}) \subsetneq C_0(\overline{\Omega})$.

We consider the three above operators in the following setting.

$$\mathscr{G}: C^0(\overline{\Omega}) \to C_d(\overline{\Omega}), \quad \mathscr{K}: C^0(\partial\Omega) \to C^0(\overline{\Omega}), \quad \mathscr{P}: C_d(\overline{\Omega}) \to C^0(\partial\Omega).$$

We also define the embedding

$$\mathscr{I}_d : C_d(\overline{\Omega}) \to C^0(\overline{\Omega}). \tag{5.57}$$

The space $C_d(\overline{\Omega})$ is a Banach lattice, that is, a Banach space with the ordering such that $|u| \le |v|$ implies $\|u\|_{C_d(\overline{\Omega})} \le \|v\|_{C_d(\overline{\Omega})}$, see Definition 3.2 or [13, 310, 360]. The positive cone

$$C_d(\overline{\Omega})^+ = \left\{ u \in C_d(\overline{\Omega}); \ u(x) \ge 0 \text{ in } \overline{\Omega} \right\}$$

is solid (namely, it has nonempty interior) and reproducing (that is, every $w \in C_d(\overline{\Omega})$ can be written as $w = u - v$ for some $u, v \in C_d(\overline{\Omega})^+$). Similarly, we define $C^0(\partial\Omega)^+$ and $C^0(\overline{\Omega})^+$.

Note that the interiors of the cones in these spaces are as follows:

$$C^0(\partial\Omega)^{+,\circ} = \left\{ v \in C^0(\partial\Omega); \ v(x) \ge c \text{ for some } c > 0 \right\},$$
$$C^0(\overline{\Omega})^{+,\circ} = \left\{ u \in C^0(\overline{\Omega}); \ u(x) \ge c \text{ for some } c > 0 \right\},$$
$$C_d(\overline{\Omega})^{+,\circ} = \left\{ u \in C_d(\overline{\Omega}); \ u(x) \ge c\,d(x) \text{ for some } c > 0 \right\}.$$

Definition 5.29. The operator $\mathscr{F} : \mathscr{C}_1 \to \mathscr{C}_2$ is called

- *nonnegative*, $\mathscr{F} \ge 0$, when $g \in \mathscr{C}_1^+ \Rightarrow \mathscr{F}g \in \mathscr{C}_2^+$;
- *strictly positive*, $\mathscr{F} \gneq 0$, when $g \in \mathscr{C}_1^+ \setminus \{0\} \Rightarrow \mathscr{F}g \in \mathscr{C}_2^+ \setminus \{0\}$;
- *strongly positive*, $\mathscr{F} > 0$, when $g \in \mathscr{C}_1^+ \setminus \{0\} \Rightarrow \mathscr{F}g \in \mathscr{C}_2^{+,\circ}$.

If $\mathscr{F} \ge 0$ and $\mathscr{F} \ne 0$, that is, for some $g \in \mathscr{C}_1^+$ we find $\mathscr{F}g \gneq 0$, we call $\mathscr{F}$ positive. Similarly, two operators are ordered through $\ge$ (respectively $\gneq$ or $>$) whenever their difference is nonnegative (respectively strictly or strongly positive).

We now prove a positivity result.

Proposition 5.30. *Suppose that $\partial\Omega \in C^2$ and $a \in C^0(\partial\Omega)$. Let $\mathscr{G}$, $\mathscr{K}$ and $\mathscr{P}$ be defined as above. Then $\mathscr{G}\mathscr{K}a\mathscr{P} : C_d(\overline{\Omega}) \to C_d(\overline{\Omega})$ is a well-defined compact linear operator. If in addition $a \gneq 0$, then $\mathscr{G}\mathscr{K}a\mathscr{P}$ is positive and even such that*

$$u \in C_d(\overline{\Omega})^+ \quad \text{implies either} \quad \mathscr{G}\mathscr{K}a\mathscr{P}u = 0 \quad \text{or} \quad \mathscr{G}\mathscr{K}a\mathscr{P}u \in C_d(\overline{\Omega})^{+,\circ}. \tag{5.58}$$

Proof. Take $\gamma \in (0,1)$, $p > n(1-\gamma)^{-1}$ and fix the embeddings $I_1 : C^0(\overline{\Omega}) \to L^p(\Omega)$, $I_2 : W^{2,p}(\Omega) \to C^{1,\gamma}(\overline{\Omega})$ and $I_3 : C^{1,\gamma} \cap C_0(\overline{\Omega}) \to C_d(\overline{\Omega})$. Since $\partial\Omega \in C^2$, for every $p \in (1,\infty)$ there exists a bounded linear operator $\mathscr{G}_p : L^p(\Omega) \to W^{2,p} \cap W_0^{1,p}(\Omega)$ such that $-\Delta \mathscr{G}_p f = f$ for all $f \in L^p(\Omega)$, see Theorem 2.20. If $\mathscr{I}_d$ is as in (5.57), then the Green operator from $C_d(\overline{\Omega})$ to $C_d(\overline{\Omega})$ should formally be denoted $\mathscr{G}\mathscr{I}_d$, where $\mathscr{G} = I_3 I_2 \mathscr{G}_p I_1$. Note that the embedding $I_1 : C^0(\overline{\Omega}) \to L^p(\Omega)$ is bounded and the embedding $I_2 : W^{2,p}(\Omega) \to C^{1,\gamma}(\overline{\Omega})$ is compact, see Theorem 2.6. Since

$W^{2,p} \cap W_0^{1,p}(\Omega) \subset C^{1,\gamma} \cap C_0(\overline{\Omega})$ and $I_3 : C^{1,\gamma} \cap C_0(\overline{\Omega}) \to C_d(\overline{\Omega})$ is bounded, $\mathscr{G}$ is not only well-defined but even compact. The strong maximum principle and Hopf's boundary point lemma allow then to conclude that $\mathscr{G} : C^0(\overline{\Omega}) \to C_d(\overline{\Omega})$ is a compact linear operator and it is strongly positive.

Since $\partial\Omega \in C^2$ and Ω is bounded all boundary points are regular. According to Perron's method [198, Theorem 2.14] the Dirichlet boundary value problem is solvable for arbitrary continuous boundary values by

$$(\mathscr{K}\phi)(x) = \sup\{v(x); v \leq \phi \text{ on } \partial\Omega \text{ and } v \text{ subharmonic in } \Omega\}.$$

For $\phi \in C^0(\partial\Omega)$ one obtains $\mathscr{K}\phi \in C^0(\overline{\Omega}) \cap C^2(\Omega)$ and by the maximum principle

$$\sup_{x\in\Omega} (\mathscr{K}\phi)(x) = \max_{x\in\partial\Omega} \phi(x) \quad \text{and} \quad \inf_{x\in\Omega} (\mathscr{K}\phi)(x) = \min_{x\in\partial\Omega} \phi(x)$$

implying not only that $\|\mathscr{K}\phi\|_{L^\infty(\Omega)} = \|\phi\|_{L^\infty(\partial\Omega)}$, but also that $\mathscr{K} : C^0(\partial\Omega) \to C^0(\overline{\Omega})$ is a strictly positive bounded linear operator.

Finally, from the fact that every function $u \in C_d(\overline{\Omega})$ can be written as $u = dw$ for some $w \in C^0(\overline{\Omega})$ and $\mathscr{P}dw = w|_{\partial\Omega}$, we infer that $\mathscr{P} : C_d(\overline{\Omega}) \to C^0(\partial\Omega)$ is a positive bounded linear operator.

From the just proved properties of $\mathscr{G}$, $\mathscr{K}$ and $\mathscr{P}$ we infer compactness and positivity of $\mathscr{G}\mathscr{K}a\mathscr{P}$ when $a \geq 0$ and that $\mathscr{K}a\mathscr{P}u \gneq 0$ implies that $\mathscr{G}\mathscr{K}a\mathscr{P}u \in C_d(\overline{\Omega})^{+,\circ}$. $\qquad\square$

Proposition 5.30 enables us to compare (using the notations of Definition 5.29) some of the operators involved in the Steklov problem.

Proposition 5.31. *Let $\Omega \subset \mathbb{R}^n$ be a bounded domain with $\partial\Omega \in C^2$ and let $\mathscr{I}_d$ be as in (5.57). Then there exists a constant $M_\Omega > 0$ such that*

$$\mathscr{G}\mathscr{K}\mathscr{P}\mathscr{G}\mathscr{I}_d\mathscr{G} \leq M_\Omega \mathscr{G}\mathscr{I}_d\mathscr{G} \text{ and } \mathscr{G}\mathscr{K}\mathscr{P}\mathscr{G}\mathscr{K} \leq M_\Omega \mathscr{G}\mathscr{K}.$$

Proof. We know that the integral kernel which corresponds to $\mathscr{G}\mathscr{K}\mathscr{P}\mathscr{G}\mathscr{I}_d\mathscr{G}$ satisfies the estimates in Lemma 4.16. By Proposition 4.13 we know estimates from below for $\mathscr{G}\mathscr{I}_d\mathscr{G}$. We have to compare these estimates. To this end, we use the following trivial fact

$$\min(1,\alpha)\min(1,\beta) \leq \min(1,\alpha\beta) \quad \text{for all } \alpha,\beta \geq 0,$$

combined with (4.53) and (4.52). Considering the different dimensions separately we then have the following. For $n \geq 5$, if $x^* \in \partial\Omega$ is such that $|x - x^*| = d(x)$,

$$(d(x)+d(y)+|x^*-y^*|)^{2-n} d(x)d(y) \preceq |x-y|^{4-n} \min\left(1, \frac{d(x)d(y)}{|x-y|^2}\right).$$

This, combined with Lemma 4.16 and (4.48), proves the statement for $n \geq 5$.

For $n = 4$ we argue as for $n = 5$ to find

$$(d(x) + d(y) + |x^* - y^*|)^{-2} d(x)d(y) \preceq \min\left(1, \frac{d(x)d(y)}{|x-y|^2}\right) \preceq \log\left(1 + \frac{d(x)d(y)}{|x-y|^2}\right).$$

This, combined with Lemma 4.16 and (4.49), proves the statement for $n = 4$.

For $n = 3$ we have

$$\begin{aligned}
&(d(x) + d(y) + |x^* - y^*|)^{-1} d(x)d(y) \\
&\preceq \sqrt{d(x)d(y)\min\left(1, \frac{d(x)d(y)}{|x-y|^2}\right)} = \sqrt{d(x)d(y)}\min\left(1, \frac{\sqrt{d(x)d(y)}}{|x-y|}\right).
\end{aligned}$$

This, combined with Lemma 4.16 and (4.50), proves the statement for $n = 3$.

For $n = 2$, by using (4.52) we find as a variation of (4.53) that

$$\log\left(2 + \frac{1}{d(x) + d(y) + |x^* - y^*|}\right) \preceq \log\left(2 + \frac{1}{|x-y|^2 + d(x)d(y)}\right).$$

This, combined with Lemma 4.16 and (4.51), proves the statement for $n = 2$. $\qquad\square$

5.4.3 Relation Between Hilbert and Schauder Setting

In this section we complete the proof of Theorem 5.22 and we give the proof of Theorem 5.27. For Theorem 5.22, it remains to prove statements (a), (b) and (c) in Item 3, see Theorem 5.37 below. In these situations it is more convenient to set the problem in spaces of continuous functions. This forces us to argue in a Schauder setting and we rewrite (5.43) as an integral equation. Then we proceed by approximation.

As in (5.57), let $\mathscr{I}_d : C_d(\overline{\Omega}) \to C^0(\overline{\Omega})$ denote the embedding operator, then (5.43) is equivalent to

$$u = \mathscr{G}\mathscr{K}a\mathscr{P}u + \mathscr{G}\mathscr{I}_d\mathscr{G}f. \tag{5.59}$$

Definition 5.32. For $f \in C^0(\overline{\Omega})$ we say that u is a $\mathscr{C}$-solution of (5.43) if $u \in C_d(\overline{\Omega})$ satisfies (5.59).

Proposition 5.33. *Suppose that Ω is a bounded domain in $\mathbb{R}^n$ ($n \geq 2$) with $\partial\Omega \in C^2$ and let $a \in C^0(\partial\Omega)$. If $f \in C^0(\overline{\Omega})$, then a $\mathscr{C}$-solution of (5.43) is also a weak solution in the sense of Definition 5.21.*

Proof. If $f \in C^0(\overline{\Omega})$ and $u \in C_d(\overline{\Omega})$ then by (5.59) it follows that $w = \mathcal{K}a\mathcal{P}u + \mathcal{I}_d\mathcal{G}f \in C^0(\overline{\Omega}) \subset L^2(\Omega)$ and hence $u = \mathcal{G}w \in H^2 \cap H_0^1(\Omega)$. Moreover, for such u and for any $v \in H^2 \cap H_0^1(\Omega)$ we have

$$\int_\Omega \Delta u \, \Delta v \, dx = \int_\Omega (\mathcal{K}a\mathcal{P}u + \mathcal{G}f) \, \Delta v \, dx = \int_{\partial\Omega} au_\nu v_\nu \, d\omega_x + \int_\Omega f \, v \, dx,$$

which is precisely (5.44). $\qquad\square$

Next, we note that (possibly by changing its sign) the minimiser $u_{1,b}$ for (5.46) lies in $C_d(\overline{\Omega})^{+,\circ}$.

Lemma 5.34. *Let $\partial\Omega \in C^2$ and suppose that $a \in C^0(\partial\Omega)$ is such that $a \lneqq \delta_{1,b}b$. Then*

$$\mathcal{E}_{\mathcal{G}}^a := (\mathcal{I} - \mathcal{G}\mathcal{K}a\mathcal{P})^{-1}\mathcal{G}\mathcal{I}_d\mathcal{G} : C^0(\overline{\Omega}) \to C_d(\overline{\Omega}),$$
$$\mathcal{E}_{\mathcal{K}}^a := (\mathcal{I} - \mathcal{G}\mathcal{K}a\mathcal{P})^{-1}\mathcal{G}\mathcal{K} : C^0(\partial\Omega) \to C_d(\overline{\Omega}),$$

are well-defined operators. Moreover, the following holds.

- *For $f \in C^0(\overline{\Omega})$ the unique $\mathcal{C}$-solution of problem (5.43) is $u = \mathcal{E}_{\mathcal{G}}^a f$.*
- *The function $u_{1,b}$ defined in Theorem 5.22 (Item 2) is a positive eigenfunction of $\mathcal{E}_{\mathcal{K}}^a (\delta_{1,b}b - a) \mathcal{P} : C_d(\overline{\Omega}) \to C_d(\overline{\Omega})$ with eigenvalue 1. Any other nonnegative eigenfunction $\tilde{u}$ of $\mathcal{E}_{\mathcal{K}}^a (\delta_{1,b}b - a) \mathcal{P}$ satisfies $(\delta_{1,b}b - a) \mathcal{P}\tilde{u} = 0$ on $\partial\Omega$.*

Proof. By Theorem 5.22 (Item 3) one finds for $a \lneqq \delta_{1,b}b$ that $\mu = 1$ is not an eigenvalue of the (compact) operator $\mathcal{G}\mathcal{K}a\mathcal{P}$. Therefore, the operator $(\mathcal{I} - \mathcal{G}\mathcal{K}a\mathcal{P})$ is invertible in $L^2(\Omega)$ and hence in $C_d(\overline{\Omega})$.

- Equation (5.59) reads as $u = (\mathcal{I} - \mathcal{G}\mathcal{K}a\mathcal{P})^{-1}\mathcal{G}\mathcal{I}_d\mathcal{G}f$.
- One directly checks that $u_{1,b}$ is an eigenfunction of $\mathcal{E}_{\mathcal{K}}^a (\delta_{1,b}b - a) \mathcal{P}$ with $\lambda = 1$ for all $a \lneqq \delta_{1,b}b$. By Theorem 5.22 (Item 2), up to its multiples, it is the unique eigenfunction with $\lambda = 1$. Let $\tilde{u}$ be another nonnegative eigenfunction of $\mathcal{E}_{\mathcal{K}}^a (\delta_{1,b}b - a) \mathcal{P}$ corresponding to some eigenvalue $\lambda \neq 1$. One finds that $\lambda = 0$ if and only if $(\delta_{1,b}b - a)\mathcal{P}\tilde{u} = 0$. For $\lambda \neq 0$ it holds that

$$\tilde{u} - \mathcal{G}\mathcal{K}\delta_{1,b}b\mathcal{P}\tilde{u} = (\lambda^{-1} - 1)\mathcal{G}\mathcal{K}(\delta_{1,b}b - a)\mathcal{P}\tilde{u}. \tag{5.60}$$

We have $u_{1,b}, \tilde{u} \in H^2 \cap H_0^1(\Omega)$. This fact allows us to use (5.60) and to find a contradiction in the case that $(\delta_{1,b}b - a)\mathcal{P}\tilde{u} \gneqq 0$. Indeed,

$$0 = \int_\Omega \Delta u_{1,b} \, \Delta \tilde{u} \, dx - \int_{\partial\Omega} \delta_{1,b}b \, (u_{1,b})_\nu \tilde{u}_\nu \, d\omega$$
$$= \int_\Omega \Delta u_{1,b} \, \Delta \left(\tilde{u} - \mathcal{G}\mathcal{K}\delta_{1,b}b\mathcal{P}\tilde{u}\right) dx$$
$$= (\lambda^{-1} - 1) \int_\Omega \Delta u_{1,b} \, \mathcal{G}\mathcal{K}(\delta_{1,b}b - a)\mathcal{P}\tilde{u}\,dx$$
$$= (1 - \lambda^{-1}) \int_\Omega u_{1,b} \, \mathcal{K}(\delta_{1,b}b - a)\mathcal{P}\tilde{u} \, dx,$$

and this last expression has a sign if $\lambda \neq 1$. $\qquad\square$

Lemma 5.35. *Let $\partial\Omega \in C^2$ and suppose that $a \in C^0(\partial\Omega)$ is such that $a \lneqq \delta_{1,b}b$. Let $\mathscr{E}_{\mathscr{G}}^a$ and $\mathscr{E}_{\mathscr{K}}^a$ be as in Lemma 5.34 and suppose that $\mathscr{E}_{\mathscr{G}}^a$ is a positive operator.*

1. *Then $\mathscr{E}_{\mathscr{G}}^a$, $\mathscr{E}_{\mathscr{K}}^a$, $\mathscr{P}\mathscr{E}_{\mathscr{G}}^a$ and $\mathscr{P}\mathscr{E}_{\mathscr{K}}^a$ are strictly positive operators.*
2. *If $\tilde{a} \in C^0(\partial\Omega)$ is such that $a \leq \tilde{a} \lneqq \delta_{1,b}b$, then $\mathscr{E}_{\mathscr{G}}^{\tilde{a}} \geq \mathscr{E}_{\mathscr{G}}^a$, $\mathscr{E}_{\mathscr{K}}^{\tilde{a}} \geq \mathscr{E}_{\mathscr{K}}^a$, $\mathscr{P}\mathscr{E}_{\mathscr{G}}^{\tilde{a}} \geq \mathscr{P}\mathscr{E}_{\mathscr{G}}^a$ and $\mathscr{P}\mathscr{E}_{\mathscr{K}}^{\tilde{a}} \geq \mathscr{P}\mathscr{E}_{\mathscr{K}}^a$.*
3. *If $\tilde{a} \in C^0(\partial\Omega)$ is such that $a < \tilde{a} \lneqq \delta_{1,b}b$, then $\mathscr{E}_{\mathscr{G}}^{\tilde{a}} > \mathscr{E}_{\mathscr{G}}^a$, $\mathscr{E}_{\mathscr{K}}^{\tilde{a}} > \mathscr{E}_{\mathscr{K}}^a$, $\mathscr{P}\mathscr{E}_{\mathscr{G}}^{\tilde{a}} > \mathscr{P}\mathscr{E}_{\mathscr{G}}^a$ and $\mathscr{P}\mathscr{E}_{\mathscr{K}}^{\tilde{a}} > \mathscr{P}\mathscr{E}_{\mathscr{K}}^a$.*

Proof. Assume that $0 \lneqq f \in C^0(\overline{\Omega})$ and $0 \lneqq \varphi \in C^0(\partial\Omega)$. Writing $u_a = \mathscr{E}_{\mathscr{G}}^a f$ and $v_a = \mathscr{E}_{\mathscr{K}}^a \varphi$ one gets

$$(\mathscr{I} - \mathscr{G}\mathscr{K}a\mathscr{P})u_a = \mathscr{G}\mathscr{I}_d\mathscr{G}f \quad \text{and} \quad (\mathscr{I} - \mathscr{G}\mathscr{K}a\mathscr{P})v_a = \mathscr{G}\mathscr{K}\varphi.$$

1. If $u_a = \mathscr{E}_{\mathscr{G}}^a f = 0$ for $f \gneqq 0$, then

$$u_a = \mathscr{G}\mathscr{K}a\mathscr{P}u_a + \mathscr{G}\mathscr{I}_d\mathscr{G}f = \mathscr{G}\mathscr{I}_d\mathscr{G}f > 0$$

by the maximum principle, a contradiction. So $\mathscr{E}_{\mathscr{G}}^a$ positive implies that $\mathscr{E}_{\mathscr{G}}^a$ is strictly positive. Since $K(x,y^*) = \lim_{t\searrow 0} G(x,y^* - tv)/t$ for $x \in \Omega$, $y^* \in \partial\Omega$ and v the exterior normal at y^*, we find that positivity of $\mathscr{E}_{\mathscr{G}}^a$ implies that $\mathscr{E}_{\mathscr{K}}^a$ is positive. We even have strict boundary positivity. Indeed, if $\mathscr{P}u_a = 0$ then $u_a = \mathscr{G}\mathscr{I}_d\mathscr{G}f$ and Hopf's boundary point lemma gives $\mathscr{P}u_a > 0$, a contradiction. A similar argument holds for v_a. This proves the first set of claims.

2. Let $a \leq \tilde{a} \lneqq \delta_{1,b}b$. We have

$$(\mathscr{I} - \mathscr{G}\mathscr{K}a\mathscr{P})u_{\tilde{a}} = \mathscr{G}\mathscr{K}(\tilde{a} - a)\mathscr{P}u_{\tilde{a}} + \mathscr{G}\mathscr{I}_d\mathscr{G}f$$

and, in turn, since $(\mathscr{I} - \mathscr{G}\mathscr{K}a\mathscr{P})$ is invertible in view of Lemma 5.34,

$$(\mathscr{I} - \mathscr{E}_{\mathscr{K}}^a(\tilde{a} - a)\mathscr{P})u_{\tilde{a}} = u_a. \tag{5.61}$$

For $\|\tilde{a} - a\|_{L^\infty(\partial\Omega)}$ small enough (say $\|\tilde{a} - a\|_{L^\infty(\partial\Omega)} < \varepsilon$) one may invert the operator in (5.61) and find an identity with a convergent series, that is

$$\mathscr{E}_{\mathscr{G}}^{\tilde{a}} = \mathscr{E}_{\mathscr{G}}^a + \sum_{k=1}^{\infty} (\mathscr{E}_{\mathscr{K}}^a(\tilde{a} - a)\mathscr{P})^k \mathscr{E}_{\mathscr{G}}^a. \tag{5.62}$$

Since $\mathscr{E}_{\mathscr{K}}^a(\tilde{a} - a)\mathscr{P} \geq 0$ holds, one finds that $u_{\tilde{a}} = \mathscr{E}_{\mathscr{G}}^{\tilde{a}}f \geq \mathscr{E}_{\mathscr{G}}^a f = u_a$. The series formula (5.62) holds for $\|\tilde{a} - a\|_{L^\infty(\partial\Omega)} < \varepsilon$. However, if $\|\tilde{a} - a\|_{L^\infty(\partial\Omega)} \geq \varepsilon$ then the above argument can be repeated by considering some intermediate $a := a_0 \lneqq a_1 \lneqq \ldots \lneqq a_k := \tilde{a}$ such that $\|a_{i+1} - a_i\|_{L^\infty(\partial\Omega)} < \varepsilon$ for all i. A similar reasoning applies to $v_{\tilde{a}}, v_a$. This proves the second set of claims.

3. Let us consider the sequence $(\varphi_m) \subset C_d(\overline{\Omega})$, defined by

$$\varphi_0 = \mathscr{E}_{\mathscr{G}}^a f,$$
$$\varphi_{m+1} = \mathscr{E}_{\mathscr{K}}^a(\delta_{1,b}b - a)\mathscr{P}\varphi_m \quad \text{for } m \geq 0.$$

Since $\mathscr{E}^a_{\mathscr{G}} f \gneq 0$ we find that $\varphi_m \geq 0$ for all $m \geq 0$. Moreover, since $\mathscr{E}^a_{\mathscr{K}}\left(\delta_{1,b}b - a\right)\mathscr{P}$ is compact, two cases may occur;

(i) there exists $m_0 > 0$ such that $\varphi_m \gneq 0$ for $m < m_0$ and $\varphi_m = 0$ for all $m \geq m_0$;

(ii) $\varphi_m / \|\varphi_m\|_{C_d(\overline{\Omega})} \to \varphi_\infty$ where φ_∞ is a nonnegative eigenfunction (with $\lambda = 1$) of the operator

$$\mathscr{E}^a_{\mathscr{K}}\left(\delta_{1,b}b - a\right)\mathscr{P}\varphi_\infty = \lambda\,\varphi_\infty.$$

If (i) occurs, then $\mathscr{E}^a_{\mathscr{K}}\left(\delta_{1,b}b - a\right)\mathscr{P}\varphi_{m_0} = 0$ so that by Item 1 we infer $\left(\delta_{1,b}b - a\right)\mathscr{P}\varphi_{m_0} = 0$ and hence $\mathscr{P}\varphi_{m_0} = 0$. We find a contradiction since as in the proof of Item 1 it follows that $\varphi_{m_0} = \mathscr{G}\mathscr{I}_d\mathscr{G}\varphi_{m_0-1}$ and $\mathscr{P}\varphi_{m_0} > 0$ holds by Hopf's boundary point lemma.

Therefore, case (ii) occurs. Then φ_∞ is a multiple of the unique positive eigenfunction $u_{1,b}$, see Lemma 5.34. So for m_1 large enough we find that there exists $c_2 > c_1 > 0$ such that

$$c_1 u_{1,b} \leq \frac{\varphi_m}{\|\varphi_m\|_{C_d(\overline{\Omega})}} \leq c_2 u_{1,b} \qquad \text{for all } m \geq m_1.$$

Now set

$$\psi_0 = \mathscr{E}^a_{\mathscr{G}} f, \qquad \psi_{m+1} = \mathscr{E}^a_{\mathscr{K}}\left(\tilde{a} - a\right)\mathscr{P}\psi_m \quad \text{for } m \geq 0. \tag{5.63}$$

Since for some $\varepsilon > 0$ it holds that

$$\varepsilon\left(\delta_{1,b}b - a\right) \leq \tilde{a} - a \leq \delta_{1,b}b - a,$$

we obtain $\psi_m \geq \varepsilon^m \varphi_m$ for all m and by (5.63)

$$\psi_m \geq \varepsilon^m \varphi_m \geq c_1 \varepsilon^m \|\varphi_m\|_{C_d(\overline{\Omega})} u_{1,b} \qquad \text{for all } m \geq m_1.$$

Then from (5.62) it follows that there exists $c_3 > 0$ such that

$$\mathscr{E}^{\tilde{a}}_{\mathscr{G}} f \geq \mathscr{E}^a_{\mathscr{G}} f + c_3 u_{1,b}.$$

In a similar way we proceed with $\mathscr{E}_{\mathscr{K}}$, $\mathscr{P}\mathscr{E}_{\mathscr{G}}$ and $\mathscr{P}\mathscr{E}_{\mathscr{K}}$. $\qquad\square$

With the result in Lemma 5.34 it will be sufficient to have positivity preserving for a negative $a \in C^0(\partial\Omega)$ in order to ensure that this property will hold for any sign-changing $\tilde{a}$ with $a \leq \tilde{a} \lneq \delta_{1,b}b$. So we may restrict ourselves to $a \leq 0$. We now prove a crucial "comparison" statement in the case where $\mathscr{G}\mathscr{K}a\mathscr{P}$ has a small spectral radius.

Lemma 5.36. *Let $\partial\Omega \in C^2$ and assume that $0 \geq a \in C^0(\partial\Omega)$ is such that*

$$r_\sigma(\mathscr{G}\mathscr{K}a\mathscr{P}) < 1.$$

If there exists $M > 0$ such that

$$\mathscr{G}\mathscr{K}\mathscr{P}\mathscr{G}\mathscr{I}_d\mathscr{G} \leq M\,\mathscr{G}\mathscr{I}_d\mathscr{G}, \tag{5.64}$$

and if $\|a\|_{L^\infty(\partial\Omega)} < M^{-1}$, then $\mathscr{E}^a_{\mathscr{G}} > 0$.

Proof. Clearly, $a = -a^-$. Since $r_\sigma(\mathscr{G}\mathscr{K}a^-\mathscr{P}) < 1$ the equation (5.59) can be rewritten as a Neumann series

$$u = \left(\mathscr{I} + \mathscr{G}\mathscr{K}a^-\mathscr{P}\right)^{-1}\mathscr{G}\mathscr{I}_d\mathscr{G}f = \sum_{k=0}^{\infty}\left(-\mathscr{G}\mathscr{K}a^-\mathscr{P}\right)^k\mathscr{G}\mathscr{I}_d\mathscr{G}f,$$

which reads

$$u = \left(\sum_{k=0}^{\infty}\left(\mathscr{G}\mathscr{K}a^-\mathscr{P}\right)^{2k}\right)\left(\mathscr{I} - \mathscr{G}\mathscr{K}a^-\mathscr{P}\right)\mathscr{G}\mathscr{I}_d\mathscr{G}f \tag{5.65}$$

after joining the odd and even powers. Next, notice that in view of (5.65) it suffices to show that the operator $(\mathscr{I} - \mathscr{G}\mathscr{K}a^-\mathscr{P})\mathscr{G}\mathscr{I}_d\mathscr{G}$ is strongly positive. This fact is a direct consequence of (5.64) and $\|a^-\|_{L^\infty(\partial\Omega)} \leq M^{-1}$. $\qquad\square$

By combining the previous statements, we obtain the following result, which completes the proof of Item 3 of Theorem 5.22. The proof uses estimates for the kernels involved and for this reason it seems more suitable to employ a Schauder setting and to approximate.

Theorem 5.37. *There exists $\delta_{c,b} := \delta_{c,b}(\Omega) \in [-\infty, 0)$ such that the following holds for a weak solution u of (5.43):*

1. *for $\delta_{c,b}b \leq a \lneqq \delta_{1,b}b$ it follows that if $0 \lneqq f \in L^2(\Omega)$, then $u \gneqq 0$;*
2. *for $\delta_{c,b}b < a \lneqq \delta_{1,b}b$ it follows that if $0 \lneqq f \in L^2(\Omega)$, then $u \geq c_f d$ for some $c_f > 0$ (depending on f), d being the distance function from (4.1);*
3. *for $a < \delta_{c,b}b$ there are $0 \lneqq f \in L^2(\Omega)$ with u somewhere negative.*

Proof. Let M_Ω be as in Proposition 5.31 and put $\delta := -\left(M_\Omega \max_{x\in\partial B} b(x)\right)^{-1} < 0$. Then by Lemmas 5.34 and 5.36 we infer that

$$\text{if } \delta b \leq a \lneqq \delta_{1,b}b \text{ and } 0 \lneqq f \in C^0(\overline{\Omega}) \quad \text{then} \quad u \gneqq 0 \text{ in } \Omega, \tag{5.66}$$

where u is the unique $\mathscr{C}$-solution of (5.43). Let $\delta_{c,b}$ be the (negative) infimum of all such δ which satisfy (5.66). We have so proved that there exists $\delta_{c,b} := \delta_{c,b}(\Omega) \in [-\infty, 0)$ such that, if $\delta_{c,b}b \leq a \lneqq \delta_{1,b}b$ and $0 \lneqq f \in C^0(\overline{\Omega})$, then $u \gneqq 0$, where u is the $\mathscr{C}$-solution of (5.43). Moreover, if $\delta_{c,b}b < a$ and $0 \lneqq f \in C^0(\overline{\Omega})$, then Lemma 5.35 yields the existence of c_f such that $u \geq c_f d$. Finally, the above definition of $\delta_{c,b}$ shows that, if $a < \delta_{c,b}b$, then there are $0 \lneqq f \in C^0(\overline{\Omega})$ with u somewhere negative. In view of Proposition 5.33, this proves Item 3.

For Item 1 we use a density argument. Assume that $\delta_{c,b}b \leq a \lneqq \delta_{1,b}b$ and $0 \lneqq f \in L^2(\Omega)$. Let $u \in H^2 \cap H_0^1(\Omega)$ be the unique weak solution of (5.43), according to Item 3 in Theorem 5.22. Consider a sequence of functions $(f_k) \subset C^0(\overline{\Omega})$ such that $f_k \gneqq 0$ for all $k \in \mathbb{N}$ and $f_k \to f$ in $L^2(\Omega)$ as $k \to \infty$. Let u_k be the $\mathscr{C}$-solution to

$$\Delta^2 u_k = f_k \text{ in } \Omega, \qquad u_k = \Delta u_k - a\frac{\partial u_k}{\partial \nu} = 0 \text{ on } \partial\Omega.$$

Then, by (5.66), $u_k \gneqq 0$ in Ω for all k. Moreover, by Corollary 2.23, the sequence (u_k) is bounded in $H^2(\Omega)$ so that, up to a subsequence, it converges weakly and pointwise to some $u \in H^2 \cap H_0^1(\Omega)$. By Definition 5.21, we know that

$$\int_\Omega \Delta u_k \Delta v\, dx - \int_{\partial\Omega} a\, (u_k)_\nu v_\nu \, d\omega = \int_\Omega f_k v\, dx \qquad \text{for all } v \in H^2 \cap H_0^1(\Omega).$$

Therefore, letting $k \to \infty$, we deduce that u is a weak solution to the original problem and

$$u \gneqq 0 \qquad \text{in } \Omega. \tag{5.67}$$

The proof of Item 2 is more delicate. Assume that $\delta_{c,b} b < a \lneqq \delta_{1,b} b$ and $0 \lneqq f \in L^2(\Omega)$. Let $u \in H^2 \cap H_0^1(\Omega)$ be the unique weak solution to (5.43). Let

$$g(x) := \min\{1, f(x)\}, \qquad x \in \Omega,$$

and let $v \in H^2 \cap H_0^1(\Omega)$ be the unique weak solution to

$$\begin{cases} \Delta^2 v = g & \text{in } \Omega, \\ v = \Delta v - a v_\nu = 0 & \text{on } \partial\Omega. \end{cases}$$

Since $g \le f$, we deduce by Lemma 5.35 and a density argument that

$$u \ge v \qquad \text{in } \Omega. \tag{5.68}$$

Moreover, since $g \in L^\infty(\Omega)$, by Corollary 2.23 and Theorem 2.6 we infer that $v \in C^1(\overline{\Omega})$.

Let $\delta_{c,b}$ be as at the beginning of this proof, take a function $a_0 \in C^0(\partial\Omega)$ such that $\delta_{c,b} b < a_0 < a$ (if $\delta_{c,b} > -\infty$ one can also take $a_0 = \delta_{c,b} b$) and consider also the unique weak solution w to

$$\begin{cases} \Delta^2 w = g & \text{in } \Omega, \\ w = \Delta w - a_0 w_\nu = 0 & \text{on } \partial\Omega. \end{cases}$$

Again, we have $w \in C^1(\overline{\Omega})$. Since $w \ge 0$ in Ω in view of Item 1, we know that $w_\nu \le 0$ on $\partial\Omega$. Moreover, it cannot be that $w_\nu \equiv 0$ since otherwise the boundary condition would imply $-\Delta w = 0$ on $\partial\Omega$ with $-\Delta w$ superharmonic in Ω. This would imply first that $-\Delta w > 0$ in Ω and next, by Hopf's lemma, that $w_\nu < 0$ on $\partial\Omega$, a contradiction. Therefore,

$$\psi := (a_0 - a) w_\nu \gneqq 0, \qquad \psi \in C^0(\partial\Omega).$$

Finally, let $z := v - w$. Then $z \in C^1(\overline{\Omega})$ and z is the unique weak solution to

$$\begin{cases} \Delta^2 z = 0 & \text{in } \Omega, \\ z = 0 & \text{on } \partial\Omega, \\ \Delta z - a z_\nu = -\psi & \text{on } \partial\Omega. \end{cases}$$

In fact, by Corollary 2.23 and Theorem 2.6, we have that $z \in C_d(\overline{\Omega})^+$ and $z = \mathscr{E}_{\mathscr{H}}^a \psi$. By Lemma 5.35 we know that $\mathscr{E}_{\mathscr{H}}^a > \mathscr{E}_{\mathscr{H}}^{a_0} \gneqq 0$ so that there exists $c > 0$ with

$$z(x) \geq c\,d(x). \tag{5.69}$$

Note that c depends on ψ and therefore also on w. Hence, it depends on f so that $c = c_f$. By combining (5.68) with (5.69) we obtain

$$u(x) \geq v(x) = z(x) + w(x) \geq z(x) \geq c_f\,d(x)$$

and Item 2 follows. $\qquad\qquad\qquad\qquad\qquad\qquad\qquad\qquad\qquad\qquad\qquad\qquad\qquad\quad\square$

Proof of Theorem 5.27. We first assume that $f \in C_c^\infty(B)$. In this case, by Corollary 2.23 we know that the weak solution u satisfies $u \in W^{2,p}(B)$ for all $p > 1$. In turn, by Theorem 2.6, this proves that $u \in C^1(\overline{B})$ and hence $\Delta u = au_v \in C^0(\partial B)$. Therefore, Theorems 2.19 and 2.25 yield $u \in C^\infty(B) \cap C^2(\overline{B})$. In particular, by Lemma 5.34, u is a $\mathscr{C}$-solution.

Consider the auxiliary function $\phi \in C^\infty(B) \cap C^0(\overline{B})$ defined by

$$\phi(x) = (|x|^2 - 1)\Delta u(x) - 4x \cdot \nabla u(x) - 2(n-4)u(x) \quad \text{for } x \in \overline{B}.$$

Since $x = v$ and $u = 0$ on ∂B, we have

$$\phi = -4u_v \qquad \text{on } \partial B. \tag{5.70}$$

Moreover, for $x \in B$ we have

$$\nabla \phi = (2\Delta u)x + (|x|^2 - 1)\nabla \Delta u + 2(2-n)\nabla u - 4D^2 u \cdot x, \tag{5.71}$$

$$-\Delta \phi = (1 - |x|^2)f(x) \geq 0, \tag{5.72}$$

where $D^2 u$ denotes the Hessian matrix of u. By (5.71) we find

$$\phi_v = 2\Delta u + 2(2-n)u_v - 4\langle v, D^2 u \cdot v \rangle \qquad \text{on } \partial B.$$

Since $\langle v, D^2 u \cdot v \rangle = u_{vv}$, by recalling that $u = 0$ on ∂B and using the expression of Δu on the boundary, the previous equation reads $\phi_v = -2\Delta u + 2nu_v$. Finally, taking into account the second boundary condition in (5.43), we obtain

$$\phi_v = 2(n-a)u_v \qquad \text{on } \partial B. \tag{5.73}$$

So, combining (5.70), (5.72) and (5.73) we find that ϕ satisfies the second order Steklov boundary value problem

$$\begin{cases} -\Delta \phi = (1 - |x|^2)f \geq 0 \text{ in } B, \\ \phi_v + \frac{1}{2}(n-a)\phi = 0 \qquad \text{on } \partial B. \end{cases}$$

As $a \leq n$, by the maximum principle (for this second order problem!) we infer that $\phi > 0$ in $\overline{B}$ and hence by (5.70) that $u_\nu \leq 0$ on ∂B. By (2.65) and Proposition 5.11, we deduce that $u > 0$ in B whenever $0 \lneqq f \in C_c^\infty(B)$.

Assume now that $0 \lneqq f \in L^2(B)$ and let $u \in H^2 \cap H_0^1(B)$ be the unique weak solution to (5.43), according to Definition 5.21. Then the same density argument leading to (5.67) shows that $u \gneqq 0$ in B. Hence, by Corollary 5.24 part 3(c), we infer that $\delta_c = -\infty$. In turn the lower bound $u(x) \geq c d(x)$ in B follows from Corollary 5.24, part 3(b). $\qquad\qquad\square$

5.5 Bibliographical Notes

The lower order perturbation theory of positivity was developed in [211], see also [205]. These results are based on Green function estimates, 3-G-theorems and Neumann series. This strategy was used e.g. by Chung, Cranston, Fabes, Hueber, Sieveking, Zhao [97, 114, 234, 421, 422] in the context of Schrödinger operators and conditioned Brownian motion and e.g. by Mitidieri-Sweers [310, 384, 385] to study positivity in noncooperatively coupled second order systems.

The discussion of the local maximum principle and of the role of boundary data follows [210, 212]. The underlying formulae for the Poisson kernels for the biharmonic Dirichlet problem go back to Lauricella-Volterra [269, 403] and were collected in the book [324] by Nicolesco. For the polyharmonic Poisson kernels we refer to Edenhofer [158, 159]. Estimates as in Lemmas 5.8 and 5.9 were proved in a more general setting but under more restrictive assumptions on the coefficients by Krasovskiĭ [256, 257]. A local maximum principle for fourth order operators was first deduced by Tomi [396].

For the positivity preserving property of the biharmonic Steklov boundary value problem (5.43), we follow Gazzola-Sweers [191], where one can also find a discussion on existence and positivity of the solution to (5.43) when $a - \delta_1$ changes sign on $\partial\Omega$. Moreover, in [191] one can also find a different proof of Theorem 5.27 which is strongly based on the behaviour of the biharmonic Green function for the Dirichlet problem, see (2.65). Corollary 5.23 is due to Parini-Stylianou [332]. We also refer to [42, 44] for some related nonlinear problems and for a first attempt to describe the positivity preserving property for (5.43).

Chapter 6
Dominance of Positivity in Linear Equations

In Section 1.2 we mentioned that although the Green function $G_{\Delta^2,\Omega}$ for the clamped plate boundary value problem

$$\begin{cases} \Delta^2 u = f & \text{in } \Omega, \\ u = |\nabla u| = 0 & \text{on } \partial\Omega, \end{cases} \tag{6.1}$$

is in general sign changing, it is very hard to display its negative part in numerical simulations or in real world experiments. Moreover, numerical work in nonlinear elliptic fourth order equations suggests that maximum or comparison principles are violated only to a "small extent". Nevertheless, we do not yet have tools at hand to give this feeling a precise form and, in particular, a quantitative form which might prove to be useful also for nonlinear higher order equations.

This chapter may be considered as a first preliminary step in this direction. We study the negative part of the biharmonic Green function $G^-_{\Delta^2,\Omega}$ and show that it is small when compared with its positive part $G^+_{\Delta^2,\Omega}$. For a precise formulation see Theorems 6.15 and 6.24 and the subsequent interpretations. We emphasise that these are not just continuous dependence on data results. Green's functions are *families* of functions with the position of the pole as a parameter and the main problem consists in gaining uniformity with respect to the position of the pole when it approaches the boundary. In proving these results, one has to distinguish between the dimensions $n = 2$ and $n \geq 3$. The second case seems to be much simpler and is carried out in detail. We are confident that the arguments can be extended to fourth order operators where the principal part is a square of a second order operator and which may contain also lower order perturbations. Uniformity with respect to unbounded families of such perturbations can, however, in general not be expected. The proof ($n \geq 3$) heavily relies on uniform Krasovskiĭ-type estimates for biharmonic Green's functions $G_{\Delta^2,\Omega}$ in general domains, which are deduced in Section 4.5. Local positivity results from Section 6.3 are used as an essential first step which, in the particular case $n = 3$, were observed first by Nehari [323]. Although in the two-dimensional case one has holomorphic maps at hand, the result there requires a much more involved proof, which we sketch in Section 6.2.2 and where for details we refer to

F. Gazzola et al., *Polyharmonic Boundary Value Problems*,
Lecture Notes in Mathematics 1991, DOI 10.1007/978-3-642-12245-3_6,
© Springer-Verlag Berlin Heidelberg 2010

the literature [117]. This proof is based on the explicit biharmonic Green functions
in the "limaçons de Pascal", on carefully putting together parts of boundaries of
several prototype domains and delicate asymptotic estimates.

A second main objective of this chapter is to show that positivity of the bihar-
monic Green function $G_{\Delta^2,B}$ in the unit ball $B \subset \mathbb{R}^n$ is *not* just a singular event
but remains true under small $C^{4,\gamma}$-smooth perturbations Ω of B. For $n \geq 3$ see
Theorem 6.29; its proof is quite similar to that of the small negative part result
mentioned before. For $n = 2$, see Theorem 6.3. Here we build on the lower order
perturbation theory developed in Theorem 5.1 and benefit from holomorphic maps
and reduction to normal form. These tools are special for $n = 2$ and allow for considering also any m-th power of a regular second order elliptic operator being close
enough to the polyharmonic prototype $(-\Delta)^m$ in domains Ω close enough to the
unit disk. Having such a perturbation theory of positivity is remarkable since, again,
this is not just a continuous dependence on data result.

6.1 Highest Order Perturbations in Two Dimensions

In two dimensions also perturbations of highest order of the polyharmonic prototype
may be taken into account. Here we consider

$$
\begin{cases}
Lu = f & \text{in } \Omega, \\
D^\alpha u|_{\partial\Omega} = 0 & \text{for } |\alpha| \leq m-1,
\end{cases}
\tag{6.2}
$$

with

$$
Lu := \left(-\sum_{i,j=1}^{2} \tilde{a}_{ij}(x)\frac{\partial^2}{\partial x_i \partial x_j} \right)^m u + \sum_{|\beta|\leq 2m-1} a_\beta(x)D^\beta u,
\tag{6.3}
$$

where $\tilde{a}_{ij} = \tilde{a}_{ji} \in C^{2m-1,\gamma}(\overline{\Omega})$, $a_\beta \in C^{0,\gamma}(\overline{\Omega})$. In view of the maximum principle for
second order operators we assume throughout the whole chapter that

$$
m \geq 2.
$$

First we define an appropriate notion of closeness for domains and operators.

Definition 6.1. We assume that $\Omega^\natural$ and Ω are bounded $C^{k,\gamma}$-smooth domains. Let
$\varepsilon \geq 0$. We call Ω ε-*close* to $\Omega^\natural$ in $C^{k,\gamma}$-sense, if there exists a $C^{k,\gamma}$-mapping $g : \overline{\Omega^\natural} \to$
$\overline{\Omega}$ such that $g(\overline{\Omega^\natural}) = \overline{\Omega}$ and

$$
\|g - Id\|_{C^{k,\gamma}(\overline{\Omega^\natural})} \leq \varepsilon.
$$

We remark that if $k \geq 1$, $\Omega^\natural$ is convex and ε is sufficiently small, then $g^{-1} \in C^{k,\gamma}(\overline{\Omega})$
exists and $\|g^{-1} - Id\|_{C^{k,\gamma}(\overline{\Omega})} = O(\varepsilon)$ as $\varepsilon \to 0$.

Definition 6.2. Let $\varepsilon \geq 0$ and assume that L is as in (6.3). We call the operator L ε-*close* to $(-\Delta)^m$ in $C^{k,\gamma}$-sense, if (in the case $k \geq 2m$) additionally $\tilde{a}_{ij} \in C^{k,\gamma}(\overline{\Omega})$ and

$$\|\tilde{a}_{ij} - \delta_{ij}\|_{C^{k,\gamma}(\overline{\Omega})} \leq \varepsilon,$$

$$\|a_\beta\|_{C^0(\overline{\Omega})} \leq \varepsilon \text{ for } |\beta| \leq 2m - 1.$$

If $\varepsilon \geq 0$ is small, then L is uniformly elliptic.

The following is the main perturbation result if $n = 2$.

Theorem 6.3. *There exists $\varepsilon_0 = \varepsilon_0(m) > 0$ such that we have for $0 \leq \varepsilon \leq \varepsilon_0$:*
If the bounded $C^{2m,\gamma}$-smooth domain $\Omega \subset \mathbb{R}^2$ is ε-close to the unit disk B in $C^{2m,\gamma}$-sense and if the differential operator L is ε-close to $(-\Delta)^m$ in $C^{2m-1,\gamma}$-sense, then for every $f \in C^{0,\gamma}(\overline{\Omega})$ satisfying $f \gneqq 0$ the solution $u \in C^{2m,\gamma}(\overline{\Omega})$ to the Dirichlet problem (6.2) is strictly positive, namely

$$u > 0 \text{ in } \Omega.$$

Remark 6.4. 1. Let $E_{a,b}$ be an ellipse with half axes $a, b > 0$. In case of small eccentricity, i.e. $\frac{a}{b} \approx 1$, Green's function for Δ^2 in $E_{a,b}$ is positive. For larger eccentricity, e.g. $\frac{a}{b} \approx 1.2$, it changes sign according to the example of Garabedian [176] and the refined version by Hedenmalm, Jakobsson, and Shimorin in [227].
2. The proof of Theorem 6.3 suggests that $\varepsilon_0(m) \searrow 0$ for increasing $m \nearrow \infty$.
3. As long as one restricts to the polyharmonic operator $(-\Delta)^m$ in perturbed domains, it was shown by Sassone [359] that $C^{m,\gamma}$-closeness to the disk is sufficient. In case of the clamped plate equation this means that positivity is governed by closeness of the boundary curvature to a constant with respect to a Hölder norm. We think that also in the case of perturbed principal parts, the required closeness to the polyharmonic operator may be relaxed. But we expect that such a relaxation will require a big technical effort. In particular, all problems should be written in divergence form and one should refer to $C^{m,\gamma}$-Schauder-theory for operators in divergence form.
4. We recall that Theorem 6.3 cannot be proved by just referring to continuous dependence on data.
5. According to Jentzsch's [237] or Kreĭn-Rutman's [258] theorem, see Theorem 3.3, positivity of the Green function implies existence of a positive first eigenfunction. A somehow stronger result was proved in [213], which was already briefly mentioned in Section 3.1.3. Assume that $(\Omega_t)_{t \geq 0}$ is a C^{2m+1}-smooth family of domains with $\Omega_0 = B$. Assume further that transition from positivity of $G_{(-\Delta)^m, \Omega_t}$ to sign change may be observed and let t_g be the largest parameter such that $G_{(-\Delta)^m, \Omega_t} > 0$ for $t \in [0, t_g)$. Then for some $\varepsilon > 0$ and for all $t \in [0, t_g + \varepsilon)$, the first polyharmonic eigenvalue in Ω_t is still simple and the corresponding eigenfunction may be chosen strictly positive.

In order to prove Theorem 6.3 we proceed in three steps.

1. First, we consider $\tilde{a}_{ij} = \delta_{ij}$ and domains Ω which are close to the disk in a conformal sense. In this case the claim can be proved by using conformal maps which leave the principal part $(-\Delta)^m$ invariant. The pulled back differential equation is a lower order perturbation of the polyharmonic equation and Theorem 5.4 is applicable. See Lemma 6.5 below.
2. Next we employ a quantitative version of the Riemannian mapping theorem. Conformal maps $B \to \Omega$ enjoy a representation based on the harmonic Green function in Ω. This representation allows to apply elliptic theory in order to conclude "conformal closeness" from "differentiable closeness". See Lemma 6.6.
3. The theory of normal forms for second order elliptic operators allows to transform the leading coefficients $\tilde{a}_{ij}$ into δ_{ij} thereby giving rise to a further "small" perturbation of the domain Ω. See Lemma 6.7.

Only in two dimensions, the theory of normal forms is available and, moreover, sufficiently many conformal maps exist to deform suitable domains into the unit disk. In higher dimensions, the only conformal maps are Möbius transforms, which map balls onto balls or half spheres.

6.1.1 Domain Perturbations

Lemma 6.5. *There exists $\varepsilon_1 = \varepsilon_1(m) > 0$ such that the following statement holds true. Let Ω be a simply connected bounded $C^{2m,\gamma}$-smooth domain. For the differential operator L in (6.3), we assume that $\tilde{a}_{ij} = \delta_{ij}$. Moreover, let $h : \overline{B} \to \overline{\Omega}$ be a biholomorphic map with $h \in C^{2m,\gamma}(\overline{B})$, $h^{-1} \in C^{2m,\gamma}(\overline{\Omega})$.*

If $\|h - Id\|_{C^{2m-1}(\overline{B})} \le \varepsilon_1$ and $\|a_\beta\|_{C^0(\overline{B})} \le \varepsilon_1$ for all $|\beta| \le 2m - 1$, then the Green function $G_{L,\Omega}$ for the boundary value problem (6.2) in Ω exists and is positive.

Proof. In the disk B, the corresponding result is given in Theorem 5.4. In order to apply this theorem also to the boundary value problem (6.2) in Ω, it has to be "pulled back" to the disk. The crucial point is that conformal maps leave the principal part $(-\Delta)^m$ invariant and yield only additional terms of lower order.

Let $\varepsilon := \max\left\{ \max_{|\beta| \le 2m-1} \|a_\beta\|_{C^0(\overline{\Omega})}, \|h - Id\|_{C^{2m-1}(\overline{B})} \right\}$ be sufficiently small. For the pulled back solution $v : \overline{B} \to \mathbb{R}$, $v(x) := u(h(x))$, using

$$\Delta v(x) = \frac{1}{2} |\nabla h(x)|^2 \left((\Delta u) \circ h \right)(x),$$

the boundary value problem

$$\begin{cases} \left(-\dfrac{2}{|\nabla h|^2} \Delta \right)^m v + \displaystyle\sum_{|\beta| \le 2m-1} \hat{a}_\beta D^\beta v = f \circ h \ \text{in } B, \\ D^\alpha v|_{\partial B} = 0 \hspace{5cm} \text{for } |\alpha| \le m - 1, \end{cases}$$

has to be considered with suitable coefficients $\hat{a}_\beta \in C^{0,\gamma}(\overline{B})$, $\|\hat{a}_\beta\|_{C^0(\overline{B})} = O(\varepsilon)$. Computing $\left(-\frac{2}{|\nabla h|^2}\Delta\right)^m$ yields additional coefficients $D^\mu\left(\frac{1}{|\nabla h|^2}\right)$ with $0 < |\mu| \le 2m-2$ for the lower order terms. The leading term becomes $\left(\frac{2}{|\nabla h|^2}\right)^m(-\Delta)^m v$. Here, $\left\|\frac{2}{|\nabla h|^2} - 1\right\|_{C^{2m-2}(\overline{B})} = O(\varepsilon)$. So, we obtain the boundary value problem

$$\begin{cases} (-\Delta)^m v + \displaystyle\sum_{|\beta| \le 2m-1} \tilde{a}_\beta D^\beta v = \tilde{f} \text{ in } B, \\ D^\alpha v|_{\partial B} = 0 & \text{for } |\alpha| \le m-1, \end{cases}$$

with $\tilde{f} := \left(\frac{|\nabla h|^2}{2}\right)^m f \circ h$ and suitable coefficients $\tilde{a}_\beta \in C^{0,\gamma}(\overline{B})$, which obey the estimate $\|\tilde{a}_\beta\|_{C^0(\overline{B})} = O(\varepsilon)$. Obviously, $f \gneqq 0$ in Ω is equivalent to $\tilde{f} \gneqq 0$ in B. Hence, for sufficiently small ε all statements of Theorem 5.4 carry over to the boundary value problem (6.2). $\qquad\square$

The Riemannian mapping theorem, combined with the Kellogg-Warschawski theorem, see e.g. [345], shows existence of conformal maps which satisfy the qualitative assumptions of Lemma 6.5. Observe that here the assumptions on the domain in Lemma 6.5 are to be used. However, even in very simple domains it may be extremely difficult to give an explicit expression for the conformal map $h : B \to \Omega$ and even more difficult to check explicitly the smallness condition imposed on $\|h - Id\|_{C^{2m-1}(\overline{B})}$. For ellipses such maps were constructed in [367] by means of elliptic functions.

So, Lemma 6.5 is not very useful yet. However, the next lemma gives a general abstract result that "differentiable closeness" always implies "conformal closeness".

Lemma 6.6. *For any $\varepsilon_1 > 0$ there exists $\varepsilon_2 = \varepsilon_2(m, \varepsilon_1) > 0$ such that for $0 \le \varepsilon \le \varepsilon_2$ we have:*

If the $C^{2m,\gamma}$-smooth domain Ω is ε-close to B in C^{2m}-sense, then there exists a biholomorphic map $h : B \to \Omega$, $h \in C^{2m,\gamma}(\overline{B})$, $h^{-1} \in C^{2m,\gamma}(\overline{\Omega})$ with

$$\|h - Id\|_{C^{2m-1}(\overline{B})} \le \varepsilon_1.$$

Proof. Let $g : \overline{B} \to \overline{\Omega}$ be a map according to Definition 6.1 with $\varepsilon := \|g - Id\|_{C^{2m}(\overline{B})}$. In what follows we always assume $\varepsilon \ge 0$ to be sufficiently small. In particular, Ω is then simply connected and bounded, and $0 \in \Omega$.

According to [112], see also [384, Sect. 4.2], a biholomorphic map $h : B \to \Omega$ such that $h \in C^{2m,\gamma}(\overline{B})$, $h^{-1} \in C^{2m,\gamma}(\overline{\Omega})$ may be constructed as follows.

Let $G_{-\Delta,\Omega}$ be the Green function of $-\Delta$ in Ω under Dirichlet boundary conditions. For $x \in \Omega$, we set

$$w(x) := 2\pi G_{-\Delta,\Omega}(x,0)$$

and introduce the conjugate harmonic function

$$w^*(x) := \int_{1/2}^x \left(-\frac{\partial}{\partial \xi_2} w(\xi)\, d\xi_1 + \frac{\partial}{\partial \xi_1} w(\xi)\, d\xi_2\right).$$

Here, the integral is taken along any curve from the complex number $\frac{1}{2}$ to $x = x_1 + ix_2$ in $\Omega \setminus \{0\}$. The function w^* is well-defined up to integer multiples of 2π. Identifying $\mathbb{R}^2$ and $\mathbb{C}$, by means of

$$h^{-1}(x) := \exp\left(-w(x) - iw^*(x)\right), \quad x \in \overline{\Omega},$$

we obtain a well-defined holomorphic map $\overline{\Omega} \to \overline{B}$ enjoying the required qualitative properties. Moreover, $h^{-1}(0) = 0$ and $\frac{1}{2}$ is mapped onto the positive real half axis.

The Green function $G_{-\Delta,\Omega}$ is given by

$$G_{-\Delta,\Omega}(x,0) = -\frac{1}{2\pi}\left(\log|x| - r(x)\right), \quad x \in \overline{\Omega},$$

where $r : \overline{\Omega} \to \mathbb{R}$ solves the boundary value problem

$$\begin{cases} \Delta r = 0 & \text{in } \Omega, \\ r(x) = \varphi(x) \text{ on } \partial\Omega, & \varphi(x) := \log|x|. \end{cases}$$

It is sufficient to show that

$$\|r\|_{C^{2m-1}(\overline{\Omega})} = O(\varepsilon), \tag{6.4}$$

because by virtue of

$$h^{-1}(x) = x \cdot \exp\left(-r(x) - ir^*(x)\right), \quad x \in \overline{\Omega},$$

one obtains $\|h^{-1} - Id\|_{C^{2m-1}(\overline{\Omega})} = O(\varepsilon)$ and finally $\|h - Id\|_{C^{2m-1}(\overline{B})} = O(\varepsilon)$. Here, the estimate $\|r\|_{C^0(\overline{\Omega})} = O(\varepsilon)$ is an obvious consequence of the maximum principle.

We assume first that $\varphi|_{\partial\Omega}$ may be extended by $\hat{\varphi} \in C^{2m}(\overline{\Omega})$ in such a way that

$$\|\hat{\varphi}\|_{C^{2m}(\overline{\Omega})} = O(\varepsilon) \tag{6.5}$$

holds true. The Schauder estimates of Theorem 2.19 then give $\|r\|_{C^{2m-1,\gamma}(\overline{\Omega})} = O(\varepsilon)$ and (6.4) is proved. Here, one should observe that thanks to the ε–closeness of Ω to B in C^{2m}-sense and $m \geq 2$, for all small enough $\varepsilon > 0$ the estimation constants can be chosen independently of ε.

Hence it remains to show that extensions $\hat{\varphi}$ of $\varphi|_{\partial\Omega}$ satisfying (6.5) indeed exist. For this purpose, only the "tangential derivatives" of $\varphi|_{\partial\Omega}$ have to be estimated. This means that it is enough to consider the boundary data being parametrised with the help of the maps $g|_{\partial B} : \partial B \to \partial\Omega$ and $\mathbb{R} \ni t \mapsto (\cos t, \sin t) \in \partial B$:

$$\psi(t) := \varphi\left(g(\cos t, \sin t)\right).$$

For this map, we show that

$$\max_{j=0,\ldots,2m} \max_{t \in \mathbb{R}} \left|\left(\frac{d}{dt}\right)^j \psi\right| = O(\varepsilon). \tag{6.6}$$

Indeed, for $j = 0$ this is due to $\|g - Id\|_{C^0(\overline{B})} = O(\varepsilon)$ and $|\log(1+\varepsilon)| = O(\varepsilon)$. We set $\tilde{g}(t) = g(\cos t, \sin t)$, $\tilde{g} : \mathbb{R} \to \partial\Omega$. For $j \geq 1$ a general chain rule shows that

$$\left(\frac{d}{dt}\right)^j \psi = \left(\frac{d}{dt}\right)^j (\varphi \circ \tilde{g})$$

$$= \sum_{|\beta|=1}^{j} \left(\left(D^\beta \varphi\right) \circ \tilde{g}\right) \left(\sum_{\substack{p_1 + \ldots + p_{|\beta|} = j \\ p_1, \ldots, p_{|\beta|} \geq 1}} d_{j,\beta,\mathbf{p}} \prod_{\ell=1}^{|\beta|} \left(\frac{d}{dt}\right)^{p_\ell} \tilde{g}^{(\mu_\ell)} \right)$$

with suitable coefficients $d_{j,\beta,\mathbf{p}}$, $\mathbf{p} = (p_1, \ldots, p_{|\beta|})$. The coefficient μ_ℓ refers to the component of $\tilde{g}$ and is chosen as $\mu_\ell = 1$ for $\ell = 1, \ldots, \beta_1$ and $\mu_\ell = 2$ for $\ell = \beta_1 + 1, \ldots, |\beta| = \beta_1 + \beta_2$. To show that this huge sum is indeed $O(\varepsilon)$, we observe that it is equal to 0, provided $\Omega = B$ and $g = Id$. So we put $\tilde{g}_0(t) = Id \circ (\cos t, \sin t) = (\cos t, \sin t)$ and compare corresponding terms. We obtain

$$\left(\frac{d}{dt}\right)^j \psi = \sum_{|\beta|=1}^{j} \left(\left((D^\beta \varphi) \circ \tilde{g} - (D^\beta \varphi) \circ \tilde{g}_0 \right) + (D^\beta \varphi) \circ \tilde{g}_0 \right)$$

$$\times \left(\sum_{\substack{p_1 + \ldots + p_{|\beta|} = j \\ p_1, \ldots, p_{|\beta|} \geq 1}} d_{j,\beta,\mathbf{p}} \prod_{\ell=1}^{|\beta|} \left(\left(\left(\frac{d}{dt}\right)^{p_\ell} \tilde{g}^{(\mu_\ell)} - \left(\frac{d}{dt}\right)^{p_\ell} \tilde{g}_0^{(\mu_\ell)} \right) \right. \right.$$

$$\left. \left. + \left(\frac{d}{dt}\right)^{p_\ell} \tilde{g}_0^{(\mu_\ell)} \right) \right).$$

As already mentioned, thanks to $\varphi(\tilde{g}_0(t)) = \log|(\cos t, \sin t)| \equiv 0$ the sum taken over all the terms which contain only $\tilde{g}_0$ and no differences, equals 0. In the remaining sum, each term contains at least one factor of the type

$$(D^\beta \varphi) \circ \tilde{g} - (D^\beta \varphi) \circ \tilde{g}_0 \qquad \text{or} \qquad \left(\frac{d}{dt}\right)^{p_\ell} \left(\tilde{g}^{(\mu_\ell)} - \tilde{g}_0^{(\mu_\ell)} \right).$$

For $\varepsilon \to 0$ each of these factors is $O(\varepsilon)$, and the remaining factors are uniformly bounded independently of ε. This proves (6.6) and the claim of the lemma. $\qquad \square$

6.1.2 Perturbations of the Principal Part

We define

$$L_0 u := -\sum_{i,j=1}^{2} \tilde{a}_{ij}(x) \frac{\partial^2 u}{\partial x_i \partial x_j}, \qquad \tilde{a}_{ij} = \tilde{a}_{ji} \in C^{2m-1,\gamma}(\overline{\Omega}), \tag{6.7}$$

the second order elliptic operator whose m-th power forms the principal part of the operator L in (6.3) under investigation. By means of a suitable coordinate transformation $(x_1, x_2) \mapsto (\xi_1, \xi_2)$, $\Omega \to \Omega^\natural$, (6.7) can be reduced to normal form

$$\tilde{L}_0 v = -A(\xi)\Delta v - B_1(\xi)\frac{\partial v}{\partial \xi_1} - B_2(\xi)\frac{\partial v}{\partial \xi_2},$$

see e.g. [177, pp. 66–68]. In this way, the operator L is transformed into an operator $\tilde{L}$ where Lemma 6.5 becomes applicable. We check that $\tilde{L}$ remains "close" to $(-\Delta)^m$ and $\Omega^\natural$ "close" to B, if the same holds for L and Ω, respectively. The new coordinates $\xi_1 = \varphi(x_1, x_2)$, $\xi_2 = \psi(x_1, x_2)$ satisfy the Beltrami system in Ω, namely

$$\frac{\partial \varphi}{\partial x_1} = \frac{\tilde{a}_{21}\psi_{x_1} + \tilde{a}_{22}\psi_{x_2}}{\sqrt{\tilde{a}_{11}\tilde{a}_{22} - \tilde{a}_{12}^2}}, \qquad \frac{\partial \varphi}{\partial x_2} = -\frac{\tilde{a}_{11}\psi_{x_1} + \tilde{a}_{12}\psi_{x_2}}{\sqrt{\tilde{a}_{11}\tilde{a}_{22} - \tilde{a}_{12}^2}}. \tag{6.8}$$

Assume that we have already found a bijective, at least twice differentiable transformation

$$\Phi = (\varphi, \psi) : \Omega \to \Omega^\natural. \tag{6.9}$$

Then as in [177] one finds that

$$L_0 u = \left(\tilde{L}_0 v\right) \circ \Phi, \tag{6.10}$$

where we put

$$\begin{cases} v(\xi_1, \xi_2) &= u \circ \Phi^{-1}(\xi_1, \xi_2), \\ A\left(\Phi(x)\right) &= \tilde{a}_{11}(x)\varphi_{x_1}^2 + 2\tilde{a}_{12}(x)\varphi_{x_1}\varphi_{x_2} + \tilde{a}_{22}(x)\varphi_{x_2}^2 \\ &= \tilde{a}_{11}(x)\psi_{x_1}^2 + 2\tilde{a}_{12}(x)\psi_{x_1}\psi_{x_2} + \tilde{a}_{22}(x)\psi_{x_2}^2 > 0, \\ B_1\left(\Phi(x)\right) &= \tilde{a}_{11}(x)\varphi_{x_1 x_1} + 2\tilde{a}_{12}(x)\varphi_{x_1 x_2} + \tilde{a}_{22}(x)\varphi_{x_2 x_2}, \\ B_2\left(\Phi(x)\right) &= \tilde{a}_{11}(x)\psi_{x_1 x_1} + 2\tilde{a}_{12}(x)\psi_{x_1 x_2} + \tilde{a}_{22}(x)\psi_{x_2 x_2}. \end{cases} \tag{6.11}$$

We determine ψ as solution of the boundary value problem

$$\begin{cases} \dfrac{\partial}{\partial x_1}\left(\dfrac{\tilde{a}_{11}\psi_{x_1} + \tilde{a}_{12}\psi_{x_2}}{\sqrt{\tilde{a}_{11}\tilde{a}_{22} - \tilde{a}_{12}^2}}\right) + \dfrac{\partial}{\partial x_2}\left(\dfrac{\tilde{a}_{21}\psi_{x_1} + \tilde{a}_{22}\psi_{x_2}}{\sqrt{\tilde{a}_{11}\tilde{a}_{22} - \tilde{a}_{12}^2}}\right) = 0 & \text{in } \Omega, \\ \psi(x) = x_2 & \text{on } \partial\Omega, \end{cases} \tag{6.12}$$

and then construct φ with the help of the Beltrami equations (6.8) and the normalisation $\varphi(0) = 0$.

In this special situation, the required results for the transformation Φ can be easily proved directly.

Lemma 6.7. *Let $\varepsilon_2 > 0$. Then there exists $\varepsilon_3 = \varepsilon_3(m, \varepsilon_2)$ such that for $0 \leq \varepsilon \leq \varepsilon_3$ the following holds true.*

Assume that the domain Ω is $C^{2m,\gamma}$-smooth and ε-close to B in C^{2m}-sense. Let the operator L of (6.3) be ε-close to $(-\Delta)^m$ in $C^{2m-1,\gamma}$-sense. Then we have for the transformation $\Phi : \Omega \to \Omega^\natural = \Phi(\Omega)$ defined in (6.8), (6.9) and (6.12), that

- Φ *is bijective,* $\Phi \in C^{2m,\gamma}(\overline{\Omega})$, $\Phi^{-1} \in C^{2m,\gamma}(\overline{\Omega^\natural})$,
- $\Omega^\natural$ *is ε_2-close to B in C^{2m}-sense.*

Putting $v := u \circ \Phi^{-1}$, see (6.11), the boundary value problem

$$\begin{cases} Lu = f & \text{in } \Omega, \\ D^\alpha u|_{\partial\Omega} = 0 & \text{for } |\alpha| \leq m-1, \end{cases}$$

is transformed into

$$\begin{cases} \hat{L}v = A^{-m}\left(f \circ \Phi^{-1}\right) \text{ in } \Omega^\natural, \\ D^\alpha v|_{\partial\Omega^\natural} = 0 & \text{for } |\alpha| \leq m-1. \end{cases}$$

Here $\hat{L}v = (-\Delta)^m v + \sum_{|\beta|\leq 2m-1} \hat{a}_\beta(.)D^\beta v$ with suitable coefficients $\hat{a}_\beta \in C^{0,\gamma}(\overline{\Omega^\natural})$ such that for all $|\beta| \leq 2m-1$ the smallness condition

$$\|\hat{a}_\beta\|_{C^0(\overline{\Omega^\natural})} \leq \varepsilon_2$$

is satisfied.

Proof. We may assume ε to be sufficiently small and in particular Ω to be bounded and uniformly convex. First we consider the boundary value problem (6.12), which is uniformly elliptic thanks to $\|\tilde{a}_{ij} - \delta_{ij}\|_{C^{2m-1,\gamma}(\overline{\Omega})} \leq \varepsilon$ with coefficients in the space $C^{2m-1,\gamma}(\overline{\Omega})$. Since Ω is $C^{2m,\gamma}$-smooth, elliptic theory (see Theorem 2.19) shows the existence of a solution $\psi \in C^{2m,\gamma}(\overline{\Omega})$ to (6.12). At the same time, this differential equation is the integrability condition for (6.8) in the convex domain Ω. This shows that a solution $\varphi \in C^{2m,\gamma}(\overline{\Omega})$ of the Beltrami system (6.8) with $\varphi(0) = 0$ exists.

Next we investigate $\Phi = (\varphi, \psi)$ quantitatively. For this purpose we consider the auxiliary function $\Psi(x) := \psi(x) - x_2$ that solves the boundary value problem

$$\begin{cases} \dfrac{\partial}{\partial x_1}\left(\dfrac{\tilde{a}_{11}\Psi_{x_1} + \tilde{a}_{12}\Psi_{x_2}}{\sqrt{\tilde{a}_{11}\tilde{a}_{22} - \tilde{a}_{12}^2}}\right) + \dfrac{\partial}{\partial x_2}\left(\dfrac{\tilde{a}_{21}\Psi_{x_1} + \tilde{a}_{22}\Psi_{x_2}}{\sqrt{\tilde{a}_{11}\tilde{a}_{22} - \tilde{a}_{12}^2}}\right) \\ \qquad = -\dfrac{\partial}{\partial x_1}\left(\dfrac{\tilde{a}_{12}}{\sqrt{\tilde{a}_{11}\tilde{a}_{22} - \tilde{a}_{12}^2}}\right) - \dfrac{\partial}{\partial x_2}\left(\dfrac{\tilde{a}_{22}}{\sqrt{\tilde{a}_{11}\tilde{a}_{22} - \tilde{a}_{12}^2}}\right) =: F(x_1, x_2) \text{ in } \Omega, \\ \Psi|_{\partial\Omega} = 0, \end{cases}$$

$$\tag{6.13}$$

where

$$F = O(\varepsilon) \text{ in } C^{2m-2,\gamma}(\overline{\Omega}).$$

Schauder estimates for higher order norms as in Theorem 2.19 yield

$$\|\psi - x_2\|_{C^{2m,\gamma}(\overline{\Omega})} = \|\Psi\|_{C^{2m,\gamma}(\overline{\Omega})} \le C\|F\|_{C^{2m-2,\gamma}(\overline{\Omega})},$$
$$\|\psi - x_2\|_{C^{2m}(\overline{\Omega})} \le C\|F\|_{C^{2m-2,\gamma}(\overline{\Omega})}. \tag{6.14}$$

Here one should observe that the $C^{2m-1,\gamma}(\overline{\Omega})$-norms of the coefficients in (6.13) are bounded independently of ε; the ellipticity constants are uniformly close to 1. Finally, by means of the uniform C^{2m}-closeness of the domains to the disk B we may choose an estimation constant in (6.14) being independent of Ω. Taking also (6.8) into account we conclude that

$$\|\Phi - Id\|_{C^{2m}(\overline{\Omega})} = O(\varepsilon), \tag{6.15}$$

thereby proving the bijectivity of Φ, the qualitative statements on Φ^{-1} and $\Omega^{\natural} = \Phi(\Omega)$, as well as

$$\|\Phi^{-1} - Id\|_{C^{2m}(\overline{\Omega^{\natural}})} = O(\varepsilon). \tag{6.16}$$

We still have to study the properties of the transformed differential operator $\hat{L}$. From (6.10) it follows that

$$\begin{aligned}
Lu &= L_0^m u + \sum_{|\beta| \le 2m-1} a_\beta D^\beta u \\
&= \left\{ \tilde{L}_0^m v + \sum_{|\beta| \le 2m-1} \left(a_\beta \circ \Phi^{-1}\right) \left(D^\beta(v \circ \Phi)\right) \circ \Phi^{-1} \right\} \circ \Phi \\
&= \left\{ \tilde{L}_0^m v + \sum_{|\beta| \le 2m-1} \tilde{a}_\beta D^\beta v \right\} \circ \Phi =: \left(A^m \, \hat{L} v\right) \circ \Phi.
\end{aligned}$$

Here the new coefficients $\tilde{a}_\beta$ contain additional derivatives of Φ of order at most $(2m-1)$ and hence $\|\tilde{a}_\beta\|_{C^0(\overline{\Omega^{\natural}})} = O(\varepsilon)$. Finally, $\tilde{L}_0 v = -A\Delta v - B_1 \frac{\partial v}{\partial \xi_1} - B_2 \frac{\partial v}{\partial \xi_2}$, so we still need to show that

$$\|A - 1\|_{C^{2m-2}(\overline{\Omega^{\natural}})} = O(\varepsilon), \qquad \|B_j\|_{C^{2m-2}(\overline{\Omega^{\natural}})} = O(\varepsilon).$$

Observing the definition (6.11) of A, B_1, B_2, this follows from the properties (6.15) and (6.16) of Φ and the assumptions on the coefficients $\tilde{a}_{ij}$. $\square$

Proof of Theorem 6.3. It follows by combining Lemmas 6.5–6.7. $\square$

Remark 6.8. Similarly as in Section 5.2, also here one has results on the qualitative boundary behaviour of solutions. Using the theory of maps described above and referring to Theorem 5.7 instead of Theorem 5.4, the claim of Theorem 5.7

on the m-th order boundary derivatives of the solution remains true also under the assumptions of Theorem 6.3, while the Dirichlet boundary data have to be prescribed homogeneously.

On the other hand, if one wants to study the influence of the sign of $D^{m-1}u|_{\partial\Omega}$ on the sign of the solution in Ω, while the first $(m-2)$ derivatives on $\partial\Omega$ are prescribed homogeneously, one has to ensure that the assumptions of Theorem 6.3 are satisfied by the (formally) adjoint operator L^*. This means that if we assume that Ω is close to B in C^{2m}-sense, L close to $(-\Delta)^m$ in C^{2m}-sense, $a_\beta \in C^{|\beta|}(\overline{\Omega})$ and $\|a_\beta\|_{C^{|\beta|}(\overline{\Omega})}$ small, then the conclusions of Theorem 5.6 remain true.

Our methods are not suitable to carry over further statements of Section 5.2 on the influence of $u|_{\partial B}$ on the sign of u in B to the situation of Theorem 6.3. This is because in the relevant result Theorem 5.15, only perturbations of order $(m-2)$ can be treated, while terms of order $(m-1)$ may indeed arise. However, in the special case of the polyharmonic operator, using Sassone's paper [359], we expect that a positivity result with respect to the two highest order boundary data may also be shown in domains Ω being a sufficiently small perturbation of the disk.

6.2 Small Negative Part of Biharmonic Green's Functions in Two Dimensions

We come back to the question raised in Section 1.2 whether the negative part of the biharmonic Green function is small in a suitable sense when compared with its positive part. In two dimensions, we have a family of domains – among which are even nonconvex ones – with positive Green's functions. These limaçons de Pascal are discussed first and serve as a basis to give a first answer to the question just mentioned.

6.2.1 The Biharmonic Green Function on the Limaçons de Pascal

Lemma 6.5 does not supply us with a reasonable bound for the perturbation in order to have a positive Green function. Hadamard found an explicit formula for the biharmonic Green function on any limaçon. As already mentioned in Section 1.2 he claimed in [223] that these Green functions were all positive. Although this claim is wrong, his formula allowed Dall'Acqua and Sweers [120] to show that the Green functions for a sufficiently large class of limaçons are indeed positive. We define the filled limaçon by

$$\Omega_a = (-a,0) + \left\{ (\rho\cos\varphi, \rho\sin\varphi) \in \mathbb{R}^2;\ 0 \le \rho < 1 + 2a\cos\varphi \right\}. \qquad (6.17)$$

For $a \in \left[0, \frac{1}{2}\right]$ the boundary is defined by $\rho = 1 + 2a \cos \varphi$; for $a = 0$ it is the unit circle and for $a = \frac{1}{2}$ one finds the cardioid. In Figure 1.2 in Section 1.2, images are shown of these limaçons which are rotated by $\frac{1}{2}\pi$.

Proposition 6.9. *The biharmonic Green function for Ω_a with $a \in \left[0, \frac{1}{2}\right]$ is positive if and only if $a \in \left[0, \frac{1}{6}\sqrt{6}\right]$.*

Before proving this result we fix some preliminaries that will give us additional information on what happens when positivity breaks down. To do so, we fix the conformal map $h_a : B = \Omega_0 \to \Omega_a$ that maps the unit disk on the perturbed domains and that keeps the horizontal axis on the horizontal axis. In complex coordinates it is defined as follows

$$h_a(\eta) = \eta + a\eta^2.$$

The Green function on Ω_a is defined through the coordinates on B; see Figure 6.1 for these curvilinear coordinates. We remark that according to Loewner [279] the only conformal maps which leave the biharmonic equation invariant are the Möbius transforms.

Let us write

$$h_a(\eta) = x_1 + ix_2 \quad \text{and} \quad h_a(\xi) = y_1 + iy_2$$

and

$$r = |\eta - \xi|, \quad R = |1 - \eta\bar{\xi}| \quad \text{and} \quad s = \left|\eta + \xi + \frac{1}{a}\right|.$$

The formula Hadamard presents in [223] using these coordinates is

$$G_{\Omega_a}(x,y) = \frac{1}{16\pi} a^2 s^2 r^2 \left(\frac{R^2}{r^2} - 1 - \log\left(\frac{R^2}{r^2}\right) - \frac{a^2}{1 - 2a^2} \frac{R^2}{s^2} \frac{r^2}{R^2} \left(\frac{R^2}{r^2} - 1\right)^2 \right).$$

To verify that this is indeed the biharmonic Green function for Ω_a requires some tedious calculations which can be found in [120]. By setting

$$F(\beta, q) = q - 1 - \log q - \beta \frac{(q-1)^2}{q} \tag{6.18}$$

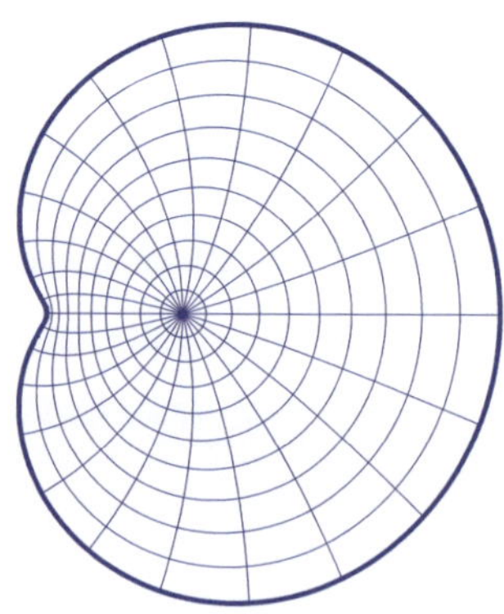

Fig. 6.1 Transformed polar coordinates corresponding to h_a.

one obtains that

$$G_{\Omega_a}(x,y) = \frac{1}{16\pi} a^2 s^2 r^2 F(\beta,q),$$

and that the Green function $G_{\Omega_a}(x,y)$ is positive if and only if $F(\beta,q)$ is positive for all the appropriate values of

$$\beta = \frac{a^2}{1-2a^2}\frac{R^2}{s^2} \quad \text{and} \quad q = \frac{R^2}{r^2}.$$

One has $q \in [1,\infty]$ and moreover, for $x \to y \in \Omega_a$ one finds that $q \to \infty$ and for $x \to x^*$ or $y \to y^*$ with $x^*,y^* \in \partial\Omega_a$ and $x^* \neq y^*$ it follows that $q \to 1$.

Lemma 6.10. *The function $q \mapsto F(\beta,q) : [1,\infty) \to \mathbb{R}$ has the following properties. It has a double zero for $q = 1$ and*

- *if $0 < \beta \leq \frac{1}{2}$, then it is positive on $(1,\infty)$ and convex;*
- *if $\frac{1}{2} < \beta < 1$, then there is $q_\beta > 1$ such that the function is negative on $(1,q_\beta)$ and positive on (q_β,∞);*
- *if $\beta \geq 1$, then the function is negative on $(1,\infty)$.*

For a plot of $q \mapsto F(\beta,q)$ see Fig. 6.2.

Proof. Let us first give the derivatives

$$\frac{\partial}{\partial q}F(\beta,q) = \frac{q-1}{q}\left(1 - \beta\frac{q+1}{q}\right) \quad \text{and} \quad \frac{\partial^2}{\partial q^2}F(\beta,q) = \frac{q-2\beta}{q^3}.$$

One finds stationary points $q = 1$ and $q = q_\beta := \beta/(1-\beta)$ with $q_\beta > 1$ only if $\beta \in \left(\frac{1}{2},1\right)$. From $\frac{\partial^2}{\partial q^2}F(\beta,q) < 0$ for $q < 2\beta$ and $\frac{\partial^2}{\partial q^2}F(\beta,q) > 0$ for $q > 2\beta$ the claim follows. $\qquad\square$

Remark 6.11. For $\eta = \bar{\xi} = -2a + i\sqrt{1-4a^2}$ one finds

$$\beta = \frac{4a^4}{(1-2a^2)(1-4a^2)} \quad \text{and} \quad q = 1. \tag{6.19}$$

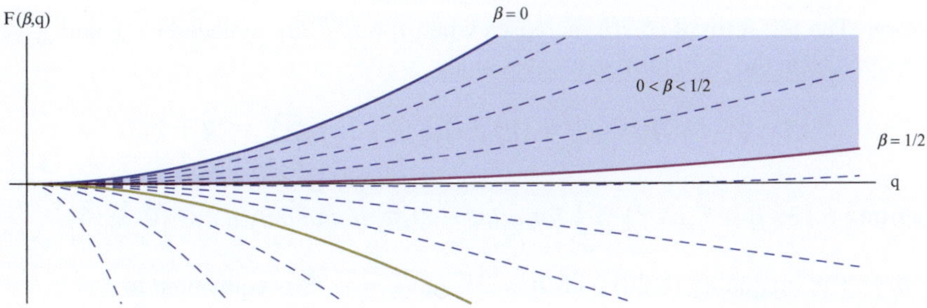

Fig. 6.2 Only the functions $q \mapsto F(\beta,q)$ in the shaded area produce positive Green functions.

Since β and q are continuous functions of η and ξ for $\xi \neq \eta \in B$, a necessary condition for positivity is

$$\frac{4a^4}{(1-2a^2)(1-4a^2)} \leq \frac{1}{2}.$$

For $a \in \left[0, \frac{1}{2}\right]$ this last inequality is equivalent to $a \in \left[0, \frac{1}{6}\sqrt{6}\right]$.

For $a \in (\frac{1}{6}\sqrt{6}, 1/2]$ this disproves a claim by Hadamard [223], according to which the clamped plate problem is positivity preserving for all limaçons.

In order to find a positive Green function it is hence sufficient to show that $\beta \leq \frac{1}{2}$, which is in its turn guaranteed when, for $a \in \left[0, \frac{1}{6}\sqrt{6}\right]$, the following holds for all $\eta, \xi \in B$:

$$2a^4 \left|1 - \eta\bar{\xi}\right|^2 \leq (1-2a^2) \left|1 + a(\eta + \xi)\right|^2.$$

Setting $\eta_1 = \mathrm{Re}\,\eta$, $\eta_2 = \mathrm{Im}\,\eta$ and $\xi_1 = \mathrm{Re}\,\xi$, $\eta_2 = \mathrm{Im}\,\eta$ and $\langle \eta, \xi \rangle = \eta_1 \xi_1 + \eta_2 \xi_2$, this condition can be rewritten as

$$\tilde{\beta}(\eta, \xi) \leq 0, \tag{6.20}$$

where $\tilde{\beta}(\eta, \xi)$ is defined by

$$\tilde{\beta}(\eta, \xi) := 2a^4 \left(1 + |\eta|^2\right)\left(1 + |\xi|^2\right)$$
$$- \left((1-2a^2)(1 + 2a(\eta_1 + \xi_1)) + a^2 |\eta + \xi|^2\right).$$

In order to show $\beta \leq \frac{1}{2}$ on B^2 we may restrict ourselves to check (6.20) only for certain pairs $(\eta, \xi) \in B^2$. For the limaçons the next lemma confirms the conjecture that sign change under smooth perturbations of the domain starts from the boundary.

Lemma 6.12. *Let $\eta, \xi \in B$. It holds that $\tilde{\beta}(\eta, \xi) \leq \tilde{\beta}(\chi, \bar{\chi})$ for $\chi \in B$ satisfying*

$$|\chi| \geq \max(|\eta|, |\xi|) \text{ and } \mathrm{Re}\,\chi = \frac{1}{2}\mathrm{Re}(\eta + \xi).$$

Proof. The left term of (6.20) increases when η and ξ are replaced by χ and $\bar{\chi}$. At the same time the right term decreases since

$$|\eta + \xi|^2 = (\eta_1 + \xi_1)^2 + (\eta_2 + \xi_2)^2 \geq (\eta_1 + \xi_1)^2 = |\chi + \bar{\chi}|^2.$$

$\square$

Lemma 6.13. $\beta\left(e^{i\varphi}, e^{-i\varphi}\right) \leq \frac{1}{2}$ *for all $\varphi \in [0, \pi]$ if and only if $a \in \left[0, \frac{1}{6}\sqrt{6}\right]$.*

Proof. The inequality (6.20) with $\eta = e^{i\varphi}$ and $\xi = e^{-i\varphi}$ is equivalent to

$$(1-2a^2)(1 + 4a\cos\varphi) + 4a^2(\cos\varphi)^2 - 8a^4 \geq 0.$$

The minimum of $f_a(t) = \left(1 - 2a^2\right)\left(1 + 4a\,t\right) + 4a^2t^2 - 8a^4$ is assumed for

$$t_a = \frac{1 - 2a^2}{-2a}.$$

Only when $a \in \left[\frac{1}{2}\left(\sqrt{3} - 1\right), \frac{1}{2}\right]$ one finds $t_a \geq -1$. Hence for $a \in \left[0, \frac{1}{2}\left(\sqrt{3} - 1\right)\right]$ the function f_a on $[-1, 1]$ has its minimum in -1 and

$$\left(1 - 2a^2\right)\left(1 + 4a\cos\varphi\right) + 4a^2\left(\cos\varphi\right)^2 - 8a^4 \geq \left(2a - 1\right)^2\left(1 - 2a^2\right) > 0.$$

For $a \in \left[\frac{1}{2}\left(\sqrt{3} - 1\right), \frac{1}{2}\right]$ one finds that f_a on $[-1, 1]$ has its minimum in t_a:

$$\min_{0 \leq \varphi \leq \pi} \left(1 - 2a^2\right)\left(1 + 4a\cos\varphi\right) + 4a^2\left(\cos\varphi\right)^2 - 8a^4 = f_a\left(t_a\right) = 2a^2\left(1 - 6a^2\right).$$

This last expression is positive if and only if $a \in \left[0, \frac{1}{6}\sqrt{6}\right]$. $\qquad\square$

Proof of Proposition 6.9. By Lemma 6.10 a necessary and sufficient condition for $G_{\Omega_a}\left(.,.\right)$ to be positive is $\beta \leq \frac{1}{2}$, which is guaranteed by $\tilde{\beta} \leq 0$. By Lemma 6.12 one finds that it suffices to prove

$$\sup_{\chi \in \partial B} \beta\left(\chi, \bar{\chi}\right) \leq \frac{1}{2}.$$

The last step follows from Lemma 6.13. $\qquad\square$

6.2.2 Filling Smooth Domains with Perturbed Limaçons

As mentioned before, the Green function for the biharmonic Dirichlet problem is in general not positive. The few exceptions that we know can be classified as follows:

- Domains for which an explicit biharmonic Green function can be constructed and for which this Green function is positive. Typical examples are the limaçons of the previous section.
- Domains which are small perturbations, in the sense of Theorem 6.3, of a domain that is known to have a positive biharmonic Green function with G satisfying similar estimates as the Green function on the disk.

We have formulated this second class of domains more widely than just perturbations of the disk. Indeed, the proof of Theorem 6.3 can be copied starting with any limaçon Ω_a as in (6.17) with $a < \frac{1}{6}\sqrt{6}$. One could even start with domains which are images of those limaçons under a Möbius mapping. Although here we only need Möbius transforms in $\mathbb{R}^2$, we briefly discuss them in $\mathbb{R}^n$. This shows that it is not a lack of Möbius transforms which prevents us from extending the present reasoning to any space dimension.

Let us recall that every Möbius transformation ϕ of $\mathbb{R}^n$ can be written as

$$\phi = \phi_1 \circ j \circ \phi_2$$

with j an inversion:

$$j(x) = |x|^{-2} x$$

and ϕ_i a similarity:

$$\phi_i(x) = a + c\, M\, x, \tag{6.21}$$

where M is an orthogonal matrix, c a nonzero multiplication and a a shift. They possess the following property.

Lemma 6.14. *Let ϕ be a Möbius transformation in $\mathbb{R}^n$. Then*

$$\Delta^k \left(J_\phi^{\frac{1}{2}-\frac{k}{n}}\, u \circ \phi \right) = J_\phi^{\frac{1}{2}+\frac{k}{n}} \left(\Delta^k u \right) \circ \phi$$

with $J_\phi = \left| \det \left(\frac{\partial \phi_i}{\partial x_j} \right) \right|$, the modulus of the Jacobian determinant.

For a proof see for example [119]. Only due to the inversion the shape of a domain will change. This lemma allows one to pull back the Dirichlet biharmonic problem from $\phi(\Omega_a)$ to Ω_a and to use the Green function on Ω_a for the problem on $\phi(\Omega_a)$. Of course, a Möbius transformed ball is still a ball. Examples of Möbius transformed limaçons can be seen in [119].

Having seen the relatively limited possibilities to have a positive Green function, one is curious to see how big the negative contribution of the Dirichlet biharmonic Green function might be.

We will give a first and still preliminary answer to the following question which was raised already in Section 1.2 and which may be considered as an adequate reformulation of the Boggio-Hadamard conjecture:

Is positivity preserving in any bounded smooth domain Ω possibly "almost true" in the sense that the negative part $G_\Omega^-(x,y) := \max\{-G_\Omega(x,y), 0\}$ of the biharmonic Green function under Dirichlet boundary conditions is "small relatively" to the singular positive part $G_\Omega^+(x,y) := \max\{G_\Omega(x,y), 0\}$?

In this section we will give an outline for the two-dimensional case, while for general dimensions, we refer to Section 6.4.

The main result is as follows.

Theorem 6.15. *Let $\Omega \subset \mathbb{R}^2$ be a bounded $C^{4,\gamma}$-smooth domain and let G_Ω denote the biharmonic Green function in Ω under Dirichlet boundary conditions.*

Then there exists a positive minimal distance $\delta = \delta(\Omega) > 0$ such that for any two points $x, y \in \Omega$, $x \neq y$,

$$|x - y| < \delta \text{ implies that } G_\Omega(x,y) > 0. \tag{6.22}$$

In particular, there exists a constant $C = C(\Omega) > 0$ such that for all $x, y \in \Omega$, $x \neq y$, we have the following estimate from below:

$$G_\Omega(x,y) \geq -Cd(x)^2 d(y)^2. \tag{6.23}$$

Remark 6.16. Together with the estimate from Theorem 4.28 (case $|\alpha| = |\beta| = 0$) we obtain that with a constant $C = C(\Omega)$

$$-Cd(x)^2 d(y)^2 \leq G_\Omega(x,y) \leq Cd(x)d(y)\min\left\{1, \frac{d(x)d(y)}{|x-y|^2}\right\}.$$

Both estimates are sharp in general. For the lower bound we refer to Garabedian's example [176], while for the upper bound one may see Theorem 4.6. For x or y closer to the boundary than to each other, the estimate from below is by a factor $|x-y|^2$ smaller than the estimate from above.

The key step in proving Theorem 6.15 consists in decomposing the Green function into a singular positive part modeled along sums of Green functions in suitably perturbed limaçons and a subordinate smooth possibly negative regular part.

Lemma 6.17. *Assume that $\Omega \subset \mathbb{R}^2$ is a bounded domain with $\partial\Omega \in C^{4,\gamma}$. Then the biharmonic Green function G_Ω for (6.1) can be written as*

$$G_\Omega(x,y) = G_\Omega^{sing}(x,y) + G_\Omega^{reg}(x,y)$$

such that for some $c_2, c_1 > 0$:

$$0 \leq G_\Omega^{sing}(x,y) \leq c_1 d(x)d(y)\min\left\{1, \frac{d(x)d(y)}{|x-y|^2}\right\}, \tag{6.24}$$

and

$$\left|G_\Omega^{reg}(x,y)\right| \leq c_2 d(x)^2 d(y)^2$$

for all $(x,y) \in \Omega^2$. Moreover:

1. $G_\Omega^{sing} \in C^{1,\tilde{\gamma}}(\bar{\Omega}^2) \cap C_0^1(\bar{\Omega}^2) \cap C^{4,\gamma}(\bar{\Omega}^2 \setminus \{(x,x); x \in \Omega\})$ for all $\tilde{\gamma} \in (0,1)$;
2. $G_\Omega^{reg} \in C^{4,\gamma}(\bar{\Omega}^2) \cap C_0^1(\bar{\Omega}^2)$.

For a compact set K, the space $C_0^1(K)$ is defined to consist of all functions $g \in C^1(K)$ satisfying $g = |\nabla g| = 0$ on ∂K.

The proof of this lemma is rather technical and lengthy. A detailed proof can be found in [117]. The main ingredients are first the positivity of the Green functions for the limaçon and its ε-close $C^{2,\gamma}$-perturbations, and secondly filling the domain from the interior with finitely many of those limaçons. See Figure 6.3.

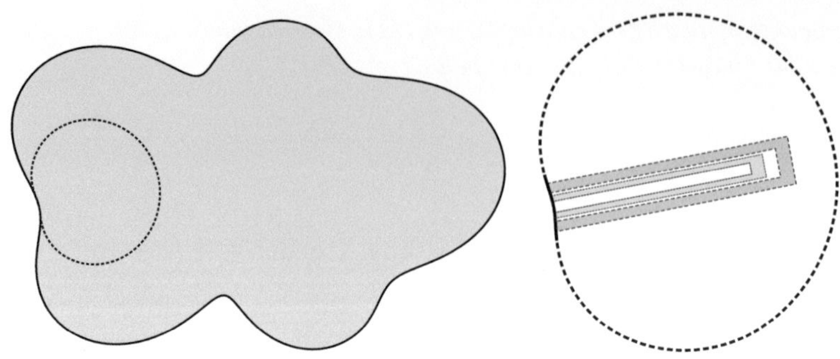

Fig. 6.3 On the left a domain Ω with one of the finitely many subdomains Ω_i close to limaçons for which the union is exactly this domain. On the right this subdomain with the supports of $\nabla \chi_i$ and $\nabla \psi_i$.

Sketch of proof of Lemma 6.17. In the first step one proves that for the domain Ω in Lemma 6.17 there exist finitely many, say k, domains Ω_i, which are, after a similarity transformation, sufficiently ε-close in $C^{3,\gamma}$-sense to a limaçon Ω_{a_i} with $a_i < \frac{1}{6}\sqrt{6}$, that is

$$\Omega = \bigcup_{1 \le i \le k} \Omega_i.$$

Here, sufficiently close is such that the individual biharmonic Green functions on Ω_i satisfy for some $c_1, c_2 \in \mathbb{R}^+$

$$c_1 d_i(x)d_i(y) \min\left\{1, \frac{d_i(x)d_i(y)}{|x-y|^2}\right\} \le G_{\Omega_i}(x,y) \le c_2 d_i(x)d_i(y) \min\left\{1, \frac{d_i(x)d_i(y)}{|x-y|^2}\right\},$$

for all $x, y \in \Omega_i$ where

$$d_i(x) = \inf\left\{|x-x^*|\,;x^* \in \partial\Omega_i\right\}.$$

The boundary $\partial\Omega$ is then covered by finitely many arcs $\Gamma_i \subset \partial\Omega_i$, say $i = 1,\ldots,$ $k' \le k$:

$$\partial\Omega = \bigcup_{1 \le i \le k'} \Gamma_i.$$

For the second step one constructs a partition of unity $\left\{\chi_i \in C^\infty\left(\bar{\Omega};[0,1]\right)\right\}_{i=1}^k$ such that $\sum_{i=1}^k \chi_i = 1$ on $\bar{\Omega}$ and

$$\operatorname{support}(\chi_i) \subset \Omega_i \cup \Gamma_i.$$

One defines cut-off functions $\left\{\psi_i \in C^\infty\left(\bar{\Omega};[0,1]\right)\right\}_{i=1}^k$, related to this partition of unity, with

$$\operatorname{support}(\psi_i) \subset \Omega_i \cup \Gamma_i$$
$$\psi_i(x) = 1 \text{ for } x \in \operatorname{support}(\chi_i),$$
$$\operatorname{support}(\nabla\psi_i) \cap \operatorname{support}(\chi_i) = \emptyset.$$

See Figure 6.3. Next, one defines

$$G_\Omega^{\mathrm{sing}}(x,y) = \sum_{i=1}^{k} \psi_i(x) G_{\Omega_i}(x,y) \chi_i(y),$$

$$G_\Omega^{\mathrm{reg}}(x,y) = G_\Omega(x,y) - G_\Omega^{\mathrm{sing}}(x,y),$$

with $\psi_i(x) G_{\Omega_i}(x,y)\chi_i(y)$ extended by 0 if $x \notin \Omega_i$ or $y \notin \Omega_i$.

Take some $y \in \Omega$, keep it fixed in what follows and define $I_y := \{i : y \in \mathrm{support}(\chi_i)\}$. Then

$$G_\Omega^{\mathrm{sing}}(x,y) = \sum_{i \in I_y} \psi_i(x) G_{\Omega_i}(x,y) \chi_i(y)$$

and one finds in distributional sense that

$$\Delta^2 G_\Omega^{\mathrm{sing}}(x,y) = \begin{cases} 0 & \text{if } x \notin \bigcup_{i \in I_y} \mathrm{support}(\psi_i) =: A_{I_y}, \\ \delta_x(y) & \text{if } x \in \bigcup_{i \in I_y} \mathrm{support}(\chi_i) =: B_{I_y}, \\ \sum_{i \in I_y} \Delta_x^2\left(\psi_i(x) G_{\Omega_i}(x,y)\right)\chi_i(y) & \text{if } x \in A_{I_y} \backslash B_{I_y}. \end{cases} \tag{6.25}$$

Note that $\Delta_x^2\left(\psi_i(x) G_{\Omega_i}(x,y)\right)\chi_i(y) = 0$ on $\left(A_{I_y} \backslash B_{I_y}\right) \cap \left(A_{I_y} \backslash \mathrm{support}(\nabla \psi_i)\right)$. In other words, G_Ω^{sing} takes care of the appropriate singularity at y and is positive through the construction that uses the G_{Ω_i}. The support of G_Ω^{sing} lies in a band around the diagonal $\{(x,x)\,;\,x \in \Omega\}$. See Figure 6.4.

For the biharmonic operator applied to the difference $G_\Omega^{\mathrm{reg}}(x,y) = G_\Omega(x,y) - G_\Omega^{\mathrm{sing}}(x,y)$ only the third term in (6.25) remains. Since $\Delta^2 G_\Omega^{\mathrm{reg}}(x,y)$ is zero near the diagonal $\{(x,x)\,;\,x \in \Omega\}$ it can be shown that G_Ω^{reg} is a smooth as the boundary allows:

$$x \mapsto \sum_{i \in I_y} \psi_i(x) G_{\Omega_i}(x,y)\chi_i(y) \in C^{4,\gamma}\left(\overline{A_{I_y} \backslash B_{I_y}}\right). \tag{6.26}$$

The support of G_Ω^{reg} will in general be the full $\bar{\Omega} \times \bar{\Omega}$.

The estimates for $G_\Omega^{\mathrm{sing}}(x,y)$ are inherited from those for $G_{\Omega_i}(x,y)$; those for $G_\Omega^{\mathrm{reg}}(x,y)$ from the boundary conditions and (6.26). $\qquad \square$

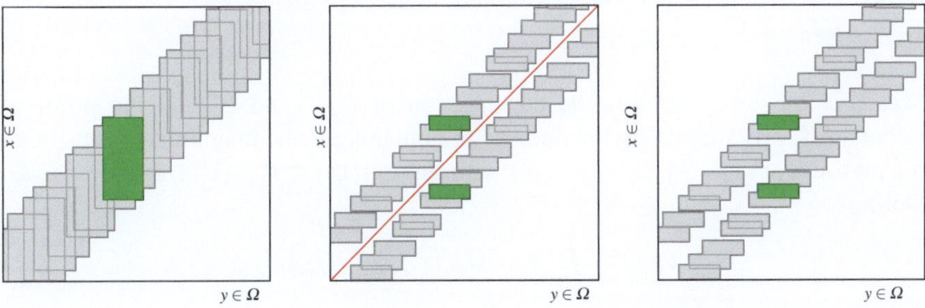

Fig. 6.4 Scheme for the support of G_Ω^{sing}, $\Delta^2 G_\Omega^{\mathrm{sing}}$ and $\Delta^2 G_\Omega^{\mathrm{reg}}$, highlighting $\mathrm{support}(\chi_i) \times \mathrm{support}(\psi_i)$ and twice $\mathrm{support}(\chi_i) \times \mathrm{support}(\nabla \psi_i)$ for one specific i.

Sketch of the proof of Theorem 6.15. By carefully checking the proof of Lemma 6.17 one finds a $r_\Omega > 0$ and a positive constant $c > 0$ such that there is the following estimate:

$$G_\Omega^{\text{sing}}(x,y) \geq c d(x)d(y)\min\left\{1,\frac{d(x)d(y)}{|x-y|^2}\right\} \quad \text{for } |x-y| < r_\Omega.$$

As a consequence one concludes that near the boundary the regular term G_Ω^{reg} is dominated by the singular term G_Ω^{sing}. Hence, there exists $\varepsilon > 0$ such that if $d(x) < \varepsilon$, then $G_\Omega(x,y) > 0$ for $y \in B_\varepsilon(x)$. In other words, we have local positivity near the boundary.

In order to prove local positivity also in the interior we observe first that for $x \in \Omega$

$$G_\Omega(x,x) = \int_\Omega (\Delta_y G_\Omega(x,y))^2\, dy > 0.$$

Moreover, in two dimension, $G_\Omega(\cdot,\cdot)$ is continuous on $\overline{\Omega} \times \overline{\Omega}$. Hence, there exists $0 < \delta \leq \varepsilon$ such that $x \in \Omega$, $d(x) \geq \varepsilon$, $|x-y| < \delta$ implies that $G_\Omega(x,y) > 0$. Together with the local positivity near the boundary we may conclude that for any $x \in \Omega$, $y \in B_\delta(x) \cap \Omega$ one has $G_\Omega(x,y) > 0$.

The estimate $G_\Omega(x,y) > -Cd(x)^2 d(y)^2$ on $\overline{\Omega} \times \overline{\Omega}$ follows directly from the local positivity just proven and the estimate from Theorem 4.28. $\square$

Lemma 6.17 allows for proving one-sided pointwise a priori estimates of solutions in terms of the data and a very weak negative Sobolev norm of the solution itself.

Theorem 6.18. *Assume that $\Omega \subset \mathbb{R}^2$ is a bounded domain with $\partial\Omega \in C^{4,\gamma}$. Then for any $q > 2$ and $\varepsilon > 0$ there exists a constant $c_{q,\Omega,\varepsilon} > 0$ such that for $f \in L^p(\Omega)$ with $p \in (1,\infty)$ the solution $u \in W^{4,p} \cap W_0^{2,p}(\Omega)$ of (6.1) satisfies for all $x \in \Omega$:*

$$u(x) \leq c_{q,\Omega,\varepsilon}\left(\|f^+\|_{L^1(B_\varepsilon(x)\cap\Omega)} + \|u\|_{W^{-1,q}(\Omega)}\right).$$

Here $f^+ = \max\{f,0\}$. A similar estimate holds from below with f^+ replaced by $f^- = \max\{-f,0\}$.

Sketch of the proof. It uses the construction of G_Ω^{sing} and G_Ω^{reg} in the proof of Lemma 6.17. By increasing the number of domains Ω_i one may find a corresponding partition of unity $\{\chi_i; i = 1,\ldots,k\}$ with $\text{support}(\psi_i) \subset B_{\varepsilon/2}(x^i)$ for some $x^i \in \Omega$. Setting

$$u_i(x) = \int_{\Omega_i} \psi_i(x)G_{\Omega_i}(x,y)\chi_i(y)f(y)dy$$

and using the positivity of G_{Ω_i} one finds $u_i(x) = 0$ for $x \notin B_{\varepsilon/2}(x^i)$ and

$$u_i(x) \le \int_{B_{\varepsilon/2}(x^i) \cap \Omega_i} \psi_i(x) G_{\Omega_i}(x,y) \chi_i(y) f^+(y) dy$$
$$\le c \left\| f^+ \right\|_{L^1\left(B_{\varepsilon/2}(x^i) \cap \Omega_i\right)} \le c \left\| f^+ \right\|_{L^1\left(B_\varepsilon(x) \cap \Omega_i\right)}$$

when $x \in B_{\varepsilon/2}(x^i)$. Since $x \in \text{support}(\psi_i)$ for at most finitely many i's, one obtains

$$u^{\text{sing}}(x) := \sum_{i=1}^k u_i(x) = \int_\Omega G_\Omega^{\text{sing}}(x,y) f(y) dy \le \tilde{c} \left\| f^+ \right\|_{L^1\left(B_\varepsilon(x) \cap \Omega\right)} .$$

The function $u - u^{\text{sing}}$ satisfies

$$\left| \Delta^2 \left(u - u^{\text{sing}} \right)(x) \right|$$
$$= \left| f(x) - \sum_{i=1}^k \Delta^2 \left(\psi_i(x) \int_{\Omega_i} G_{\Omega_i}(x,y) \chi_i(y) f(y) dy \right) \right|$$
$$= \left| \sum_{i=1}^k \sum_{\substack{|\alpha+\beta|=4 \\ |\alpha| \ge 1}} c_\alpha D^\alpha \psi_i(x) D^\beta \int_{\Omega_i} G_{\Omega_i}(x,y) \chi_i(y) f(y) dy \right|$$
$$\le \sum_{i=1}^k \sum_{\substack{|\alpha+\beta|=4 \\ |\alpha| \ge 1}} c_\alpha \left\| D^\alpha \psi_i(.) D^\beta \int_{\Omega_i} G_{\Omega_i}(.,y) \chi_i(y) f(y) dy \right\|_{L^\infty(\text{support}(\nabla \psi_i))} .$$

By the Sobolev embedding $W^{1,q} \subset L^\infty$ for $q > 2$ and the properties of ψ_i it follows that

$$\left\| D^\alpha \psi_i(.) D^\beta \int_{\Omega_i} G_{\Omega_i}(.,y) \chi_i(y) f(y) dy \right\|_{L^\infty(\text{support}(\nabla \psi_i))}$$
$$\le \left\| D^\alpha \psi_i(.) D^\beta \int_{\Omega_i} G_{\Omega_i}(.,y) \chi_i(y) f(y) dy \right\|_{W_0^{1,q}(\text{support}(\nabla \psi_i))}$$
$$= \left\| D^\alpha \psi_i(.) D^\beta \int_{\Omega_i} G_{\Omega_i}(.,y) \chi_i(y) \left(\Delta^2 u \right)(y) dy \right\|_{W_0^{1,q}(\text{support}(\nabla \psi_i))}$$
$$= \left\| D^\alpha \psi_i(.) D^\beta \int_{\Omega_i} \Delta_y^2 \left(G_{\Omega_i}(.,y) \chi_i(y) \right) u(y) dy \right\|_{W_0^{1,q}(\text{support}(\nabla \psi_i))} . \qquad (6.27)$$

This last expression does not involve points (x,y) near the diagonal: $\text{support}(\chi_i) \times \text{support}(\nabla \psi_i)$ is as in Figure 6.4 on the right. So $(x,y) \mapsto D^\alpha \psi_i(x) D^\beta G_{\Omega_i}(.,y) \chi_i(y)$ is a C^∞-function with compact support. It allows us to estimate (6.27) by a weak norm of u like e.g. $\|u\|_{W^{-1,q}(\Omega)}$. Elliptic L^p-estimates, see Section 2.5.2, and Sobolev embeddings translate in several steps the L^∞-bound for $\Delta^2 \left(u - u^{\text{sing}} \right)$ into an L^∞-bound for $u - u^{\text{sing}}$. $\qquad \square$

6.3 Regions of Positivity in Arbitrary Domains in Higher Dimensions

The remaining part of this chapter is mainly devoted to the biharmonic Green function in dimensions

$$n \geq 3,$$

which corresponds to the Dirichlet problem

$$\begin{cases} \Delta^2 u = f & \text{in } \Omega, \\ u = |\nabla u| = 0 & \text{on } \partial\Omega. \end{cases} \tag{6.28}$$

For brevity let $G_\Omega := G_{\Delta^2,\Omega}$ denote these Green functions in what follows.

One goal will be to verify perturbation results like Theorem 6.3, another to prove "smallness" of the negative part G_Ω^- in general bounded smooth domains $\Omega \subset \mathbb{R}^n$ also if $n \geq 3$, i.e. to extend Theorem 6.15 to higher dimensions. Our strategy to achieve the second goal is to identify subsets

$$\mathscr{P} \subset \Omega \times \Omega \setminus \{(x,x)\}$$

where one can directly and explicitly verify that

$$G_\Omega(x,y) > 0 \qquad \text{for all } (x,y) \in \mathscr{P}.$$

Further steps to be done in Sections 6.4 and 6.5 will be a blow-up analysis and a discussion of several limiting situations, which all together will imply also perturbation results to achieve the first goal.

A first step to identify positivity sets $\mathscr{P}$ was done by Nehari [323] in dimensions $n = 2$ and $n = 3$. Developing Nehari's idea, in Section 6.3.1 we prove the following result.

Theorem 6.19. *Let $n \geq 3$. Then there exists a constant $\delta_n > 0$, which depends only on the dimension n, such that the following holds true.*

Assume $\Omega \subset \mathbb{R}^n$ to be a $C^{4,\gamma}$-smooth bounded domain and let $G_\Omega := G_{\Delta^2,\Omega}$ denote the Green function for the biharmonic operator under Dirichlet boundary conditions. If

$$|x - y| < \delta_n \max\{d(x), d(y)\},$$

then we have

$$G_\Omega(x,y) > 0.$$

For the constant δ_n, one may achieve that

$$\delta_3 \geq 4 - 2\sqrt{3} \approx 0.53,$$
$$\delta_4 \geq 0.59,$$
$$\delta_n \geq 0.6 \qquad \text{for } n \geq 5,$$

and

$$\lim_{n\to\infty} \delta_n = \frac{\sqrt{5}-1}{2} \approx 0.618.$$

We recall that $d(x) = \mathrm{dist}(x,\partial\Omega)$. For two-dimensional domains, only a much more restricted statement seems to be available, where also the *maximal* distance of x,y to boundary points of $\partial\Omega$ is involved, see [323].

By making use of $d(x) \leq d(y) + |x-y|$ one may observe that the condition $|x-y| < \delta_n \max\{d(x),d(y)\}$ implies that $|x-y| < \frac{\delta_n}{1-\delta_n} \min\{d(x),d(y)\}$.

The preceding theorem shows that the negative part of the Green function is uniformly bounded and hence *relatively small* compared with the singular positive part, as long as x and y stay uniformly away from the boundary $\partial\Omega$. It is remarkable that the constant δ_n can be found independently of the domains Ω. In this form, the result cannot be deduced from Schauder-type estimates.

With slightly more complicated but similar techniques as in the proof of Theorem 6.19, one may also cover the Green function for the Dirichlet problem $G_{(-\Delta)^m,\Omega}$ for the polyharmonic operator. By means of the formula

$$u(x) = \int_\Omega G_{(-\Delta)^m,\Omega}(x,y)\,f(y)\,dy$$

we find solutions of the polyharmonic Dirichlet problem

$$\begin{cases} (-\Delta)^m u = f & \text{in } \Omega, \\ D^\alpha u|_{\partial\Omega} = 0 & \text{for } |\alpha| \leq m-1, \end{cases} \tag{6.29}$$

provided f and Ω are smooth enough.

In order to avoid distinctions and too many technicalities we prove in Section 6.3.2 the following result only for large dimensions.

Theorem 6.20. *Let $m \in \mathbb{N}^+$, $n > 2m$. Then there exists a constant $\delta_{m,n} > 0$, which depends only on the dimension n and the order $2m$ of the polyharmonic operator, such that the following holds true.*

Assume $\Omega \subset \mathbb{R}^n$ to be a $C^{2m,\gamma}$-smooth bounded domain and let $G_{(-\Delta)^m,\Omega}$ denote the polyharmonic Green function under Dirichlet boundary conditions. If

$$|x-y| < \delta_{m,n} \max\{d(x),d(y)\},$$

then

$$G_{(-\Delta)^m,\Omega}(x,y) > 0.$$

For the constant $\delta_{m,n}$, one may achieve that

$$\delta_{m,2m+1} \geq 1 + \frac{\Gamma(m)\Gamma\left(\frac{3}{2}\right)}{\Gamma\left(m+\frac{1}{2}\right)} - \sqrt{1 + \frac{\Gamma(m)^2\Gamma\left(\frac{3}{2}\right)^2}{\Gamma\left(m+\frac{1}{2}\right)^2}}$$

and, for fixed m, that

$$\lim_{n \to \infty} \delta_{m,n} = \frac{\sqrt{5}-1}{2} \approx 0.618.$$

Remark 6.21. 1. A similar result was obtained for $n = 2m$ by Köckritz in a seminar thesis [251]. Checking carefully the proof below indicates that it can be extended to $n \geq 2m - 1$ but will presumably fail for $n \leq 2m - 2$.

2. Numerical evidence indicates the following for the constants $\delta_{m,n}$.

- For each m the sequence $(\delta_{m,n})_{n=2m+1}^{\infty}$ is increasing to $\frac{\sqrt{5}-1}{2}$.
- The sequence $(\delta_{m,2m+1})_{m=2}^{\infty}$ is decreasing to 0.
- We provide lower bounds for $\delta_{m,n}$, the limit of which for $n \to \infty$ is $\frac{\sqrt{5}-1}{2}$, independently of m. The speed of convergence, however, seems to depend strongly on m.

6.3.1 The Biharmonic Operator

In this section we prove Theorem 6.19. We consider the situation where, for some $R > 1$, we have

$$B := B_1(0) \subset \Omega \subset B_R := B_R(0)$$

and write, for suitable $f : \mathbb{R}^n \to \mathbb{R}$,

$$\mathcal{G}_\Omega f(x) := \int_\Omega G_\Omega(x,y) f(y) \, dy$$

which yields the solution $u(x) := \mathcal{G}_\Omega f(x)$ to the Dirichlet problem (6.28).

Let us recall that a fundamental solution for Δ^2 on $\mathbb{R}^n$ is given by

$$F_n(x) = \begin{cases} c_n |x|^{4-n} & \text{if } n \notin \{2,4\}, \\ -2c_4 \log |x| & \text{if } n = 4, \\ 2c_2 |x|^2 \log |x| & \text{if } n = 2, \end{cases}$$

where

$$c_n = \begin{cases} \dfrac{1}{2(n-4)(n-2)ne_n} & \text{if } n \notin \{2,4\}, \\[4mm] \dfrac{1}{8ne_n} & \text{if } n \in \{2,4\}. \end{cases}$$

The Green function may be decomposed into the fundamental solution plus a regular part

$$G_\Omega(x,y) = F_n(|x-y|) + H_\Omega(x,y),$$

where $H_\Omega \in C^{4,\gamma}\left(\overline{\Omega}^2\right)$. We will also use

$$\mathscr{H}_\Omega f(x) := \int_\Omega H_\Omega(x,y) f(y)\, dy.$$

Lemma 6.22. *Let f, g be smooth and supported in B. Then*

$$4\int_\Omega (\Delta\mathscr{G}_\Omega f)(\Delta\mathscr{G}_\Omega g)\, dx \geq \int_B \left(f\left(\mathscr{H}_B f - \mathscr{H}_{B_R} f\right) + g\left(\mathscr{H}_B g - \mathscr{H}_{B_R} g\right)\right) dx$$
$$+ \int_B \left(f\left(\mathscr{G}_B g + \mathscr{G}_{B_R} g\right) + g\left(\mathscr{G}_B f + \mathscr{G}_{B_R} f\right)\right) dx.$$

Proof. We consider the quadratic form

$$\mathbb{R}^2 \ni (\beta,\gamma) \mapsto \int_\Omega \left(\beta\Delta\mathscr{G}_\Omega f + \gamma\Delta\mathscr{G}_\Omega g\right)^2 dx$$

and show that this is non-decreasing with respect to domain inclusion. For this purpose one considers $\omega \subset \Omega$ and one gets

$$\int_\Omega \left(\beta\Delta\mathscr{G}_\Omega f + \gamma\Delta\mathscr{G}_\Omega g\right)^2 dx - \int_\omega \left(\beta\Delta\mathscr{G}_\omega f + \gamma\Delta\mathscr{G}_\omega g\right)^2 dx$$
$$= \int_\Omega \left(\beta\Delta\mathscr{G}_\Omega f + \gamma\Delta\mathscr{G}_\Omega g\right)^2 dx + \int_\omega \left(\beta\Delta\mathscr{G}_\omega f + \gamma\Delta\mathscr{G}_\omega g\right)^2 dx$$
$$- 2\int_\omega \left(\beta\mathscr{G}_\omega f + \gamma\mathscr{G}_\omega g\right)\left(\beta\Delta^2\mathscr{G}_\omega f + \gamma\Delta^2\mathscr{G}_\omega g\right) dx$$
$$= \int_\Omega \left(\beta\Delta\mathscr{G}_\Omega f + \gamma\Delta\mathscr{G}_\Omega g\right)^2 dx + \int_\omega \left(\beta\Delta\mathscr{G}_\omega f + \gamma\Delta\mathscr{G}_\omega g\right)^2 dx$$
$$- 2\int_\omega \left(\beta\mathscr{G}_\omega f + \gamma\mathscr{G}_\omega g\right)\left(\beta f + \gamma g\right) dx$$
$$= \int_\Omega \left(\beta\Delta\mathscr{G}_\Omega f + \gamma\Delta\mathscr{G}_\Omega g\right)^2 dx + \int_\omega \left(\beta\Delta\mathscr{G}_\omega f + \gamma\Delta\mathscr{G}_\omega g\right)^2 dx$$
$$- 2\int_\omega \left(\beta\mathscr{G}_\omega f + \gamma\mathscr{G}_\omega g\right)\left(\beta\Delta^2\mathscr{G}_\Omega f + \gamma\Delta^2\mathscr{G}_\Omega g\right) dx$$
$$= \int_\Omega \left(\beta\Delta\mathscr{G}_\Omega f + \gamma\Delta\mathscr{G}_\Omega g\right)^2 dx + \int_\omega \left(\beta\Delta\mathscr{G}_\omega f + \gamma\Delta\mathscr{G}_\omega g\right)^2 dx$$
$$- 2\int_\omega \left(\beta\Delta\mathscr{G}_\omega f + \gamma\Delta\mathscr{G}_\omega g\right)\left(\beta\Delta\mathscr{G}_\Omega f + \gamma\Delta\mathscr{G}_\Omega g\right) dx$$
$$= \int_{\Omega\setminus\omega} \left(\beta\Delta\mathscr{G}_\Omega f + \gamma\Delta\mathscr{G}_\Omega g\right)^2 dx$$
$$+ \int_\omega \left(\beta(\Delta\mathscr{G}_\Omega f - \Delta\mathscr{G}_\omega f) + \gamma(\Delta\mathscr{G}_\Omega g - \Delta\mathscr{G}_\omega g)\right)^2 dx$$
$$\geq 0.$$

In a first step we exploit this monotonicity in $B \subset \Omega$ with $\beta = \gamma = 1$, that is

$$\int_{\Omega} (\Delta \mathscr{G}_{\Omega} f + \Delta \mathscr{G}_{\Omega} g)^2 \, dx \geq \int_{B} (\Delta \mathscr{G}_B f + \Delta \mathscr{G}_B g)^2 \, dx = \int_{B} (f+g)\mathscr{G}_B (f+g) \, dx. \tag{6.30}$$

In a second step it is used in $\Omega \subset B_R$ with $\beta = -\gamma = 1$, that is

$$\int_{\Omega} (\Delta \mathscr{G}_{\Omega} f - \Delta \mathscr{G}_{\Omega} g)^2 \, dx \leq \int_{B_R} (\Delta \mathscr{G}_{B_R} f - \Delta \mathscr{G}_{B_R} g)^2 \, dx = \int_{B} (f-g)\mathscr{G}_{B_R} (f-g) \, dx. \tag{6.31}$$

Subtracting (6.31) from (6.30) yields

$$4\int_{\Omega} (\Delta \mathscr{G}_{\Omega} f)(\Delta \mathscr{G}_{\Omega} g) \, dx \geq \int_{B} f\,(\mathscr{G}_B f - \mathscr{G}_{B_R} f) \, dx + \int_{B} g\,(\mathscr{G}_B g - \mathscr{G}_{B_R} g) \, dx$$
$$+ \int_{B} f\,(\mathscr{G}_B g + \mathscr{G}_{B_R} g) \, dx + \int_{B} g\,(\mathscr{G}_B f + \mathscr{G}_{B_R} f) \, dx.$$

Since $\mathscr{G}_B - \mathscr{G}_{B_R} = \mathscr{H}_B - \mathscr{H}_{B_R}$, the claim follows. $\qquad\square$

Lemma 6.23. *For $x,y \in B$, $x \neq y$, we have the following estimate from below for the biharmonic Green function of Ω,*

$$G_{\Omega}(x,y) \geq \frac{1}{4} \left(H_B(x,x) - H_{B_R}(x,x) + H_B(y,y) - H_{B_R}(y,y) \right)$$
$$+ \frac{1}{2} \left(G_B(x,y) + G_{B_R}(x,y) \right).$$

Proof. The statement follows directly from Lemma 6.22 by taking smooth approximations of the Dirac delta distribution concentrated in x and y respectively as f and g. One also uses the symmetry of the Green function $G_{\Omega}(x,y) = G_{\Omega}(y,x)$. $\quad\square$

Proof of Theorem 6.19. We recall from Lemma 2.27 that for $n > 4$ and $n = 3$, observing that $c_3 < 0$,

$$G_B(x,y) = c_n \left\{ |x-y|^{4-n} - \frac{n-2}{2} \left| |x|y - \frac{x}{|x|} \right|^{4-n} + \frac{n-4}{2} \left| |x|y - \frac{x}{|x|} \right|^{2-n} |x-y|^2 \right\}, \tag{6.32}$$

$$G_{B_R}(x,y) = R^{4-n} G_B \left(\frac{1}{R}x, \frac{1}{R}y \right),$$

$$H_{B_R}(x,x) = -c_n \frac{n-2}{2} \left(R - \frac{|x|^2}{R} \right)^{4-n}, \tag{6.33}$$

while for $n = 4$

$$G_B(x,y) = c_4 \left\{ -2\log|x-y| + 2\log\left| |x|y - \frac{x}{|x|} \right| - 1 + \left| |x|y - \frac{x}{|x|} \right|^{-2} |x-y|^2 \right\},$$
(6.34)

$$G_{B_R}(x,y) = G_B\left(\frac{1}{R}x, \frac{1}{R}y \right),$$

$$H_{B_R}(x,x) = 2c_4 \log\left(1 - \frac{|x|^2}{R^2} \right) - c_4 + 2c_4 \log R.$$
(6.35)

In order to prove Theorem 6.19, by rescaling and translation, it is enough to consider $x = 0$, $y \in B_{\delta_n}(0)$, where $\delta_n \in (0,1)$ has to be suitably specified below.

We consider first the case $n > 4$, where Lemma 6.23 and formulae (6.32)-(6.33) yield

$$\frac{4}{c_n}G_\Omega(0,y) \geq -\frac{n-2}{2} + \frac{n-2}{2}R^{4-n} - \frac{n-2}{2}\left(1 - |y|^2\right)^{4-n} + \frac{n-2}{2}\left(R - \frac{|y|^2}{R} \right)^{4-n}$$

$$+ 4|y|^{4-n} - (n-2) + (n-4)|y|^2 - (n-2)R^{4-n} + (n-4)R^{2-n}|y|^2.$$

Letting $R \to \infty$, we obtain

$$\frac{4}{c_n}G_\Omega(0,y) \geq 4|y|^{4-n} + (n-4)|y|^2 - \frac{n-2}{2}\left(1 - |y|^2\right)^{4-n} - \frac{3}{2}(n-2). \quad (6.36)$$

If $n = 5$ one has to check whether

$$0 < 4 - 6|y| - 4|y|^2 + \frac{11}{2}|y|^3 - |y|^5.$$

The right hand side is strictly decreasing in $|y| \in [0,0.6]$ and takes on a positive value for $|y| = 0.6$. According to maple^{TM} the above inequality is satisfied for $|y| \in (0, 0.612865\ldots)$.

If $n \geq 6$, we drop the term $(n-4)|y|^2$ in (6.36) and have to determine δ_n such that

$$4\delta_n^{4-n} - \frac{n-2}{2}\left(1 - \delta_n^2\right)^{4-n} - \frac{3}{2}(n-2) \geq 0. \quad (6.37)$$

Asymptotically, δ_n should be chosen close to the positive root δ_∞ of

$$\delta = 1 - \delta^2,$$

i.e. to $\delta_\infty = (\sqrt{5}-1)/2$. We show that (6.37) is satisfied with $\delta_n = 0.6$, i.e. that

$$4\left(\frac{3}{5}\right)^{4-n} - \frac{n-2}{2}\left(\frac{16}{25}\right)^{4-n} - \frac{3}{2}(n-2) \geq 0$$

$$\Leftrightarrow 8 - (n-2)\left(\frac{15}{16}\right)^{n-4} - 3(n-2)\left(\frac{3}{5}\right)^{n-4} \geq 0.$$

The left hand side of the last expression is increasing for $n \geq 18$ and attains positive values for $n = 6, \ldots, 18$, thereby showing that (6.37) holds true for $\delta_n = 0.6$.

We discuss now the case $n = 4$ and choose $R \geq 1$, where Lemma 6.23 and formulae (6.34)-(6.35) yield

$$\frac{2}{c_4}G_\Omega(0,y) \geq -\log\left(1 - \frac{|y|^2}{R^2}\right) + \log\left(1 - |y|^2\right) - 2\log R - 4\log|y|$$

$$+ 2\log R - 2 + |y|^2 + \frac{|y|^2}{R^2}$$

$$\geq -\log\left(1 - \frac{|y|^2}{R^2}\right) + \log\left(1 - |y|^2\right) - 4\log|y| - 2 + |y|^2 + \frac{|y|^2}{R^2}.$$

Letting $R \to \infty$, we conclude that

$$\frac{2}{c_4}G_\Omega(0,y) \geq -4\log|y| + \log\left(1 - |y|^2\right) - 2 + |y|^2.$$

The right hand side is certainly decreasing in $|y| \in [0,0.6]$ and takes a positive value for $|y| = \delta_4 = 0.59$. With the help of maple^{TM} we see that it is positive for $|y| \in (0, 0.594160\ldots)$.

Finally, we discuss the case $n = 3$, where $c_3 = -\frac{1}{8\pi} < 0$. Proceeding as before we find

$$-\frac{2}{c_3}G_\Omega(0,y) \geq \frac{1}{4}\left(1 - R + 1 - |y|^2 - R + \frac{|y|^2}{R}\right)$$

$$+ \left(-2|y| + \frac{1}{2} + \frac{R}{2} + (\frac{1}{2} + \frac{1}{2R})|y|^2\right)$$

$$= 1 - 2|y| + \frac{1}{4}|y|^2 + \frac{3}{4R}|y|^2.$$

Letting $R \to \infty$ yields

$$-\frac{2}{c_3}G_\Omega(0,y) \geq 1 - 2|y| + \frac{1}{4}|y|^2,$$

where the right hand side is positive for $|y| < 4 - 2\sqrt{3}$. $\qquad\square$

6.3.2 *Extensions to Polyharmonic Operators*

Here, we prove Theorem 6.20. The arguments are very similar to Section 6.3.1 and we may be very brief and focus mainly on what is different. Throughout this section, according to Theorem 6.20, we confine ourselves to the case

$$n > 2m.$$

We consider

$$B = B_1(0) \subset \Omega \subset B_R = B_R(0)$$

and the Green function $G_{(-\Delta)^m,\Omega}$ corresponding to (6.29) in Ω. Again, this Green function may be decomposed into a singular and a regular part

$$G_{(-\Delta)^m,\Omega}(x,y) = c_{m,n}|x-y|^{2m-n} + H_{(-\Delta)^m,\Omega}(x,y),$$

where $H_{(-\Delta)^m,\Omega} \in C^{2m,\gamma}(\overline{\Omega}^2)$ denotes the regular part and $c_{m,n} > 0$ is a suitable positive constant. Lemma 6.23 directly generalises to the polyharmonic situation and we may perform the

Proof of Theorem 6.20. According to Lemma 2.27 we have with a suitable positive constant $k_{m,n}$,

$$G_{(-\Delta)^m,B}(x,y) = k_{m,n}|x-y|^{2m-n} \int_1^{\left| |x|y - \frac{x}{|x|} \right| / |x-y|} \left(v^2 - 1 \right)^{m-1} v^{1-n} \, dv, \quad (6.38)$$

$$G_{(-\Delta)^m,B_R}(x,y) = R^{2m-n} G_{(-\Delta)^m,B} \left(\frac{1}{R}x, \frac{1}{R}y \right), \quad (6.39)$$

$$H_{(-\Delta)^m,B_R}(x,x) = -\frac{k_{m,n}}{n-2m} \left(R - \frac{|x|^2}{R} \right)^{2m-n}. \quad (6.40)$$

The constants $c_{m,n}$ and $k_{m,n}$ are related by

$$c_{m,n} = k_{m,n} \int_1^\infty \left(v^2 - 1 \right)^{m-1} v^{1-n} \, dv = k_{m,n}(-1)^m \sum_{j=0}^{m-1} \frac{(-1)^j \binom{m-1}{j}}{2j+2-n}$$

$$= k_{m,n} \frac{2^{m-1}(m-1)!}{\displaystyle\prod_{j=1}^m (n-2j)}.$$

The latter identity is verified by induction.

By the generalisation of Lemma 6.23, formulae (6.38)–(6.40) and letting $R \to \infty$, we obtain

$$G_{(-\Delta)^m,\Omega}(0,y) \geq c_{m,n}|y|^{2m-n} - \frac{k_{m,n}}{2}|y|^{2m-n} \int_{1/|y|}^\infty \left(v^2 - 1 \right)^{m-1} v^{1-n} \, dv$$

$$- \frac{k_{m,n}}{4(n-2m)} \left(1 + (1 - |y|^2)^{2m-n} \right)$$

$$\geq \frac{c_{m,n}}{2}|y|^{2m-n} - \frac{k_{m,n}}{4(n-2m)} \left(1 + (1 - |y|^2)^{2m-n} \right) \quad (6.41)$$

so that

$$\frac{4(n-2m)}{k_{m,n}} G_{(-\Delta)^m,\Omega}(0,y) \geq \frac{2(m-1)!}{\prod_{j=1}^{m-1}\left(\frac{n}{2}-j\right)} |y|^{2m-n} - 1 - \left(1-|y|^2\right)^{2m-n}. \tag{6.42}$$

Certainly, one finds $\delta_{m,n} > 0$ such that the right hand side is positive for $|y| < \delta_{m,n}$. For m fixed and $n \to \infty$, the powers $2m - n$ dominate all the other terms and $\delta_{m,n}$ may be chosen such that they approach, as in the biharmonic case, the positive zero δ_∞ of

$$\delta = 1 - \delta^2,$$

which is precisely $\frac{\sqrt{5}-1}{2}$. In the case $n = 2m+1$, (6.42) reads

$$\frac{4}{k_{m,2m+1}} G_{(-\Delta)^m,\Omega}(0,y) \geq 2\frac{\Gamma(m)\Gamma\left(\frac{3}{2}\right)}{\Gamma\left(m+\frac{1}{2}\right)} \frac{1}{|y|} - 1 - \frac{1}{1-|y|^2} \tag{6.43}$$

$$\geq 2\frac{\Gamma(m)\Gamma\left(\frac{3}{2}\right)}{\Gamma\left(m+\frac{1}{2}\right)} \frac{1}{|y|} - 1 - \frac{1}{1-|y|}. \tag{6.44}$$

The right hand side (6.44) is positive if and only if

$$|y| < 1 + \frac{\Gamma(m)\Gamma\left(\frac{3}{2}\right)}{\Gamma\left(m+\frac{1}{2}\right)} - \sqrt{1 + \frac{\Gamma(m)^2\Gamma\left(\frac{3}{2}\right)^2}{\Gamma\left(m+\frac{1}{2}\right)^2}}. \tag{6.45}$$

One might wonder whether dropping a positive term in (6.41) gives rise to a very rough estimate. The previous estimate (6.45) would still allow for choosing $\delta_{2,5} = 0.46$, while the right hand side of (6.43) is positive for $|y| < \delta_{2,5} = 0.54$. On the other hand, according to Theorem 6.19, $\delta_{2,5} = 0.59$ is admissible. This shows that one has not lost much in (6.41). In any case, our proof shows that we cannot do better than a constant $\delta_{m,n}$ with $\lim_{m\to\infty} \delta_{m,2m+1} = 0$, even if one had kept the second term in (6.41). $\qquad\square$

6.4 Small Negative Part of Biharmonic Green's Functions in Higher Dimensions

The goal of this section is to extend Theorem 6.15 to any dimension $n \geq 2$.

6.4.1 Bounds for the Negative Part

Theorem 6.24. *Let $\Omega \subset \mathbb{R}^n$ ($n \geq 2$) be a bounded $C^{4,\gamma}$-smooth domain. Let G_Ω denote the biharmonic Green function in Ω under Dirichlet boundary conditions.*

Then there exists a positive minimal distance $\delta = \delta(\Omega) > 0$ such that for any two points $x, y \in \Omega$, $x \neq y$,

$$|x - y| < \delta \text{ implies that } G_\Omega(x, y) > 0. \tag{6.46}$$

In particular, there exists a constant $C = C(\Omega) > 0$ such that for all $x, y \in \Omega$, $x \neq y$, we have the following estimate from below

$$G_\Omega(x, y) \geq -C d(x)^2 d(y)^2. \tag{6.47}$$

For $n = 2$, the statement just recalls Theorem 6.15. The proof for the case $n \geq 3$ will be given in Section 6.4.2.

Remark 6.25. So for any bounded $C^{4,\gamma}$-smooth domain $\Omega \subset \mathbb{R}^n$ there exists a constant $C = C(\Omega) > 0$ such the biharmonic Green function G_Ω satisfies the following estimate:

$$-C d(x)^2 d(y)^2 \leq G_\Omega(x, y) \leq \begin{cases} C|x - y|^{4-n} \min\left\{1, \dfrac{d(x)^2 d(y)^2}{|x - y|^4}\right\} & \text{if } n > 4, \\[3mm] C \log\left(1 + \dfrac{d(x)^2 d(y)^2}{|x - y|^4}\right) & \text{if } n = 4, \\[3mm] C d(x)^{2-n/2} d(y)^{2-n/2} \min\left\{1, \dfrac{d(x)^{n/2} d(y)^{n/2}}{|x - y|^n}\right\} \\[2mm] \qquad\qquad\qquad\qquad \text{if } n = 2, 3. \end{cases} \tag{6.48}$$

The estimates from above follow from the two-sided estimates in Theorem 4.28. See also (4.8). Even for $n = 2, 3$, where the Green function is bounded, the bound from below in 6.48 is a strong statement because in the case where x or y is closer to the boundary than they are to each other, we gain a factor of order $|x - y|^n$ compared with the estimate in Theorem 4.28.

Remark 6.26. A functional analytic approach by Malyshev [284] shows for $n = 2, 3$ this estimate from below:

$$G_{\Delta^2,\Omega}(x, y) \geq -C(\Omega)\sqrt{G_{\Delta^2,\Omega}(x, x) G_{\Delta^2,\Omega}(y, y)} \tag{6.49}$$

with a constant $C(\Omega) \in [0, 1)$. Starting from (4.8) and using $C(\Omega) < 1$ it follows that the optimal constant for $G_{\Delta^2,\Omega}$ from below in (4.8) for x near y is smaller than the one from above. From (6.49) and (4.8) alone one does not get a qualitative better estimate from below as in (6.48).

Since Theorem 6.24 in the case $n = 2$ is already proved by Theorem 6.15, we restrict ourselves in what follows to

$$n \geq 3.$$

And as we already mentioned, completely different techniques than for $n = 2$ have to be used.

6.4.2 A Blow-Up Procedure

The proof of Theorem 6.24 as well as the perturbation result Theorem 6.29 below are deduced from the following statement that for smooth perturbations of a fixed domain, pairs of interior zeros of the biharmonic Green functions cannot get arbitrarily close to each other. Theorem 6.29 will extend Theorem 6.3 for the biharmonic operator to dimensions $n \geq 3$ and will be proved below in Section 6.5.

Theorem 6.27. *We assume that $n \geq 3$ and that (Ω_k) is a $C^{4,\gamma}$-perturbation of the bounded $C^{4,\gamma}$-smooth domain Ω according to Definition 4.19. Let $G_k = G_{\Delta^2,\Omega_k}$ denote the biharmonic Green functions in Ω_k under Dirichlet boundary conditions. Then there exists a positive number $\delta > 0$, which is uniform for the whole family of domains, such that for all $k \in \mathbb{N}$ it holds that*

$$x, y \in \Omega_k, \ x \neq y, \ G_k(x,y) = 0 \ \text{implies that} \ |x - y| \geq \delta.$$

Proof. 1. The proof will be performed by contradiction making use of a rescaling or blow-up analysis which was developed by Grunau and Robert in [208].

After possibly passing to a subsequence, possibly replacing Ω with some fixed Ω_{k_0} and possibly relabeling the sequence we may assume by contradiction that

$$\text{there exist } x_k, y_k \in \Omega_k, \ x_k \neq y_k, \ \text{such that } |x_k - y_k| \to 0 \text{ and } G_k(x_k, y_k) = 0. \quad (6.50)$$

In view of local positivity – see Theorem 6.19 – there exists $\rho > 0$ such that

$$|x_k - y_k| \geq \rho \max\{d(x_k, \partial\Omega_k), d(y_k, \partial\Omega_k)\}. \quad (6.51)$$

In particular, x_k and y_k are approaching the boundary $\partial\Omega$. After possibly passing to a further subsequence we may assume that

$$\text{there exists } x_\infty \in \partial\Omega : x_k, y_k \to x_\infty.$$

By means of a translation and rotation we may achieve that

$$x_\infty = 0 \text{ and the first unit vector } \mathbf{e}_1 \text{ is the exterior unit normal in } x_\infty = 0 \text{ of } \partial\Omega.$$

According to Definition 4.19 there exist $C^{4,\gamma}$-diffeomorphisms $\Psi_k : \overline{\Omega} \to \overline{\Omega_k}$ such that $\|Id - \Psi_k\|_{C^{4,\gamma}(\overline{\Omega})} \to 0$. We introduce

$$x'_k := \Psi_k^{-1}(x_k) \in \Omega, \quad y'_k := \Psi_k^{-1}(y_k) \in \Omega,$$

where obviously $x'_k \to 0$, $y'_k \to 0$, $|x'_k - y'_k| \to 0$ as $k \to \infty$. We introduce furthermore $\tilde{x}_k \in \partial\Omega$ as the closest boundary point to x'_k:

$$|x'_k - \tilde{x}_k| = d(x'_k).$$

By virtue of $x'_k \to 0 \in \partial\Omega$ and the smoothness of $\partial\Omega$, this is uniquely determined for k large enough and also satisfies $\tilde{x}_k \to 0$ for $k \to \infty$.

2. For $\xi, \eta \in \frac{1}{|x'_k - y'_k|}(-\tilde{x}_k + \overline{\Omega})$ we introduce the following rescalings of the Green functions G_k

$$\tilde{G}_k(\xi, \eta) := |x'_k - y'_k|^{n-4} G_k(\Psi_k(\tilde{x}_k + |x'_k - y'_k|\xi), \Psi_k(\tilde{x}_k + |x'_k - y'_k|\eta)).$$

The rescaled domains $\frac{1}{|x'_k - y'_k|}(-\tilde{x}_k + \Omega)$ approach the half space

$$\mathscr{H} := \{x \in \mathbb{R}^n : x_1 < 0\}.$$

The goal is to prove that in a suitable sense $(\tilde{G}_k)$ approaches the biharmonic Green function $G_{\Delta^2, \mathscr{H}}$, while from our assumption (6.50) we deduce some degenerate behaviour of $G_{\Delta^2, \mathscr{H}}$. In view of Boggio's formula, the latter cannot occur and we obtain a contradiction.

First we see that the bounds from Theorem 4.20 yield uniformly in k, ξ and η

$$|\tilde{G}_k(\xi, \eta)| \leq C|\xi - \eta|^{4-n}, \quad \text{provided that } n > 4. \tag{6.52}$$

If $n = 3, 4$ we conclude first that

$$|\nabla \tilde{G}_k(\xi, \eta)| \leq C \begin{cases} 1 & \text{if } n = 3, \\ |\xi - \eta|^{-1} & \text{if } n = 4. \end{cases}$$

Upon integration we obtain that

$$|\tilde{G}_k(\xi, \eta)| \leq C \begin{cases} 1 + |\xi| + |\eta| & \text{if } n = 3, \\ 1 + |\log|\xi - \eta|| + \log(1 + |\xi| + |\eta|) & \text{if } n = 4. \end{cases} \tag{6.53}$$

We observe, defining

$$g_k(\xi) = \Psi_k^*(\mathscr{E})(\tilde{x}_k + |x_k' - y_k'|\xi)$$

with the pulled back euclidean metric $\mathscr{E}$ in rescaled variables, that for any $\xi \in \frac{1}{|x_k' - y_k'|}\left(-\tilde{x}_k + \overline{\Omega}\right)$, we have in the sense of distributions

$$\Delta^2_{g_k,\eta}\tilde{G}_k(\xi, \,.\,) = \delta_\xi. \tag{6.54}$$

This family of differential equations, the coefficients of which obey uniform bounds, together with the uniform bounds (6.52) and (6.53) allows to apply local interior and boundary Schauder estimates (see Theorem 2.19). For each $\xi \in \mathscr{H}$ we find a function $G(\xi, \,.\,) \in C^4(\overline{\mathscr{H}} \setminus \{\xi\})$ such that $\tilde{G}_k(\xi, \,.\,) \to G(\xi, \,.\,)$ in $C^4_{loc}(\overline{\mathscr{H}} \setminus \{\xi\})$. Since $\Psi_k \to Id$ in C^4 implies that $g_k(\xi) \to (\delta_{ij})$ locally uniformly in C^3, this and (6.54) give in turn that for any $\xi \in \mathscr{H}$

$$\Delta^2_\eta G(\xi, \,.\,) = \delta_\xi$$

in the sense of distributions. Finally, in order to explain carefully the boundary conditions to be attained by $G(\xi, \,.\,)$, we consider a suitable local $C^{4,\gamma}$-smooth coordinate chart Φ for Ω near 0 with $\Phi(\xi) = \xi + O(|\xi|^2)$, which maps in particular a neighbourhood of 0 in $\partial\mathscr{H}$ onto a neighbourhood of 0 in $\partial\Omega$. Defining

$$\Phi_k(\xi) := \frac{1}{|x_k' - y_k'|}\left(\Phi\left(\Phi^{-1}(\tilde{x}_k) + |x_k' - y_k'|\xi\right) - \tilde{x}_k\right), \tag{6.55}$$

we obtain local coordinate charts for $\frac{1}{|x_k' - y_k'|}\left(-\tilde{x}_k + \Omega\right)$ with the coordinates ξ in bounded neighbourhoods of 0 in $\overline{\mathscr{H}}$. Since $\tilde{x}_k \to 0$ and $\Phi^{-1}(\tilde{x}_k) \to 0$, we have uniformly for ξ in such sets and for $k \to \infty$ that:

$$\Phi\left(\Phi^{-1}(\tilde{x}_k) + |x_k' - y_k'|\xi\right)$$
$$= \Phi\left(\Phi^{-1}(\tilde{x}_k)\right) + |x_k' - y_k'|D\Phi\left(\Phi^{-1}(\tilde{x}_k)\right)\xi \; + \; O(|x_k' - y_k'|^2 |\xi|^2)$$
$$= \tilde{x}_k + |x_k' - y_k'|\xi + o(1)|x_k' - y_k'||\xi| + O(|x_k' - y_k'|^2 |\xi|^2)$$

with $o(1) \to 0$ for $k \to \infty$. This shows that $\tilde{\Phi}_k \to Id$ locally uniformly. The chain rule directly yields also convergence of the derivatives so that we may conclude that $\Phi_k \to Id$ in C^4_{loc}. This allows us to infer that

$$G(\xi, \eta) = \partial_{\eta_1} G(\xi, \eta) = 0 \;\text{ for all }\; \xi \in \mathscr{H} \;\text{ and all }\; \eta \in \partial\mathscr{H}.$$

Finally, (6.52) and (6.53) yield the following bounds for G (where $\xi, \eta \in \mathcal{H}$, $\xi \neq \eta$)

$$|G(\xi,\eta)| \leq C \begin{cases} 1 + |\xi| + |\eta| & \text{if } n = 3, \\ 1 + |\log|\xi - \eta|| + \log(1 + |\xi| + |\eta|) & \text{if } n = 4, \\ |\xi - \eta|^{4-n} & \text{if } n \geq 5. \end{cases} \qquad (6.56)$$

3. We prove that

$$G(\xi,\eta) = G_{\Delta^2,\mathcal{H}}(\xi,\eta) \qquad \text{for all } \xi, \eta \in \mathcal{H} \text{ such that } \xi \neq \eta. \qquad (6.57)$$

To this end we keep $\xi \in \mathcal{H}$ fixed in what follows. As for the biharmonic Green function in the half space we recall Boggio's formula

$$G_{\Delta^2,\mathcal{H}}(\xi,\eta) = \frac{1}{4ne_n} |\xi - \eta|^{4-n} \int_1^{|\xi^* - \eta|/|\xi - \eta|} (v^2 - 1)v^{1-n}\,dv; \qquad (6.58)$$

where $\xi, \eta \in \mathcal{H}$, $\xi^* = (-\xi_1, \xi_2, \ldots, \xi_n)$, see Lemma 2.27. We comment on the uniqueness issue of $G_{\Delta^2,\mathcal{H}}$ in dimensions $n = 3, 4$ at the end of the present proof.

Both $G(\xi,.)$ and $G_{\Delta^2,\mathcal{H}}(\xi,.)$ satisfy the biharmonic equation with the δ-distribution δ_ξ as right hand side and zero Dirichlet boundary conditions on $\{\eta_1 = 0\}$. We let $\psi := \psi_\xi := G(\xi,.) - G_{\Delta^2,\mathcal{H}}(\xi,.)$. Hence, $\psi \in C^\infty(\overline{\mathcal{H}})$ solves

$$\begin{cases} \Delta^2 \psi = 0 & \text{in } \mathcal{H}, \\ \psi = \frac{\partial}{\partial \eta_1} \psi = 0 & \text{on } \{\eta_1 = 0\}. \end{cases} \qquad (6.59)$$

Moreover, according to (6.56) and (6.58), for all $\eta \in \mathcal{H}$ we have that

$$|\psi(\eta)| \leq C \begin{cases} 1 + |\eta| & \text{if } n = 3, \\ 1 + |\log|\eta|| & \text{if } n = 4, \\ |\eta|^{4-n} & \text{if } n \geq 5; \end{cases} \qquad (6.60)$$

$$|\nabla \psi(\eta)| \leq C \begin{cases} 1 & \text{if } n = 3, \\ |\eta|^{-1} & \text{if } n = 4; \end{cases} \qquad (6.61)$$

where $C = C(\xi)$. According to [151, 233], writing $\bar{\eta} = (\eta_2, \ldots, \eta_n)$,

$$\psi^*(\eta) := \begin{cases} \psi(\eta) & \text{if } \eta_1 \leq 0, \\ -\psi(-\eta_1, \bar{\eta}) - 2\eta_1 \frac{\partial}{\partial y_1} \psi(-\eta_1, \bar{\eta}) - \eta_1^2 \Delta \psi(-\eta_1, \bar{\eta}) & \text{if } \eta_1 > 0, \end{cases}$$

$\psi^* \in C^4(\mathbb{R}^n)$ is an entire biharmonic function. We consider now first the case $n > 4$. Below we will prove that (6.59) and (6.60) imply that also

$$|\nabla^j \psi(\eta)| \le C|\eta|^{4-n-j} \qquad \text{for all } \eta \in \mathscr{H} \text{ and for } j = 1, 2, \tag{6.62}$$

where $C = C(\xi)$. This immediately gives that $|\psi^*(\eta)| \le C|\eta|^{4-n}$ and in particular that ψ^* is a bounded entire biharmonic function. Liouville's theorem for biharmonic functions [324, p.19] yields that $\psi^*(\eta) \equiv 0$ so that the claim of the lemma follows, provided $n > 4$.

If $n = 3, 4$ we shall prove below that for $j = 0, 1, 2$ and for all $\eta \in \mathscr{H}$

$$|D^{2+j} \psi(\eta)| \le C|\eta|^{2-n-j}, \quad \text{where } C = C(\xi). \tag{6.63}$$

As above ψ^* is an entire biharmonic function and so are $D\psi^*$ and $D^2\psi^*$. We have $|D^2\psi^*(\eta)| \le C(1+|\eta|)^{2-n}$ and so, it follows that $D^2\psi^*(\eta) \equiv 0$. In view of the boundary conditions in (6.59) we come up with $\psi^*(\eta) \equiv 0$ also in the case $n = 3, 4$.

It remains to prove (6.62) and (6.63). We consider first $n > 4$. Assume by contradiction that there exists a sequence $(\eta_\ell) \subset \mathscr{H}$ such that $|\nabla^j \psi(\eta_\ell)||\eta_\ell|^{n+j-4} \to \infty$ for $\ell \to \infty$. Then, taking $\eta_{\ell,1}$ as the first component of η_ℓ,

$$\tilde{\psi}_\ell(\eta) := |\eta_\ell|^{n-4} \psi\left(\eta_\ell - \eta_{\ell,1}\mathbf{e}_1 + |\eta_\ell|\eta\right)$$

would solve

$$\begin{cases} \Delta^2 \tilde{\psi}_\ell = 0 & \text{in } \mathscr{H}, \\ \tilde{\psi}_\ell = \frac{\partial}{\partial \eta_1} \tilde{\psi}_\ell = 0 & \text{on } \{\eta_1 = 0\}. \end{cases}$$

From the assumption we conclude that

$$\left|\nabla^j \tilde{\psi}_\ell\left(\frac{\eta_{\ell,1}}{|\eta_\ell|}\mathbf{e}_1\right)\right| = |\eta_\ell|^{n+j-4}|\nabla^j \psi(\eta_\ell)| \to \infty. \tag{6.64}$$

On the other hand,

$$|\tilde{\psi}_\ell(\eta)| \le C|\eta_\ell|^{n-4}\left|\eta_\ell - \eta_{\ell,1}\mathbf{e}_1 + |\eta_\ell|\eta\right|^{4-n} \le C\left|\frac{\eta_\ell}{|\eta_\ell|} + \eta - \frac{\eta_{\ell,1}}{|\eta_\ell|}\mathbf{e}_1\right|^{4-n}, \tag{6.65}$$

so that $\tilde{\psi}_\ell$ remains bounded in a neighbourhood of $\frac{\eta_{\ell,1}}{|\eta_\ell|}\mathbf{e}_1$ in $\overline{\mathscr{H}}$. Local Schauder estimates (see Theorem 2.19) yield

$$\left|\nabla^j \tilde{\psi}_\ell\left(\frac{\eta_{\ell,1}}{|\eta_\ell|}\mathbf{e}_1\right)\right| \le C,$$

thereby contradicting (6.64). This proves (6.62).

As for (6.63), i.e. in particular $n = 3, 4$, the proof is quite similar since we can already make use of the gradient estimates (6.61). Instead of (6.65) one has to employ

$$|\nabla \tilde{\psi}_\ell(\eta)| \le C|\eta_\ell|^{n-3} \left|\eta_\ell - \eta_{\ell,1}\mathbf{e}_1 + |\eta_\ell|\eta\right|^{3-n} \le C \left| \frac{\eta_\ell}{|\eta_\ell|} + \eta - \frac{\eta_{\ell,1}}{|\eta_\ell|}\mathbf{e}_1 \right|^{3-n},$$

so that $\nabla \tilde{\psi}_\ell$ remains bounded uniformly outside $\frac{\eta_\ell}{|\eta_\ell|} - \frac{\eta_{\ell,1}}{|\eta_\ell|}\mathbf{e}_1$. Therefore, since $\tilde{\psi}_\ell$ vanishes on $\partial \mathcal{H}$, we get that $\tilde{\psi}_\ell$ is bounded in a neighbourhood of $\frac{\eta_{\ell,1}}{|\eta_\ell|}\mathbf{e}_1$ in $\overline{\mathcal{H}}$. The proof of (6.57) is complete.

4. From assumption (6.50) we obtain

$$\tilde{G}_k \left(\frac{x'_k - \tilde{x}_k}{|x'_k - y'_k|}, \frac{y'_k - \tilde{x}_k}{|x'_k - y'_k|} \right) = 0. \tag{6.66}$$

Since $\frac{x'_k - \tilde{x}_k}{|x'_k - y'_k|}, \frac{y'_k - \tilde{x}_k}{|x'_k - y'_k|}$ are bounded by virtue of $|x'_k - \tilde{x}_k| = d(x'_k)$ and boundedness of $d(x'_k)/|x'_k - y'_k|$ according to (6.51), we may assume after possibly passing to a further subsequence that

$$\text{there exist } \xi_0, \eta_0 \in \overline{\mathcal{H}} : \quad \frac{x'_k - \tilde{x}_k}{|x'_k - y'_k|} \to \xi_0, \quad \frac{y'_k - \tilde{x}_k}{|x'_k - y'_k|} \to \eta_0.$$

Observe that $|\xi_0 - \eta_0| = 1$. We recall from (6.55) the construction of local $C^{4,\gamma}$-smooth coordinate charts $\xi \mapsto \Phi_k(\xi)$ for $\frac{1}{|x'_k - y'_k|}(-\tilde{x}_k + \Omega)$ with coordinates ξ in bounded neighbourhoods of 0 in $\overline{\mathcal{H}}$. These converge in C^4_{loc} to the identity. Instead of $\tilde{G}_k$ we consider $\hat{G}_k := \tilde{G}_k \circ (\Phi_k, \Phi_k)$ in $\overline{\mathcal{H}} \times \overline{\mathcal{H}} \setminus \{\xi = \eta\}$. As in Step 2 of this proof, elliptic theory (see Theorem 2.19) yields that $\hat{G}_k \to G_{\Delta^2, \mathcal{H}}$ in $C^4((\overline{\mathcal{H}} \cap B_{1/2}(\xi_0)) \times (\overline{\mathcal{H}} \cap B_{1/2}(\eta_0)))$. Exploiting this convergence, the boundary data of the Green functions and using Taylor's expansion we conclude from (6.66) that one of the following three possibilities occurs

1. $\xi_0 \in \mathcal{H}, \eta_0 \in \mathcal{H}, G_{\Delta^2, \mathcal{H}}(\xi_0, \eta_0) = 0$;
2. $\xi_0 \in \mathcal{H}, \eta_0 \in \partial \mathcal{H}, \Delta_\eta G_{\Delta^2, \mathcal{H}}(\xi_0, \eta_0) = 0$, or vice versa;
3. $\xi_0 \in \partial \mathcal{H}, \eta_0 \in \partial \mathcal{H}, \Delta_\xi \Delta_\eta G_{\Delta^2, \mathcal{H}}(\xi_0, \eta_0) = 0$.

From Boggio's formula (6.58) for the biharmonic Green function $G_{\Delta^2, \mathcal{H}}$ in the half space $\mathcal{H}$, it follows by direct calculation that

$$\text{for all } \xi, \eta \in \mathcal{H} \text{ such that } x \ne y: \quad G_{\Delta^2, \mathcal{H}}(\xi, \eta) > 0;$$
$$\text{for all } \xi \in \mathcal{H}, \eta \in \partial \mathcal{H}: \quad \Delta_\eta G_{\Delta^2, \mathcal{H}}(\xi, \eta) > 0;$$
$$\text{for all } x, y \in \partial \mathcal{H}, x \ne y: \quad \Delta_\xi \Delta_\eta G_{\Delta^2, \mathcal{H}}(\xi, \eta) > 0.$$

This shows that none of the possibilities mentioned above may occur. Hence we have achieved a contradiction and the theorem is proved. $\qquad \square$

Remark 6.28. The arguments in Step 3 of the previous proof show that if $n = 3, 4$, the "boundary conditions" at infinity $|G_{\Delta^2, \mathcal{H}}(x, y)| \le C(1 + |x| + |y|)$ yield a uniquely determined biharmonic Green function in $\mathcal{H}$.

Finally, basing upon Theorem 6.27 it is straightforward to prove Theorem 6.24.

Proof of Theorem 6.24. In order to prove (6.46) we assume by contradiction that there exist sequences $(x_k), (y_k) \subset \Omega$, $x_k \ne y_k$ such that $G_\Omega(x_k, y_k) \le 0$ and $\lim_{k \to \infty} |x_k - y_k| = 0$. In view of the smoothness assumptions made on Ω one may construct sequences $(\xi_k), (\eta_k) \subset \Omega$, $\xi_k \ne \eta_k$ such that $G_\Omega(\xi_k, \eta_k) = 0$ as well as $\lim_{k \to \infty} |\xi_k - \eta_k| = 0$. Application of Theorem 6.27 with all $\Omega_k \equiv \Omega$ directly yields a contradiction.

Formula (6.47) now follows from (6.46) and Theorem 4.28. $\qquad\qquad\qquad\square$

6.5 Domain Perturbations in Higher Dimensions

In Section 6.1 we studied the question whether the positivity of the biharmonic Green function in the two-dimensional (unit) disk remains under sufficiently small smooth domain perturbations. Here, we prove the same result in dimensions $n \ge 3$. For domain closeness we refer to Definition 6.1.

Theorem 6.29. *Let B be the unit ball of $\mathbb{R}^n$, $n \ge 2$. Then there exists $\varepsilon_0 = \varepsilon_0(n) > 0$ such that for $\varepsilon \in [0, \varepsilon_0]$ the following holds true.*

Assume that $\Omega \subset \mathbb{R}^n$ is a $C^{4, \gamma}$-smooth domain which is ε-close to the ball B in the $C^{4, \gamma}$-sense. Then the Green function $G_{\Delta^2, \Omega}$ for Δ^2 in Ω under Dirichlet boundary conditions is strictly positive, that is

$$G_{\Delta^2, \Omega}(x, y) > 0 \qquad \text{for all } x, y \in \Omega, \quad x \ne y.$$

The proof given in Section 6.1 for the case $n = 2$ is direct and explicit, based on perturbation series, Green's function estimates and conformal maps. This means that in principle ε_0 there may be calculated explicitly. Moreover, in the case $n = 2$, according to the observation by Sassone [359], closeness has to be assumed only in a weaker norm.

In what follows, again, we restrict ourselves to considering

$$n \ge 3.$$

In this case, the proof is somehow more indirect since a number of proofs by contradiction are involved so that it will be impossible to calculate ε_0 for $n \ge 3$ from the proofs.

Recalling Remark 6.4 we emphasise once more that Theorem 6.29 is by no means just a continuous dependence on data result.

Under suitable assumptions on the dimension n, the same techniques will allow for proving a result like Theorem 6.29 also for polyharmonic operators $(-\Delta)^m$.

Presumably one will have to require the condition $n \geq 2m - 1$ on the dimension, see Remark 6.21.

The proof of Theorem 6.29 for $n \geq 3$ will basically follow from the next result which describes in which way the transition from positivity to sign change may occur within a smooth family of domains, see Definition 4.19.

Theorem 6.30. *We assume that $n \geq 3$ and that (Ω_k) is a $C^{4,\gamma}$-perturbation of the bounded $C^{4,\gamma}$-smooth domain Ω. Let $G_k = G_{\Delta^2,\Omega_k}$ and $G = G_{\Delta^2,\Omega}$ denote the biharmonic Green functions in Ω_k and Ω respectively under Dirichlet boundary conditions.*

We suppose that there exist two sequences $(x_k), (y_k)$ such that $x_k, y_k \in \Omega_k$ and

$$x_k \neq y_k, \quad G_k(x_k, y_k) = 0 \text{ for all } k \in \mathbb{N}.$$

Up to a subsequence, let $x_\infty := \lim_{k \to +\infty} x_k$ and $y_\infty := \lim_{k \to +\infty} y_k$. Then $x_\infty, y_\infty \in \overline{\Omega}$, $x_\infty \neq y_\infty$ and we are in one of the following situations

- *if $x_\infty, y_\infty \in \Omega$, then $G(x_\infty, y_\infty) = 0$;*
- *if $x_\infty \in \Omega$ and $y_\infty \in \partial\Omega$, then $\Delta_y G(x_\infty, y_\infty) = 0$;*
- *if $x_\infty \in \partial\Omega$ and $y_\infty \in \Omega$, then $\Delta_x G(x_\infty, y_\infty) = 0$;*
- *if $x_\infty, y_\infty \in \partial\Omega$, then $\Delta_x \Delta_y G(x_\infty, y_\infty) = 0$.*

Proof. The crucial statement is that $x_\infty \neq y_\infty$, which already follows from Theorem 6.27. In view of the smoothness and convergence properties of the Green functions G_k, G, which are proved in Section 4.4, the claim follows directly from the assumption $G_k(x_k, y_k) = 0$ by means of Taylor's expansion. $\square$

Theorem 6.29 now follows with a simple proof by contradiction since in the unit ball $B \subset \mathbb{R}^n$ Boggio's formula (2.65) shows that none of the above mentioned degeneracies may indeed occur.

Lemma 6.31. *The biharmonic Green function in $B = B_1(0) \subset \mathbb{R}^n$, which is given by*

$$G_{\Delta^2,B}(x,y) = \frac{1}{4ne_n} |x-y|^{4-n} \int_1^{\left| |x|y - \frac{x}{|x|} \right|/|x-y|} (v^2 - 1)v^{1-n} \, dv \qquad (x, y \in B, \, x \neq y)$$

enjoys the following properties

$$\begin{aligned}
&\text{for all } x, y \in B, \, x \neq y: &&G_{\Delta^2,B}(x,y) > 0; \\
&\text{for all } x \in B, y \in \partial B: &&\Delta_y G_{\Delta^2,B}(x,y) > 0; \\
&\text{for all } x, y \in \partial B, \, x \neq y: &&\Delta_x \Delta_y G_{\Delta^2,B}(x,y) > 0.
\end{aligned}$$

Proof. These properties are verified by explicit calculations. $\square$

6.6 Bibliographical Notes

The presentation of the perturbation theory of positivity and smallness of the negative parts is based on [117, 120, 205, 207–209, 215]. The results in Section 6.1 were first proved in [209], while the exposition is adapted from [205]. Section 6.2 outlines and develops results by Dall'Acqua, Meister, and Sweers which originally appeared in [117, 120]. The presentation of Section 6.3 follows [215]. The material in Sections 6.4 and 6.5 is based on [207, 208], which are joint works of Grunau and Robert.

In two dimensions, the required notion of domain closeness to have positivity as in Theorem 6.3 was relaxed by Sassone [359]. Local positivity was considered first by Nehari [323] for the clamped plate equation in dimensions $n = 2, 3$, where Theorem 6.19 was proved for $n = 3$ and a restrictive version of it for $n = 2$. A first mathematical indication that the negative part of biharmonic Green's functions is smaller than their positive part was obtained by Malyshev [284] by means of a functional analytic approach. There, the estimate from below uses the same expressions as from above but with a smaller constant while we use smaller expressions and do not discuss the magnitude of the constants. Hedenmalm, Jakobsson, and Shimorin [226, 227] exhibit further examples of fourth order operators with a positive Green function and outline a strategy how to reduce the positivity discussion of Green's functions to that of suitable Poisson kernels. This approach supports the feeling that when transition from positivity to sign change occurs, this will happen most likely via the last variant in Theorem 6.30. Corresponding results for boundary value problems for fourth order ordinary differential equations were achieved by Schröder [364–366].

Further contributions to the Boggio-Hadamard conjecture are discussed in detail in Section 1.2. One may also see [226, 227, 284, 295].

Chapter 7
Semilinear Problems

We study the elliptic polyharmonic reaction-diffusion-type model equation

$$(-\Delta)^m u = f(u) \tag{7.1}$$

in bounded domains $\Omega \subset \mathbb{R}^n$ and in most cases together with Dirichlet boundary conditions

$$D^\alpha u|_{\partial\Omega} = 0 \quad \text{for } |\alpha| \leq m-1. \tag{7.2}$$

These boundary conditions prevent (7.1) from being written as a system of second order boundary value problems. However, in some cases, also (homogeneous) Navier boundary conditions (2.21) or Steklov boundary conditions (2.22) may be particularly interesting. We assume that the nonlinearity $f : \mathbb{R} \to \mathbb{R}$ is continuous and, in most cases, non-decreasing. The latter condition will be crucial in proving existence and further properties of positive solutions. A large part of this chapter will be devoted to special superlinear nonlinearities of polynomial growth such as

$$f(u) = \lambda u + |u|^{p-1}u, \qquad p > 1, \tag{7.3}$$

where $\lambda \in \mathbb{R}$ is a parameter.

As was pointed out in Section 1.5, problem (7.1)-(7.3) with $m = 2$ and $p = \frac{n+4}{n-4}$ may be considered as prototype for the Paneitz equation from fourth order conformal geometry. We come back to this geometrical issue in Section 7.10. However, the main focus of the present chapter is different. In higher order equations no general maximum principle is available (see Section 1.2), and truncation methods are not admissible. Therefore, many well established techniques from second order equations fail. One goal is to study the model problem (7.1)-(7.2) in order to establish some new and general techniques which may be useful also in other situations and which may outline future developments. A second goal is to find out, in how far it is still possible to show qualitative properties of solutions in spite of the lack of a maximum principle.

If $m = 1$ and $\Omega = B$ is the unit ball, a celebrated result by Gidas-Ni-Nirenberg [195] states radial symmetry of positive solutions to (7.1)-(7.2) under mild smoothness assumptions on the nonlinearity f. The proof is strongly based on the use of maximum principles and reflection methods. In Section 7.1 we explain how to

F. Gazzola et al., *Polyharmonic Boundary Value Problems*,
Lecture Notes in Mathematics 1991, DOI 10.1007/978-3-642-12245-3_7,
© Springer-Verlag Berlin Heidelberg 2010

substitute these by kernel estimates and monotonicity properties of the biharmonic Green function. Only an additional sign and a monotonicity condition have to be imposed on f in order to enable us to extend the result to all $m \geq 1$.

Concerning existence and nonexistence of nontrivial solutions of the Dirichlet problem (7.1)-(7.2)-(7.3) the results strongly depend – as in the case $m = 1$ of second order equations – on the value of the exponent p. In order to avoid technical distinctions we assume that $n > 2m$ in what follows. To explain the meaning of "critical growth" we start with the observation that (7.1)-(7.2)-(7.3) may be considered – at least formally – as Euler-Lagrange equations of the functional

$$
E_\lambda(v) := \begin{cases} \frac{1}{2} \int_\Omega \left(\Delta^k v \right)^2 dx - \frac{\lambda}{2} \int_\Omega v^2 dx - \frac{1}{p+1} \int_\Omega |v|^{p+1} dx & \text{if } m = 2k \text{ is even,} \\[2ex] \frac{1}{2} \int_\Omega \left| \nabla \Delta^k v \right|^2 dx - \frac{\lambda}{2} \int_\Omega v^2 dx - \frac{1}{p+1} \int_\Omega |v|^{p+1} dx & \text{if } m = 2k+1 \text{ is odd.} \end{cases}
$$

$$(7.4)$$

The highest order term forces us to work in the space $H_0^m(\Omega)$. Thanks to the embedding

$$
H_0^m(\Omega) \subset L^{s+1}(\Omega), \qquad s = \frac{n+2m}{n-2m}, \qquad n > 2m,
$$

the functional (7.4) is well-defined for $p \leq s$ and enjoys compactness properties if $p < s$. The latter so-called subcritical case is relatively simple and in Section 7.2 we show for any $\lambda \in \mathbb{R}$ the existence of infinitely many solutions.

The situation is completely different and much more difficult in the critical case where $p = s$, which is the case of geometrical relevance. This issue reflects the conformal covariance property of the Paneitz operator. The polyharmonic model problem at critical growth reads as follows,

$$
\begin{cases} (-\Delta)^m u = \lambda u + |u|^{s-1}u, & u \not\equiv 0 \text{ in } \Omega, \\ D^\alpha u|_{\partial\Omega} = 0 & \text{for } |\alpha| \leq m-1 \end{cases}
$$

and is studied in detail in Sections 7.4 to 7.9 also under different boundary conditions. A first systematical investigation of the partial loss of compactness and – more generally – the compactness properties of variational functionals with critical growth is due to Struwe [381] when $m = 1$. In Section 7.9.5 we prove an extension of his result to the present case, namely $m = 2$.

In the supercritical regime $p > s$ $(n > 2m)$, the variational formulation (7.4) breaks down completely. Instead, in Section 7.11 different techniques like a super-subsolution method or – in the case of radial solutions – dynamical systems techniques have to be developed and are applied to study the slightly different model problem

$$
\begin{cases} \Delta^2 u = \lambda(1+u)^p, & u > 0 \text{ in } B, \\ u = |\nabla u| = 0 & \text{on } \partial B, \end{cases}
$$

where $p > (n+4)/(n-4)$ and $n > 4$. Although similar results may be expected also for the analogous polyharmonic problem, technical difficulties prevent us here from developing this issue further.

7.1 A Gidas-Ni-Nirenberg Type Symmetry Result

In this section we extend the symmetry result by Gidas-Ni-Nirenberg [195], which holds for $m = 1$, to higher order elliptic problems with $m \geq 2$. We consider both Dirichlet and Navier boundary conditions. Let us start with the first case.

Theorem 7.1. *Assume that $f : [0,\infty) \to \mathbb{R}$ is a continuous, non-decreasing function with $f(0) \geq 0$ and that $u \in H_0^m \cap L^\infty(B)$ is a nonnegative nontrivial weak solution to the Dirichlet problem*

$$\begin{cases} (-\Delta)^m u = f(u) & \text{in } B, \\ D^\alpha u|_{\partial B} = 0 & \text{for } |\alpha| \leq m - 1. \end{cases} \tag{7.5}$$

Then u is radially symmetric and strictly decreasing in the radial variable.

By elliptic regularity (see Theorem 2.20) we readily infer that a bounded weak solution u to (7.5) satisfies $u \in C^{2m-1,\gamma}(\overline{B})$, see also Lemma 7.5 below. Moreover, if f is smooth, Theorem 2.19 shows that $u \in C^{2m}(\overline{B})$ is a classical solution of (7.5). In fact, u is even more regular. From Boggio's formula it is immediate that $u > 0$ in B.

The proof of Theorem 7.1 is given as follows. In Section 7.1.1 we establish some inequalities for the polyharmonic Green function relative to Dirichlet boundary conditions and its derivatives. In Section 7.1.2 we carry out the moving plane procedure on the corresponding integral equation and complete the proof of Theorem 7.1.

Remark 7.2. 1. The monotonicity assumption on f crucially enters the proof. If we assume instead that f is differentiable and satisfies for every $s \geq 0$ that $f'(s) < \Lambda_{m,1} :=$ first Dirichlet eigenvalue of $(-\Delta)^m$, then (7.5) admits at most one weak solution which is then necessarily radially symmetric. Indeed, assuming by contradiction that $u, v \in H_0^m \cap L^\infty(B)$ are different weak solutions of (7.5), we find

$$(u - v, u - v)_{H_0^m} = \int_B (f(u) - f(v))(u - v)\, dx < \Lambda_{m,1} \int_B (u - v)^2\, dx$$

contrary to the variational characterisation of $\Lambda_{m,1}$. Here, we used the scalar product defined in (2.10). In fact, symmetry is also ensured if $\overline{\{f'(s) : s \in \mathbb{R}\}}$ does not contain any nonradial Dirichlet eigenvalue of $(-\Delta)^m$ (see [270]) or if $f'(s) < \Lambda_{m,2}$ (see [124]). Uniqueness is guaranteed for sublinear f (see [127]).
2. An inspection of the proof shows that Theorem 7.1 is also valid for (7.5) with $f(u)$ replaced by the nonautonomous radial nonlinearity $f(|x|, u)$ provided that $f : [0,1] \times [0,\infty) \to [0,\infty)$ is continuous, non-increasing in the first variable and non-decreasing in the second one. If f is increasing in the first variable, the statement may become false, see [45, Theorem 3].

On the other hand, as long as Navier boundary conditions are involved, the scalar equation (7.1) may be rewritten as a system of m equations and the classical moving plane procedure applies. In this respect, the following statement is a direct consequence of a result by Troy [397].

Theorem 7.3. *Assume that $f : [0,\infty) \to \mathbb{R}$ is a continuously differentiable, non-decreasing function with $f(0) \geq 0$ and that $u \in H^m \cap L^\infty(B)$ is a nonnegative nontrivial weak solution to the Navier problem*

$$\begin{cases} (-\Delta)^m u = f(u) & \text{in } B, \\ \Delta^j u|_{\partial B} = 0 & \text{for } j = 0, \ldots, m-1. \end{cases}$$

Then the maps $r \mapsto (-\Delta)^k u(r)$ are radially symmetric and strictly decreasing for all $k = 0, \ldots, m-1$.

We point out that under the assumptions of Theorem 7.3, the solution u satisfies $u \in C^{2m}(\overline{B})$, see Theorem 2.19.

The following example shows that the *sign assumption* on f, in both Theorems 7.1 and 7.3, is necessary in order to have the monotonicity of positive radial solutions.

Example 7.4. Let J_0 and I_0 denote respectively the Bessel function and the so-called modified Bessel function of the first kind. Some computations show that the function

$$v(r) := i\left(J_0\left(e^{\frac{1}{4}\pi i} r\right) - I_0\left(e^{\frac{1}{4}\pi i} r\right) \right)$$

$$= i \sum_{k=0}^\infty \left(\frac{\left(-\frac{1}{4}e^{\frac{1}{2}\pi i} r^2\right)^k}{(k!)^2} - \frac{\left(\frac{1}{4}e^{\frac{1}{2}\pi i} r^2\right)^k}{(k!)^2} \right) = \sum_{k=0}^\infty \frac{2(-1)^k}{((2k+1)!)^2} \left(\frac{r}{2}\right)^{4k+2}$$

solves $\Delta^2 v = -v$ in $\mathbb{R}^2$ and oscillates with increasing amplitude.

Let $r_0 > 0$ denote the first nonzero minimum of v and let $m_0 = -v(r_0) > 0$. Numerically one finds that $r_0 = 8.28\ldots$ and $m_0 = 72.33\ldots$. Putting $u_0(r) = v(r) + m_0$ and $f(u) = m_0 - u$, one finds that $u_0 > 0$ satisfies

$$\begin{cases} \Delta^2 u_0 = f(u_0) & \text{in } B_{r_0}, \\ u_0 = |\nabla u_0| = 0 & \text{on } \partial B_{r_0}, \end{cases} \tag{7.6}$$

but the radially symmetric solution u_0 is not decreasing, see Figure 7.1.

Now let $r_1 > 0$ denote the first zero of $\Delta v(r) = v''(r) + v'(r)/r$. Define the number $m_1 = -v(r_1) > 0$. Numerically one finds $r_1 = 7.23\ldots$ and $m_1 = 50.15\ldots$. By setting $u_1(r) = v(r) + m_1$ and $f(u) = m_1 - u$ one finds that $u_1 > 0$ satisfies

$$\begin{cases} \Delta^2 u_1 = f(u_1) & \text{in } B_{r_1}, \\ u_1 = \Delta u_1 = 0 & \text{on } \partial B_{r_1}, \end{cases} \tag{7.7}$$

u_1 is radial but not decreasing, see Figure 7.1.

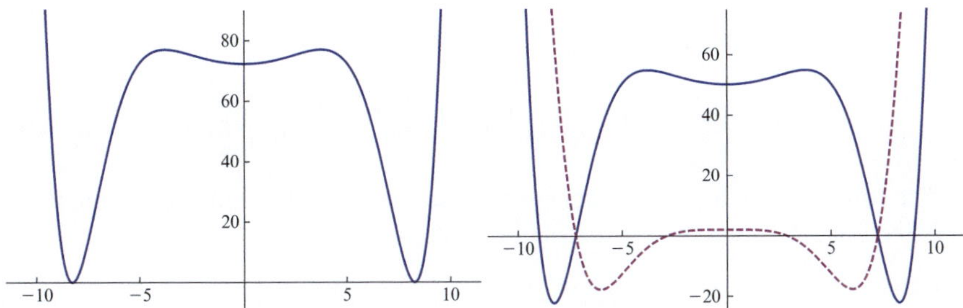

Fig. 7.1 The graphs of u_0 (left picture) and of u_1 and Δu_1 (right picture).

The above two-dimensional construction may be extended to higher dimensions. In dimensions $n \geq 3$ one replaces v by

$$v_n(x) = 2 \sum_{k=0}^{\infty} \frac{(-1)^k}{(2k+1)! \, \Gamma\left(2k+1+\frac{1}{2}n\right)} \left(\frac{r}{2}\right)^{4k+2}$$

and proceeds in a similar way. This formula even applies in one dimension and $v_1(x) = \frac{2}{\sqrt{\pi}} \sin(\frac{\sqrt{2}}{2}x) \sinh(\frac{\sqrt{2}}{2}x)$ may be used for the construction of a counterexample.

7.1.1 Green Function Inequalities

In this section we derive some pointwise inequalities for the Green function of $(-\Delta)^m$ on $B \subset \mathbb{R}^n$ relative to Dirichlet boundary conditions. It is convenient to introduce the quantity

$$\theta(x,y) = \begin{cases} (1 - |x|^2)(1 - |y|^2) & \text{if } x,y \in B, \\ 0 & \text{if } x \notin B \text{ or } y \notin B. \end{cases}$$

Then for $x,y \in B$, $x \neq y$ we use the following representation of the Green function due to Boggio, see Lemma 2.27:

$$G(x,y) = G_{m,n}(x,y) = k_{m,n}|x-y|^{2m-n} \int_1^{\left(1 + \frac{\theta(x,y)}{|x-y|^2}\right)^{1/2}} \frac{(z^2 - 1)^{m-1}}{z^{n-1}} \, dz$$

$$= \frac{k_{m,n}}{2}|x-y|^{2m-n} \int_0^{\frac{\theta(x,y)}{|x-y|^2}} \frac{z^{m-1}}{(z+1)^{n/2}} \, dz = \frac{k_{m,n}}{2} H(|x-y|^2, \theta(x,y)). \quad (7.8)$$

Here $k_{m,n}$ is the positive constant from Lemma 2.27 and

$$H : (0,\infty) \times [0,\infty) \to \mathbb{R}, \quad H(s,t) = s^{m-\frac{n}{2}} \int_0^{\frac{t}{s}} \frac{z^{m-1}}{(z+1)^{n/2}} \, dz. \tag{7.9}$$

The following lemma is a direct consequence of formula (7.8), elliptic regularity (see Section 2.5.2) and the estimates in Chapter 4.

Lemma 7.5. *Let $h \in L^\infty(B)$, and let $u \in H_0^m(B)$ satisfy*

$$(u,v)_{H_0^m} = \int_B hv \, dx \qquad \text{for all } v \in H_0^m(B),$$

i.e., u is a weak solution of $(-\Delta)^m u = h$ in B under Dirichlet boundary conditions. Then $u \in C^{2m-1,\gamma}(\overline{B})$, and u satisfies

$$D^\alpha u(x) = \int_B D_x^\alpha G(x,y)h(y) \, dy \qquad \text{for every } x \in \overline{B},$$

where D^α stands for any partial derivative of order $|\alpha| < 2m$. In particular, $D^\alpha u \equiv 0$ on ∂B for $|\alpha| \leq m-1$.

We need the following inequalities for the derivatives of the function H defined in (7.9).

Lemma 7.6. *For all $s,t > 0$ we have*

$$H_s(s,t) < 0, \qquad H_t(s,t) > 0, \qquad H_{st}(s,t) < 0.$$

Proof. We have

$$H_t(s,t) = \frac{t^{m-1}}{(t+s)^{n/2}}, \qquad H_{st}(s,t) = -\frac{nt^{m-1}}{2(t+s)^{n/2+1}}$$

and

$$H_s(s,t) = \left(m - \frac{n}{2}\right)s^{m-\frac{n}{2}-1} \int_0^{\frac{t}{s}} \frac{z^{m-1}}{(z+1)^{n/2}} \, dz - \frac{t^m}{s(t+s)^{n/2}}.$$

Hence the last two inequalities follow. Also the first inequality follows in case $n \geq 2m$ while in the remaining case $n < 2m$, we rewrite $H_s(s,t)$ as

$$H_s(s,t) = \left(m - \frac{n}{2}\right) \int_0^t \frac{x^{m-1}}{s(x+s)^{n/2}} \, dx - \frac{t^m}{s(t+s)^{n/2}}$$

$$= \left(m - \frac{n}{2}\right) \int_0^t \frac{x^{m-\frac{n}{2}-1}}{s} \left(\frac{x}{x+s}\right)^{n/2} dx - \frac{t^m}{s(t+s)^{n/2}}$$

$$< \left(m - \frac{n}{2}\right) \left(\frac{t}{t+s}\right)^{n/2} \int_0^t \frac{x^{m-\frac{n}{2}-1}}{s} \, dx - \frac{t^m}{s(t+s)^{n/2}} = 0.$$

This completes the proof. $\qquad\qquad\qquad\qquad\qquad\qquad\qquad\qquad\qquad\qquad\qquad\square$

In the following, we assume that G is trivially extended to $\mathbb{R}^n \times \mathbb{R}^n \setminus \{(x,x);\ x \in \mathbb{R}^n\}$, i.e., $G(x,y) = 0$ if $|x| \geq 1$ or $|y| \geq 1$. Then formula (7.8) is valid for all $x, y \in \mathbb{R}^n$, $x \neq y$. We introduce some more notation. For all $\lambda \in [0,1]$ we put

$$T_\lambda := \{x = (x_1,\ldots,x_n) \in \mathbb{R}^n;\ x_1 = \lambda\} \quad \text{and} \quad \Sigma_\lambda := \{x = (x_1,\ldots,x_n) \in B;\ x_1 < \lambda\}. \tag{7.10}$$

Moreover, for any $x \in \mathbb{R}^n$ let $\bar{x}$ denote its reflection about T_λ.

Lemma 7.7. *Let $\lambda \in [0,1)$. Then for every $x \in B \cap T_\lambda$ and $y \in \Sigma_\lambda$ we have*

$$G_{x_1}(x,y) < 0 \qquad \text{and} \tag{7.11}$$
$$G_{x_1}(x,y) + G_{x_1}(x,\bar{y}) \leq 0. \tag{7.12}$$

Moreover, the second inequality is strict if $\lambda > 0$.

Proof. For abbreviation, we put $d := |x-y|^2 = |x-\bar{y}|^2 > 0$, $\theta = \theta(x,y) > 0$, and $\bar{\theta} = \theta(x,\bar{y}) \geq 0$. Then

$$G_{x_1}(x,y) = k_{m,n}\Big(H_s(d,\theta)(x_1 - y_1) - H_t(d,\theta)(1 - |y|^2)x_1\Big) < 0,$$

by Lemma 7.6, since $x_1 \geq 0$ and $x_1 > y_1$. Moreover

$$\begin{aligned}
G_{x_1}(x,y) + G_{x_1}(x,\bar{y}) &= k_{m,n}\Big(H_s(d,\theta)(x_1 - y_1) + H_s(d,\bar{\theta})(x_1 - (\bar{y})_1)\Big) \\
&\quad - k_{m,n}\Big([H_t(d,\theta)(1 - |y|^2) + H_t(d,\bar{\theta})(1 - |\bar{y}|^2)]x_1\Big) \\
&\leq k_{m,n}[H_s(d,\theta) - H_s(d,\bar{\theta})](x_1 - y_1), \tag{7.13}
\end{aligned}$$

where we used Lemma 7.6 and the fact that $x_1 - \bar{y}_1 = y_1 - x_1$. Since moreover $\bar{\theta} \leq \theta$ and $H_{st} < 0$ in $(0,\infty)^2$ by Lemma 7.6, we conclude that $H_s(d,\theta) - H_s(d,\bar{\theta}) \leq 0$. Hence, (7.12) follows from (7.13). Finally, if $\lambda > 0$, then we have the strict inequality $\bar{\theta} < \theta$, so that we obtain a strict inequality in (7.13). $\qquad\square$

Lemma 7.8. *Let $\lambda \in (0,1)$. For all $x,y \in \Sigma_\lambda$, $x \neq y$, we have*

$$G(x,y) > \max\{G(x,\bar{y}), G(\bar{x},y)\}, \tag{7.14}$$
$$G(x,y) - G(\bar{x},\bar{y}) > |G(x,\bar{y}) - G(\bar{x},y)|. \tag{7.15}$$

Proof. Concerning (7.14), it suffices to prove $G(x,y) > G(x,\bar{y})$ due to symmetry and to consider the case $\bar{y} \in B$. We first observe that

$$|x-y| = |\bar{x}-\bar{y}| < |x-\bar{y}| = |\bar{x}-y|. \tag{7.16}$$

Moreover, since $|\bar{x}| > |x|$, $|\bar{y}| > |y|$, we have that

$$\theta(x,y) > \max\{\theta(x,\bar{y}), \theta(\bar{x},y)\} \geq \min\{\theta(x,\bar{y}), \theta(\bar{x},y)\} > \theta(\bar{x},\bar{y}) \tag{7.17}$$

and may conclude that

$$G(x,y) = \frac{k_{m,n}}{2} H\left(|x-y|^2, \theta(x,y)\right) > \frac{k_{m,n}}{2} H\left(|x-\bar{y}|^2, \theta(x,y)\right)$$

$$> \frac{k_{m,n}}{2} H\left(|x-\bar{y}|^2, \theta(x,\bar{y})\right) = G(x,\bar{y}).$$

Here we used that H is strictly decreasing in the first argument and increasing in the second one, see Lemma 7.6.

In view of (7.14), in order to prove (7.15) we may restrict ourselves to $\bar{x}, \bar{y} \in B$ and observe first that (7.15) is equivalent to

$$H\left(|x-y|^2, \theta(x,y)\right) - H\left(|\bar{x}-\bar{y}|^2, \theta(\bar{x},\bar{y})\right)$$

$$> \left|H\left(|x-\bar{y}|^2, \theta(x,\bar{y})\right) - H\left(|\bar{x}-y|^2, \theta(\bar{x},y)\right)\right|.$$

This means that, in view of (7.16) and (7.17), we have to show that

$$0 < s_1 < s_2, \quad 0 < t_1 < t_2 < t_4, \quad 0 < t_1 < t_3 < t_4$$

$$\Rightarrow \quad H(s_1,t_4) - H(s_1,t_1) > |H(s_2,t_2) - H(s_2,t_3)|.$$

Indeed, the latter follows again from Lemma 7.6 since

$$H(s_1,t_4) - H(s_1,t_1) = \int_{t_1}^{t_4} H_t(s_1,t)\, dt > \int_{t_1}^{t_4} H_t(s_2,t)\, dt$$

$$> \int_{\min\{t_2,t_3\}}^{\max\{t_2,t_3\}} H_t(s_2,t)\, dt = |H(s_2,t_2) - H(s_2,t_3)|.$$

This completes the proof. $\square$

7.1.2 The Moving Plane Argument

In this section we complete the proof of Theorem 7.1. Consider a fixed nonnegative nontrivial weak solution $u \in H_0^m \cap L^\infty(B)$ of (7.5). From Corollary 2.21 we know that $u \in C^{2m-1,\gamma}(\overline{B})$, see also Proposition 7.15 below. Hence, from Lemma 2.27 we know that $u > 0$ in B, see also Theorem 3.6 for a direct statement without regularity. Let T_λ, Σ_λ and $\bar{x}$ be defined as in (7.10). We first provide some crucial estimates for directional derivatives which are related to the Hopf boundary lemma for second order problems. We recall the following statement as a special case of Theorem 5.7.

Lemma 7.9. *If $x_0 \in \partial B$ and μ is a unit vector with $\mu \cdot x_0 < 0$, then $\dfrac{\partial^m u}{\partial \mu^m}(x_0) > 0$.*

In the following we extend u by zero outside of B so that it is defined on the whole of $\mathbb{R}^n$ and we put

$$\tilde{f}(s) = \begin{cases} f(s) & \text{if } s > 0 \\ 0 & \text{if } s = 0 \end{cases} \tag{7.18}$$

so that $\tilde{f}: [0, \infty) \to [0, \infty)$ is still non-decreasing while it may lose continuity at $s = 0$.

For the next estimate we need the following technical result.

Lemma 7.10. *Let $0 < \lambda < 1$, and suppose that $u(x) \geq u(\bar{x})$ for all $x \in \Sigma_\lambda$. Then $f(u(y)) \geq \tilde{f}(u(\bar{y})) \geq 0$ for all $y \in \Sigma_\lambda$, and there exists a nonempty open set $\mathcal{O}_\lambda \subset \Sigma_\lambda$ such that $f(u(y)) > \tilde{f}(u(\bar{y}))$ or $\tilde{f}(u(\bar{y})) > 0$ for all $y \in \mathcal{O}_\lambda$.*

Proof. The inequalities $f(u(y)) \geq \tilde{f}(u(\bar{y})) \geq 0$ for all $y \in \Sigma_\lambda$ follow from the monotonicity and positivity assumption on f. For the second statement it then suffices to show that $f(u(y)) \not\equiv 0$ in Σ_λ since then one of the two above inequalities would become strict in a nonempty open set $\mathcal{O}_\lambda \subset \Sigma_\lambda$. By contradiction, if $f(u) \equiv 0$ in Σ_λ then the above inequalities would imply $f(u) \equiv 0$ in B. In turn, this implies $(-\Delta)^m u \equiv 0$ which contradicts the positivity of u. $\qquad\square$

Thanks to the above lemmas we obtain a statement on the sign of $\frac{\partial u}{\partial x_1}$.

Lemma 7.11. *Let $0 < \lambda < 1$, and suppose that $u(x) \geq u(\bar{x})$ for all $x \in \Sigma_\lambda$. Then there exists $\gamma \in (0, \lambda)$ such that $\frac{\partial u}{\partial x_1} < 0$ on $T_\ell \cap B$ for all $\ell \in (\lambda - \gamma, \lambda)$.*

Proof. By Lemma 7.5, for all $x \in T_\lambda \cap B$ we have

$$\frac{\partial u}{\partial x_1}(x) = \int_B G_{x_1}(x,y) f(u(y))\, dy = \int_{\Sigma_\lambda} \left(G_{x_1}(x,y) f(u(y)) + G_{x_1}(x,\bar{y}) \tilde{f}(u(\bar{y})) \right) dy.$$

$$\tag{7.19}$$

According to Lemma 7.10 we have $f(u(y)) \geq \tilde{f}(u(\bar{y})) \geq 0$ for all $y \in \Sigma_\lambda$ and two cases may occur. In the first case, $f(u(y)) > \tilde{f}(u(\bar{y}))$ for all $y \in \mathcal{O}_\lambda$; in this case, (7.19) yields

$$\frac{\partial u}{\partial x_1}(x) < \int_{\Sigma_\lambda} \left(G_{x_1}(x,y) + G_{x_1}(x,\bar{y}) \right) \tilde{f}(u(\bar{y}))\, dy \leq 0 \quad \text{for all } x \in T_\lambda \cap B,$$

where in the first inequality we used (7.11) and in the second we used (7.12). In the second case, $\tilde{f}(u(\bar{y})) > 0$ for all $y \in \mathcal{O}_\lambda$; in this case,

$$\frac{\partial u}{\partial x_1}(x) \leq \int_{\Sigma_\lambda} \left(G_{x_1}(x,y) + G_{x_1}(x,\bar{y}) \right) \tilde{f}(u(\bar{y}))\, dy < 0 \quad \text{for all } x \in T_\lambda \cap B,$$

where in the first inequality we used (7.12) which is strict for $\lambda > 0$. In any case,

$$\frac{\partial u}{\partial x_1}(x) < 0 \quad \text{for all } x \in T_\lambda \cap B. \tag{7.20}$$

For any $y \in \mathbb{R}^n$ and any $a > 0$ consider now the cube centered at y, namely

$$\mathcal{U}_a(y) := \left\{ x \in \mathbb{R}^n;\ \max_{1 \le i \le n} |x_i - y_i| < a \right\}.$$

In view of Lemma 7.9, for any $x_0 \in T_\lambda \cap \partial B$ we know that

$$(-1)^m \left(\frac{\partial}{\partial x_1} \right)^{m-1} \frac{\partial u}{\partial x_1}(x_0) = \left(-\frac{\partial}{\partial x_1} \right)^m u(x_0) > 0 .$$

Since from the boundary conditions we also know that $(\frac{\partial}{\partial x_1})^k u(x_0) = 0$ for all $k = 0, \ldots, m-1$, there exists $a = a(x_0) > 0$ such that

$$\frac{\partial u}{\partial x_1}(x) < 0 \qquad \text{for all } x \in \mathcal{U}_a(x_0) \cap B. \tag{7.21}$$

Then by the compactness of $T_\lambda \cap \partial B$ there exists $\bar{a} > 0$ such that

$$\frac{\partial u}{\partial x_1}(x) < 0 \quad \text{for all } x \in A := \bigcup_{x_0 \in T_\lambda \cap \partial B} (\mathcal{U}_{\bar{a}}(x_0) \cap B). \tag{7.22}$$

Consider now the compact set $K := (T_\lambda \cap B) \setminus A$ and for $d > 0$ consider $K_d := K - de_1$. In view of (7.20), there exists $\delta > 0$ such that

$$\frac{\partial u}{\partial x_1} < 0 \quad \text{on} \quad K_d \qquad \text{for all } d \in [0, \delta]. \tag{7.23}$$

Let $\gamma := \min\{\bar{a}, \delta\} > 0$. Then the statement follows from (7.22)-(7.23). $\quad\square$

We are now ready to start the moving plane procedure by shifting the plane T_λ from the initial tangent position T_1 towards the interior of B.

Lemma 7.12. *There exists $\varepsilon > 0$ such that for all $\lambda \in [1 - \varepsilon, 1)$ we have*

$$u(x) > u(\bar{x}) \quad \text{for } x \in \Sigma_\lambda , \qquad \frac{\partial u}{\partial x_1}(x) < 0 \quad \text{for } x \in T_\lambda \cap B. \tag{7.24}$$

Proof. Note that $T_1 \cap \partial B = \{e_1\}$, where $e_1 = (1, 0, \ldots, 0)$. By arguing as for (7.21), we infer that there exists $\varepsilon > 0$ such that $\frac{\partial u}{\partial x_1}(x) < 0$ for $x \in B \setminus \Sigma_{1-2\varepsilon}$. In turn, from this we infer that (7.24) holds for all $\lambda \in [1 - \varepsilon, 1)$. $\quad\square$

Next we make sure that we can move the plane until we reach the origin.

Lemma 7.13. *We have*

$$\Lambda := \left\{ \lambda \in (0, 1);\ u(x) > u(\bar{x})\ \forall x \in \Sigma_\lambda,\ \frac{\partial u}{\partial x_1}(x) < 0\ \forall x \in T_\lambda \cap B \right\} = (0, 1).$$

Proof. By Lemma 7.12 we know that $[1-\varepsilon,1) \subset \Lambda$. Let $\bar{\lambda} \in [0,1)$ be the smallest number such that $(\bar{\lambda},1) \subset \Lambda$; the proof will be complete once we show that $\bar{\lambda} = 0$. By continuity we have

$$u(x) \geq u(\bar{x}) \quad \text{for all } x \in \Sigma_{\bar{\lambda}}. \tag{7.25}$$

By contradiction assume that $\bar{\lambda} > 0$. Then by Lemma 7.11 and (7.25) we infer that

$$\text{there exists } \gamma \in (0,\bar{\lambda}) \text{ such that } \frac{\partial u}{\partial x_1} < 0 \text{ on } T_\ell \cap B \text{ for all } \ell \in (\bar{\lambda}-\gamma,\bar{\lambda}). \tag{7.26}$$

Consider the function $\tilde{f}$ defined in (7.18). For all $x \in \Sigma_{\bar{\lambda}}$ we compute

$$\begin{aligned}
u(x) - u(\bar{x}) &= \int_B (G(x,y) - G(\bar{x},y))\, f(u(y))\, dy \\
&= \int_{\Sigma_{\bar{\lambda}}} (G(x,y) - G(\bar{x},y))\, f(u(y))\, dy \\
&\quad + \int_{\Sigma_{\bar{\lambda}}} (G(x,\bar{y}) - G(\bar{x},\bar{y}))\, \tilde{f}(u(\bar{y}))\, dy .
\end{aligned} \tag{7.27}$$

According to Lemma 7.10, two cases may occur. If $f(u(y)) > \tilde{f}(u(\bar{y}))$ for all $y \in \mathscr{O}_\lambda$, then (7.14) and (7.27) yield

$$u(x) - u(\bar{x}) > \int_{\Sigma_{\bar{\lambda}}} (G(x,y) - G(\bar{x},y) + G(x,\bar{y}) - G(\bar{x},\bar{y}))\, \tilde{f}(u(\bar{y}))\, dy \geq 0,$$

where the last inequality follows from (7.15). If $\tilde{f}(u(\bar{y})) > 0$ for all $y \in \mathscr{O}_\lambda$, then again (7.14), (7.15) and (7.27) yield

$$u(x) - u(\bar{x}) \geq \int_{\Sigma_{\bar{\lambda}}} (G(x,y) - G(\bar{x},y) + G(x,\bar{y}) - G(\bar{x},\bar{y}))\, \tilde{f}(u(\bar{y}))\, dy > 0 .$$

Hence we have shown in any case that

$$u(x) > u(\bar{x}) \qquad \text{for all} \quad x \in \Sigma_{\bar{\lambda}}. \tag{7.28}$$

From (7.26) and (7.28) we deduce by a standard compactness argument that there exists $0 < \gamma_1 < \gamma$ such that

$$u(x) > u(\bar{x}) \quad \text{for all } x \in \Sigma_\ell \quad \text{and } \ell \in (\bar{\lambda}-\gamma_1,\bar{\lambda}].$$

This, combined with (7.26), shows that $(\bar{\lambda}-\gamma_1,\bar{\lambda}] \subset \Lambda$, contrary to the characterisation of $\bar{\lambda}$. $\qquad\square$

Now we complete the *proof of Theorem 7.1*. Since $0 \in \partial \Lambda$ by Lemma 7.13, the continuity of u implies that

$$u(-x_1, x_2, \ldots, x_n) \geq u(x_1, x_2, \ldots, x_n) \text{ for } x = (x_1, \ldots, x_n) \in B \text{ with } x_1 \geq 0. \quad (7.29)$$

Since, for a given rotation $A \in SO(n)$, the function $u_A := u \circ A$ is also a positive weak solution of (7.5), inequality (7.29) also holds for u_A in place of u. This readily implies that u is symmetric with respect to every hyperplane containing the origin. Consequently, u is radially symmetric. Moreover, we have $\frac{\partial u}{\partial r} < 0$ in $B \setminus \{0\}$, since $\frac{\partial u}{\partial x_1} < 0$ in $\{x \in B, x_1 > 0\}$ by definition of Λ. $\qquad\qquad\square$

7.2 A Brief Overview of Subcritical Problems

In this section we briefly show how existence results for higher order semilinear subcritical problems exhibit no surprises when compared with second order problems. On the other hand, positivity of particular solutions (such as mountain-pass solutions) is not ensured in general domains. We start with some important facts concerning the regularity of the solution.

7.2.1 Regularity for At Most Critical Growth Problems

Let $\Omega \subset \mathbb{R}^n$ be a bounded domain and consider the equation

$$(-\Delta)^m u = \lambda u + |u|^{p-1} u \quad \text{in } \Omega, \quad (7.30)$$

where $\lambda \in \mathbb{R}$ and

$$\begin{cases} 1 < p \leq s = \frac{n+2m}{n-2m} & \text{if } n > 2m, \\ 1 < p < \infty & \text{if } n \leq 2m. \end{cases} \quad (7.31)$$

These equations are called *at most critical* with respect to the space $H^m(\Omega)$ due to the upper bound for p, which coincides with the critical Sobolev exponent s. We complement (7.30) with homogeneous boundary conditions of different kinds. We mainly focus our attention on the Dirichlet boundary conditions

$$D^\alpha u|_{\partial \Omega} = 0 \text{ for } |\alpha| \leq m - 1, \quad (7.32)$$

but we will also consider Navier boundary conditions

$$\Delta^j u|_{\partial \Omega} = 0 \text{ for } j \leq m - 1, \quad (7.33)$$

or, when $m = 2$, Steklov boundary conditions

$$u|_{\partial \Omega} = (\Delta u - au_\nu)|_{\partial \Omega} = 0 \quad (a \in \mathbb{R}). \quad (7.34)$$

Before going through existence and nonexistence results, we should clarify what is meant by a solution for each one of these problems. For the Navier problem, the suitable space

$$H_\vartheta^m(\Omega) = \{v \in H^m(\Omega);\ \Delta^j v = 0 \text{ on } \partial\Omega \text{ for } j < \tfrac{1}{2}m\} \qquad (7.35)$$

was defined in (2.35) which, thanks to elliptic estimates, is a Hilbert space when endowed with the following scalar product and corresponding norm

$$(u,v)_{H_\vartheta^m} = \begin{cases} \int_\Omega \Delta^k u \Delta^k v & \text{if } m = 2k, \\ \int_\Omega \nabla(\Delta^k u) \cdot \nabla(\Delta^k v) & \text{if } m = 2k+1, \end{cases} \qquad \|u\|_{H_\vartheta^m} = (u,u)_{H_\vartheta^m}^{1/2}. \quad (7.36)$$

Note that $H_\vartheta^2(\Omega) = H^2 \cap H_0^1(\Omega)$, the space which is needed for Steklov boundary conditions. Recalling the definition of solution in the linear case, see (2.42), it is natural to give the following definition.

Definition 7.14. Assume (7.31).

1. Let $(.,.)_{H_0^m}$ denote the bilinear form in (2.10). We say that $u \in H_0^m(\Omega)$ is a *weak solution* to (7.30)-(7.32) if

$$(u,v)_{H_0^m} = \int_\Omega (\lambda u + |u|^{p-1}u)v\,dx \quad \text{for all } v \in H_0^m(\Omega).$$

2. We say that $u \in H_\vartheta^m(\Omega)$ is a *weak solution* to (7.30)-(7.33) if

$$(u,v)_{H_\vartheta^m} = \int_\Omega (\lambda u + |u|^{p-1}u)v\,dx \quad \text{for all } v \in H_\vartheta^m(\Omega).$$

3. We say that $u \in H^2 \cap H_0^1(\Omega)$ is a *weak solution* to (7.30)-(7.34) if

$$\int_\Omega \Delta u \Delta v\,dx = \int_\Omega (\lambda u + |u|^{p-1}u)v\,dx + a\int_{\partial\Omega} u_\nu v_\nu\,d\omega \quad \text{for all } v \in H^2 \cap H_0^1(\Omega).$$

Thanks to the at most criticality assumption (7.31), all the above definitions make sense. This definition should also be compared with (2.42) which holds for linear problems. Moreover, we refer to the discussion following Theorem 2.15 to find an explanation of how these weak formulations "contain" the boundary conditions.

Once existence of a solution (according to the previous definition) is established, one would like to find out if it has some regularity properties. The answer shows a deep difference between at most critical and supercritical problems. Let us explain this crucial fact with a simple example. In a bounded $C^{2m,\gamma}$-domain Ω, consider (7.30) with $\lambda = 0$ and complemented with Dirichlet boundary conditions

$$\begin{cases} (-\Delta)^m u = |u|^{p-1}u & \text{in } \Omega, \\ D^\alpha u|_{\partial\Omega} = 0 & \text{for } |\alpha| \le m-1. \end{cases} \qquad (7.37)$$

Assume first that $p < s$. Then, if $u \in H_0^m(\Omega)$ is a weak solution to (7.37), by Theorem 2.4 we know that $u \in L^{s+1}(\Omega)$ so that the right hand side $f(u) = |u|^{p-1}u$ of (7.37) belongs to $L^{(s+1)/p}(\Omega)$. Therefore, Theorem 2.20 implies that $u \in W^{2m,(s+1)/p}(\Omega)$. In turn, a further application of Theorem 2.4 shows that $u \in L^{n(s+1)/(np-2ms-2m)}(\Omega)$. Since $p < s$ we have $\frac{n(s+1)}{np-2ms-2m} > s+1$ so that we have gained some summability of u and $f(u)$. By repeating a finite number of times the same combination of Theorems 2.20 and 2.4, we arrive at $u \in W^{2m,q}(\Omega)$ for some $q > \frac{n}{2m}$. Hence, Theorem 2.6 applies and we infer that $f(u) \in C^{0,\gamma}(\overline{\Omega})$ for some $\gamma > 0$. Finally, Theorem 2.19 implies that $u \in C^{2m,\gamma}(\overline{\Omega})$ and that u is a classical solution to (7.37).

The same arguments apply to more general semilinear equations such as

$$(-\Delta)^m u = g(u)$$

with $g \in C^1(\mathbb{R})$, $|g(s)| \le c(1 + |s|^p)$ for all $s \in \mathbb{R}$ and for some $p < s$ and $c > 0$. Moreover, one can replace Dirichlet boundary conditions (7.32) with (7.33), (7.34), or any other set of boundary conditions which satisfy the complementing condition, see Definition 2.9.

On the other hand, this procedure fails if $p > s$. To see this, let $n \ge 3$, let $p > \frac{2(n-1)}{n-2}$ and consider the second order problem

$$\begin{cases} -\Delta u = \lambda(1+u)^p & \text{in } B, \\ u = 0 & \text{on } \partial B, \end{cases} \tag{7.38}$$

with $\lambda = \frac{2}{p-2}(n-2-\frac{2}{p-2}) > 0$. Some computations allow to verify that the function $u(x) = |x|^{-2/(p-2)} - 1$ solves (7.38) in $B \setminus \{0\}$. Moreover, u is a weak solution to (7.38) in $H_0^1(B)$, if and only if $p > s = \frac{n+2}{n-2}$. However, u is not a classical solution since $u \notin L^\infty(B)$. We study in some detail supercritical *biharmonic* equations like (7.38) in Section 7.11.

This striking difference between the two cases $p < s$ and $p > s$ is one of the reasons why $p = s$ is called the *critical exponent* with respect to the space H^m. In fact, the above described bootstrap argument also fails for $p = s$ since, in this case, there is no gain of regularity after an application of Theorems 2.20 and 2.4. However, with some different and more involved arguments, Luckhaus [282] was able to prove the optimal regularity result also for the critical growth case. By restricting our attention to the problem treated in this chapter, we quote the following general result.

Proposition 7.15. *Assume that $\Omega \subset \mathbb{R}^n$ is a bounded domain and that (7.31) holds. Let u be a weak solution to one of the above problems. Then u is an analytic function in $\{x \in \Omega; u(x) \neq 0\}$. Moreover, if $\partial\Omega \in C^{2m,\gamma}$, then $u \in C^{2m,\gamma}(\overline{\Omega})$ and u is a classical solution.*

Proof. We refer to Luckhaus [282] for the general statement under Dirichlet boundary conditions (7.32). For Navier boundary conditions (7.33), the problem may be treated as a system, see [400, Appendix B] for the details in the case $m = 2$. The same arguments apply also to Steklov boundary conditions (7.34), see [42, Proposition 23]. The function $u \mapsto \lambda u + |u|^{p-1}u$ is analytic for $u \neq 0$ so that the solution u is analytic in $\{x \in \Omega; u(x) \neq 0\}$ in view of [313, 314]. $\qquad\qquad\qquad\qquad\square$

Additional regularity (such as in $C^{k,\gamma}$ for $k > 2m$) can be obtained, provided one takes into account the regularity of the nonlinearity $s \mapsto |s|^{p-1}s$ in a neighbourhood of $s = 0$.

Remark 7.16. Since variational methods are involved, the boundary $\partial\Omega$ needs not be smooth if one is merely interested in *existence results* for (7.30). In general, solutions only enjoy local smoothness properties. If one wishes to have *global reg-ularity results*, Proposition 7.15 tells us that the solution inherits smoothness from the boundary provided $p \leq s$. Global regularity is quite important because, when $p = s$, the Pohožaev identity (see Section 7.4 below) shows nonexistence of smooth solutions in certain domains, and in order to have complementary existence results, it appears convenient to consider smooth solutions.

We conclude this section by emphasising a further difference between (strictly) subcritical problems and critical problems. In the case of reaction-type model non-linearities $f(u)$ as in (7.30), the possibility of finding a priori estimates seems to be limited to the (strictly) subcritical case $p < s$, see [126, 353, 376] and, for second order equations, the fundamental contribution [197] of Gidas and Spruck. On the other hand, in the critical regime branches of solutions to (7.30) may blow up for λ in bounded intervals, see e.g. [25]. The situation is slightly different in the coercive case, i.e. when $\lambda \geq 0$ and the right hand side in (7.30) is replaced by its negative, see e.g. [395, 396, 405]. We comment on this issue in more detail in the bibliographical notes.

7.2.2 Existence

For simplicity, we consider in detail only the superlinear subcritical problem under Dirichlet boundary conditions in a smooth bounded domain $\Omega \subset \mathbb{R}^n$

$$\begin{cases} (-\Delta)^m u = \lambda u + |u|^{p-1}u & \text{in } \Omega, \\ D^\alpha u|_{\partial\Omega} = 0 & \text{for } |\alpha| \leq m - 1. \end{cases} \tag{7.39}$$

Here λ is a parameter and the exponent p is subject to the superlinearity and (strict) subcriticality assumption

$$\begin{cases} 1 < p < s = \frac{n+2m}{n-2m} & \text{if } n > 2m, \\ 1 < p < \infty & \text{if } n \leq 2m. \end{cases} \tag{7.40}$$

Then according to Proposition 7.15 any H_0^m-solution to (7.39) is as smooth as $\partial\Omega$ and p permit.

As for "minimal" nontrivial solutions we have the following existence result.

Theorem 7.17. *Let $\Lambda_{m,1}$ denote the first Dirichlet eigenvalue of $(-\Delta)^m$ in Ω and let p be subject to condition (7.40). Then for all $\lambda < \Lambda_{m,1}$ the Dirichlet problem (7.39) has a nontrivial solution.*

Proof. We minimise the functional

$$
F_\lambda(v) := \begin{cases} \displaystyle\int_\Omega \left(\left(\Delta^{m/2} v \right)^2 - \lambda v^2 \right) dx & \text{if } m \text{ even,} \\[4mm] \displaystyle\int_B \left(\left| \nabla \Delta^{(m-1)/2} v \right|^2 - \lambda v^2 \right) dx & \text{if } m \text{ odd,} \end{cases}
\tag{7.41}
$$

in $H_0^m(\Omega)$ subject to the constraint $v \in N$ with

$$
N := \left\{ v \in H_0^m(\Omega) : \int_\Omega |v|^{p+1} \, dx = 1 \right\}.
\tag{7.42}
$$

Since $\lambda < \Lambda_{m,1}$ the functional F_λ is coercive. Hence, any minimising sequence $(v_k)_{k \in \mathbb{N}} \subset N$ is bounded in $H_0^m(\Omega)$. We may extract a subsequence which is weakly convergent to some $\tilde{u} \in H_0^m(\Omega)$. Due to the compactness of the embeddings $H_0^m(\Omega) \subset L^{p+1}(\Omega)$, $H_0^m(\Omega) \subset L^2(\Omega)$, see Theorem 2.4, and weak lower semicontinuity of F_λ, which is related to a suitable scalar product in $H_0^m(\Omega)$, we have that $\tilde{u} \in N$ is a minimiser for F_λ in N. We call this minimum S_λ. By putting $u = S_\lambda^{1/(p-1)} \tilde{u}$, we obtain a solution of (7.39). $\qquad\square$

In order to remove the restriction $\lambda < \Lambda_{m,1}$ in Theorem 7.17 we notice that the above reasoning shows that a variational approach to (7.39) is quite similar to the second order case $m = 1$, as long as only existence of solutions is investigated. As in Struwe's book [382, Ch. II, Theorem 6.6], basing upon a mountain pass lemma for symmetric functionals due to Ambrosetti and Rabinowitz [14], one may prove the following result.

Theorem 7.18. *Let $\Omega \subset \mathbb{R}^n$ be a bounded smooth domain, $\lambda \in \mathbb{R}$ and let p be subject to the strict subcriticality condition (7.40). Then the Dirichlet problem (7.39) admits an unbounded sequence of solutions $(u_k) \subset H_0^m(\Omega)$.*

Theorems 7.17 and 7.18 hold (and can be proved exactly as above) also if homogeneous Dirichlet conditions (7.32) are replaced by homogeneous Navier conditions (7.33). When $m = 2$, similar statements are also available under homogeneous Steklov conditions (7.34), because variational methods apply as well when the linear term λu in Ω is replaced by the linear term au_ν on $\partial\Omega$, see Section 3.3.

7.2.3 Positivity and Symmetry

As for positivity and symmetry, in view of Chapter 6, we confine ourselves to the case where $\Omega = B$ is the unit ball and $\lambda \in [0, \Lambda_{m,1})$.

Theorem 7.19. *Assume that $\Omega = B$ and let $0 \le \lambda < \Lambda_{m,1}$. Then any solution constructed as in the proof of Theorem 7.17 by means of constrained minimisation is necessarily of fixed sign and radially symmetric.*

Proof. We use the method of decomposition with respect to dual cones which was explained in detail in Section 3.1.2. As in the proof of Theorem 7.17 let $\tilde{u}$ denote a minimum of F_λ on N. We assume that $\tilde{u}$ is sign changing. Let $\mathscr{K} \subset H_0^m(B)$ be the cone of nonnegative functions and $\mathscr{K}^*$ its dual cone. We decompose

$$\tilde{u} = u_1 + u_2, \quad u_1 \in \mathscr{K}, \quad u_2 \in \mathscr{K}^*, \quad u_1 \perp u_2 \text{ in } H_0^m(B).$$

By Proposition 3.6, our assumption implies that

$$0 \not\equiv u_1 \ge 0 \text{ and } u_2 < 0.$$

Recalling (2.11) and putting $\hat{u} = u_1 - u_2$ we see that

$$\begin{aligned}
F_\lambda(\tilde{u}) &= \|\tilde{u}\|_{H_0^m}^2 - \lambda \int_B \tilde{u}^2 \, dx \\
&= \|\hat{u}\|_{H_0^m}^2 - \lambda \int_B (u_1^2 + 2u_1 u_2 + u_2^2) \, dx \\
&\ge \|\hat{u}\|_{H_0^m}^2 - \lambda \int_B (u_1^2 - 2u_1 u_2 + u_2^2) \, dx = F_\lambda(\hat{u}),
\end{aligned}$$

while

$$|\tilde{u}| = |u_1 + u_2| \le u_1 - u_2 = \hat{u} = |\hat{u}|$$

with strict inequality on the set $\{u_1 > 0\}$ of positive measure. Hence

$$1 = \|\tilde{u}\|_{L^{p+1}} < \|\hat{u}\|_{L^{p+1}}$$

so that

$$\frac{\hat{u}}{\|\hat{u}\|_{L^{p+1}}} \in N, \qquad F_\lambda\left(\frac{\hat{u}}{\|\hat{u}\|_{L^{p+1}}}\right) < F_\lambda(\tilde{u}).$$

We achieved a contradiction to the minimality of $\tilde{u}$ so that $\tilde{u}$ is of fixed sign. A suitable multiple of $\tilde{u}$ solves (7.39). We may assume that $0 \not\equiv \tilde{u} \ge 0$ so that thanks to Boggio's principle, see Lemma 2.27, and $\lambda \ge 0$ we even have $\tilde{u} > 0$. Finally, Theorem 7.1 shows radial symmetry of $\tilde{u}$. $\square$

Conversely, if $u \in H_0^m(B)$ is a strictly positive solution to (7.39) then necessarily $\lambda < \Lambda_{m,1}$. To see this, notice that thanks to Theorem 3.7 we know that the first Dirichlet eigenfunction φ to $(-\Delta)^m$ in B is of fixed sign and the corresponding eigenvalue $\Lambda_{m,1}$ is simple. We may assume that $\varphi > 0$. Multiplying (7.39) by φ and integrating by parts yields

$$\lambda \int_B u\varphi\,dx < \int_B u^p\varphi\,dx + \lambda \int_B u\varphi\,dx = \int_B (-\Delta)^m u\,\varphi\,dx$$

$$= \int_B (-\Delta)^m \varphi\,u\,dx = \Lambda_{m,1} \int_B u\varphi\,dx. \tag{7.43}$$

Since by assumption $\int_B u\varphi\,dx > 0$, we necessarily have that $\lambda < \Lambda_{m,1}$.

Remark 7.20. In view of Theorem 7.3, a statement similar to Theorem 7.19 also holds if the Dirichlet conditions in (7.39) are replaced by the Navier conditions. Moreover, when $m = 2$ (biharmonic operator), $\lambda = 0$ and $\Omega = B$, both the Dirichlet problem (7.39) and the corresponding Navier problem admit a unique positive solution. This follows by combining Theorems 7.1 and 7.3 with a result by Dalmasso [125].

7.3   The Hilbertian Critical Embedding

The existence results stated in the previous section are obtained thanks to the compactness of the embedding $H_0^m \subset L^{p+1}$ for $p < s$. In this section we analyse the continuous (but not compact) embedding $H_0^m \subset L^{s+1}$.

As before, for all $n > 2m$ we define

$$s := \frac{n+2m}{n-2m}$$

to be the *critical Sobolev exponent*. Theorem 2.3 states that there exists a constant $S > 0$ such that $S\|u\|_{L^{s+1}}^2 \leq \|u\|_{\mathscr{D}^{m,2}}^2$ for all $u \in \mathscr{D}^{m,2}(\mathbb{R}^n)$. The optimal (largest) constant S may be characterised variationally as

$$S = \inf_{\mathscr{D}^{m,2}(\mathbb{R}^n)\setminus\{0\}} \frac{\|v\|_{\mathscr{D}^{m,2}}^2}{\|v\|_{L^{s+1}}^2}, \tag{7.44}$$

where

$$\|v\|_{\mathscr{D}^{m,2}} := \begin{cases} \|\Delta^k v\|_{L^2} & \text{if } m = 2k, \\ \|\nabla(\Delta^k v)\|_{L^2} & \text{if } m = 2k+1. \end{cases}$$

The corresponding scalar product $(\,.\,,.\,)_{\mathscr{D}^{m,2}}$, see the bilinear form in (2.13), gives an Hilbertian structure to $\mathscr{D}^{m,2}(\mathbb{R}^n)$ in view of Theorem 2.2.

Obviously, $S = S(n,m)$ and $s = s(n,m)$. Since it will be clear from the context we prefer to use the short notation.

Up to a multiplier, the Euler-Lagrange equation relative to the minimisation problem (7.44) reads

$$(-\Delta)^m u = |u|^{s-1} u \qquad \text{in } \mathbb{R}^n. \tag{7.45}$$

We now put

$$c_{m,n} := \left[\prod_{h=-m}^{m-1} (n+2h) \right]^{(n-2m)/4m}$$

and for any $x_0 \in \mathbb{R}^n$ and $\varepsilon > 0$ we consider the family of entire functions

$$u_{\varepsilon,x_0}(x) = \frac{c_{m,n}\,\varepsilon^{(n-2m)/2}}{(\varepsilon^2 + |x-x_0|^2)^{(n-2m)/2}}. \tag{7.46}$$

We first state

Theorem 7.21. *Let $m \in \mathbb{N}^+$ and let $n > 2m$. Then $\mathscr{D}^{m,2}(\mathbb{R}^n) \subset L^{s+1}(\mathbb{R}^n)$ and*

$$S\|u\|_{L^{s+1}}^2 \le \|u\|_{\mathscr{D}^{m,2}}^2 \qquad \text{for all } u \in \mathscr{D}^{m,2}(\mathbb{R}^n). \tag{7.47}$$

Equality in (7.47) holds if and only if, up to multiples, $u = u_{\varepsilon,x_0}$ for some $x_0 \in \mathbb{R}^n$ and some $\varepsilon > 0$, where u_{ε,x_0} is as in (7.46). Moreover, the functions in (7.46) are the only positive solutions of (7.45).

Proof. The estimate (7.47) follows directly from the variational characterisation (7.44). The rest of the proof is quite delicate, involving a number of deep results. Whence we briefly sketch it by giving the main points and referring to the literature for the details.

In [278, 383] it is shown that the functions u_{ε,x_0} achieve equality in (7.47). In order to show that these are the only functions with this property, we first remark that a suitable multiple of any function achieving the equality in (7.47) satisfies (7.45). Then from Lemma 7.22 below we know that any minimiser of (7.44) is of constant sign. Hence, minimisers for the Sobolev ratio are positive (or negative) solutions of (7.45). Finally, [411, Theorem 1.3] ensures that, up to scaling, translations and nontrivial multiples, they have precisely the form of u_{ε,x_0}. See also earlier results in [275] for the case $m = 2$ and subsequent stronger results in [95]. $\qquad\square$

In view of [32], we know that (7.45) admits infinitely many nodal finite energy solutions. Therefore, it is of some interest to prove the following "energy doubling property" of nodal solutions to (7.45).

Lemma 7.22. *Let $u \in \mathscr{D}^{m,2}(\mathbb{R}^n)$ be a nodal solution of the equation (7.45). Then $\|u\|_{\mathscr{D}^{m,2}}^2 \ge 2^{(s-1)/(s+1)} S\|u\|_{L^{s+1}}^2$.*

Proof. Consider the convex closed cone

$$\mathscr{K} = \left\{ u \in \mathscr{D}^{m,2}(\mathbb{R}^n) : u \ge 0 \text{ a.e. in } \mathbb{R}^n \right\},$$

and its dual cone

$$\mathscr{K}^* = \left\{ u \in \mathscr{D}^{m,2}(\mathbb{R}^n) : (u,v)_{\mathscr{D}^{m,2}} \le 0 \text{ for all } v \in \mathscr{K} \right\}.$$

By Proposition 3.6 we know that $\mathscr{K}^* \subset -\mathscr{K}$. Moreover, by the dual cones decomposition, see Theorem 3.4, for each $u \in \mathscr{D}^{m,2}(\mathbb{R}^n)$ there exists a unique pair (u_1, u_2) in $\mathscr{K} \times \mathscr{K}^*$ such that

$$u = u_1 + u_2, \quad (u_1, u_2)_{\mathscr{D}^{m,2}} = 0. \tag{7.48}$$

Let u be a nodal solution of (7.45) and let $u_1 \in \mathscr{K}$ and $u_2 \in \mathscr{K}^*$ be the components of u according to this decomposition. We obtain that $u_i \neq 0$ and

$$|u(x)|^{s-1} u(x) u_i(x) \leq |u_i(x)|^{s+1}, \quad i = 1, 2, \tag{7.49}$$

for a.e. $x \in \mathbb{R}^n$. Indeed, if $i = 1$ and $u(x) \leq 0$ then (7.49) is trivial, while if $u(x) \geq 0$, since $u_2 \in -\mathscr{K}$, one has $u(x) = u_1(x) + u_2(x) \leq u_1(x)$ and again (7.49) holds. The case $i = 2$ is similar. By combining the Sobolev inequality (7.47) with (7.48) and (7.49), we get for $i = 1, 2$

$$S\|u_i\|_{L^{s+1}}^2 \leq \|u_i\|_{\mathscr{D}^{m,2}}^2 = (u, u_i)_{\mathscr{D}^{m,2}} = \int_{\mathbb{R}^n} (-\Delta)^m u\, u_i\, dx$$
$$= \int_{\mathbb{R}^n} |u|^{s-1} u\, u_i\, dx \leq \int_{\mathbb{R}^n} |u_i|^{s+1}\, dx = \|u_i\|_{L^{s+1}}^{s+1},$$

which implies $\|u_i\|_{L^{s+1}}^2 \geq S^{2/(s-1)}$ for $i = 1, 2$. Hence, again by (7.47) and (7.48), one obtains

$$\frac{\|u\|_{\mathscr{D}^{m,2}}^2}{\|u\|_{L^{s+1}}^2} = \|u\|_{\mathscr{D}^{m,2}}^{2(s-1)/(s+1)} = \left(\|u_1\|_{\mathscr{D}^{m,2}}^2 + \|u_2\|_{\mathscr{D}^{m,2}}^2\right)^{(s-1)/(s+1)}$$
$$\geq \left(S\|u_1\|_{L^{s+1}}^2 + S\|u_2\|_{L^{s+1}}^2\right)^{(s-1)/(s+1)} \geq 2^{(s-1)/(s+1)} S,$$

which concludes the proof. $\qquad\square$

Notice that Theorem 7.21 enables us to compute explicitly the optimal constant S defined in (7.44). By Theorem 7.21, we know that $u_{1,0}$ (see (7.46)) solves (7.45). If we multiply (7.45) by $u_{1,0}$ and integrate by parts m times we obtain that $\|u_{1,0}\|_{\mathscr{D}^{m,2}}^2 = \|u_{1,0}\|_{L^{s+1}}^{s+1}$. Therefore,

$$S = \frac{\|u_{1,0}\|_{\mathscr{D}^{m,2}}^2}{\|u_{1,0}\|_{L^{s+1}}^2} = \|u_{1,0}\|_{L^{s+1}}^{s-1} = c_{m,n}^{4m/(n-2m)} \left(n e_n \int_0^\infty \frac{r^{n-1}}{(1+r^2)^n}\, dr\right)^{2m/n}.$$

Recalling the definition of the Gamma function, some tedious computations yield

$$S = \pi^m \left(\frac{\Gamma(\frac{n}{2})}{\Gamma(n)}\right)^{2m/n} \prod_{h=-m}^{m-1} (n + 2h). \tag{7.50}$$

In fact, the constant S serves as an embedding constant for $H_0^m(\Omega) \subset L^{s+1}(\Omega)$ for any domain $\Omega \subset \mathbb{R}^n$.

Theorem 7.23. *Let $m \in \mathbb{N}^+$ and let $\Omega \subset \mathbb{R}^n$ ($n > 2m$) be a bounded (respectively unbounded) domain. Then $H_0^m(\Omega) \subset L^{s+1}(\Omega)$ (respectively $\mathscr{D}^{m,2}(\Omega) \subset L^{s+1}(\Omega)$) and*

$$S = \inf_{H_0^m(\Omega)\setminus\{0\}} \frac{\|v\|_{H_0^m}^2}{\|v\|_{L^{s+1}}^2} \qquad \left(\text{respectively } S = \inf_{\mathscr{D}^{m,2}(\Omega)\setminus\{0\}} \frac{\|v\|_{\mathscr{D}^{m,2}}^2}{\|v\|_{L^{s+1}}^2}\right). \qquad (7.51)$$

Moreover, the minimum in (7.51) is not attained if $\mathbb{R}^n \setminus \overline{\Omega}$ has positive (possibly infinite) measure.

Proof. We only give the proof when Ω is bounded. In the case of an unbounded Ω one just has to replace the norms in H_0^m with the norms in $\mathscr{D}^{m,2}$.

Let $S(\Omega)$ denote the infimum in (7.51). Since any function in $H_0^m(\Omega)$ can be trivially extended to $\mathbb{R}^n$ as a function in $\mathscr{D}^{m,2}(\mathbb{R}^n)$, it is clear that

$$S(\Omega) \geq S. \qquad (7.52)$$

To see the converse we consider for $\varepsilon \searrow 0$ the following one-parameter family of functions of the kind (7.46),

$$v_\varepsilon(x) = \gamma_{m,n} \frac{\varepsilon^{(n-2m)/2}}{(\varepsilon^2 + |x|^2)^{(n-2m)/2}} = \varepsilon^{-(n-2m)/2} v_1\left(\frac{x}{\varepsilon}\right).$$

Here $\gamma_{m,n}$ is chosen in such a way that, independently of ε, we have

$$\int_{\mathbb{R}^n} |v_\varepsilon(x)|^{s+1}\, dx = 1. \qquad (7.53)$$

By Theorem 7.21 we know that for every $\varepsilon > 0$

$$S = \|v_\varepsilon\|_{\mathscr{D}^{m,2}}^2. \qquad (7.54)$$

By [383, Lemma 2] we also know that for $j = 0, \dots, m-1$, $|x| \geq 1$, we have

$$|\Delta^j v_1(x)| \leq C \frac{|x|^{2j}}{(1 + |x|^2)^{(n-2m+4j)/2}},$$

$$|\nabla \Delta^j v_1(x)| \leq C \frac{|x|^{2j+1}}{(1 + |x|^2)^{(n-2m+4j+2)/2}},$$

implying that for $0 < \varepsilon \leq \frac{1}{2}$, $|x| \geq \frac{1}{2}$, we have

$$\left|\Delta^j v_\varepsilon(x)\right| \le C \frac{\varepsilon^{(n-2m)/2}|x|^{2j}}{(\varepsilon^2+|x|^2)^{(n-2m+4j)/2}},$$

$$\left|\nabla\Delta^j v_\varepsilon(x)\right| \le C \frac{\varepsilon^{(n-2m)/2}|x|^{2j+1}}{(\varepsilon^2+|x|^2)^{(n-2m+4j+2)/2}}. \tag{7.55}$$

For $0 < \varepsilon \le \frac{1}{2}$, $\frac{1}{2} \le |x| \le 1$ we conclude

$$\left|\Delta^j v_\varepsilon(x)\right| \le C\varepsilon^{(n-2m)/2}, \qquad \left|\nabla\Delta^j v_\varepsilon(x)\right| \le C\varepsilon^{(n-2m)/2}.$$

Since v_ε is radially symmetric, for all $|\alpha| \le m$, $0 < \varepsilon \le \frac{1}{2}$, $\frac{1}{2} \le |x| \le 1$ one finds that, in particular,

$$\left|D^\alpha v_\varepsilon(x)\right| \le C\varepsilon^{(n-2m)/2}. \tag{7.56}$$

After scaling and translation we may assume that $B \subset \Omega$. Let $\xi \in C_c^\infty(B)$ be a fixed cut-off function, $0 \le \xi \le 1$, $\xi(x) = 1$ on $\{|x| \le \frac{1}{2}\}$. We proceed by localising the minimisers v_ε and put

$$w_\varepsilon := \xi \cdot v_\varepsilon \in C_c^\infty(B). \tag{7.57}$$

For even m we conclude with the help of (7.55) and (7.56) that

$$\left|\int_{\mathbb{R}^n} \left(\Delta^{m/2} v_\varepsilon\right)^2 dx - \int_B \left(\Delta^{m/2} w_\varepsilon\right)^2 dx\right|$$

$$\le \left|\int_{\frac{1}{2}\le|x|\le 1} \left(\left(\Delta^{m/2} v_\varepsilon\right)^2 - \left(\Delta^{m/2}(\xi v_\varepsilon)\right)^2\right) dx\right| + \int_{|x|\ge 1} \left(\Delta^{m/2} v_\varepsilon\right)^2 dx$$

$$\le C\varepsilon^{n-2m}\left(1 + \int_1^\infty r^{2m-n-1}\, dr\right) = O(\varepsilon^{n-2m}).$$

For odd m an analogous estimate holds true. By virtue of (7.53) and (7.54) it follows that

$$\|w_\varepsilon\|^2_{H_0^m} = S + O(\varepsilon^{n-2m}); \tag{7.58}$$

$$\left|\int_{\mathbb{R}^n} v_\varepsilon^{s+1}\, dx - \int_B w_\varepsilon^{s+1}\, dx\right| \le \int_{|x|\ge\frac{1}{2}} v_\varepsilon^{s+1}\, dx \le C\int_{|x|\ge\frac{1}{2}} \frac{\varepsilon^n}{(\varepsilon^2+|x|^2)^n}\, dx = O(\varepsilon^n);$$

$$\int_B w_\varepsilon^{s+1}\, dx = 1 + O(\varepsilon^n). \tag{7.59}$$

Since $w_\varepsilon \ge 0$ and

$$\tilde w_\varepsilon := \frac{w_\varepsilon}{(\int_B w_\varepsilon^{s+1}\, dx)^{1/(s+1)}} \in C_c^\infty(B), \qquad \|\tilde w_\varepsilon\|_{L^{s+1}} = 1,$$

by making use of (7.58) and (7.59) we obtain

$$\|\tilde w_\varepsilon\|^2_{H_0^m} = \frac{\|w_\varepsilon\|^2_{H_0^m}}{(\int_B w_\varepsilon^{s+1}\, dx)^{2/(s+1)}} = \frac{S + O(\varepsilon^{n-2m})}{(1 + O(\varepsilon^n))^{2/(s+1)}} = S + O(\varepsilon^{n-2m}),$$

so that $S(\Omega) \le S$ by letting $\varepsilon \to 0$. Together with (7.52), this proves that S is independent of the domain.

In order to show that the minimum in (7.51) is not attained, one argues that an optimal function $v_0 \in H_0^m(\Omega)$ may be extended by 0 to a function $v_0 \in H_0^m(\mathbb{R}^n)$ which vanishes identically on the nonempty open set $\mathbb{R}^n \setminus \overline{\Omega}$. Since we have just shown that the infimum is domain independent, this extended function would be a minimiser of the Sobolev ratio in $\mathbb{R}^n$ and would therefore solve (7.45), up to a Lagrange multiplier. This is in contradiction with the unique continuation principle [337, 346] which excludes the existence of H_0^m-solutions to (7.45) which vanish identically on some open set. Let us also mention that if Ω is bounded, the simplest way to prove nonattainment is to apply Theorem 7.58 below. $\qquad\square$

Let us also emphasise that

$$S = \inf_{H^m(\mathbb{R}^n)\setminus\{0\}} \frac{\|v\|_{H^m}^2}{\|v\|_{L^{s+1}}^2} \tag{7.60}$$

but, contrary to (7.44), the infimum in (7.60) is not attained since the additional L^2-norms, see (2.8), make the corresponding inequality strict. In order to prove (7.60), consider the functions u_{ε,x_0} in (7.46) with $x_0 = 0$. By Theorem 7.21 they satisfy $S\|u_{\varepsilon,0}\|_{L^{s+1}}^2 = \|u_{\varepsilon,0}\|_{\mathscr{D}^{m,2}}^2$. A computation shows that $\|D^k u_{\varepsilon,0}\|_{L^2} \to 0$ as $\varepsilon \to 0$ for all $k = 0, \ldots, m-1$, see also the proof of Theorem 7.23 above. This proves (7.60).

One may then wonder whether S and its attainment remain the same if in (7.51) we replace the functional space $H_0^m(\Omega)$ with a different (larger) subspace of $H^m(\Omega)$. Restricting our attention to the case of Navier boundary conditions, we consider the space $H_\vartheta^m(\Omega)$ introduced in (7.35) endowed with the norm (7.36). We have the following positive answer from van der Vorst [400] and Ge [192], see also [187].

Theorem 7.24. *Let $\Omega \subset \mathbb{R}^n$ $(n > 2m)$ be a bounded C^m-smooth domain. Then $H_\vartheta^m(\Omega) \subset L^{s+1}(\Omega)$ and*

$$S = \inf_{H_\vartheta^m(\Omega)\setminus\{0\}} \frac{\|v\|_{H_\vartheta^m}^2}{\|v\|_{L^{s+1}}^2}. \tag{7.61}$$

Moreover, the infimum in (7.61) is not achieved.

Theorems 7.23 and 7.24 tell us that not only the best embedding constant is independent of the domain Ω, but it is also independent of the kernels of the trace operators defined in (2.4). We also refer to [41] for embeddings in case of "free" boundary conditions.

Proof. 1. By scaling we may assume that $|\Omega| = e_n = |B|$. Since $H_0^m(\Omega) \subset H_\vartheta^m(\Omega)$ it is obvious that

$$S \ge \inf_{H_\vartheta^m(\Omega)\setminus\{0\}} \frac{\|v\|_{H_\vartheta^m}^2}{\|v\|_{L^{s+1}}^2}.$$

In order to prove equality we assume by contradiction that the above inequality is strict. This means that there exists $u \in H^m_\vartheta(\Omega) \setminus \{0\}$ such that

$$S > \frac{\|u\|^2_{H^m_\vartheta}}{\|u\|^2_{L^{s+1}}}. \tag{7.62}$$

In a first step we prove that there exists a radially symmetric function $v \in H^m_\vartheta(B) \setminus \{0\}$ satisfying

$$(-\Delta)^j v \text{ is positive and radially decreasing for all } j \le \frac{m}{2} \tag{7.63}$$

and

$$\frac{\|u\|^2_{H^m_\vartheta}}{\|u\|^2_{L^{s+1}}} \ge \frac{\|v\|^2_{H^m_\vartheta}}{\|v\|^2_{L^{s+1}}}. \tag{7.64}$$

In the case that $m = 2k$ is even, we consider $v \in H^{m,p}_\vartheta(B)$ given by

$$\begin{cases} (-\Delta)^k v = ((-\Delta)^k u)^* & \text{in } B \\ \Delta^j v = 0 & \text{on } \partial B, \quad j = 0,\dots,k-1, \end{cases} \tag{7.65}$$

where $((-\Delta)^k u)^*$ denotes the spherical rearrangement of $(-\Delta)^k u$ according to Definition 3.10. Then $(-\Delta)^j v$ is positive, radially symmetric and radially decreasing for all $j = 0,\dots,k$. Moreover, Theorem 3.12 yields $v \ge u^*$ so that

$$\|v\|_{L^{s+1}(B)} \ge \|u^*\|_{L^{s+1}(B)} = \|u\|_{L^{s+1}(\Omega)}, \tag{7.66}$$

where the last equality follows from standard properties of the spherical rearrangement, see Theorem 3.11. For the same reason, we also have

$$\left\|\Delta^k v\right\|_{L^2(B)} = \left\|(\Delta^k u)^*\right\|_{L^2(B)} = \left\|\Delta^k u\right\|_{L^2(\Omega)}.$$

Altogether we see that v has the claimed properties.

To prove (7.64) in the case that $m = 2k+1$ is odd we consider again $v \in H^m_\vartheta(B)$ defined by (7.65). Then $(-\Delta)^j v$ is positive, radially symmetric and radially decreasing for all $j = 0,\dots,k$. Moreover, we obtain again (7.66). Finally, by standard properties of the spherical rearrangement, see again Theorem 3.11, we infer

$$\left\|\nabla(\Delta^k v)\right\|_{L^2(B)} = \left\|\nabla(\Delta^k u)^*\right\|_{L^2(B)} \le \left\|\nabla(\Delta^k u)\right\|_{L^2(\Omega)}$$

and v is as stated.

2. Take now v as constructed in the previous step. Below, we shall introduce an extension procedure and prove that there exists a radially symmetric function

$w \in \mathscr{D}^{m,2}\left(\mathbb{R}^n\right)$ with

$$\frac{\|v\|_{H^m_\vartheta}^2}{\|v\|_{L^{s+1}}^2} > \frac{\|w\|_{\mathscr{D}^{m,2}}^2}{\|w\|_{L^{s+1}}^2}. \tag{7.67}$$

Since by definition of the optimal Sobolev constant one has that

$$\frac{\|w\|_{\mathscr{D}^{m,2}}^2}{\|w\|_{L^{s+1}}^2} \geq S.$$

Combining this with (7.62), (7.64), and (7.67) yields a contradiction. Quite similarly one shows that S is not achieved in $H^m_\vartheta(\Omega)$. This means that up to showing (7.67), the proof of Theorem 7.24 is complete. $\qquad\square$

In order to prove (7.67) one needs to distinguish between even and odd m.
Even m, $m = 2k$ for some $k \geq 1$.
    For $g : [0,1] \to \mathbb{R}$ let us define

$$(\mathscr{G}_0 g)(r) := \int_r^1 \int_0^\rho \left(\frac{s}{\rho}\right)^{n-1} g(s)\,ds\,d\rho.$$

Hence, $\mathscr{G}_0$ is the solution operator for the radially symmetric Poisson problem in the unit ball of $\mathbb{R}^n$, that is, it satisfies

$$\begin{cases} -\Delta\left(\mathscr{G}_0 g\right)(|x|) = g\left(|x|\right) & \text{for } |x| < 1, \\ \left(\mathscr{G}_0 g\right)(|x|) = 0 & \text{for } |x| = 1. \end{cases}$$

Let us also define for $g : [0,\infty) \to \mathbb{R}$ with appropriate integrability conditions

$$(\mathscr{G} g)(r) := \int_r^\infty \int_0^\rho \left(\frac{s}{\rho}\right)^{n-1} g(s)\,ds\,d\rho.$$

If g also goes to 0 fast enough for $r \to \infty$ (e.g. like $r^{-\gamma}$ with $\gamma > 2$), then an integration by parts gives

$$(\mathscr{G} g)(r) = \frac{1}{n-2} r^{2-n} \int_0^r s^{n-1} g(s)\,ds + \frac{1}{n-2} \int_r^\infty s g(s)\,ds, \tag{7.68}$$

and

$$-\Delta\left(\mathscr{G} g\right)(|x|) = g\left(|x|\right) \text{ for } x \in \mathbb{R}^n.$$

Note that

$$g \geq 0 \implies \mathscr{G} g \geq \mathscr{G}_0 g \text{ in } B. \tag{7.69}$$

We now describe the inductive procedure which we will use in order to suitably extend radial functions in $H^m_\vartheta(B)$ satisfying (7.63).

Lemma 7.25. *Let $n, \gamma > 2$, $\gamma \neq n$ and let $\ell \geq 0$. If $f \in H^\ell_{loc}(\mathbb{R}^n)$ is radially symmetric, positive and such that*

$$f(x) \leq c_f |x|^{-\gamma} \text{ for } |x| > 1,$$

then there is a unique radially symmetric solution $z \in H^{\ell+2}_{loc}(\mathbb{R}^n)$ of

$$\begin{cases} -\Delta z = f \text{ in } \mathbb{R}^n, \\ \lim_{|x| \to \infty} z(x) = 0. \end{cases}$$

Moreover, $z = \mathcal{G} f$ implies that z is positive and

$$z(x) \leq c_2 |x|^{2-n} + \frac{c_f}{(\gamma-2)(n-\gamma)} |x|^{2-\gamma} \text{ for } |x| > 1.$$

This inequality becomes an equality if $f(x) = c_f |x|^{-\gamma}$ for $|x| > 1$.

Proof. In view of the "boundary condition" at infinity, uniqueness follows from Weyl's lemma and Liouville's theorem.

Suppose first that f is continuous. We have

$$-r^{1-n} \frac{\partial}{\partial r} \left(r^{n-1} \frac{\partial z}{\partial r}(r) \right) = f(r).$$

Since $\frac{\partial z}{\partial r}$ is bounded in 0 we find

$$r^{n-1} \frac{\partial z}{\partial r}(r) = -\int_0^r s^{n-1} f(s)\, ds$$

and since z goes to 0 at ∞, it follows that

$$\begin{aligned}
z(r) &= \int_r^\infty \rho^{1-n} \int_0^\rho s^{n-1} f(s)\, ds\, d\rho \\
&= \frac{-1}{n-2} \left[\rho^{2-n} \int_0^\rho s^{n-1} f(s)\, ds \right]_r^\infty + \frac{1}{n-2} \int_r^\infty s f(s)\, ds \\
&= \frac{1}{n-2} r^{2-n} \int_0^r s^{n-1} f(s)\, ds + \frac{1}{n-2} \int_r^\infty s f(s)\, ds. \tag{7.70}
\end{aligned}$$

If $f \geq 0$ is not identically 0, then $z > 0$. For $r > 1$ it follows from (7.70) that

$$\begin{aligned}
z(r) &\leq \frac{1}{n-2} r^{2-n} \left(\int_0^1 s^{n-1} f(s)\, ds + c_f \int_1^r s^{n-1-\gamma}\, ds \right) + \frac{c_f}{(n-2)(\gamma-2)} r^{2-\gamma} \\
&= \frac{1}{n-2} \left(\int_0^1 s^{n-1} f(s)\, ds - \frac{c_f}{n-\gamma} \right) r^{2-n} + \frac{c_f}{(\gamma-2)(n-\gamma)} r^{2-\gamma}.
\end{aligned}$$

Equality holds if $f(x) = c_f |x|^{-\gamma}$ for $|x| > 1$. The formula in (7.70) also holds for $f \in H^{\ell}_{loc}(\mathbb{R}^n)$. The claim that $z \in H^{\ell+2}_{loc}(\mathbb{R}^n)$ is direct. $\qquad\square$

The second tool is a variation of an extension result which enables us to modify radial functions in $H^m_\vartheta(B)$ to functions on the whole space with no increase of the Dirichlet norm.

Lemma 7.26. *Let* $m = 2k$ *and let* $v \in H^m_\vartheta(B) \setminus \{0\}$ *be radially symmetric and satisfy* (7.63). *Let* $w(r) = \left(\mathscr{G}^k f\right)(r)$ *for*

$$f(r) = \begin{cases} (-\Delta)^k v(r) & \text{for } r \leq 1, \\ 0 & \text{for } r > 1. \end{cases}$$

Then $w \in \mathscr{D}^{m,2}(\mathbb{R}^n)$ *and*

1. $\|w\|_{\mathscr{D}^{m,2}} = \|v\|_{H^m_\vartheta(B)}$
2. $\|w\|_{L^{s+1}(\mathbb{R}^n)} > \|v\|_{L^{s+1}(B)}.$

Proof. From Lemma 7.25 we find that

$$w(r) = c_1 r^{2-n} + c_2 r^{4-n} + \cdots + c_m r^{m-n} \text{ for } r > 1$$

which implies with $w \in H^m_{loc}(\mathbb{R}^n)$ that $w \in \mathscr{D}^{m,2}(\mathbb{R}^n)$. Since

$$f(r) = (-\Delta)^k \left(\mathscr{G}^k f\right)(r) = (-\Delta)^k w(r)$$

it even follows that

$$\|w\|_{\mathscr{D}^{m,2}} = \left\|\Delta^k w\right\|_{L^2(\mathbb{R}^n)} = \|f\|_{L^2(\mathbb{R}^n)} = \|f\|_{L^2(B)} = \left\|\Delta^k v\right\|_{L^2(\Omega)} = \|v\|_{H^m_\vartheta(B)}.$$

Moreover, by (7.68) it follows that $\mathscr{G} f(1) > 0 = \left((-\Delta)^{k-1} v\right)(1)$ and hence by the maximum principle and by (7.69):

$$\mathscr{G} f \geq \mathscr{G}_0 f = (-\Delta)^{k-1} v \text{ in } B. \tag{7.71}$$

Since $\mathscr{G}^2 f(1) > 0 = \left((-\Delta)^{k-2} v\right)(1)$ and since (7.71) holds, a further iteration of the maximum principle and (7.69) implies

$$\mathscr{G}^2 f \geq \mathscr{G}_0^2 f = (-\Delta)^{k-2} v \text{ in } B.$$

Repeating this argument we find

$$w = \mathscr{G}^k f \geq \mathscr{G}_0^k f = v \text{ in } B.$$

Hence $\|w\|_{L^{s+1}(\mathbb{R}^n)} > \|w\|_{L^{s+1}(B)} \geq \|v\|_{L^{s+1}(B)}.$ $\qquad\square$

If $m = 2k$ is even, the function $w \in \mathscr{D}^{m,2}(\mathbb{R}^n) \setminus \{0\}$ constructed in Lemma 7.26 satisfies (7.67).

Odd m, $m = 2k+1$ for some $k \geq 1$. In this case, we take advantage of what has just been proved for the even exponent $2k$. Since $H_\vartheta^{2k+1}(B) \subset H_\vartheta^{2k}(B)$, by Lemma 7.26 we know that any radially symmetric function $v \in H_\vartheta^{2k+1}(B) \setminus \{0\}$ satisfying (7.63) allows to define an entire function w such that

$$w > v \text{ in } B, \quad \Delta^k(w-v) = 0 \text{ in } B, \quad \Delta^k w = 0 \text{ in } \mathbb{R}^n \setminus B.$$

In particular, this implies that also

$$\nabla(\Delta^k(w-v)) = 0 \text{ in } B, \quad \nabla(\Delta^k w) = 0 \text{ in } \mathbb{R}^n \setminus B.$$

The construction for the $2k$-case also enables us to conclude that $w \in C^{2k-1}(\mathbb{R}^n)$, a regularity which is not enough to obtain $w \in \mathscr{D}^{2k+1,2}(\mathbb{R}^n)$, we need here one more degree of regularity. This is obtained by recalling the extra boundary condition that appears by going from $H_\vartheta^{2k}(B)$ to $H_\vartheta^{2k+1}(B)$, namely $\Delta^k v = 0$ on ∂B, and that $\Delta^k w = 0$ in $\mathbb{R}^n \setminus B$. Hence, (7.67) follows also for odd m.

7.4 The Pohožaev Identity for Critical Growth Problems

In what follows we assume again that $n > 2m$ and we consider the case where the exponent p in (7.39) is critical,

$$p = s = \frac{n+2m}{n-2m}.$$

When trying to carry over the methods from Section 7.2 to this case, one is immediately confronted with the fact that the embedding $H_0^m(\Omega) \subset L^{s+1}(\Omega)$ is no longer compact although still continuous. Not only the arguments break down, but a huge number of new phenomena can be observed, some of which will be explained below. On the other hand, according to Proposition 7.15, all weak H_0^m–solutions of critical growth equations are as smooth as the domain and the nonlinearity permit. So in what follows, in smooth domains we need not specify the type of solution we are dealing with.

The starting point of our analysis is a celebrated identity due to Pohožaev [340, 341], see also Pucci-Serrin [348]. It is deduced from the equation by means of the testing functions $x \cdot \nabla u$ and u itself, and by partial integration. It reflects translation and scaling equivariance of the differential equation.

Theorem 7.27 (Pohožaev identity). *Assume that $\Omega \subset \mathbb{R}^n$ $(n > 2m)$ is a bounded smooth domain with exterior unit normal ν, and let $u \in C^{2m}(\overline{\Omega})$ be a solution of*

$$\begin{cases} (-\Delta)^m u = \lambda u + |u|^{s-1} u \ \text{ in } \Omega, \\ D^\alpha u|_{\partial\Omega} = 0 \qquad\qquad \text{for } |\alpha| \le m-1, \end{cases} \tag{7.72}$$

where $\lambda \in \mathbb{R}$. Then one has the following variational identity

$$2m\lambda \int_\Omega u^2 \, dx = \begin{cases} \int_{\partial\Omega} \left(\Delta^{m/2} u \right)^2 (x \cdot \nu) \, d\omega & \text{if } m \text{ is even,} \\[2em] \int_{\partial\Omega} \left(\dfrac{\partial}{\partial\nu} \Delta^{(m-1)/2} u \right)^2 (x \cdot \nu) \, d\omega & \text{if } m \text{ is odd.} \end{cases} \tag{7.73}$$

Proof. We consider first the case where $m = 2k$ is even. Using $x \cdot \nabla u$ as testing function we obtain for each term in (7.72)

$$\int_\Omega |u|^{s-1} u \left(\sum_{j=1}^n x_j u_{x_j} \right) dx = -s \sum_{j=1}^n \int_\Omega |u|^{s-1} u u_{x_j} x_j \, dx - n \int_\Omega |u|^{s+1} \, dx$$

$$\Rightarrow \int_\Omega |u|^{s-1} u \left(\sum_{j=1}^n x_j u_{x_j} \right) dx = -\frac{n}{s+1} \int_\Omega |u|^{s+1} \, dx = -\frac{n-2m}{2} \int_\Omega |u|^{s+1} \, dx;$$

$$\int_\Omega u \left(\sum_{j=1}^n x_j u_{x_j} \right) dx = -\frac{n}{2} \int_\Omega u^2 \, dx;$$

$$\int_\Omega \left((-\Delta)^{2k} u \right) \left(\sum_{j=1}^n x_j u_{x_j} \right) dx$$

$$= \int_\Omega \left(\Delta^{k+1} u \right) \Delta^{k-1} \left(\sum_{j=1}^n x_j u_{x_j} \right) dx$$

$$= 2(k-1) \int_\Omega \left(\Delta^{k+1} u \right) \left(\Delta^{k-1} u \right) dx + \sum_{j=1}^n \int_\Omega \left(\Delta^{k+1} u \right) x_j \left(\Delta^{k-1} u \right)_{x_j} dx$$

$$= 2(k-1) \int_\Omega \left(\Delta^k u \right)^2 dx - \sum_{i,j=1}^n \int_\Omega \left(\Delta^k u \right)_{x_i} \delta_{ij} \left(\Delta^{k-1} u \right)_{x_j} dx$$

$$- \sum_{i,j=1}^n \int_\Omega \left(\Delta^k u \right)_{x_i} x_j \left(\Delta^{k-1} u \right)_{x_i x_j} dx$$

$$= (2k-1) \int_\Omega \left(\Delta^k u \right)^2 dx - \sum_{i,j=1}^n \int_{\partial\Omega} \left(\Delta^k u \right) \left(\Delta^{k-1} u \right)_{x_i x_j} (\nu_i x_j) \, d\omega$$

$$+ \sum_{j=1}^n \int_\Omega \left(\Delta^k u \right) x_j \left(\Delta^k u \right)_{x_j} dx + \sum_{i,j=1}^n \int_\Omega \left(\Delta^k u \right) \delta_{ij} \left(\Delta^{k-1} u \right)_{x_i x_j} dx.$$

Since $\left(\Delta^{k-1}u\right)_{x_j}\big|_{\partial\Omega}=0$, one has that on $\partial\Omega$

$$\nabla\left(\left(\Delta^{k-1}u\right)_{x_j}\right)=\frac{\partial}{\partial v}\left(\left(\Delta^{k-1}u\right)_{x_j}\right)\cdot v=\sum_{\ell=1}^{n}\left(\left(\Delta^{k-1}u\right)_{x_\ell x_j}v_\ell\right)\cdot v$$

so that we may proceed as follows.

$$\int_\Omega\left((-\Delta)^{2k}u\right)\left(\sum_{j=1}^{n}x_ju_{x_j}\right)dx$$

$$=2k\int_\Omega\left(\Delta^ku\right)^2 dx-\sum_{i,j=1}^{n}\int_{\partial\Omega}\left(\Delta^ku\right)\left(\sum_{\ell=1}^{n}\left(\left(\Delta^{k-1}u\right)_{x_\ell x_j}v_\ell\right)\cdot v_i\right)(v_ix_j)\,d\omega$$

$$+\frac{1}{2}\int_{\partial\Omega}\left(\Delta^ku\right)^2(x\cdot v)\,d\omega-\frac{n}{2}\int_\Omega\left(\Delta^ku\right)^2 dx$$

$$=\frac{2m-n}{2}\int_\Omega\left(\Delta^ku\right)^2 dx-\sum_{j,\ell=1}^{n}\int_{\partial\Omega}\left(\Delta^ku\right)\left(\Delta^{k-1}u\right)_{x_\ell x_j}(v_\ell x_j)\,d\omega$$

$$+\frac{1}{2}\int_{\partial\Omega}\left(\Delta^ku\right)^2(x\cdot v)\,d\omega$$

$$=\frac{2m-n}{2}\int_\Omega\left(\Delta^ku\right)^2 dx-\sum_{i,j,\ell=1}^{n}\int_{\partial\Omega}\left(\Delta^ku\right)\left(\Delta^{k-1}u\right)_{x_\ell x_i}(v_iv_jv_\ell x_j)\,d\omega$$

$$+\frac{1}{2}\int_{\partial\Omega}\left(\Delta^ku\right)^2(x\cdot v)\,d\omega$$

$$=\frac{2m-n}{2}\int_\Omega\left(\Delta^ku\right)^2 dx-\sum_{j=1}^{n}\int_{\partial\Omega}\left(\Delta^ku\right)\left(\frac{\partial^2}{\partial v^2}\Delta^{k-1}u\right)(v_jx_j)\,d\omega$$

$$+\frac{1}{2}\int_{\partial\Omega}\left(\Delta^ku\right)^2(x\cdot v)\,d\omega$$

$$=\frac{2m-n}{2}\int_\Omega\left(\Delta^ku\right)^2 dx-\int_{\partial\Omega}\left(\Delta^ku\right)\left(\Delta^ku\right)(x\cdot v)\,d\omega$$

$$+\frac{1}{2}\int_{\partial\Omega}\left(\Delta^ku\right)^2(x\cdot v)\,d\omega$$

$$=-\frac{n-2m}{2}\int_\Omega\left(\Delta^ku\right)^2 dx-\frac{1}{2}\int_{\partial\Omega}\left(\Delta^ku\right)^2(x\cdot v)\,d\omega,$$

where we exploited again that $\left(\Delta^{k-1}u\right)_{x_j}\big|_{\partial\Omega}=0$. Making use of the differential equation yields

$$\frac{n-2m}{2}\int_\Omega\left(\Delta^ku\right)^2 dx+\frac{1}{2}\int_{\partial\Omega}\left(\Delta^ku\right)^2(x\cdot v)\,d\omega$$

$$=\lambda\frac{n}{2}\int_\Omega u^2\,dx+\frac{n-2m}{2}\int_\Omega|u|^{s+1}\,dx.$$

Using now $\frac{n-2m}{2}u$ as a testing function gives

$$\frac{n-2m}{2}\int_{\Omega}\left(\Delta^k u\right)^2 dx = \lambda\frac{n-2m}{2}\int_{\Omega}u^2\,dx + \frac{n-2m}{2}\int_{\Omega}|u|^{s+1}\,dx.$$

We subtract both equations and come up with

$$\frac{1}{2}\int_{\partial\Omega}\left(\Delta^k u\right)^2 (x\cdot\nu)\,d\omega = m\lambda\int_{\Omega}u^2\,dx,$$

thereby proving the claim for even $m=2k$.

If $m=2k+1$ is odd, only the differential operator needs some extra consideration.

$$\int_{\Omega}\left((-\Delta)^{2k+1}u\right)\left(\sum_{j=1}^{n}x_j u_{x_j}\right)dx$$

$$= -\int_{\Omega}\left(\Delta^{k+1}u\right)\Delta^k\left(\sum_{j=1}^{n}x_j u_{x_j}\right)dx$$

$$= -2k\int_{\Omega}\left(\Delta^{k+1}u\right)\left(\Delta^k u\right)dx - \sum_{j=1}^{n}\int_{\Omega}\left(\Delta^{k+1}u\right)x_j\left(\Delta^k u\right)_{x_j}dx$$

$$= 2k\int_{\Omega}\left|\nabla\Delta^k u\right|^2 dx - \sum_{i,j=1}^{n}\int_{\partial\Omega}\left(\Delta^k u\right)_{x_i}\left(\Delta^k u\right)_{x_j}(x_j\nu_i)\,d\omega$$

$$+ \sum_{i,j=1}^{n}\int_{\Omega}\left(\Delta^k u\right)_{x_i}\delta_{ij}\left(\Delta^k u\right)_{x_j}dx + \sum_{i,j=1}^{n}\int_{\Omega}\left(\Delta^k u\right)_{x_i}x_j\left(\Delta^k u\right)_{x_i x_j}dx$$

$$= (2k+1)\int_{\Omega}\left|\nabla\Delta^k u\right|^2 dx - \sum_{i,j=1}^{n}\int_{\partial\Omega}\left(\frac{\partial}{\partial\nu}\Delta^k u\right)^2(\nu_i\nu_j x_j\nu_i)\,d\omega$$

$$+ \frac{1}{2}\sum_{i,j=1}^{n}\int_{\partial\Omega}\left(\Delta^k u\right)_{x_i}^2(x_j\nu_j)\,d\omega - \frac{n}{2}\int_{\Omega}\left|\nabla\Delta^k u\right|^2 dx$$

$$= -\frac{n-2m}{2}\int_{\Omega}\left|\nabla\Delta^k u\right|^2 dx - \frac{1}{2}\int_{\partial\Omega}\left(\frac{\partial}{\partial\nu}\Delta^k u\right)^2(x\cdot\nu)\,d\omega.$$

The remaining arguments are exactly the same as for even m. $\qquad\square$

Remark 7.28. A careful analysis of the proof shows that the statement remains true if we merely assume that $u\in C^{2m}(\Omega)\cap C^{2m-1}(\overline{\Omega})$. This is also true for the next two identities.

A similar identity is also available for critical growth problems under Navier boundary conditions. The following result, which can be proved essentially as Theorem 7.27, is due to Mitidieri [308] and van der Vorst [399].

Theorem 7.29. *Assume that $\Omega \subset \mathbb{R}^n$ $(n > 2m)$ is a bounded smooth domain with exterior unit normal ν, and let $u \in C^{2m}(\overline{\Omega})$ be a solution of*

$$\begin{cases} (-\Delta)^m u = \lambda u + |u|^{s-1}u & \text{in } \Omega, \\ \Delta^j u|_{\partial\Omega} = 0 & \text{for } j = 0, \ldots m-1, \end{cases}$$

where $\lambda \in \mathbb{R}$. Then one has the following variational identity

$$2m\lambda \int_\Omega u^2 \, dx = (-1)^{m+1} \sum_{j=1}^m \int_{\partial\Omega} (\Delta^{j-1}u)_\nu (\Delta^{m-j}u)_\nu (x \cdot \nu) \, d\omega. \qquad (7.74)$$

Proof. As for Theorem 7.27, one uses again $x \cdot \nabla u$ and $\frac{n-2m}{2} u$ as testing functions and integrates by parts by replacing the Dirichlet conditions with the Navier conditions $\Delta^j u|_{\partial\Omega} = 0$ for $j = 0, \ldots m-1$. This gives rise to different boundary integrals. And, again, the cases where m is even or odd need to be considered separately. $\qquad\square$

There are several equivalent ways of writing the identity (7.74). We chose the one which appears most elegant since it does not require to distinguish between even and odd m. Note that if m is even every summand in the right hand side of (7.74) appears twice while if m is odd the summand indexed with $j = \frac{m+1}{2}$ appears only once and all the others summands appear twice. We refer to [376, Lemma 2] for a different form of (7.74).

Finally, we consider the case of Steklov boundary conditions, where the situation is slightly different. For later purposes we only need to deal with the case $\lambda = 0$.

Theorem 7.30. *Assume that $\Omega \subset \mathbb{R}^n$ $(n > 4)$ is a bounded smooth domain with exterior unit normal ν, let $a \in \mathbb{R}$ and let $u \in C^4(\overline{\Omega})$ be a solution of*

$$\begin{cases} \Delta^2 u = |u|^{8/(n-4)}u & \text{in } \Omega, \\ u = \Delta u - a u_\nu = 0 & \text{on } \partial\Omega. \end{cases} \qquad (7.75)$$

Then

$$\int_{\partial\Omega} \left(2(x \cdot \nabla \Delta u) - a^2 (x \cdot \nu) u_\nu + n a u_\nu \right) u_\nu \, d\omega = 0. \qquad (7.76)$$

Proof. The starting point is a Rellich-type identity [308, (2.6)] of Mitidieri for arbitrary $C^4(\overline{\Omega})$-functions, which can be obtained as follows.

$$\int_\Omega (\Delta^2 u)(x \cdot \nabla u) \, dx = -\int_\Omega \nabla \Delta u \cdot \nabla u \, dx - \sum_{i,j=1}^n \int_\Omega (\Delta u)_{x_i x_j} u_{x_i x_j} \, dx$$

$$+ \int_{\partial\Omega} (x \cdot \nabla u)(\Delta u)_\nu \, d\omega$$

$$= (n-1) \int_\Omega \nabla \Delta u \cdot \nabla u \, dx + \sum_{i,j=1}^n \int_\Omega (\Delta u)_{x_i x_j} x_j u_{x_i} \, dx$$

$$+ \int_{\partial\Omega} \left((x \cdot \nabla u)(\Delta u)_\nu - (\nabla \Delta u \cdot \nabla u)(x \cdot \nu) \right) d\omega$$

$$= (n-2) \int_\Omega \nabla \Delta u \cdot \nabla u \, dx - \sum_{j=1}^n \int_\Omega (\Delta u)_{x_j} x_j \Delta u \, dx$$

$$+ \int_{\partial \Omega} \left((x \cdot \nabla u)(\Delta u)_v - (\nabla \Delta u \cdot \nabla u)(x \cdot v) + (x \cdot \nabla \Delta u) u_v \right) d\omega$$

$$= (n-2) \int_\Omega \nabla \Delta u \cdot \nabla u \, dx + \frac{n}{2} \int_\Omega (\Delta u)^2 \, dx$$

$$+ \int_{\partial \Omega} \left((x \cdot \nabla u)(\Delta u)_v - (\nabla \Delta u \cdot \nabla u)(x \cdot v) \right.$$

$$\left. + (x \cdot \nabla \Delta u) u_v - \frac{(\Delta u)^2}{2} (x \cdot v) \right) d\omega.$$

By separating interior terms from boundary terms we finally get

$$\int_\Omega (\Delta^2 u) x \cdot \nabla u \, dx - \frac{n}{2} \int_\Omega (\Delta u)^2 \, dx - (n-2) \int_\Omega \nabla \Delta u \cdot \nabla u \, dx$$

$$= \int_{\partial \Omega} \left((x \cdot \nabla u)(\Delta u)_v - (\nabla \Delta u \cdot \nabla u)(x \cdot v) + (x \cdot \nabla \Delta u) u_v - \frac{(\Delta u)^2}{2} (x \cdot v) \right) d\omega.$$

$$(7.77)$$

Since $u = 0$ on $\partial \Omega$, (7.77) reduces to

$$\int_\Omega (\Delta^2 u) x \cdot \nabla u \, dx - \frac{n}{2} \int_\Omega (\Delta u)^2 \, dx + (n-2) \int_\Omega (\Delta^2 u) u \, dx$$

$$= -\frac{1}{2} \int_{\partial \Omega} (\Delta u)^2 (x \cdot v) \, d\omega + \int_{\partial \Omega} (x \cdot \nabla \Delta u) u_v \, d\omega. \qquad (7.78)$$

Here we integrated by parts to get the third term in (7.78). Now, for solutions of (7.75) the left hand side of (7.78) becomes

$$\int_\Omega (\Delta^2 u) x \cdot \nabla u \, dx - \frac{n}{2} \int_\Omega (\Delta u)^2 \, dx + (n-2) \int_\Omega (\Delta^2 u) u \, dx$$

$$= \int_\Omega |u|^{8/(n-4)} u x \cdot \nabla u \, dx - \frac{n}{2} \left(\int_\Omega |u|^{2n/(n-4)} \, dx + a \int_{\partial \Omega} u_v^2 \, d\omega \right)$$

$$+ (n-2) \int_\Omega |u|^{2n/(n-4)} \, dx$$

$$= \frac{n-4}{2n} \int_\Omega (x \cdot \nabla |u|^{2n/(n-4)}) \, dx + \frac{n-4}{2} \int_\Omega |u|^{2n/(n-4)} \, dx - \frac{na}{2} \int_{\partial \Omega} u_v^2 \, d\omega$$

$$= -\frac{na}{2} \int_{\partial \Omega} u_v^2 \, d\omega.$$

Combining the latter with (7.78) and using that $\Delta u = a u_v$ on $\partial \Omega$, we get (7.76). $\quad \square$

7.5 Critical Growth Dirichlet Problems

In this section, we consider the semilinear polyharmonic problem

$$
\begin{cases}
(-\Delta)^m u = \lambda u + |u|^{s-1} u, \ u \not\equiv 0 & \text{in } \Omega, \\
D^\alpha u|_{\partial\Omega} = 0 & \text{for } |\alpha| \le m-1,
\end{cases}
\tag{7.79}
$$

where $\Omega \subset \mathbb{R}^n$ is a bounded smooth domain and $\lambda \in \mathbb{R}$. As before, $n > 2m$ and $s = (n+2m)/(n-2m)$ is the critical Sobolev exponent. We will prove both nonexistence results – essentially based on the Pohožaev identity in Theorem 7.27 – and existence results.

7.5.1 Nonexistence Results

Restricting ourselves to starshaped domains, we draw here the following conclusion from the Pohožaev identity (7.73). In Section 7.9 we discuss whether nontrivial solutions exist in geometrically or topologically more complicated domains.

Theorem 7.31. *Assume that $\Omega \subset \mathbb{R}^n$ $(n > 2m)$ is a bounded smooth starshaped domain. Then the Dirichlet problem (7.72) only has the trivial solution $u \equiv 0$, if*

1. *$\lambda \le 0$ in the case $m = 1$,*
2. *$\lambda < 0$ in the case $m \ge 2$.*

Proof. With no loss of generality we may assume that Ω is starshaped with respect to the origin so that $x \cdot \nu \ge 0$ on $\partial\Omega$. Therefore, nonexistence of nontrivial solution for $\lambda < 0$ follows from (7.73). Moreover, by the divergence theorem one has $\int_{\partial\Omega} (x \cdot \nu) \, d\omega = n|\Omega| > 0$. Hence, there exists a relatively open part of $\partial\Omega$ where $x \cdot \nu > 0$ so that there, also $\nabla^m u = 0$. If we assume now in addition that $m = 1$, then this is enough to extend u by 0 as a solution of the differential equation (7.72) beyond this part of $\partial\Omega$. By means of the unique continuation principle for $-\Delta$ we may conclude that $u(x) \equiv 0$ in particular in Ω. See [248] for the details. $\qquad\square$

Remark 7.32. If $m > 1$, one still has a unique continuation principle [337, 346]. Knowing $\nabla^m u = 0$ on a part of $\partial\Omega$, however, does not suffice to extend u as a solution beyond $\partial\Omega$. Instead, one would need $\nabla^j u = 0$ for $j = 0,\ldots,2m-1$.

If $m \ge 2$ and $\lambda = 0$, the nonexistence of any nontrivial solution to (7.72) has to be left open, only more restricted results are available. Concerning the biharmonic problem, i.e. $m = 2$, by combining (7.73) with the classical Hopf's lemma, Oswald [328] could prove the following.

Theorem 7.33. *Assume that $\Omega \subset \mathbb{R}^n$ $(n > 4)$ is a strictly starshaped bounded smooth domain. Then the problem*

$$\begin{cases} \Delta^2 u = u^{(n+4)/(n-4)} & \text{in } \Omega, \\ u = |\nabla u| = 0 & \text{on } \partial\Omega \end{cases} \tag{7.80}$$

admits no nontrivial nonnegative solution.

Proof. With no loss of generality we may assume that Ω is strictly starshaped with respect to the origin so that $x \cdot \nu > 0$ on $\partial\Omega$. Assume that u is a nonnegative nontrivial solution to (7.80). Since $\partial\Omega$ is smooth, by Proposition 7.15 we know that $u \in C^4(\overline{\Omega})$. Then from (7.73) (recall $\lambda = 0$) we deduce that $\Delta u = 0$ on $\partial\Omega$. But $-\Delta(-\Delta u) = u^{(n+4)/(n-4)} \gneqq 0$ in Ω so that $-\Delta u$ is superharmonic in Ω and vanishes on its boundary. The maximum principle then yields $-\Delta u > 0$ in Ω. Since $u = 0$ on $\partial\Omega$, Hopf's boundary lemma for strictly superharmonic functions implies that $u_\nu < 0$ on $\partial\Omega$, a contradiction. $\qquad\square$

However, in general domains Ω, no result excludes the existence of simplest possible nontrivial solutions – such as mountain pass solutions or constrained minima – as they are discussed in the subcritical regime in the proof of Theorem 7.17. Assuming that such a solution exists would not directly yield a contradiction because due to the lack of positivity preserving – see Chapter 6 – there does not seem to be a way to prove its positivity. Such a conclusion would be obvious only in domains with positive biharmonic Green's function.

Let us now briefly discuss the situation for general $m \geq 2$ whenever $\lambda = 0$. Here, it seems that, so far, nonexistence results are available only in balls where, in view of Theorem 7.1, the class of radial solutions contains the class of nonnegative solutions. Hence, the following statement, due to Lazzo and Schmidt [273], also excludes the existence of nontrivial nonnegative solutions.

Theorem 7.34. *Let $B \subset \mathbb{R}^n$ $(n > 2m)$ be the unit ball, $s = (n+2m)/(n-2m)$. Then the problem*

$$\begin{cases} (-\Delta)^m u = |u|^{s-1} u & \text{in } B, \\ D^\alpha u|_{\partial B} = 0 & \text{for } |\alpha| \leq m-1 \end{cases} \tag{7.81}$$

admits no nontrivial radial solution. In particular, (7.81) admits no nontrivial nonnegative solution.

Proof. Assume that $r \mapsto u(r) \not\equiv 0$ is a radial solution to (7.81). By putting $v_i := \Delta^{i-1} u$ and $w_i := r^{n-1} v_i'$ for $i \in \{1,\dots,m\}$, equation (7.81) is equivalent to the system

$$r^{n-1} v_i' = w_i, \quad w_i' = r^{n-1} v_{i+1}, \quad i \in \{1,\dots,m\},$$

where we have set $v_{m+1} := (-1)^m |v_1|^{s-1} v_1$. Note that the w-components satisfy the initial conditions

$$w_i(0) = 0 \quad \text{for all } i \in \{1,\dots,m\}. \tag{7.82}$$

It is clear that u (and so v_1) cannot have an accumulation point of zeros since otherwise all its derivatives would have the same accumulation point of zeros and, by continuity, they would all vanish at that point. Then unique solvability of the Cauchy problem would imply that $u \equiv 0$, against the assumption.

For each $i \in \{1,\ldots,m\}$ let $\sigma(v_i)$ be the (finite) number of interior zeros of v_i, namely the number of values $r \in (0,1)$ where $v_i(r)$ vanishes. Also let $\beta(v_i)$ be the number of boundary zeros, namely the zeros of $v_i(r)$ when $r \in \{0,1\}$. Clearly, $\beta(v_i) \in \{0,1,2\}$ for all i. Similarly, we define $\sigma(w_i)$ and $\beta(w_i)$.

Let $\phi : [0,1] \to \mathbb{R}$ be a continuously differentiable function such that both ϕ and ϕ' have a finite number of interior zeros. Between any two zeros of ϕ there is a point where ϕ attains a positive maximum or a negative minimum which is an interior zero of ϕ'. It follows, in particular, that $\sigma(\phi') \geq \sigma(\phi) + \beta(\phi) - 1$. If we apply this remark to (v, w), for all $i \in \{1,\ldots,m\}$ we have

$$\sigma(w_i) = \sigma(r^{n-1}v_i') = \sigma(v_i') \geq \sigma(v_i) + \beta(v_i) - 1$$

and

$$\sigma(v_{i+1}) = \sigma(r^{n-1}v_{i+1}) = \sigma(w_i') \geq \sigma(w_i) + \beta(w_i) - 1,$$

which imply

$$\sigma(v_{i+1}) \geq \sigma(w_i) + \beta(w_i) - 1 \geq \sigma(v_i) + \beta(v_i) + \beta(w_i) - 2. \tag{7.83}$$

We now put $k := m/2$ if m is even and $k := (m-1)/2$ if m is odd. We claim that

1. $\sigma(v_{i+1}) - 1 \geq \sigma(w_i) \geq \sigma(v_i)$ for all $i \in \{1,\ldots,k\}$,
2. $\sigma(v_{i+1}) \geq \sigma(w_i) \geq \sigma(v_i) - 1$ for all $i \in \{m-k+1,\ldots,m\}$,
3. if m is odd, $\sigma(v_{k+2}) \geq \sigma(w_{k+1}) \geq \sigma(v_{k+1})$,
4. $\sigma(v_1) \leq \sigma(v_2) - 1 \leq \cdots \leq \sigma(v_k) - (k-1) \leq \sigma(v_{k+1}) - k \leq$
   $\leq \sigma(v_{m-k+1}) - k \leq \sigma(v_{m-k+2}) - (k-1) \leq \cdots \leq \sigma(v_m) - 1 \leq \sigma(v_{m+1})$.

To see this, notice that if $i \in \{1,\ldots,k\}$, both v_i and w_i vanish for $r = 1$, so that $\beta(v_i) \geq 1$ and $\beta(w_i) = 2$ in view of (7.82). If $i \in \{m-k+1,\ldots,m\}$, there is no condition at $r = 1$ for v_i or w_i, and all we know is that $\beta(v_i) \geq 0$ and $\beta(w_i) \geq 1$. If m is odd, then also v_{k+1} vanishes at 1, but w_{k+1} needs not, so that $\beta(v_{k+1}) \geq 1$ and $\beta(w_{k+1}) \geq 1$. Items 1-2-3 now follow directly from (7.83). Furthermore, it follows from Item 1 that

$$\sigma(v_1) \leq \sigma(v_2) - 1 \leq \cdots \leq \sigma(v_k) - (k-1) \leq \sigma(v_{k+1}) - k$$

and from Item 2 that

$$\sigma(v_{m+1}) \geq \sigma(v_m) - 1 \geq \cdots \geq \sigma(v_{m-k+2}) - (k-1) \geq \sigma(v_{m-k+1}) - k.$$

If m is even, $k = m - k$ and so $\sigma(v_{k+1}) = \sigma(v_{m-k+1})$. If m is odd, $\sigma(v_{k+1}) \leq \sigma(v_{k+2}) = \sigma(v_{m-k+1})$ due to Item 3. Altogether, these inequalities prove Item 4 and complete the proof of the claim.

In fact, since $\sigma(v_{m+1}) = \sigma(|v_1|^{s-1}v_1) = \sigma(v_1)$ all the inequalities in Item 4 are necessarily equalities. But then, also the inequalities in Items 1-2-3 are equalities. From this we finally conclude that also (7.83) are equalities so that, in particular (for $i = k+1$),

$$\beta(v_{k+1}) + \beta(w_{k+1}) = 1 \quad \text{if } m \text{ is even}, \qquad \beta(v_{k+1}) + \beta(w_{k+1}) = 2 \quad \text{if } m \text{ is odd}.$$

In turn, since $\beta(w_{k+1}) \geq 1$ in view of (7.82) and since $\beta(v_{k+1}) \geq 1$ whenever m is odd, the latter implies

$$\beta(v_{k+1}) = 0 \quad \text{if } m \text{ is even} \qquad \beta(w_{k+1}) = 1 \quad \text{if } m \text{ is odd}.$$

From this, using again (7.82) when m is odd, we infer that

$$v_{k+1}(1) \neq 0 \quad \text{if } m \text{ is even} \qquad w_{k+1}(1) \neq 0 \quad \text{if } m \text{ is odd}.$$

On the other hand, rewritten in our context, (7.73) reads

$$v_{k+1}(1) = 0 \quad \text{if } m \text{ is even} \qquad w_{k+1}(1) = 0 \quad \text{if } m \text{ is odd}.$$

which gives a contradiction. $\qquad\qquad\qquad\qquad\qquad\qquad\qquad\qquad\qquad\qquad$ $\square$

7.5.2 *Existence Results for Linearly Perturbed Equations*

In the previous section several nonexistence results are collected for the critical growth polyharmonic problem (7.72) in starshaped domains. However, the discussion of the case $\lambda = 0$ made clear that an exhaustive treatment of nonexistence results seems to be out of reach. Only in the ball the nonexistence situation is fairly well understood although one should also find complementing existence results. For this reason we mainly restrict our attention in this section to the discussion of existence for radial solutions to

$$\begin{cases} (-\Delta)^m u = \lambda u + |u|^{s-1}u, \ u \neq 0 & \text{in } B, \\ D^\alpha u|_{\partial B} = 0 & \text{for } |\alpha| \leq m - 1. \end{cases} \tag{7.84}$$

As before, $n > 2m$ and $s = (n+2m)/(n-2m)$ is the critical Sobolev exponent. If one is merely interested in nontrivial solutions, one can also consider the problem in general domains, see Theorem 7.40 below.

According to Theorems 7.31 and 7.34 one has nonexistence of radial solutions to (7.84) for all $\lambda \leq 0$. The goal should now be to discuss (7.84) for $\lambda > 0$ in the class of radial functions. However, for general $m \geq 1$, as we shall explain below, this class is still too large for having complementary existence and nonexistence results. Since we are concerned with the case $\lambda > 0$, not only Boggio's comparison principle Lemma 2.27 is available but also the symmetry result in Theorem 7.1 becomes applicable. It will turn out that the class of positive radial functions is suitable for a satisfactory discussion of (7.84) for $\lambda > 0$. The argument in (7.43) then shows that a first natural restriction is $\lambda < \Lambda_{m,1}$, the first Dirichlet eigenvalue. But we will show more serious obstructions than just $\lambda \in (0, \Lambda_{m,1})$ in order to have positive radial solutions to (7.84).

In the case $m = 1$, Brezis and Nirenberg [72] discovered an interesting phenomenon. They showed that there exists a positive radial solution to (7.84) for every $\lambda \in (0, \Lambda_{1,1})$ if $n \geq 4$, and for every $\lambda \in (\frac{1}{4}\Lambda_{1,1}, \Lambda_{1,1})$ if $n = 3$. Moreover, in the latter case, problem (7.84) has no nontrivial radial solution if $\lambda \leq \frac{1}{4}\Lambda_{1,1}$. Pucci-Serrin [349] raised the question in which way this critical behaviour of certain dimensions depends on the order $2m$ of the semilinear polyharmonic eigenvalue problem (7.84), if m increases arbitrarily. In order to have a suggestive name for those dimensions we define according to Pucci and Serrin:

Definition 7.35. The dimension n is called *critical* (with respect to the boundary value problem (7.84)) if and only if there is a positive bound $\Lambda > 0$ such that a necessary condition for a nontrivial radial solution to (7.84) to exist is $\lambda > \Lambda$.

Pucci and Serrin [349] showed that for any m the dimension $n = 2m+1$ is critical and, moreover, that $n = 5,6,7$ are critical in the fourth order problem $m = 2$. They also had a clear idea about the precise range of critical dimensions.

Conjecture 7.36. [Pucci-Serrin] The critical dimensions for the boundary value problem (7.84) are precisely $n = 2m+1, \ldots, 4m-1$.

By the work of Brezis and Nirenberg [72], this conjecture is true for $m = 1$. Subsequently, Pucci-Serrin [349] and Edmunds-Fortunato-Jannelli [160] showed that the conjecture is also true for $m = 2$ while, according to Theorem 7.38 below, for arbitrary m the critical dimensions are *at most* $n = 2m+1, \ldots, 4m-1$. Finally, Bernis and Grunau [53, 202] proved the Pucci-Serrin conjecture for $m = 3$ and $m = 4$ while for any $m \geq 5$ they proved that there are at least five critical dimensions $n = 2m+1, \ldots, 2m+5$. But in general, the conjecture of Pucci and Serrin is still open. Since in view of Theorem 7.1, for $\lambda \geq 0$, it is equivalent to consider positive or positive radial solutions, a possible relaxation of the original notion of critical dimension is the following.

Definition 7.37. The dimension n is called *weakly critical* (with respect to the boundary value problem (7.84)) if and only if there is a positive bound $\tilde{\Lambda} > 0$ such that a necessary condition for a positive solution to (7.84) to exist is $\lambda > \tilde{\Lambda}$.

We recall that the case $\lambda \leq 0$ was already studied in Section 7.4. Here we prove the following result.

Theorem 7.38. *Let $n > 2m$, $s = (n+2m)/(n-2m)$, $B \subset \mathbb{R}^n$ the unit ball. Let $\Lambda_{m,1}$ denote the first Dirichlet eigenvalue of $(-\Delta)^m$ in B.*

1. *If $n \geq 4m$, then for every $\lambda \in (0, \Lambda_{m,1})$ there exists a positive solution $u \in C^{\infty}(B) \cap C^{2m+1}(\overline{B})$ to the Dirichlet problem (7.84). This solution is radially symmetric, $u = u(r)$, and strictly decreasing in $r = |x| \in (0,1)$.*
2. *If $2m + 1 \leq n \leq 4m - 1$, then there exist $0 < \underline{\Lambda} \leq \bar{\Lambda} < \Lambda_{m,1}$ such that for every $\lambda \in (\bar{\Lambda}, \Lambda_{m,1})$ the Dirichlet problem (7.84) has a solution u as above and for every $\lambda \in (0, \underline{\Lambda})$ it has no positive solution.*
3. *If $\lambda \geq \Lambda_{m,1}$, then (7.84) has no positive solution.*

In particular, Theorem 7.38 – together with Theorems 7.1, 7.31, and 7.34 – shows a relaxed version of Conjecture 7.36.

Corollary 7.39. *The weakly critical dimensions for the boundary value problem (7.84) are precisely $n = 2m + 1, \ldots, 4m - 1$*

Before starting the proof of Theorem 7.38, let us mention that similar but somehow weaker results hold in general domains.

Theorem 7.40. *Let $n > 2m$, $s = (n+2m)/(n-2m)$, let $\Omega \subset \mathbb{R}^n$ a bounded domain. Let $\Lambda_{m,1}$ denote the first Dirichlet eigenvalue of $(-\Delta)^m$ in Ω.*

1. *If $n \geq 4m$, then for every $\lambda \in (0, \Lambda_{m,1})$ there exists a solution $u \in H_0^m(\Omega)$ to the Dirichlet problem (7.79).*
2. *If $2m + 1 \leq n \leq 4m - 1$, then there exists $0 < \bar{\Lambda} < \Lambda_{m,1}$ such that for every $\lambda \in (\bar{\Lambda}, \Lambda_{m,1})$ the Dirichlet problem (7.79) has a solution $u \in H_0^m(\Omega)$.*

In order to prove Theorem 7.38 we proceed as in Section 7.2. We minimise the functional $v \mapsto F_\lambda(v)$ in (7.41) subject to the constraint

$$v \in N = \left\{ v \in H_0^m(B) : \int_B |v|^{s+1}\, dx = 1 \right\}$$

and we define

$$S_\lambda := \inf_{v \in N} F_\lambda(v). \tag{7.85}$$

In Section 7.2, the required compactness was a consequence of the Rellich-Kondrachov embedding theorem 2.4. Here, in general, compactness may fail. However, we will show that in a suitable energy range, one still has compactness. The threshold of this range is related to the optimal Sobolev constant defined in (7.51) and which is independent of the domain Ω, see Theorem 7.23.

Lemma 7.41. *Let $0 < \lambda < \Lambda_{m,1}$ and $(v_k) \subset H_0^m(B) \cap N$ be a minimising sequence for $F_\lambda|_N$. Moreover we assume that $S_\lambda < S$. Then $F_\lambda|_N$ attains its minimum in a function $v_0 \in H_0^m(B) \cap N$, namely $F_\lambda(v_0) = S_\lambda$. Up to a subsequence, one has strong convergence $v_k \to v_0$ in $H_0^m(B)$.*

Proof. Since $\lambda < \Lambda_{m,1}$ we have that $S_\lambda > 0$ and the minimising sequence (v_k) is bounded in $H_0^m(B)$. After possibly passing to a subsequence we may assume that there exists $v_0 \in H_0^m(B)$ with

$$v_k \rightharpoonup v_0 \text{ in } H_0^m(B), \ v_k \rightharpoonup v_0 \text{ in } L^{s+1}(B), \ v_k \to v_0 \text{ in } L^2(B), \ v_k \to v_0 \text{ a.e. in } B.$$

By Vitali's convergence theorem, we know that for $k \to \infty$

$$\int_B \left(|v_k|^{s+1} - |v_k - v_0|^{s+1} \right) dx = \int_B |v_0|^{s+1} dx + o(1)$$

so that, since $v_k \in N$, we have

$$1 - \int_B |v_k - v_0|^{s+1} dx = \int_B |v_0|^{s+1} dx + o(1). \tag{7.86}$$

As for the quadratic functionals, by weak convergence it follows that

$$\|v_k\|_{H_0^m}^2 - \|v_k - v_0\|_{H_0^m}^2 = \|v_0\|_{H_0^m}^2 + o(1),$$
$$F_\lambda(v_k) - \|v_k - v_0\|_{H_0^m}^2 = F_\lambda(v_0) + o(1). \tag{7.87}$$

In order to conclude for compactness we need that a *positive* multiple of v_0 obeys the constraint N. So, we show first that

$$\int_B |v_0|^{s+1} dx > 0.$$

Indeed, since by assumption $S_\lambda < S$, (7.86)-(7.87) and $F_\lambda(v_k) = S_\lambda + o(1)$ show that

$$\int_B |v_0|^{s+1} dx = 1 - \int_B |v_k - v_0|^{s+1} dx + o(1)$$
$$\geq 1 - S^{-(s+1)/2} \|v_k - v_0\|_{H_0^m}^{s+1} + o(1)$$
$$= 1 - S^{-(s+1)/2} \left(F_\lambda(v_k) - F_\lambda(v_0) + o(1) \right)^{(s+1)/2} + o(1)$$
$$\geq 1 - S^{-(s+1)/2} \left(F_\lambda(v_k) + o(1) \right)^{(s+1)/2} + o(1)$$
$$= 1 - S^{-(s+1)/2} \left(S_\lambda + o(1) \right)^{(s+1)/2} + o(1) = 1 - \left(\frac{S_\lambda}{S} \right)^{(s+1)/2} + o(1) > 0$$

for sufficiently large k. Hence,

$$\tilde{v}_0 := \frac{v_0}{\|v_0\|_{L^{s+1}}} \in N;$$

and by making use of $F_\lambda(\tilde{v}_0) \geq S_\lambda$ we conclude that

$$F_\lambda(v_0) \geq S_\lambda \left(\int_B |v_0|^{s+1} dx \right)^{2/(s+1)}. \tag{7.88}$$

We now prove strong convergence $v_k \to v_0$ in $L^{s+1}(\Omega)$. Since $2/(s+1) \leq 1$ we employ the simple but very useful inequality

$$(a+b)^{2/(s+1)} \le a^{2/(s+1)} + b^{2/(s+1)} \qquad \text{for all } a,b \ge 0$$

to obtain

$$S_\lambda = F_\lambda(v_k) + o(1) = \|v_k - v_0\|_{H_0^m}^2 + F_\lambda(v_0) + o(1) \quad \text{due to (7.87)}$$

$$\ge S\|v_k - v_0\|_{L^{s+1}}^2 + S_\lambda \left(\int_B |v_0|^{s+1} \, dx \right)^{2/(s+1)} + o(1) \quad \text{due to (7.51), (7.88)}$$

$$= (S - S_\lambda)\|v_k - v_0\|_{L^{s+1}}^2$$

$$+ S_\lambda \left(\left(\int_B |v_k - v_0|^{s+1} \, dx \right)^{2/(s+1)} + \left(\int_B |v_0|^{s+1} \, dx \right)^{2/(s+1)} \right) + o(1)$$

$$\ge (S - S_\lambda)\|v_k - v_0\|_{L^{s+1}}^2 + S_\lambda \left(\int_B |v_k - v_0|^{s+1} \, dx + \int_B |v_0|^{s+1} \, dx \right)^{2/(s+1)} + o(1)$$

$$\ge (S - S_\lambda)\|v_k - v_0\|_{L^{s+1}}^2 + S_\lambda (1 + o(1))^{2/(s+1)} + o(1) \quad \text{due to (7.86)}$$

$$= (S - S_\lambda)\|v_k - v_0\|_{L^{s+1}}^2 + S_\lambda + o(1).$$

Since by assumption $S - S_\lambda > 0$, we obtain $v_k \to v_0$ in L^{s+1} so that $v_0 \in N$. Weak lower semicontinuity of F_λ (see (7.87)) gives

$$S_\lambda \ge F_\lambda(v_0) \text{ and because of } v_0 \in N \text{ also } S_\lambda = F_\lambda(v_0).$$

Combining this with (7.87) finally proves that $v_k \to v_0$ in $H_0^m(B)$. $\qquad \square$

In order to obtain compactness we verify the assumption $S_\lambda < S$ of Lemma 7.41. To this end, we need to take into account the calculations of Theorem 7.23.

Lemma 7.42.
1. *Assume $n \ge 4m$. Then for any $\lambda > 0$ we have $S_\lambda < S$.*
2. *Let $2m+1 \le n \le 4m-1$. Then there exists a number $\bar\Lambda = \bar\Lambda(n,m) < \Lambda_{m,1}$ such that for all $\lambda > \bar\Lambda$ one has $S_\lambda < S$.*

Proof. In view of Theorem 7.23 we know that $S = \inf_{v \in N} \|v\|_{H_0^m}^2$. By using the same notations as in that proof, we see that there exist constants $c_1, c_2, c_3, c_4 > 0$ such that

$$\int_B w_\varepsilon^2 \, dx \ge \gamma_{m,n}^2 \varepsilon^{n-2m} \int_{|x| \le \varepsilon} \frac{dx}{(2\varepsilon^2)^{n-2m}} + \gamma_{m,n}^2 \varepsilon^{n-2m} \int_{\varepsilon \le |x| \le \frac{1}{2}} \frac{dx}{(2|x|^2)^{n-2m}}$$

$$= c_1 \varepsilon^{2m} + c_2 \varepsilon^{n-2m} \int_\varepsilon^{1/2} r^{4m-n-1} \, dr$$

$$= \begin{cases} c_3 \varepsilon^{2m} + O(\varepsilon^{n-2m}) & \text{if } n > 4m, \\ c_2 \varepsilon^{2m} |\log \varepsilon| + O(\varepsilon^{2m}) & \text{if } n = 4m, \\ c_4 \varepsilon^{n-2m} + O(\varepsilon^{2m}) & \text{if } n < 4m. \end{cases} \qquad (7.89)$$

If $n > 4m$, we use (7.89) to get

$$F_\lambda(\tilde{w}_\varepsilon) \leq \frac{S + O(\varepsilon^{n-2m}) - \lambda c_3 \varepsilon^{2m}}{(1 + O(\varepsilon^n))^{2/(s+1)}} = S - \lambda c_3 \varepsilon^{2m} + O(\varepsilon^{n-2m}),$$

while in the case $n = 4m$, we obtain

$$F_\lambda(\tilde{w}_\varepsilon) \leq S - \lambda c_2 \varepsilon^{2m} |\log \varepsilon| + O(\varepsilon^{2m}).$$

By choosing $\varepsilon > 0$ small enough, the proof of Item 1 is so complete and $S_\lambda < S$.

To prove Item 2, we consider a positive radially symmetric eigenfunction

$$(-\Delta)^m \varphi = \Lambda_{m,1} \varphi, \quad \varphi > 0 \text{ in } B, \quad D^\alpha \varphi|_{\partial B} = 0 \text{ for } |\alpha| \leq m - 1,$$

see Theorem 3.7. Alternatively, the Kreĭn-Rutman theorem may be applied. We may assume $\|\varphi\|_{L^{s+1}}^{s+1} = 1$ so that $\varphi \in N$. Putting $\bar{\Lambda} = \Lambda_{m,1} - S e_n^{-2m/n}$, we find for $\lambda > \bar{\Lambda}$:

$$S_\lambda \leq F_\lambda(\varphi) = (\Lambda_{m,1} - \lambda) \int_B \varphi^2 \, dx < (\Lambda_{m,1} - \bar{\Lambda}) \left(\int_B \varphi^{s+1} \, dx \right)^{2/(s+1)} e_n^{2m/n} = S$$

and thereby also proving Item 2. $\qquad\square$

Proof of Theorem 7.38. The existence part (Items 1 and 2) is proved exactly as in Theorem 7.17, while only the reference to the Rellich-Kondrachov theorem has to be replaced by referring to the compactness Lemma 7.41 and to Lemma 7.42. Like in Theorem 7.19 one sees that the constructed solution is of one sign and, hence, may be chosen to be positive. The symmetry result Theorem 7.1 applies and shows that these solutions are necessarily radially symmetric and strictly decreasing in the radial variable. The limiting exponent for Luckhaus's [282] regularity result (see Proposition 7.15) is also s. The proof of Item 3, i.e. nonexistence of positive solutions for $\lambda \geq \Lambda_{m,1}$ follows as in the subcritical case, see (7.43).

Therefore, in order to complete the proof of Theorem 7.38, it remains to show the nonexistence part of Item 2, that is, the following statement.

Lemma 7.43. *If $2m + 1 \leq n \leq 4m - 1$, then there exists $\underline{\Lambda} \in (0, \Lambda_{m,1})$ such that for every $\lambda \in (0, \underline{\Lambda})$ problem (7.84) has no positive solution.*

Proof. We assume that there exists a positive solution $u \in C^{2m}(\overline{B})$ to (7.84). From Theorem 7.31 we know that necessarily $\lambda \geq 0$ so that, from the differential equation (7.84), we immediately conclude that

$$(-\Delta)^m u > 0 \text{ in } B. \tag{7.90}$$

Theorem 7.1 shows that u is radial, so, we write $r = |x|$, $u = u(r)$, $u'(r) = \frac{du}{dr}(r) = \nabla(u(|x|)) \cdot \frac{x}{|x|}$. Moreover, $[0,1] \ni r \mapsto u(r)$ is strictly decreasing.

1. We introduce the positive weight function $w(r) := (1 - r^2)^{m-1}$. With the help of the simple observations

$$
\begin{cases}
(-\Delta)^m w = 0, \\
\left(\dfrac{d}{dr}\right)^j w|_{r=1} = 0 \text{ for } j = 0, \ldots, m-2, \\
\left(-\dfrac{d}{dr}\right)^{m-1} w|_{r=1} > 0,
\end{cases}
\tag{7.91}
$$

integration by parts yields, if m is even,

$$
\begin{aligned}
\int_B w(-\Delta)^m u \, dx &= \int_B \nabla\left((-\Delta)^{(m/2)-1}w\right) \cdot \nabla\left((-\Delta)^{m/2}u\right) dx \\
&= (-1)^{m-1} \int_{\partial B} \left(\frac{d}{dr}\left(\Delta^{(m/2)-1}w\right)\right)\left(\Delta^{m/2}u\right) d\omega \\
&= C(n,m) \int_{\partial B} \Delta^{m/2}u \, d\omega,
\end{aligned}
$$

and, if m is odd,

$$
\begin{aligned}
\int_B w(-\Delta)^m u \, dx &= \int_B \left((-\Delta)^{(m-1)/2}w\right)\left((-\Delta)^{(m+1)/2}u\right) dx \\
&= (-1)^m \int_{\partial B} \left(\Delta^{(m-1)/2}w\right)\left(\frac{d}{dr}\Delta^{(m-1)/2}u\right) d\omega \\
&= C(n,m) \int_{\partial B} \left(-\frac{d}{dr}\Delta^{(m-1)/2}u\right) d\omega.
\end{aligned}
$$

From (7.91) we see that the constants $C(n,m)$ are strictly positive. Applying the Cauchy-Schwarz inequality we find

$$
\left(\int_B w(-\Delta)^m u \, dx\right)^2 \leq
\begin{cases}
C(n,m) \displaystyle\int_{\partial B} \left(\Delta^{m/2}u\right)^2 d\omega & \text{if } m \text{ is even,} \\
C(n,m) \displaystyle\int_{\partial B} \left(\frac{d}{dr}\Delta^{(m-1)/2}u\right)^2 d\omega & \text{if } m \text{ is odd.}
\end{cases}
\tag{7.92}
$$

2. The crucial information, which we take from the Dirichlet problem (7.84), is the following Pohožaev identity, see Theorem 7.27

$$
2m\lambda \int_B u^2 \, dx =
\begin{cases}
\displaystyle\int_{\partial B} \left(\Delta^{m/2}u\right)^2 d\omega & \text{if } m \text{ even,} \\
\displaystyle\int_{\partial B} \left(\frac{d}{dr}\Delta^{(m-1)/2}u\right)^2 d\omega & \text{if } m \text{ odd.}
\end{cases}
\tag{7.93}
$$

Combining (7.92) and (7.93) we have

$$\left(\int_B w(-\Delta)^m u\, dx\right)^2 \leq C(n,m)\, \lambda \int_B u^2\, dx. \qquad (7.94)$$

3. We want to show that the weighted L^1-norm $\int_B w(-\Delta)^m u\, dx$ is equivalent to the L^1-norm $\int_B (-\Delta)^m u\, dx = \int_B |(-\Delta)^m u|\, dx$, see also (7.90). For this purpose we need some additional property of $(-\Delta)^m u$. Employing the strict monotonicity of u in the radial variable r, we may estimate as follows

$$0 < \int_B (-\Delta)^m u\, dx = \int_{|x|\leq 1/2} (-\Delta)^m u\, dx + \int_{1/2\leq |x|\leq 1} (-\Delta)^m u\, dx$$

$$\leq \int_{|x|\leq 1/2} (-\Delta)^m u\, dx + \int_{1/2\leq |x|\leq 1} dx\ \left((-\Delta)^m u|_{r=1/2}\right)$$

$$\leq \int_{|x|\leq 1/2} (-\Delta)^m u\, dx + 2^n \int_{|x|\leq 1/2} dx\ \left((-\Delta)^m u|_{r=1/2}\right)$$

$$\leq (2^n+1)\int_{|x|\leq 1/2} (-\Delta)^m u\, dx \leq (2^n+1)\left(\frac{4}{3}\right)^{m-1}\int_{|x|\leq 1/2} w(-\Delta)^m u\, dx,$$

$$\Rightarrow 0 < \int_B (-\Delta)^m u\, dx \leq C(n,m)\int_B w(-\Delta)^m u\, dx.$$

Together with (7.94) we come up with

$$\|(-\Delta)^m u\|_{L^1}^2 \leq \lambda\, C(n,m)\|u\|_{L^2}^2. \qquad (7.95)$$

4. It remains to estimate the L^1-norm of $(-\Delta)^m u$ from below by the L^2-norm of u itself. We use a duality argument, which was proposed by Brezis-Nirenberg [72, Theorem 1.2″].

Let $\varphi \in W^{2m,2}\cap W_0^{m,2}(B)$ be a solution of $(-\Delta)^m\varphi = u$. Since $n < 4m$, we have a continuous embedding $W^{2m,2}(B) \subset C^0(\overline{B})$. By means of elliptic estimates (see Theorem 2.20) we find

$$\|\varphi\|_{L^\infty} \leq C\|\varphi\|_{W^{2m,2}} \leq C\|(-\Delta)^m\varphi\|_{L^2} = \|u\|_{L^2};$$

$$\|u\|_{L^2}^2 = \int_B (-\Delta)^m\varphi\, u\, dx = \int_B \varphi\,(-\Delta)^m u\, dx$$
$$\leq \|\varphi\|_{L^\infty}\|(-\Delta)^m u\|_{L^1} \leq C\|u\|_{L^2}\|(-\Delta)^m u\|_{L^1};$$

$$\|u\|_{L^2}^2 \leq C(n,m)\|(-\Delta)^m u\|_{L^1}^2.$$

Combining this estimate with (7.95) we finally have $\lambda \geq \Lambda := \frac{1}{C(n,m)} > 0.$ $\square$

It seems that a full proof of the original Pucci-Serrin conjecture 7.36, if possible at all, will be significantly more difficult. The most general question in this context, whether we have nonexistence of any nontrivial (not necessarily radial) solution to (7.84) for λ close to 0, is completely open as far as the authors know, even in the second order case ($m = 1$).

7.5.3 Nontrivial Solutions Beyond the First Eigenvalue

Here, we address the question whether, in a bounded domain Ω, the critical growth problem (7.79) has a solution for all positive $\lambda > 0$. Let us recall once more that weak solutions are smooth in smooth domains, see Proposition 7.15.

We remark that according to a bifurcation result of Böhme [64], every Dirichlet eigenvalue $\Lambda_{m,j}$, $j \in \mathbb{N}^+$, of $(-\Delta)^m$ in Ω is a bifurcation point for the semilinear problem (7.79). Since the problem is variational, this holds true irrespective of the multiplicity of the eigenvalues. The following statement suggests a picture how these branches may look like, having the global bifurcation result of Rabinowitz [350] in mind.

Theorem 7.44. *Let $\Omega \subset \mathbb{R}^n$ ($n > 2m$) be a bounded domain, let $(\Lambda_{m,j})_{j \in \mathbb{N}^+}$ denote the ordered sequence of Dirichlet eigenvalues of $(-\Delta)^m$ and let μ_j denote their multiplicity, the eigenvalues being repeated according to their multiplicity.*

1. *For $j \in \mathbb{N}^+$ and for all $\lambda \in (\Lambda_{m,j} - S|\Omega|^{-2m/n}, \Lambda_{m,j})$ problem (7.79) has μ_j (pairs of) solutions.*
2. *If $4m \leq n \leq (2\sqrt{2}+2)m$, then (7.79) has a pair of solutions for any $\lambda > 0$ such that $\lambda \notin (\Lambda_{m,j})_{j \in \mathbb{N}^+}$.*
3. *If $n > (2+2\sqrt{2})m$, then (7.79) has a pair of solutions for every $\lambda > 0$. Moreover, for any $j \in \mathbb{N}^+$ there exists $\delta_j > 0$ such that (7.79) has $\mu_j + 1$ pairs of solutions for any $\lambda \in (\Lambda_{m,j} - \delta_j, \Lambda_{m,j})$.*

Direct methods in the calculus of variations like constrained minimisation as used in Section 7.5.2 are suitable only for constructing the simplest nontrivial solutions. In order to understand global properties of (7.79) we have to study the "free" functional

$$E_\lambda(v) := \frac{1}{2}F_\lambda(v) - \frac{1}{s+1}\int_\Omega |v|^{s+1}\,dx \tag{7.96}$$

$$= \begin{cases} \int_\Omega \left(\frac{1}{2}\left(\Delta^{m/2}v\right)^2 - \frac{1}{2}\lambda v^2 - \frac{1}{s+1}|v|^{s+1}\right)dx & \text{if } m \text{ is even,} \\ \int_\Omega \left(\frac{1}{2}\left|\nabla\Delta^{(m-1)/2}v\right|^2 - \frac{1}{2}\lambda v^2 - \frac{1}{s+1}|v|^{s+1}\right)dx & \text{if } m \text{ is odd,} \end{cases}$$

in $H_0^m(\Omega)$. To explain the main features of E_λ we need several basic notions.

Definition 7.45. Let H be a Hilbert space and let $E : H \to \mathbb{R}$ be a continuous functional. We say that E is Fréchet differentiable if for all $u \in H$ there exists a linear continuous operator $L_u \in H'$ such that

$$E(u+h) - E(u) = \langle L_u, h \rangle + o(\|h\|_H) \qquad \text{as } \|h\|_H \to 0.$$

If the map $u \mapsto L_u$ is continuous from H to H', we say that E is a continuously Fréchet differentiable functional and we write $dE(u) = L_u$. In this case, a sequence $(u_k) \subset H$ is called a *Palais-Smale sequence* for E, if

$$\lim_{k \to \infty} E(u_k) \in \mathbb{R} \text{ exists}, \qquad dE(u_k) \to 0 \text{ strongly in } H'.$$

We say that E satisfies a *local Palais-Smale condition* below the level c_0, if every Palais-Smale sequence with $\lim_{k \to \infty} E(u_k) < c_0$ has a strongly convergent subsequence in H.

The functional E_λ is continuously Fréchet differentiable. Let us first make precise the meaning of "critical energy level" below which a suitable compactness holds true. Since Struwe's pioneering work [381] such a local compactness is a well understood phenomenon in critical growth problems. The functional E_λ defined in (7.96) satisfies indeed a local Palais-Smale condition. Like in Section 7.5.2 a key role is played by the optimal constant S for the Sobolev embedding $H_0^m \subset L^{s+1}$, see (7.51).

Lemma 7.46. *In $H_0^m(\Omega)$ the functional E_λ satisfies a local Palais-Smale condition below the level $c_0 = \frac{m}{n} S^{n/2m}$.*

Proof. Let $(u_k) \subset H_0^m(\Omega)$ be a Palais-Smale sequence at level below $\frac{m}{n} S^{n/2m}$, i.e.

$$\lim_{k \to \infty} E_\lambda(u_k) < c_0 = \frac{m}{n} S^{n/2m}, \tag{7.97}$$

$$dE_\lambda(u_k) \to 0 \text{ strongly in } H^{-m}(\Omega) = (H_0^m(\Omega))'. \tag{7.98}$$

1. First we show the boundedness of (u_k) in $H_0^m(\Omega)$. A suitable difference between the functional and its differential gives

$$2E_\lambda(u_k) - \langle dE_\lambda(u_k), u_k \rangle = \frac{2m}{n} \int_\Omega |u_k|^{s+1}\, dx,$$

$$\Rightarrow \|u_k\|_{L^{s+1}}^{s+1} \leq C \left\{ |E_\lambda(u_k)| + \|dE_\lambda(u_k)\|_{H^{-m}} \|u_k\|_{H_0^m} \right\}$$

$$= O(1) + o(1) \|u_k\|_{H_0^m}.$$

We combine this estimate with the assumption (7.97) of bounded E_λ-energy to obtain

$$\|u_k\|_{H_0^m}^2 = 2E_\lambda(u_k) + \lambda \int_\Omega u_k^2\, dx + \frac{2}{s+1} \int_\Omega |u_k|^{s+1}\, dx$$

$$\leq O(1) + C\|u_k\|_{L^{s+1}}^{s+1} \leq O(1) + o(1)\|u_k\|_{H_0^m}.$$

It follows that $\|u_k\|_{H_0^m} \leq O(1)$.

2. After possibly passing to a subsequence we may assume that

$$u_k \rightharpoonup u \text{ in } H_0^m(\Omega), \qquad\qquad u_k \rightharpoonup u \text{ in } L^{s+1}(\Omega),$$

$$u_k \to u \text{ in } L^p(\Omega), \ p < s+1, \ u_k \to u \text{ almost everywhere in } \Omega. \tag{7.99}$$

For each fixed testing function $\varphi \in C_c^\infty(\Omega)$ one has $\langle dE_\lambda(u_k), \varphi \rangle = o(1)$. Hence, due to the convergence properties (7.99) of (u_k), the limiting function u weakly solves the Dirichlet problem

$$\begin{cases} (-\Delta)^m u = \lambda u + |u|^{s-1}u & \text{in } \Omega, \\ D^\alpha u|_{\partial\Omega} = 0 & \text{for } |\alpha| \le m-1. \end{cases}$$

Testing the differential equation with u gives

$$F_\lambda(u) = \int_\Omega |u|^{s+1}\, dx, \qquad E_\lambda(u) = \frac{m}{n}\int_\Omega |u|^{s+1}\, dx \ge 0. \tag{7.100}$$

So, any (weak) limit of such a Palais-Smale sequence has nonnegative E_λ-energy.

3. Like in the first part of the proof of Lemma 7.41 we deduce the following identities which quantify the possible deviation from strong convergence

$$\int_\Omega |u_k|^{s+1}\, dx - \int_\Omega |u_k - u|^{s+1}\, dx = \int_\Omega |u|^{s+1}\, dx + o(1), \tag{7.101}$$

$$F_\lambda(u_k) - \|u_k - u\|_{H_0^m}^2 = F_\lambda(u) + o(1),$$

$$E_\lambda(u_k) - E_0(u_k - u) = E_\lambda(u) + o(1). \tag{7.102}$$

Using Vitali's convergence theorem we obtain further that

$$\int_\Omega \left(|u_k|^{s-1}u_k - |u|^{s-1}u \right)(u_k - u)\, dx$$

$$= \int_\Omega |u_k|^{s+1}\, dx - \int_\Omega |u_k|^{s-1}u_k u\, dx - \int_\Omega |u|^{s-1}uu_k\, dx + \int_\Omega |u|^{s+1}\, dx$$

$$= \int_\Omega |u_k|^{s+1}\, dx - \int_\Omega |u|^{s+1}\, dx + o(1).$$

Combining this with (7.101) yields

$$\int_\Omega \left(|u_k|^{s-1}u_k - |u|^{s-1}u \right)(u_k - u)\, dx = \int_\Omega |u_k - u|^{s+1}\, dx + o(1). \tag{7.103}$$

4. The assumption (7.98) on the differential $dE_\lambda(u_k)$ and equation (7.103), together with $u_k \rightharpoonup u$ in $H_0^m(\Omega)$, show that

$$o(1) = \langle dE_\lambda(u_k) - dE_\lambda(u), u_k - u \rangle$$
$$= F_\lambda(u_k - u) - \int_\Omega \left(|u_k|^{s-1} u_k - |u|^{s-1} u \right) (u_k - u)\, dx$$
$$= \|u_k - u\|^2_{H_0^m} - \int_\Omega |u_k - u|^{s+1}\, dx + o(1),$$

$$\|u_k - u\|^2_{H_0^m} = \int_\Omega |u_k - u|^{s+1}\, dx + o(1), \tag{7.104}$$

$$E_0(u_k - u) = \frac{m}{n} \|u_k - u\|^2_{H_0^m} + o(1). \tag{7.105}$$

Applying (7.104) and the optimal Sobolev embedding (7.51) yields

$$\|u_k - u\|^2_{H_0^m} = \|u_k - u\|^{s+1}_{L^{s+1}} + o(1) \leq S^{-(s+1)/2} \|u_k - u\|^{s+1}_{H_0^m} + o(1)$$

$$= S^{-(s+1)/2} \|u_k - u\|^2_{H_0^m} \left(\frac{n}{m} E_0(u_k - u) \right)^{(s-1)/2} + o(1) \qquad \text{by virtue of (7.105)}$$

$$= S^{-(s+1)/2} \|u_k - u\|^2_{H_0^m} \left(\frac{n}{m} E_\lambda(u_k) - \frac{n}{m} E_\lambda(u) + o(1) \right)^{(s-1)/2} + o(1)$$

$$\text{by virtue of (7.102)}$$

$$\leq S^{-(s+1)/2} \|u_k - u\|^2_{H_0^m} \left(\frac{n}{m} \lim_{k\to\infty} E_\lambda(u_k) + o(1) \right)^{(s-1)/2} + o(1)$$

$$\text{by virtue of (7.100)}$$

$$\leq \left[S^{-(s+1)/2} \left(\frac{n}{m} \lim_{k\to\infty} E_\lambda(u_k) \right)^{(s-1)/2} \right] \|u_k - u\|^2_{H_0^m} + o(1).$$

According to assumption (7.97) the square bracket is strictly smaller than 1, indeed

$$S^{-(s+1)/2} \left(\frac{n}{m} \lim_{k\to\infty} E_\lambda(u_k) \right)^{(s-1)/2} < S^{-n/(n-2m)} \left(S^{n/2m} \right)^{2m/(n-2m)} = 1.$$

This shows that $\|u_k - u\|_{H_0^m} \to 0$ and, therefore, the local Palais-Smale condition. $\square$

A suitable variational tool for proving Theorem 7.44 is a general "symmetric mountain pass lemma", due to Ambrosetti-Rabinowitz [14] and refined by Bartolo-Benci-Fortunato [30, Theorem 2.4]. We quote the version as used by Capozzi-Fortunato-Palmieri [84, Theorem 1.2].

Lemma 7.47. *Let H be a real Hilbert space, $E : H \to \mathbb{R}$ a continuously Fréchet differentiable functional satisfying a local Palais-Smale condition below the level $c_0 > 0$. Moreover, we assume that*

1. $E(0) = 0$.
2. E is even, i.e. $E(u) = E(-u)$ for all $u \in H$.
3. There exist two closed subspaces V^+, $V^- \subset H$ and positive numbers ρ, $\delta > 0$ such that $\mathrm{codim}\, V^+ < \infty$ *and*

$$\sup_{u \in V^-} E(u) < c_0, \tag{7.106}$$

$$E(u) \geq \delta \text{ for all } u \in V^+ \text{ with } \|u\|_H = \rho. \tag{7.107}$$

Then E has at least

$$\dim(V^+ \cap V^-) - \text{codim}(V^+ + V^-)$$

pairs of nontrivial critical points.

To prove Theorem 7.44, we seek suitable subspaces V^+, V^- such that the assumptions of Lemma 7.47, in particular (7.106), are satisfied. As we shall see, each one of Items 1-3 in Theorem 7.44 needs its own choice for these spaces.

Proof of Theorem 7.44 (Item 1). In view of Theorem 7.40 we may restrict ourselves to $j \geq 2$. Let $(\varphi_i)_{i \in \mathbb{N}^+}$ be an orthonormal system in $L^2(\Omega)$ of eigenfunctions for $(-\Delta)^m$ corresponding to the eigenvalues $(\Lambda_{m,i})_{i \in \mathbb{N}^+}$. This system is also orthogonal with respect to the scalar product in $H_0^m(\Omega)$. We put

$$V^+ = \left(\text{span}\{\varphi_1, \ldots, \varphi_{j-1}\}\right)^\perp \text{ in } H_0^m, \qquad V^- = \text{span}\{\varphi_1, \ldots, \varphi_{j+\mu_j-1}\},$$

where orthogonality $\perp$ is meant with respect to the scalar product in $H_0^m(\Omega)$. Let $\lambda \in (\Lambda_{m,j} - S|\Omega|^{-2m/n}, \Lambda_{m,j})$. For $u \in V^-$ there exist numbers $\tau_1, \ldots, \tau_{j+\mu_j-1} \in \mathbb{R}$ and $t > 0$, such that

$$u = t \sum_{i=1}^{j+\mu_j-1} \tau_i \varphi_i, \qquad \left\| \sum_{i=1}^{j+\mu_j-1} \tau_i \varphi_i \right\|_{L^{s+1}} = 1.$$

Let F_λ be as in (7.41). Due to the orthogonality of (φ_i) in $L^2(\Omega)$ as well as in $H_0^m(\Omega)$ we find

$$F_\lambda \left(\sum_{i=1}^{j+\mu_j-1} \tau_i \varphi_i \right) = \sum_{i=1}^{j+\mu_j-1} (\Lambda_{m,i} - \lambda) \tau_i^2 \|\varphi_i\|_{L^2}^2$$

$$\leq \sum_{i=1}^{j+\mu_j-1} (\Lambda_{m,j} - \lambda) \tau_i^2 \|\varphi_i\|_{L^2}^2 = (\Lambda_{m,j} - \lambda) \left\| \sum_{i=1}^{j+\mu_j-1} \tau_i \varphi_i \right\|_{L^2}^2$$

$$\leq (\Lambda_{m,j} - \lambda) |\Omega|^{(s-1)/(s+1)} \left\| \sum_{i=1}^{j+\mu_j-1} \tau_i \varphi_i \right\|_{L^{s+1}}^2 = (\Lambda_{m,j} - \lambda) |\Omega|^{2m/n}.$$

Here and in the sequel, we need the following fact from elementary calculus

$$\max_{t \geq 0} \left(\frac{a}{2} t^2 - \frac{b}{s+1} t^{s+1} \right) = \frac{(s-1)a}{2(s+1)} \left(\frac{a}{b} \right)^{2/(s-1)} \qquad \text{for any } a, b > 0. \tag{7.108}$$

Recalling the assumption $(\Lambda_{m,j} - \lambda)|\Omega|^{2m/n} < S$, it follows from (7.108) and by compactness that

$$E_\lambda(u) = \frac{t^2}{2}F_\lambda\left(\sum_{i=1}^{j+\mu_j-1}\tau_i\varphi_i\right) - \frac{t^{s+1}}{s+1} \le \frac{m}{n}\left((\Lambda_{m,j} - \lambda)|\Omega|^{2m/n}\right)^{n/2m}$$

$$\Rightarrow \sup_{u \in V^-} E_\lambda(u) < \frac{m}{n}S^{n/2m}.$$

This means that on V^-, E_λ stays below the critical energy level, under which the local Palais-Smale condition is satisfied according to Lemma 7.46. This proves (7.106).

Finally, thanks to the variational characterisation of the eigenvalues, we find for $u \in V^+$

$$E_\lambda(u) = \frac{1}{2}\|u\|^2_{H^m_0} - \frac{\lambda}{2}\|u\|^2_{L^2} - \frac{1}{s+1}\|u\|^{s+1}_{L^{s+1}}$$

$$\ge \frac{1}{2}\left(1 - \frac{\lambda}{\Lambda_{m,j}}\right)\|u\|^2_{H^m_0} - \frac{1}{s+1}S^{-(s+1)/2}\|u\|^{s+1}_{H^m_0} \ge \delta \qquad (7.109)$$

with some $\delta > 0$, provided that $\|u\|_{H^m_0} = \rho$ and $\rho > 0$ is suitably chosen.

Lemma 7.47 yields the existence of $\mu_j = \dim(V^+ \cap V^-)$ distinct pairs of nontrivial critical points and so of solutions to (7.79). □

In what follows we assume that $n \ge 4m$ and we prove Items 2 and 3 of Theorem 7.44. Let c stand for suitable positive constants, which may vary also within the same formula. In view of Theorem 7.40, we may confine ourselves to the case where $\Lambda_{m,j-1} < \Lambda_{m,j}$ and $\Lambda_{m,j-1} \le \lambda < \Lambda_{m,j}$ for some $j \ge 2$.

After scaling and translation we may assume $B \subset \Omega$. As in Section 7.5.2, minimising sequences for the optimal embedding (7.51) will play a key role. However, these functions will here be combined with suitable approximating eigenfunctions. For any $i \in \mathbb{N}^+$ let φ_i be an L^2-normalised eigenfunction relative to $\Lambda_{m,i}$. For any $h \ge 2$ the ball $B_{2/h}$ of radius $2/h$ is contained in Ω and we define

$$\zeta_h(x) := \rho(h|x|)$$

where ρ is a C^∞-function with $0 \le \rho \le 1$, $\rho(r) = 0$ for $r \le 1$ and $\rho(r) = 1$ for $r \ge 2$. Clearly,

$$\|D^\ell\zeta_h\|_{L^\infty} \le ch^\ell \qquad \text{for } \ell = 0,\dots,m. \qquad (7.110)$$

Defining $\varphi_i^h := \zeta_h\varphi_i$ for all $i = 1,\dots,j-1$, we have that $\varphi_i^h \to \varphi_i$ in $H^m_0(\Omega)$ as $h \to \infty$. In turn, if we define $H^- := \mathrm{span}\{\varphi_1^h,\dots,\varphi_{j-1}^h\}$, this shows in particular that

$$H^- \cap \left(\mathrm{span}\{\varphi_1,\dots,\varphi_{j-1}\}\right)^\perp = \{0\} \qquad (7.111)$$

for large enough h where, again, orthogonality $\perp$ is meant in $H_0^m(\Omega)$.

If $u \in H^-$ then for some $\alpha_i \in \mathbb{R}$ $(i = 1,\ldots,j-1)$

$$u = \sum_{i=1}^{j-1} \alpha_i \varphi_i^h = \zeta_h \sum_{i=1}^{j-1} \alpha_i \varphi_i =: \zeta_h u_h$$

with $u_h \in \operatorname{span}\{\varphi_1,\ldots,\varphi_{j-1}\}$. Moreover,

$$\|u_h\|_{L^2}^2 \leq \int_\Omega u^2(x)\,dx + \int_{B_{2/h}} u_h^2(x)\,dx \leq \|u\|_{L^2}^2 + c\|u_h\|_{L^\infty}^2 h^{-n}. \tag{7.112}$$

Let $u \in H^-$ be such that $\|u\|_{L^2} = 1$ and

$$\|u\|_{H_0^m}^2 = \max_{\{v \in H^-;\, \|v\|_{L^2}=1\}} \|v\|_{H_0^m}^2.$$

Then $u = \zeta_h u_h$ for some $u_h \in \operatorname{span}\{\varphi_1,\ldots,\varphi_{j-1}\}$ and, by means of (7.110)-(7.112),

$$\|u\|_{H_0^m}^2 \leq \|u_h\|_{H_0^m}^2 +
\begin{cases}
\|\Delta^k(\zeta_h u_h)\|_{L^2(B_{2/h}\setminus B_{1/h})}^2 & \text{if } m = 2k \\
\|\nabla(\Delta^k(\zeta_h u_h))\|_{L^2(B_{2/h}\setminus B_{1/h})}^2 & \text{if } m = 2k+1
\end{cases}$$

$$\leq \Lambda_{m,j-1}\|u_h\|_{L^2}^2 + ch^{2m-n} \leq \Lambda_{m,j-1}(1 + c\|u_h\|_{L^\infty}^2 h^{-n}) + ch^{2m-n}$$

$$\leq \Lambda_{m,j-1} + ch^{2m-n}$$

for sufficiently large h. We have so proved that there exists $c_j > 0$ such that for large enough h we have

$$\max_{\{u \in H^-;\, \|u\|_{L^2}=1\}} \|u\|_{H_0^m}^2 \leq \Lambda_{m,j-1} + c_j h^{2m-n}. \tag{7.113}$$

Next, we take a positive cut-off function $\xi \in C_c^\infty(B_{1/h})$ such that $\xi \equiv 1$ in $B_{1/2h}$ and

$$\|D^\ell \xi\|_{L^\infty} \leq ch^\ell \qquad \text{for } \ell = 0,\ldots,m. \tag{7.114}$$

Then we recall the definition (7.57) of $w_\varepsilon \in C_c^\infty(\Omega)$ and we estimate the energy of the family (w_ε) which now depends also on h.

Lemma 7.48. *Assume that $h\varepsilon \to 0$, then there exist $c_1,c_2 > 0$, depending on n but independent of h and ε, such that*

$$\max_{t \in \mathbb{R}} E_\lambda(tw_\varepsilon) \leq \frac{m}{n} S^{n/2m} + c_1(h\varepsilon)^{n-2m} - c_2
\begin{cases}
\varepsilon^{2m} & \text{if } n > 4m, \\
\varepsilon^{2m}|\log(h\varepsilon)| & \text{if } n = 4m.
\end{cases}$$

Proof. First, we slightly refine (7.58), (7.59) and (7.89) by emphasising the dependence on h, namely by replacing B with $B_{1/h}$. By (7.114) we obtain

$$\|w_\varepsilon\|_{H_0^m}^2 \le S + c(h\varepsilon)^{n-2m}, \qquad \|w_\varepsilon\|_{L^{s+1}}^{s+1} \ge 1 - c(h\varepsilon)^n,$$

$$\|w_\varepsilon\|_{L^2}^2 \ge c \begin{cases} \varepsilon^{2m} & \text{if } n > 4m, \\ \varepsilon^{2m}|\log(h\varepsilon)| & \text{if } n = 4m. \end{cases}$$

By combining these estimates with (7.108), we obtain the statement. $\qquad\square$

Assuming that $\Lambda_{m,j-1} < \Lambda_{m,j}$ and $\Lambda_{m,j-1} \le \lambda < \Lambda_{m,j}$ for some $j \ge 2$, we now introduce the spaces

$$V^+ := \left(\text{span}\{\varphi_1,\dots,\varphi_{j-1}\}\right)^\perp, \qquad V^- := \text{span}\{\varphi_1^h,\dots,\varphi_{j-1}^h, w_\varepsilon\}. \qquad (7.115)$$

Notice that the space V^- depends both on h and ε. We show that these parameters may be chosen in such a way that E_λ is below the compactness threshold in the subspace V^-. Here, we have to distinguish whether λ is a Dirichlet eigenvalue or not.

Lemma 7.49. *Assume that $n \ge 4m$, that $\Lambda_{m,j-1} < \Lambda_{m,j}$ and that $\lambda \in (\Lambda_{m,j-1},\Lambda_{m,j})$ for some $j \ge 2$. Let V^- be the space defined in (7.115). Then for h large enough and $\varepsilon > 0$ sufficiently small we have*

$$\sup_{u\in V^-} E_\lambda(u) < \frac{m}{n} S^{n/2m}. \qquad (7.116)$$

Proof. Take h large enough so that (7.111) holds and that

$$\lambda - \Lambda_{m,j-1} > c_j h^{2m-n} \qquad (7.117)$$

where c_j is as in (7.113). This h is now kept fixed. Take any function $u \in V^-$ so that $u = v + t w_\varepsilon$ for some $v \in H^-$ and some $t \in \mathbb{R}$. In view of (7.113) and (7.117), we have

$$E_\lambda(v) = \frac{\|v\|_{H_0^m}^2}{2} - \lambda\frac{\|v\|_{L^2}^2}{2} - \frac{\|v\|_{L^{s+1}}^{s+1}}{s+1} \le -(\lambda - \Lambda_{m,j-1} - c_j h^{2m-n})\frac{\|v\|_{L^2}^2}{2} \le 0.$$

Hence, since v and w_ε have disjoint support, by Lemma 7.48 we infer

$$E_\lambda(u) = E_\lambda(v) + E_\lambda(t w_\varepsilon) \le \frac{m}{n} S^{n/2m} + c_1(h\varepsilon)^{n-2m} - c_2 \begin{cases} \varepsilon^{2m} & \text{if } n > 4m, \\ \varepsilon^{2m}|\log(h\varepsilon)| & \text{if } n = 4m \end{cases}$$

so that (7.116) holds for sufficiently small ε. $\qquad\square$

Lemma 7.50. *Assume that $n > (2+2\sqrt{2})m$ and assume that $\lambda = \Lambda_{m,j-1} < \Lambda_{m,j}$ for some $j \ge 2$. Let V^- be the space defined in (7.115). Then (7.116) holds, provided h is sufficiently large and ε is suitably small.*

Proof. Here we cannot proceed as in Lemma 7.49 because we cannot obtain (7.117). We take again a function $u \in V^-$ so that $u = v + t w_\varepsilon$ for some $v \in H^-$ and some $t \in \mathbb{R}$. By (7.113) we have

$$E_\lambda(v) = \frac{\|v\|_{H_0^m}^2}{2} - \Lambda_{m,j-1} \frac{\|v\|_{L^2}^2}{2} - \frac{\|v\|_{L^{s+1}}^{s+1}}{s+1} \le c h^{2m-n} \|v\|_{L^2}^2 - c\|v\|_{L^{s+1}}^{s+1}.$$

Since the norms L^2 and L^{s+1} are equivalent in the finite dimensional space H^-, we obtain

$$E_\lambda(v) \le c h^{2m-n} \|v\|_{L^2}^2 - c\|v\|_{L^2}^{s+1} \le c h^{-n(n-2m)/2m}$$

where we used (7.108).

Now we let $h \to \infty$ and we choose

$$\varepsilon_h = h^{-(n+2m)/2m} \tag{7.118}$$

so that Lemma 7.48 applies and

$$\max_{t \in \mathbb{R}} E_\lambda(t w_\varepsilon) \le \frac{m}{n} S^{n/2m} + c_1 (h\varepsilon)^{n-2m} - c_2 \varepsilon^{2m}$$

$$\le \frac{m}{n} S^{n/2m} + c_1 h^{-n(n-2m)/2m} - c_2 h^{-(n+2m)}$$

where the last inequality is a consequence of the choice in (7.118). By recalling that v and w_ε have disjoint support, the two previous estimates show that

$$E_\lambda(u) = E_\lambda(v) + E_\lambda(t w_\varepsilon)$$
$$\le \frac{m}{n} S^{n/2m} - c_3 h^{-(n+2m)} + c_4 h^{-n(n-2m)/2m} = \frac{m}{n} S^{n/2m} - c h^{-(n+2m)},$$

the latter equality being a consequence of the assumption $n > (2 + 2\sqrt{2})m$. Hence, with the choice of $\varepsilon = \varepsilon_h$ in (7.118), we see that (7.116) holds for sufficiently large h. $\qquad\square$

For any $n \ge 4m$ we choose h sufficiently large so that (7.111) holds and we consider now the space V^+ defined in (7.115). If $u \in V^+$ then $\|u\|_{H_0^m}^2 \ge \Lambda_{m,j} \|u\|_{L^2}^2$ so that, recalling (7.51), we obtain again (7.109). This shows that

$$\text{there exist } \alpha, \rho > 0 \text{ such that } E_\lambda(u) \ge \alpha \text{ for all } u \in V^+, \ \|u\|_{H_0^m} = \rho. \tag{7.119}$$

Proof of Theorem 7.44 (Item 2). We first fix h as in (7.117). Then we let $\varepsilon \to 0$ so that Lemma 7.48 applies. When ε is sufficiently small, we choose V^- according to (7.115) so that Lemma 7.49 holds true. This shows (7.106) with $c_0 = \frac{m}{n} S^{n/2m}$ which is precisely the critical energy level under which E_λ satisfies a local Palais-Smale condition, see Lemma 7.46. Finally, (7.107) follows from (7.119).

Since $\operatorname{codim} V^+ = j-1$ and $\dim V^- = j$ we know that $\dim(V^+ \cap V^-) \geq 1$ so that by Lemma 7.47 we have at least a pair of critical points of E_λ, that is, a pair of nontrivial solutions to (7.79). $\qquad\square$

Proof of Theorem 7.44 (Item 3). The first statement can be proved exactly as for Item 2, it suffices to replace (7.117) with (7.118) and to Lemma 7.49 with Lemma 7.50.

In order to prove the multiplicity statement of Item 3, we modify the spaces in (7.115) by enlarging V^+. More precisely, we set

$$V^+ := \left(\operatorname{span}\{\varphi_1,\ldots,\varphi_{j-2}\}\right)^\perp, \quad V^- := \operatorname{span}\{\varphi_1^h,\ldots,\varphi_{j-2+\mu_{j-1}}^h, w_\varepsilon\}. \quad (7.120)$$

To maintain the notations of Lemma 7.50 we have chosen to shift the indices and take $\lambda \in (\Lambda_{m,j-2}, \Lambda_{m,j-1})$ for some $j \geq 2$ with the convention that $\Lambda_{m,0} = 0$ and $V^+ = H_0^m(\Omega)$ if $j = 2$. From Lemma 7.50, we know that there exists $\omega > 0$ such that

$$\sup_{u \in V^-} E_{\Lambda_{m,j-1}}(u) = \frac{m}{n} S^{n/2m} - \omega.$$

Recalling that V^- is finite dimensional and since $E_{\Lambda_{m,j-2}}(u) < 0$ for $\|u\|_{L^2}$ sufficiently large, we know that

$$\gamma := \sup\{\|u\|_{L^2}^2; u \in V^-, E_{\Lambda_{m,j-2}}(u) > 0\} < \infty.$$

Let $\delta_{j-1} = \omega/\gamma$ and take $\lambda \in (\Lambda_{m,j-1} - \delta_{j-1}, \Lambda_{m,j-1})$, then

$$E_\lambda(u) = E_{\Lambda_{m,j-1}}(u) + \frac{\Lambda_{m,j-1} - \lambda}{2} \|u\|_{L^2}^2 \leq \frac{m}{n} S^{n/2m} - \frac{\omega}{2} \qquad \text{for all } u \in V^-$$

since $E_\lambda(u) \leq 0$ whenever $\|u\|_{L^2}^2 \geq \gamma$. Moreover, since now $\lambda < \Lambda_{m,j-1}$ we still have (7.119) but with the new characterisation of V^+, see (7.120). Then the statement follows again from Lemma 7.47 once we note that $\dim(V^+ \cap V^-) \geq \mu_{j-1} + 1$. $\qquad\square$

Remark 7.51. When $m = 1$, the additional difficulty for $4m \leq n \leq (2 + 2\sqrt{2})m$ and $\lambda \in (\Lambda_{m,j})_{j \in \mathbb{N}^+}$ was observed by Zhang [419] but not by Capozzi-Fortunato-Palmieri [84]. In fact, the proof in [84] would work analogously in balls restricting to radial functions yielding the existence of radial solutions also in dimension $n = 4$. However, in [20] it is proved that for $m = 1$, $n = 4$, $\lambda = \Lambda_{1,1}$, (7.84) has no radial solution. For interesting studies of branches of radial solutions one could also see [25]. Combining [20,25] one sees that the branch of radial solutions emanating from the second radial eigenvalue of $(-\Delta)$ in B bends back to the first eigenvalue without ever crossing it. If $n = 5$, the same branch also bends back to the first eigenvalue but crossing it and approaching it from the left, see [180].

On the other hand, for the second order problem (7.79) with $m = 1$, Fortunato-Jannelli [171] constructed *nonsymmetric* solutions in the ball $B \subset \mathbb{R}^n$, $n \geq 4$, for all $\lambda > 0$. It is by no means obvious how to generalise such a result to higher order equations.

7.6 Critical Growth Navier Problems

As we have seen in the previous section, the Dirichlet problem for polyharmonic critical growth semilinear elliptic equations is quite delicate, since existence results strongly depend on the dimension, on subcritical perturbations and on the geometry of the domain. We restricted ourselves to Dirichlet boundary conditions, since these ensure the "pure higher order" character of our problems. In this and the next section, we show how critical growth problems behave when considered under Navier or Steklov boundary conditions. We recall that, in these cases, the problem can be transformed into a second order system. These boundary conditions exhibit slightly different phenomena for critical growth problems and this is the reason why we believe that they deserve some attention.

Consider the following Navier boundary value problem at critical growth

$$\begin{cases} (-\Delta)^m u = \lambda u + |u|^{s-1} u, \ u \not\equiv 0 & \text{in } \Omega, \\ \Delta^j u|_{\partial\Omega} = 0 & \text{for } j \leq m-1, \end{cases} \tag{7.121}$$

where $\Omega \subset \mathbb{R}^n$ $(n > 2m)$ is a bounded domain and $\lambda \in \mathbb{R}$ is a parameter.

The case $\lambda = 0$

In this case, no positive solution to (7.121) may exist in any starshaped domain Ω.

Theorem 7.52. *Assume that Ω is smooth and starshaped with respect to some point and that $\lambda = 0$. Then there exists no nonnegative solution to (7.121).*

Proof. With no loss of generality we may assume that Ω is starshaped with respect to the origin. Assume for contradiction that (7.121) admits a nontrivial solution u. Since Ω is smooth we have $u \in C^{2m,\gamma}(\overline{\Omega})$, see Proposition 7.15. Moreover, by applying m times the maximum principle for $-\Delta$ we infer that $(-\Delta)^j u \geq 0$ in Ω for $j = 0,\dots,m$. Therefore, if $j = 0,\dots,m-1$ then $(-\Delta)^j u$ is superharmonic in Ω and vanishes on $\partial\Omega$, so that Hopf's boundary lemma yields

$$\frac{\partial}{\partial \nu}(-\Delta)^j u < 0 \quad \text{on } \partial\Omega. \tag{7.122}$$

On the other hand, by applying (7.74), we obtain

$$0 = \sum_{j=1}^m \int_{\partial\Omega} (\Delta^{j-1} u)_\nu (\Delta^{m-j} u)_\nu (x \cdot \nu) \, d\omega. \tag{7.123}$$

Since Ω is starshaped, we know that $x \cdot \nu \geq 0$ on $\partial\Omega$. In turn, in view of (7.122) and (7.123), this yields $x \cdot \nu \equiv 0$ on $\partial\Omega$. This is impossible since by the divergence theorem $\int_{\partial\Omega} (x \cdot \nu) \, d\omega = n|\Omega| > 0$. $\qquad\square$

If we restrict ourselves to a ball, according to Theorem 7.3, positive solutions are radially symmetric so that Theorem 7.52 can be improved with the following statement.

Theorem 7.53. *Problem (7.121) has no radial solution when $\lambda = 0$ and $\Omega = B$.*

Proof. For simplicity we give the proof only in the biharmonic case $m = 2$ since it already contains the main ideas of the general proof for which we refer to Lazzo-Schmidt [273].

We use the same notations introduced in Theorem 7.34 and we assume for contradiction that (7.121) admits a nontrivial radial solution u. Then u and its derivatives only have a finite number of zeros, see Theorem 7.34. From Theorem 7.34 we also recall that for any $\phi \in C^1([0,1];\mathbb{R})$, such that both ϕ and ϕ' have a finite number of interior zeros, we have $\sigma(\phi') \geq \sigma(\phi) + \beta(\phi) - 1$. In view of the boundary conditions we have $u(1) = \Delta u(1) = 0$, whereas $u'(0) = (\Delta u)'(0) = 0$ since u is smooth at the origin. In particular, these equalities imply that $\beta(u), \beta(u'), \beta(\Delta u), \beta((\Delta u)') \geq 1$. Taking all these remarks into account, we infer that

$$\sigma(u) = \sigma(|u|^{8/(n-4)}u) = \sigma(\Delta^2 u) \geq \sigma(r^{n-1}(\Delta u)') = \sigma((\Delta u)')$$
$$\geq \sigma(\Delta u) = \sigma((r^{n-1}u')') \geq \sigma(r^{n-1}u') = \sigma(u') \geq \sigma(u).$$

This means that all the above inequalities are in fact equalities. But in view of Pohožaev's identity in Theorem 7.29 we have that $u'(1) = 0$ or $(\Delta u)'(1) = 0$ so that $\beta(u') = 2$ or $\beta((\Delta u)') = 2$. This means that at least one of the above inequalities is strict, a contradiction. $\qquad\Box$

The case $\lambda > 0$

Also under Navier boundary conditions a critical dimension phenomenon can be observed. Before stating the corresponding result, let us remark that the first Navier eigenvalue for $(-\Delta)^m$ is $\Lambda_{1,1}^m$, that is, the m-th power of the first Dirichlet eigenvalue for $-\Delta$. Therefore, arguing as in (7.43), one sees that no positive solution to (7.121) may exist if $\lambda \geq \Lambda_{1,1}^m$.

Theorem 7.54. *Let $B \subset \mathbb{R}^n$ $(n > 2m)$ be the unit ball. Then for problem (7.121) in B, the following holds:*

1. *If $2m + 1 \leq n \leq 4m - 1$, then there exist $0 < \lambda_* \leq \lambda^* < \Lambda_{1,1}^m$ such that (7.121) admits a positive solution if $\lambda \in (\lambda^*, \Lambda_{1,1}^m)$ and no positive solution if $\lambda \in (0, \lambda_*)$.*
2. *If $n \geq 4m$, then (7.121) admits a positive solution for all $\lambda \in (0, \Lambda_{1,1}^m)$.*

Hence $n \in \{2m + 1, \ldots, 4m - 1\}$ are the critical dimensions also for problem (7.121), when restricting to positive solutions in the ball. Let us also mention that Item 2 in Theorem 7.54 also holds in general bounded domains Ω, see the proof below. Again, in Item 1 nonexistence of *any nontrivial* solution has to be left open.

Proof of Theorem 7.54. 1. We assume here that $2m+1 \le n \le 4m-1$. When λ is in a left neighbourhood of $\Lambda_{1,1}^m$, the existence part may be obtained as in Item 2 of Theorem 7.38. Assume now that $\lambda > 0$ and that (7.121) has a positive solution u in B. By Theorem 7.3, $(-\Delta)^k u$ is radially symmetric and strictly decreasing for all $k = 0,\ldots,m-1$. Moreover, an iterated application of the maximum principle for $-\Delta$ yields that $(-\Delta)^j u > 0$ in B for $j = 0,\ldots,m$. Hence, if $j = 0,\ldots,m-1$, then $(-\Delta)^j u$ is superharmonic in B and vanishes on ∂B, so that Hopf's boundary lemma yields

$$\frac{\partial}{\partial \nu}(-\Delta)^j u < 0 \quad \text{on } \partial B. \tag{7.124}$$

In the radial framework, Theorem 7.29 reads

$$2m\lambda \int_B u^2 \, dx = (-1)^{m+1} \sum_{j=1}^{m} \int_{\partial B} (\Delta^{j-1} u)' (\Delta^{m-j} u)' \, d\omega. \tag{7.125}$$

In view of (7.124), counting also the factor $(-1)^{m+1}$, all the terms in the sum on the right hand side of (7.125) are strictly positive. Therefore, (7.125) implies

$$m\lambda \int_B u^2 \, dx \ge (-1)^{m+1} \int_{\partial B} u' (\Delta^{m-1} u)' \, d\omega. \tag{7.126}$$

Let ne_n denote the $(n-1)$-dimensional surface measure of the unit ball B and let $c(n)$ denote a generic positive constant which may vary from line to line. As u is radially symmetric, we may proceed as follows with the terms on the right hand side of (7.126).

$$\int_{\partial B} u' (\Delta^{m-1} u)' \, d\omega = ne_n u'(1) \cdot (\Delta^{m-1} u)'(1)$$
$$= \frac{1}{ne_n} \left(\int_{\partial B} u' \, d\omega \right) \left(\int_{\partial B} (\Delta^{m-1} u)' \, d\omega \right)$$
$$= \frac{1}{ne_n} \left(\int_B \Delta u \, dx \right) \left(\int_B \Delta^m u \, dx \right). \tag{7.127}$$

Let w be the unique (smooth radial positive) solution to

$$\begin{cases} (-\Delta)^{m-1} w = 1 & \text{in } B, \\ \Delta^j w|_{\partial B} = 0 & \text{for } j \le m-2. \end{cases}$$

Integrating by parts and using the homogeneous Navier boundary conditions both for u and w, yields

$$-\int_B \Delta u \, dx = -\int_B \Delta u (-\Delta)^{m-1} w \, dx = \int_B w (-\Delta)^m u \, dx.$$

Inserting this identity and (7.127) into (7.126) gives

$$m\lambda \int_B u^2\,dx \geq \frac{1}{ne_n}\left(\int_B(-\Delta)^m u\,dx\right)\left(\int_B w(-\Delta)^m u\,dx\right).$$

Since u is positive and strictly radially decreasing, by (7.121) also $(-\Delta)^m u$ is positive and strictly radially decreasing. This fact can be used to get rid of the weight factor w in the last integral. Indeed,

$$\begin{aligned}
\int_B(-\Delta)^m u\,dx &= \int_{|x|\leq 1/2}(-\Delta)^m u\,dx + \int_{1/2\leq|x|\leq 1}(-\Delta)^m u\,dx \\
&\leq \int_{|x|\leq 1/2}(-\Delta)^m u\,dx + c(n)(-\Delta)^m u\left(\frac{1}{2}\right) \\
&\leq \int_{|x|\leq 1/2}(-\Delta)^m u\,dx + c(n)\int_{|x|\leq 1/2}(-\Delta)^m u\,dx \\
&\leq c(n)\int_B w(-\Delta)^m u\,dx.
\end{aligned}$$

Combining this estimate with the previous inequality, we obtain

$$m\lambda \int_B u^2\,dx \geq c(n)\left(\int_B(-\Delta)^m u\,dx\right)^2 = c(n)\|(-\Delta)^m u\|_{L^1}^2. \tag{7.128}$$

Since $2m+1 \leq n \leq 4m-1$, by a duality argument and using elliptic estimates, we find

$$\|u\|_{L^2} \leq c(n)\|(-\Delta)^m u\|_{L^1}. \tag{7.129}$$

Observe that L^2-estimates for the polyharmonic operator under homogeneous Navier boundary conditions follow immediately from the L^2-estimates for the Laplacian, see Theorem 2.20.

Combining (7.128) and (7.129), we see that for any positive radial solution u of (7.121) in B, we have

$$\lambda \int_B u^2\,dx \geq c(n)\int_B u^2\,dx,$$

where the constant $c(n)$ is independent of u. Since $u > 0$, we necessarily have that $\lambda \geq c(n) =: \lambda_*$ and no solutions exists if $\lambda < \lambda_*$.

2. We consider again the functional F_λ in (7.41), but instead of minimising it over N (see (7.42)) we minimise it now over

$$M = \left\{v \in H^m_\vartheta(B) : \int_B |v|^{s+1}\,dx = 1\right\}, \tag{7.130}$$

where $H^m_\vartheta(B)$ is defined in (7.35). Correspondingly, instead of (7.85), we define

$$S_\lambda := \inf_{v\in M} F_\lambda(v).$$

Arguing as in Lemma 7.41, we see that if $S_\lambda < S$, then $F_\lambda|_M$ attains its minimum. Notice that S is the same as for the Dirichlet problem, see Theorem 7.24. Recalling that $H_0^m(B) \subset H_\vartheta^m(B)$, Lemma 7.42 ensures that indeed $S_\lambda < S$ whenever $n \geq 4m$. Hence, $F_\lambda|_M$ attains its minimum at some $v \in M$. Once we have a minimiser v, by the maximum principle for powers of $-\Delta$ (which holds in *any* domain!), we obtain a positive minimiser by taking a multiple of the unique function $u \in H_\vartheta^m(B)$ such that $(-\Delta)^{m/2}u = |\Delta^{m/2}v|$ if m is even and such that $(-\Delta)^{(m-1)/2}u = |\Delta^{(m-1)/2}v|$ if m is odd. $\qquad\square$

The case $\lambda < 0$

In the preceding nonexistence results, the positivity of a solution, the nonexistence of which had to be shown, is intensively exploited. For $\lambda \geq 0$ one can conclude from $u \geq 0$ that even $-\Delta u \geq 0$ in B. But for $\lambda < 0$, we can only argue as before, when λ is close enough to 0, see [247, 310]. If $\lambda \ll 0$, the framework of positive solutions is no longer adequate for (7.121) in B, and the argument in [399, pp. 390-391] and [401, Theorem 3] breaks down. Instead, one has to look for nonexistence of *any* nontrivial solution. Such a result seems to be still unknown, even for nontrivial radial solutions in the ball. And also the proof of Theorem 7.53 does not seem to extend to the case $\lambda < 0$.

7.7 Critical Growth Steklov Problems

Once again we restrict our attention to the case of the unit ball B and we consider the following fourth order elliptic problem with purely critical growth and Steklov boundary conditions

$$\begin{cases} \Delta^2 u = u^{(n+4)/(n-4)}, & u > 0 & \text{in } B, \\ u = \Delta u - au_\nu = 0 & & \text{on } \partial B, \end{cases} \tag{7.131}$$

where $a \in \mathbb{R}$. In this case we have

Theorem 7.55. *If $a \leq 4$ or $a \geq n$, then (7.131) admits no solution. For all $a \in (4,n)$ problem (7.131) admits a unique radial solution u_a.*

We also point out that as $a \nearrow n$ we have $u_a \to 0$ uniformly in $\overline{B}$ whereas as $a \searrow 4$ we have $u_a(0) \to +\infty$ and $u_a(x) \to 0$ for all $x \in B \setminus \{0\}$, see [189].

As for the proof of Theorem 7.55, we recall that the first Steklov eigenvalue equals n and the first eigenfunction is given by $\phi_1(x) = 1 - |x|^2$, see Section 3.3.1. Nonexistence for $a \geq n$ is obtained again by testing (7.131) with the positive first eigenfunction ϕ_1, see (7.43). The other parts of the proof are more delicate.

Proof of nonexistence for $a \le 4$. Assume by contradiction that u is a solution of (7.131) for some $a \le 4$. Consider the auxiliary function $\phi \in C^2(\overline{B})$ defined by

$$\phi(x) = (4 - a + a|x|^2)\Delta u(x) - 4ax \cdot \nabla u(x) + a(8 - 2n)u(x), \qquad x \in \overline{B}.$$

Then $\phi = 0$ on ∂B, since $u = 0$ and $\Delta u = au_\nu$ on ∂B. A short computation shows

$$\begin{aligned}
\Delta \phi &= 2an\Delta u + 4ax \cdot \nabla \Delta u + (4 - a + a|x|^2)\Delta^2 u \\
&\quad - 4a(2\Delta u + x \cdot \nabla \Delta u) + a(8 - 2n)\Delta u \\
&= (4 - a + a|x|^2)u^{(n+4)/(n-4)}.
\end{aligned}$$

If $u > 0$ solves (7.131), then $\Delta \phi \gneqq 0$, since $a \le 4$. By the maximum principle we conclude that $\phi < 0$ in B, and $\phi_\nu > 0$ on ∂B by the Hopf boundary lemma. But on ∂B we also get by direct computation, using the second boundary condition,

$$\begin{aligned}
\phi_\nu &= 2a\Delta u + 4(\Delta u)_\nu - 4a(u_\nu + u_{\nu\nu}) + a(8 - 2n)u_\nu \\
&= 2a\Delta u + 4(\Delta u)_\nu - 4a(u_\nu + \Delta u - (n-1)u_\nu) + a(8 - 2n)u_\nu \\
&= 2\left(2(\Delta u)_\nu + a(n - a)u_\nu\right),
\end{aligned}$$

so that $2(\Delta u)_\nu + a(n - a)u_\nu > 0$ on ∂B. Since $u > 0$ in B we have $u_\nu \le 0$ on ∂B. We also use (7.76) which, for solutions of (7.131), reads as follows

$$\int_{\partial B}\left(2(\Delta u)_\nu + a(n - a)u_\nu\right)u_\nu\, d\omega = 0. \tag{7.132}$$

This yields $u_\nu = 0$ on ∂B. But then $u > 0$ would be a solution of the *Dirichlet* problem in B, which is known to have no positive solutions, see Theorem 7.33. This contradiction concludes the proof of the nonexistence statement for $a \le 4$ in Theorem 7.55. $\qquad\square$

Proof of existence when $a \in (4, n)$. From Theorem 7.24 we know that the optimal Sobolev constant

$$S = \inf_{H^2 \cap H_0^1(B)\setminus\{0\}} \frac{\|\Delta v\|_{L^2}^2}{\|v\|_{L^{s+1}}^2}$$

is not achieved. We recall that the exact value of S is given in (7.50) for $m = 2$. We also recall from Theorem 7.21 that S is achieved by the radial entire functions

$$u_\varepsilon(x) := \frac{1}{(\varepsilon^2 + |x|^2)^{\frac{n-4}{2}}}$$

which satisfy

$$S = \frac{\displaystyle\int_{\mathbb{R}^n} |\Delta u_\varepsilon|^2\, dx}{\left(\displaystyle\int_{\mathbb{R}^n} |u_\varepsilon|^{2n/(n-4)}\, dx\right)^{(n-4)/n}} \qquad \text{for all } \varepsilon > 0. \tag{7.133}$$

Moreover,

$$\int_{\mathbb{R}^n} |u_\varepsilon|^{2n/(n-4)}\,dx = \int_{\mathbb{R}^n} \frac{dx}{(\varepsilon^2 + |x|^2)^n} = \frac{1}{\varepsilon^n}\int_{\mathbb{R}^n} \frac{dx}{(1+|x|^2)^n}$$

$$= \frac{n e_n}{\varepsilon^n}\int_0^\infty \frac{r^{n-1}}{(1+r^2)^n}\,dr = \frac{n e_n}{2\varepsilon^n}\int_0^\infty \frac{t^{\frac{n}{2}-1}}{(1+t)^n}\,dt$$

$$= \frac{n e_n}{2\varepsilon^n}\frac{\Gamma(\frac{n}{2})^2}{\Gamma(n)} =: \frac{K_2}{\varepsilon^n}, \tag{7.134}$$

and, by (7.133),

$$\int_{\mathbb{R}^n} |\Delta u_\varepsilon|^2\,dx = S\left(\int_{\mathbb{R}^n} |u_\varepsilon|^{2n/(n-4)}\,dx\right)^{(n-4)/n} = S\frac{K_2^{(n-4)/n}}{\varepsilon^{n-4}} =: \frac{K_1}{\varepsilon^{n-4}}. \tag{7.135}$$

Let $\mathscr{H}_r = \{u \in H^2 \cap H_0^1(B); u = u(|x|)\}$ denote the closed subspace of radially symmetric functions and for all nontrivial $u \in \mathscr{H}_r$ consider the ratio

$$Q_a(u) := \frac{\|\Delta u\|_{L^2}^2 - a\|u_v\|_{L^2(\partial B)}^2}{\|u\|_{L^{s+1}}^2}. \tag{7.136}$$

We consider the minimisation problem

$$\Sigma_a := \inf_{\mathscr{H}_r\setminus\{0\}} Q_a(v) \tag{7.137}$$

and prove the following compactness result which is the analogue of Lemma 7.41 for Steklov boundary conditions.

Lemma 7.56. *Assume that $0 < a < n$. If $\Sigma_a < S$, then the infimum in (7.137) is achieved. Moreover, up to a sign change and up to a Lagrange multiplier, any minimiser is a superharmonic radial solution of (7.131).*

Proof. Let $(u_k) \subset \mathscr{H}_r \setminus \{0\}$ be a minimising sequence for Σ_a such that

$$\|u_k\|_{L^{s+1}} = 1. \tag{7.138}$$

Then

$$\|\Delta u_k\|_{L^2}^2 - a\|(u_k)_v\|_{L^2(\partial B)}^2 = \Sigma_a + o(1) \qquad \text{as } k \to +\infty. \tag{7.139}$$

Moreover, from Theorem 3.20 we know that $\|\Delta u\|_{L^2}^2 \ge n\|u_v\|_{L^2(\partial B)}^2$ for all $u \in H^2 \cap H_0^1(B)$ and, in turn,

$$\|\Delta u_k\|_{L^2}^2 = \Sigma_a + a\|(u_k)_v\|_{L^2(\partial B)}^2 + o(1) \le \Sigma_a + \frac{a}{n}\|\Delta u_k\|_{L^2}^2 + o(1)$$

so that (u_k) is bounded in $H^2 \cap H_0^1(B)$. Hence (∇u_k) is bounded in $H^1(B)$. Exploiting the compactness of the embeddings $H^1(B) \subset L^2(\partial B)$ and $H^2 \cap H_0^1(B) \subset L^2(B)$, we deduce that there exists $u \in H^2 \cap H_0^1(B)$ such that, up to a subsequence,

$$u_k \rightharpoonup u \text{ in } H^2 \cap H_0^1(B), \quad (u_k)_v \to u_v \text{ in } L^2(\partial B), \quad u_k \to u \text{ in } L^2(B). \qquad (7.140)$$

That is, if we set $v_k := u_k - u$, then

$$v_k \rightharpoonup 0 \text{ in } H^2 \cap H_0^1(B), \quad (v_k)_v \to 0 \text{ in } L^2(\partial B), \quad v_k \to 0 \text{ in } L^2(B). \qquad (7.141)$$

On the other hand, by (7.138) we infer that $\|\Delta u_k\|_{L^2}^2 \geq S$, so that from (7.139) we also obtain

$$a\|(u_k)_v\|_{L^2(\partial B)}^2 = \|\Delta u_k\|_{L^2}^2 - \Sigma_a + o(1) \geq S - \Sigma_a + o(1)$$

which remains bounded away from 0 since $\Sigma_a < S$. From this fact we deduce that $u \neq 0$.

In view of (7.140)-(7.141) we may rewrite (7.139) as

$$\|\Delta u\|_{L^2}^2 + \|\Delta v_k\|_{L^2}^2 - a\|u_v\|_{L^2(\partial B)}^2 = \Sigma_a + o(1). \qquad (7.142)$$

Moreover, by (7.138) and Vitali's convergence theorem, we have

$$1 = \|u + v_k\|_{L^{s+1}}^{s+1} = \|u\|_{L^{s+1}}^{s+1} + \|v_k\|_{L^{s+1}}^{s+1} + o(1) \leq \|u\|_{L^{s+1}}^2 + \|v_k\|_{L^{s+1}}^2 + o(1)$$
$$\leq \|u\|_{L^{s+1}}^2 + \frac{1}{S}\|\Delta v_k\|_{L^2}^2 + o(1) ,$$

where we also used the fact that both $\|u\|_{L^{s+1}}$ and $\|v_k\|_{L^{s+1}}$ do not exceed 1. Since $\Sigma_a \geq 0$ for every $0 < a < n$, this last inequality gives

$$\Sigma_a \leq \Sigma_a\|u\|_{L^{s+1}}^2 + \frac{\Sigma_a}{S}\|\Delta v_k\|_{L^2}^2 + o(1).$$

By combining the last estimate with (7.142), we obtain

$$\|\Delta u\|_{L^2}^2 - a\|u_v\|_{L^2(\partial B)}^2 = \Sigma_a - \|\Delta v_k\|_{L^2}^2 + o(1) \leq \Sigma_a\|u\|_{L^{s+1}}^2$$
$$+ \left(\frac{\Sigma_a}{S} - 1\right)\|\Delta v_k\|_{L^2}^2 + o(1) \leq \Sigma_a\|u\|_{L^{s+1}}^2 + o(1),$$

which shows that $u \neq 0$ is a radial minimiser for (7.137). This proves the first part of Lemma 7.56.

Consider now a radial minimiser u for (7.137) and assume for contradiction that it is not superharmonic (nor subharmonic) in B. Then let $w \in H^2 \cap H_0^1(B)$ be the unique solution to

$$\begin{cases} -\Delta w = |\Delta u| & \text{in } B, \\ w = 0 & \text{on } \partial B. \end{cases}$$

By the maximum principle for superharmonic functions it follows that $w > 0$ in B and $w_\nu < 0$ on ∂B. Moreover, both $w \pm u$ are superharmonic (but not harmonic!) in B and vanish on ∂B. This proves that

$$|u| < w \quad \text{in } B, \qquad |u_\nu| < |w_\nu| \quad \text{on } \partial B .$$

In turn, these inequalities (and $-\Delta w = |\Delta u|$) prove that $Q_a(w) < Q_a(u)$ which contradicts the assumption that u minimises (7.137).

Therefore, any minimiser u for (7.137) is superharmonic or subharmonic in B. By the Lagrange multiplier method, it is readily seen that a multiple of u is a radial superharmonic solution to (7.131). $\qquad\square$

Assume that $a > 4$, fix a real number

$$0 < \delta < \sqrt[n]{\frac{a-4}{n+a-4}} \tag{7.143}$$

and consider the following two-parameters family of functions

$$U_{\varepsilon,\delta}(x) = g_\delta(|x|) u_\varepsilon(x) \equiv \frac{g_\delta(|x|)}{\left(\varepsilon^2 + |x|^2\right)^{\frac{n-4}{2}}} ,$$

where $g_\delta \in C^1[0,1] \cap W^{2,\infty}(0,1)$, $g_\delta(r) = 1$ for $0 \le r \le \delta$ and $g_\delta(1) = 0$. Then $U_{\varepsilon,\delta} \in \mathscr{H}_r$ and

$$U_{\varepsilon,\delta}(x) = u_\varepsilon(x) = \frac{1}{\left(\varepsilon^2 + |x|^2\right)^{\frac{n-4}{2}}} \quad \text{in } B_\delta = \{x \in \mathbb{R}^n; |x| < \delta\}.$$

In view of Lemma 7.56 existence for (7.131) is proved if we find $U_{\varepsilon,\delta}$ such that

$$Q_a(U_{\varepsilon,\delta}) < S \tag{7.144}$$

for suitable $\varepsilon, \delta > 0$. Hence, our purpose is now to estimate the ratio $Q_a(U_{\varepsilon,\delta})$. We proceed as in the proof of Lemma 7.42 but here the explicit expression of the cut-off function g_δ will play a major role. In view of (7.134) a lower bound for the denominator in (7.136) is readily obtained:

$$\int_B |U_{\varepsilon,\delta}|^{2^*} dx = \int_{\mathbb{R}^n} |u_\varepsilon|^{2^*} dx - \int_{\mathbb{R}^n \setminus B} |u_\varepsilon|^{2^*} dx - \int_{B \setminus B_\delta} \frac{1 - g_\delta(|x|)^{2^*}}{\left(\varepsilon^2 + |x|^2\right)^n} dx$$

$$\ge \frac{K_2}{\varepsilon^n} + O(1). \tag{7.145}$$

We seek an upper bound for the numerator. From (7.135) we get

$$\int_B |\Delta U_{\varepsilon,\delta}|^2\, dx = \int_{\mathbb{R}^n} |\Delta u_\varepsilon|^2\, dx + \int_{B\setminus B_\delta} \left(|\Delta U_{\varepsilon,\delta}|^2 - |\Delta u_\varepsilon|^2\right) dx - \int_{\mathbb{R}^n\setminus B} |\Delta u_\varepsilon|^2\, dx$$

$$= \frac{K_1}{\varepsilon^{n-4}} + o(1) - 4n(n-4)e_n$$

$$+ \int_{B\setminus B_\delta} \left(|\Delta U_{\varepsilon,\delta}|^2 - |\Delta u_\varepsilon|^2\right) dx. \tag{7.146}$$

After some computations we find in radial coordinates $r = |x|$

$$\Delta U_{\varepsilon,\delta}(r) = U''_{\varepsilon,\delta}(r) + \frac{n-1}{r} U'_{\varepsilon,\delta}(r)$$

$$= \frac{g''_\delta(r)}{(\varepsilon^2+r^2)^{(n-4)/2}} + \frac{g'_\delta(r)}{r(\varepsilon^2+r^2)^{(n-2)/2}}\left[(7-n)r^2 + (n-1)\varepsilon^2\right]$$

$$- (n-4)\frac{g_\delta(r)}{(\varepsilon^2+r^2)^{n/2}}(2r^2 + n\varepsilon^2)\,.$$

Let us recall that $g'_\delta(r) = g''_\delta(r) = 0$ for $r < \delta$. Furthermore, as $\varepsilon \to 0$, we have

$$\Delta U_{\varepsilon,\delta}(r) = \frac{g''_\delta(r)}{r^{n-4}} + (7-n)\frac{g'_\delta(r)}{r^{n-3}} - 2(n-4)\frac{g_\delta(r)}{r^{n-2}} + o(1)$$

uniformly with respect to $r \in [\delta, 1]$. By squaring, we get

$$|\Delta U_{\varepsilon,\delta}(r)|^2 = \frac{g''_\delta(r)^2}{r^{2n-8}} + (7-n)^2\frac{g'_\delta(r)^2}{r^{2n-6}} + 4(n-4)^2\frac{g_\delta(r)^2}{r^{2n-4}} + 2(7-n)\frac{g''_\delta(r)g'_\delta(r)}{r^{2n-7}}$$

$$-4(n-4)\frac{g''_\delta(r)g_\delta(r)}{r^{2n-6}} + 4(n-4)(n-7)\frac{g'_\delta(r)g_\delta(r)}{r^{2n-5}} + o(1).$$

We may now rewrite in a simplified radial form the terms contained in the last integral in (7.146). With some integrations by parts, and taking into account the behaviour of $g_\delta(r)$ for $r \in \{1,\delta\}$, we obtain

$$\int_\delta^1 \frac{g''_\delta(r)g'_\delta(r)}{r^{n-6}}\, dr = \frac{n-6}{2}\int_\delta^1 \frac{g'_\delta(r)^2}{r^{n-5}}\, dr + \frac{g'_\delta(1)^2}{2},$$

$$\int_\delta^1 \frac{g''_\delta(r)g_\delta(r)}{r^{n-5}}\, dr = -\int_\delta^1 \frac{g'_\delta(r)^2}{r^{n-5}}\, dr + (n-5)\int_\delta^1 \frac{g'_\delta(r)g_\delta(r)}{r^{n-4}}\, dr,$$

$$\int_\delta^1 \frac{g'_\delta(r)g_\delta(r)}{r^{n-4}}\, dr = \frac{n-4}{2}\int_\delta^1 \frac{g_\delta(r)^2}{r^{n-3}}\, dr - \frac{1}{2\delta^{n-4}}.$$

Collecting terms we find

$$\int_{B \setminus B_\delta} \left(|\Delta U_{\varepsilon,\delta}|^2 - |\Delta u_\varepsilon|^2 \right) dx = n e_n \int_\delta^1 \left(\frac{g_\delta''(r)^2}{r^{n-7}} + 3(n-3) \frac{g_\delta'(r)^2}{r^{n-5}} \right) dr$$
$$+ n(7-n) e_n g_\delta'(1)^2 + 4n(n-4) e_n. \quad (7.147)$$

Moreover, simple computations show that

$$\int_{\partial B} (U_{\varepsilon,\delta})_\nu^2 = n e_n g_\delta'(1)^2 + o(1)$$

which, combined with (7.146) and (7.147), yields

$$\int_B |\Delta U_{\varepsilon,\delta}|^2 \, dx - a \int_{\partial B} (U_{\varepsilon,\delta})_\nu^2 \, d\omega = \frac{K_1}{\varepsilon^{n-4}} + n e_n \int_\delta^1 \left(\frac{g_\delta''(r)^2}{r^{n-7}} + 3(n-3) \frac{g_\delta'(r)^2}{r^{n-5}} \right) dr$$
$$+ n(7-n-a) e_n g_\delta'(1)^2 + o(1). \quad (7.148)$$

Putting $f = g_\delta'$, we are so led to find negative values for the functional

$$J(f) := \int_\delta^1 \left(\frac{f'(r)^2}{r^{n-7}} + 3(n-3) \frac{f(r)^2}{r^{n-5}} \right) dr + (7-n-a)f(1)^2$$

when $f \in C^0[\delta,1] \cap W^{1,\infty}(\delta,1)$ satisfies

$$f(\delta) = 0 \qquad \text{and} \qquad \int_\delta^1 f(r) \, dr = -1. \quad (7.149)$$

The Euler-Lagrange equation relative to the functional J reads

$$r^2 f''(r) + (7-n) r f'(r) - 3(n-3) f(r) = 0 \qquad \delta \le r \le 1,$$

whose solutions have the following general form $f(r) = \alpha r^{n-3} + \beta r^{-3}$ for any $\alpha, \beta \in \mathbb{R}$. The first condition in (7.149) yields $\beta = -\alpha \delta^n$. To determine the value of α, we use the second condition in (7.149) and obtain

$$\alpha = -\frac{2(n-2)}{2 - n\delta^{n-2} + (n-2)\delta^n} \, .$$

So, let us consider the function $f(r) = \alpha(r^{n-3} - \delta^n r^{-3})$ and compute

$$\frac{J(f)}{\alpha^2} = \int_\delta^1 \left[n(n-3) r^{n-1} + \frac{3n\delta^{2n}}{r^{n+1}} \right] dr + (7-n-a)(1-\delta^n)^2$$
$$= (1-\delta^n) \left[(4-a)(1-\delta^n) + n\delta^n \right] =: \gamma < 0,$$

where the sign of γ follows from the initial choice of δ in (7.143). Summarising, with the above choice of f and recalling that $g_\delta(r) = -\int_r^1 f(s)\,ds$, we obtain from (7.148)

$$\int_B |\Delta U_{\varepsilon,\delta}|^2 \, dx - a \int_{\partial B} (U_{\varepsilon,\delta})_\nu^2 \, d\omega = \frac{K_1}{\varepsilon^{n-4}} + ne_n \alpha^2 \gamma + o(1) \,.$$

Finally, by combining this estimate with (7.145) and recalling the definition in (7.136), we find

$$Q_a(U_{\varepsilon,\delta}) \le \frac{\frac{K_1}{\varepsilon^{n-4}} + ne_n \alpha^2 \gamma + o(1)}{[\frac{K_2}{\varepsilon^n} + O(1)]^{2/2^*}} \qquad \text{as } \varepsilon \to 0$$

so that (7.144) holds for sufficiently small ε. $\square$

Proof of uniqueness. If we consider radially symmetric solutions and put $r = |x|$, then for $r \in [0,1)$ the equation in (7.131) reads

$$u^{iv}(r) + \frac{2(n-1)}{r} u'''(r) + \frac{(n-1)(n-3)}{r^2} u''(r) - \frac{(n-1)(n-3)}{r^3} u'(r) = u^{\frac{n+4}{n-4}}(r)$$

$$(7.150)$$

while the boundary conditions become

$$u(1) = 0 \,, \qquad u''(1) + (n-1-a)u'(1) = 0 \,. \tag{7.151}$$

Moreover, the Pohožaev-type identity (7.132) yields

$$2u'''(1) + 2(n-1)u''(1) + (2(1-n) + a(n-a))u'(1) = 0 \,. \tag{7.152}$$

In turn, with the change of variables

$$u(r) = r^{-\frac{n-4}{2}} v(\log r) \quad (0 < r \le 1) \,, \quad v(t) = e^{\frac{n-4}{2}t} u(e^t) \quad (t \le 0), \tag{7.153}$$

equation (7.150) may be rewritten as

$$v^{iv}(t) - C_2 v''(t) + C_1 v(t) = v^{\frac{n+4}{n-4}}(t) \qquad t \in (-\infty, 0) \,, \tag{7.154}$$

where

$$C_1 = \left(\frac{n(n-4)}{4} \right)^2 \,, \qquad C_2 = \frac{n^2 - 4n + 8}{2} > 0. \tag{7.155}$$

After some computations we find from (7.151) and (7.152) that

$$v(0) = 0, \qquad\qquad v'(0) < 0,$$
$$v''(0) = (a-2)v'(0), \quad v'''(0) = \tfrac{1}{4}\left(n^2 - 4n + 2a^2 - 8a + 16\right)v'(0).$$

$$(7.156)$$

Assume that there exist two solutions u_1 and u_2 of (7.150) and let v_1 and v_2 be the corresponding functions obtained through the change of variables (7.153). Then both v_1 and v_2 satisfy (7.154) and (7.156). Put

$$w_i(t) := \frac{v_i(t)}{|v_i'(0)|} \qquad (i = 1, 2)$$

so that

$$w_i^{iv}(t) - C_2 w_i''(t) + C_1 w_i(t) = \lambda_i w_i^{(n+4)/(n-4)}(t), \qquad t \in (-\infty, 0),$$

with

$$w_i(0) = 0, \qquad w_i'(0) = -1,$$
$$w_i''(0) = 2 - a, \qquad w_i'''(0) = -\tfrac{1}{4}\left(n^2 - 4n + 2a^2 - 8a + 16\right),$$
$$\lambda_i = |v_i'(0)|^{8/(n-4)} > 0.$$

With no loss of generality we may assume that $\lambda_1 \geq \lambda_2$. Let $w := w_1 - w_2$ so that w satisfies

$$w^{iv}(t) - C_2 w''(t) + C_1 w(t) = \lambda_1 w_1^{(n+4)/(n-4)}(t) - \lambda_2 w_2^{(n+4)/(n-4)}(t)$$

with homogeneous initial conditions at $t = 0$. This equation may be rewritten as

$$w^{iv}(t) - C_2 w''(t) + C_1 w(t) = (\lambda_1 - \lambda_2) w_1^{(n+4)/(n-4)}(t) + f(t) w(t), \qquad (7.157)$$

where, by Lagrange's theorem,

$$f(t) = \frac{n+4}{n-4} \lambda_2 \int_0^1 (s w_1(t) + (1-s) w_2(t))^{8/(n-4)} \, ds \geq 0.$$

We now prove a technical result.

Lemma 7.57. *Let $h \in C^0(-\infty, 0]$, then the unique solution $w \in C^4(-\infty, 0]$ of the Cauchy problem*

$$\begin{cases} w^{iv}(t) - C_2 w''(t) + C_1 w(t) = h(t), & t \in (-\infty, 0) \\ w(0) = w'(0) = w''(0) = w'''(0) = 0 \end{cases}$$

is given by

$$w(t) = \frac{4}{n(n-4)} \int_t^0 \sinh\frac{n(\tau - t)}{2} \int_\tau^0 h(s) \sinh\frac{(n-4)(s-\tau)}{2} \, ds \, d\tau$$
$$= \int_t^0 G(s-t) h(s) \, ds, \qquad (7.158)$$

where

$$G(\sigma) = \frac{1}{n(n-2)} \sinh \frac{n\sigma}{2} - \frac{1}{(n-2)(n-4)} \sinh \frac{(n-4)\sigma}{2}$$

is positive for $\sigma > 0$.

Proof. It follows by combining three simple facts. First, the unique solution w of the problem

$$\begin{cases} w''(t) - \frac{n^2}{4} w(t) = z(t), & t \in (-\infty, 0) \\ w(0) = w'(0) = 0 \end{cases}$$

is given by

$$w(t) = \frac{2}{n} \int_t^0 z(\tau) \sinh \frac{n(\tau - t)}{2} \, d\tau \ .$$

Second, the unique solution z of the problem

$$\begin{cases} z''(t) - \frac{(n-4)^2}{4} z(t) = h(t), & t \in (-\infty, 0) \\ z(0) = z'(0) = 0 \end{cases}$$

is given by

$$z(t) = \frac{2}{n-4} \int_t^0 h(\tau) \sinh \frac{(n-4)(\tau - t)}{2} \, d\tau \ .$$

Third, by (7.155) the left hand side of (7.157) may be written as

$$\left(\frac{d^2}{dt^2} - \frac{(n-4)^2}{4} \right) \left(\frac{d^2}{dt^2} - \frac{n^2}{4} \right) w.$$

Finally, by changing the order of integration in the second term of (7.158) we get

$$G(s-t) = \frac{4}{n(n-4)} \int_t^s \sinh \frac{(n-4)(s-\tau)}{2} \sinh \frac{n(\tau - t)}{2} \, d\tau, \quad t < s < 0,$$

and the explicit form of G follows by elementary calculations. $\square$

By Lemma 7.57 the homogeneous Cauchy problem for (7.157) is equivalent to the following integral equation

$$w(t) = (\lambda_1 - \lambda_2) \int_t^0 G(s-t) w_1^{(n+4)/(n-4)}(s) \, ds + \int_t^0 G(s-t) f(s) w(s) \, ds \ .$$

In turn, by putting

$$W(t) := (\lambda_1 - \lambda_2) \int_t^0 G(s-t) w_1^{(n+4)/(n-4)}(s) \, ds \geq 0 \ ,$$

the above integral equation reads

$$w(t) = W(t) + \int_t^0 G(s-t)f(s)w(s)\,ds.$$

The solution to this problem is obtained by iteration. One makes an "initial guess" w_0 and constructs a sequence of approximations which converges to the solution. The sequence is defined by

$$w_{k+1}(t) = W(t) + \int_t^0 G(s-t)f(s)w_k(s)\,ds \qquad \text{for all } k \geq 0.$$

We may start with a nonnegative initial guess w_0. Then by recalling that $W \geq 0$, $f \geq 0$ on $(-\infty,0]$ and that $G(s-t) \geq 0$ for $s > t$, we readily obtain

$$w(t) \geq 0 \qquad \text{for all } t \leq 0.$$

Finally, if we multiply (7.157) by $e^{nt/2}$ we may rewrite it as

$$\frac{d}{dt}\left(e^{nt/2}\left(w'''(t) - \frac{n}{2}w''(t) - \frac{(n-4)^2}{4}w'(t) + \frac{n(n-4)^2}{8}w(t)\right)\right)$$
$$= e^{nt/2}\left((\lambda_1 - \lambda_2)w_1^{(n+4)/(n-4)}(t) + f(t)w(t)\right).$$

By integrating this equation over $(-\infty,0)$ and using the homogeneous boundary conditions we obtain

$$\int_{-\infty}^0 e^{nt/2}\left((\lambda_1 - \lambda_2)w_1^{(n+4)/(n-4)}(t) + f(t)w(t)\right)dt = 0.$$

In view of the sign conditions

$$\lambda_1 - \lambda_2 \geq 0, \quad w_1 > 0, \quad f,w \geq 0,$$

this implies that $\lambda_1 = \lambda_2$, $w \equiv 0$ and $u_1 = u_2$. Uniqueness is so proved. $\qquad\square$

7.8 Optimal Sobolev Inequalities with Remainder Terms

As we have seen in Theorem 7.23, the best Sobolev constant (in the critical case) is not achieved on a bounded domain. In this section we show that the corresponding embedding inequality can be improved by adding remainder terms.

Let $\Omega \subset \mathbb{R}^n$ be a smooth bounded domain and for all $q \in (1,\infty)$ let

$$\|u\|_{L_w^q} := \sup_{A \subset \Omega,\, |A|>0} |A|^{-1/q'} \int_A |u|\,dx$$

denote the weak L^q-norm, where q' is the conjugate of q.

Recall the scalar product and norm introduced in (2.10)-(2.11) for the space $H_0^m(\Omega)$ and let S be as in (7.50). We have the following improvement of Theorem 7.23.

Theorem 7.58. *Let $m \in \mathbb{N}^+$ and let $\Omega \subset \mathbb{R}^n$ ($n > 2m$) be a bounded domain. Then there exists a constant $C = C(\mathrm{diam}(\Omega), n, m) > 0$ such that*

$$\|u\|_{H_0^m}^2 \geq S\|u\|_{L^{s+1}}^2 + C\|u\|_{L_w^{\frac{n}{n-2m}}}^2 \qquad \text{for all } u \in H_0^m(\Omega). \tag{7.159}$$

Proof. Let $R > 0$ be such that $\Omega \subset B_R$, where B_R denotes the ball centered at the origin with radius R. Extending any function $u \in H_0^m(\Omega)$ by 0 in $B_R \setminus \Omega$ shows $H_0^m(\Omega) \subset H_0^m(B_R)$. Therefore, it suffices to prove (7.159) in the case where $\Omega = B_R$ for some $R > 0$. We divide the proof into two parts.

Proof of (7.159) for positive functions. Consider the closed convex cone of non-negative functions

$$\mathscr{K} = \{v \in H_0^m(B_R);\ v \geq 0 \text{ a.e. in } B_R\}.$$

Let $g \in L^\infty(B_R)$, $g \not\equiv 0$ and $g \geq 0$ a.e. in B_R. Let v be the solution of the problem

$$\begin{cases} (-\Delta)^m v = g & \text{in } B_R, \\ D^\alpha v|_{\partial B_R} = 0 & \text{for } |\alpha| \leq m - 1. \end{cases}$$

Then $v \in H_0^m \cap L^\infty(B_R)$. Take $u \in \mathscr{K} \setminus \{0\}$ and consider the (entire) function

$$\phi = \begin{cases} u - v + \|v\|_{L^\infty} & \text{in } B_R, \\ \|v\|_{L^\infty} P_m(\frac{R}{|x|}) & \text{in } B_R^c, \end{cases}$$

where P_m is the polynomial of degree $n - m - 1$ whose lowest power is $n - 2m$ and satisfies

$$P_m(1) = 1, \qquad P_m^{(\ell)}(1) = 0 \quad \ell = 1, \dots, m - 1.$$

Since $\phi \in \mathscr{D}^{m,2}(\mathbb{R}^n)$, we may write inequality (7.47) for ϕ and obtain (for some $C = C(n, R) > 0$)

$$\|u - v\|_{H_0^m}^2 + C\|v\|_{L^\infty}^2 \geq S\|u\|_{L^{s+1}}^2,$$

where we used the fact that both $u \geq 0$ and $-v + \|v\|_{L^\infty} \geq 0$ a.e. in B_R. Therefore, we obtain

$$\|u\|_{H_0^m}^2 + \|v\|_{H_0^m}^2 - 2(u, v)_{H_0^m} + C\|v\|_{L^\infty}^2 \geq S\|u\|_{L^{s+1}}^2$$

and hence

$$\|u\|_{H_0^m}^2 \geq 2\int_{B_R} ug\,dx - (\|v\|_{H_0^m}^2 + C\|v\|_{L^\infty}^2) + S\|u\|_{L^{s+1}}^2.$$

Replacing g, v with $\lambda g, \lambda v$ the previous inequality remains true for all $\lambda > 0$. Taking into account that for all $a \geq 0$ and $b > 0$ we have $\max_{\lambda \geq 0}(a\lambda - b\lambda^2) = \frac{a^2}{4b}$, we obtain

$$\|u\|^2_{H_0^m} \geq S\|u\|^2_{L^{s+1}} + \frac{(\int_{B_R} ug)^2}{\|v\|^2_{H_0^m} + C\|v\|^2_{L^\infty}} \,. \tag{7.160}$$

Note that here we also used $g \geq 0$ and $u \geq 0$. Let $A \subset B_R$, $|A| > 0$ and take $g = 1_A$ (the characteristic function of A) in (7.160). Then we have

$$\int_{B_R} ug \, dx = \int_A u \, dx \,. \tag{7.161}$$

Moreover, if $\gamma = \sqrt[n]{\frac{1}{e_n}|A|}$ and $G_{(-\Delta)^m, B_R}$ denotes the Green function corresponding to $(-\Delta)^m$ in B_R, then for a.e. $x \in B_R$ we have by Boggio's explicit formula (2.65) for $G_{(-\Delta)^m, B_R}$ (see also (4.24))

$$|v(x)| = \left| \int_{B_R} G_{(-\Delta)^m, B_R}(x, y) \, 1_A(y) \, dy \right| \leq C \int_A |x-y|^{2m-n} \, dy$$

$$\leq C \int_{B_\gamma(x)} |x-y|^{2m-n} \, dy \leq C \int_0^\gamma \rho^{n-1+2m-n} \, d\rho$$

$$= C|A|^{2m/n},$$

where the C denote possibly different constants that only depend on n and m. This proves that

$$\|v\|_{L^\infty} \leq C(n, m)|A|^{2m/n} \,. \tag{7.162}$$

Finally, by Hölder and Sobolev inequalities, we have

$$\|v\|^2_{H_0^m} = (v, v)_{H_0^m} = \int_A v \, dx \leq \|v\|_{L^{s+1}} |A|^{(n+2m)/2n} \leq C(n, m)\|v\|_{H_0^m} |A|^{(n+2m)/2n} \,;$$

therefore,

$$\|v\|^2_{H_0^m} \leq C(n, m)|A|^{1+2m/n} \,. \tag{7.163}$$

Since $|A| \leq |B_R| = e_n R^n$, from (7.162) and (7.163) we get

$$\|v\|^2_{H_0^m} + C\|v\|^2_{L^\infty} \leq C(n, m, R)|A|^{4m/n} \,.$$

By arbitrariness of A, this inequality and (7.161) may be replaced into (7.160) to obtain

$$\|u\|^2_{H_0^m} \geq S\|u\|^2_{L^{s+1}} + \overline{C}\|u\|^2_{L_w^{\frac{n}{n-2m}}} \qquad \text{for all } u \in H_0^m(B_R) \cap \mathcal{K}, \tag{7.164}$$

for some suitable $\overline{C} > 0$.

Proof of (7.159) for sign-changing functions. Take a sign-changing function $u \in H_0^m(B_R)$. By Theorem 3.4 we infer that there exists a unique couple $(u_1, u_2) \in \mathcal{K} \times \mathcal{K}^*$ such that $u = u_1 + u_2$ and $(u_1, u_2)_{H_0^m} = 0$. Then

$$\|u\|_{H_0^m}^2 = \|u_1\|_{H_0^m}^2 + \|u_2\|_{H_0^m}^2. \tag{7.165}$$

Since $u_1 \geq 0$ and $u_2 < 0$ a.e. in B_R (by Proposition 3.6) we have $|u_1 + u_2| \leq \max\{|u_1|, |u_2|\}$ and, for all $r > 0$,

$$|u_1(x) + u_2(x)|^r \leq \max\{|u_1(x)|^r, |u_2(x)|^r\} \leq |u_1(x)|^r + |u_2(x)|^r \qquad \text{for a.e. } x \in B_R.$$

Furthermore, if $r \geq 2$ we obtain

$$\|u\|_{L^r}^2 = \left(\int_{B_R} |u_1 + u_2|^r \, dx \right)^{2/r} \leq \left(\int_{B_R} \left(|u_1|^r + |u_2|^r \right) dx \right)^{2/r}$$

$$\leq \|u_1\|_{L^r}^2 + \|u_2\|_{L^r}^2. \tag{7.166}$$

Since we already know that (7.164) holds for both u_1 and u_2, by (7.165) and (7.166), we get

$$\|u\|_{H_0^m}^2 = \|u_1\|_{H_0^m}^2 + \|u_2\|_{H_0^m}^2$$
$$\geq S(\|u_1\|_{L^{s+1}}^2 + \|u_2\|_{L^{s+1}}^2) + \overline{C}(\|u_1\|_{L_w^{\frac{n}{n-2m}}}^2 + \|u_2\|_{L_w^{\frac{n}{n-2m}}}^2)$$
$$\geq S\|u\|_{L^{s+1}}^2 + C\|u\|_{L_w^{\frac{n}{n-2m}}}^2$$

with $C = \overline{C}/2$. This completes the proof of (7.159). $\qquad\square$

Recalling the properties of the weak norms (see [235, p. 255] and [394, Section 4.2]), from Theorem 7.58 we immediately obtain

Corollary 7.59. *Let $m \in \mathbb{N}^+$ and let $\Omega \subset \mathbb{R}^n$ ($n > 2m$) be a bounded domain. Then for any $p \in [1, \frac{n}{n-2m})$ there exists a constant $C = C(\Omega, m, p) > 0$ such that*

$$\|u\|_{H_0^m}^2 \geq S\|u\|_{L^{s+1}}^2 + C\|u\|_{L^p}^2 \qquad \text{for all } u \in H_0^m(\Omega).$$

Comparing this result with Corollary 7.39 allows for the following interpretation. *On bounded domains, the Sobolev inequality for the embedding $H_0^m \subset L^{s+1}$ with optimal constant S may be improved by adding an L^2-norm if and only if the space dimension is weakly critical.*

Similar results also hold in other subspaces of $H^m(\Omega)$ such as $H_\vartheta^m(\Omega)$, see (7.35). We prove here two inequalities which are strictly related to existence and nonexistence results for critical growth problems.

Theorem 7.60. *Let $\Omega \subset \mathbb{R}^n$ ($n > 2m$) be a bounded C^m-smooth domain. Then for all $p \in [1, \frac{n}{n-2m})$ there exists a constant $C = C(n, p, |\Omega|) > 0$, such that*

$$\|u\|_{H_\vartheta^m}^2 \geq S\|u\|_{L^{s+1}}^2 + C\|u\|_{L^p}^2 \qquad \text{for all } u \in H_\vartheta^m(\Omega).$$

Let $B \subset \mathbb{R}^n$ ($n > 4$) denote the unit ball. Then

$$\|\Delta u\|_{L^2}^2 \geq S\|u\|_{L^{2n/(n-4)}}^2 + 4\int_{\partial B} u_\nu^2 \, d\omega \qquad \text{for all } u \in H^2 \cap H_0^1(B).$$

Proof. We keep $p \in [1, \frac{n}{n-2m})$ fixed as in the first statement. Let B denote the unit ball. By scaling and Talenti's comparison principle, see Theorem 3.12, it suffices to prove that there exists $C_B > 0$ with

$$\|u\|_{H_\vartheta^m}^2 \geq S\|u\|_{L^{s+1}}^2 + \frac{1}{C_B}\|u\|_{L^p}^2 \tag{7.167}$$

for each $u \in H_\vartheta^m(B)$, which is radially symmetric and radially decreasing with respect to the origin (and hence positive). Let us set

$$S_{\lambda,p} = \inf\left\{ \frac{\|v\|_{H_\vartheta^m}^2 - \lambda\|v\|_{L^p}^2}{\|v\|_{L^{s+1}}^2} ; v \in H_\vartheta^m(B) \text{ and } v \text{ radially decreasing} \right\}.$$

If by contradiction (7.167) does not hold, one gets $S_{\lambda,p} < S$ for all $\lambda > 0$. Let us fix some $\lambda \in (0, \lambda_{1,p})$, where

$$\lambda_{1,p} := \inf_{H_\vartheta^m(B)\setminus\{0\}} \frac{\|v\|_{H_\vartheta^m}^2}{\|v\|_{L^p}^2}$$

is the first positive eigenvalue of

$$\begin{cases} (-\Delta)^m u = \lambda\|u\|_{L^p}^{2-p}|u|^{p-2}u & \text{in } B, \\ \Delta^j u|_{\partial B} = 0 & \text{for } j = 0, \ldots, m-1. \end{cases}$$

Arguing along the lines of Sections 7.5.1 and 7.6 one finds a smooth positive radial strictly decreasing solution of

$$\begin{cases} (-\Delta)^m u = u^s + \lambda\|u\|_{L^p}^{2-p}|u|^{p-2}u & \text{in } B, \\ \Delta^j u|_{\partial B} = 0 & \text{for } j = 0, \ldots, m-1. \end{cases} \tag{7.168}$$

On the other hand, for any positive radially decreasing solution of (7.168) one has the following variant of the Pohožaev identity (7.74):

$$2\lambda \left(\frac{n}{p} - \frac{n-2m}{2} \right) \|u\|_{L^p}^2 = (-1)^{m+1} \sum_{j=1}^{m} \int_{\partial B} (\Delta^{j-1} u)_\nu (\Delta^{m-j} u)_\nu \, d\omega.$$

As in the proof of Theorem 7.54, we conclude

$$\lambda \|u\|_{L^p}^2 \geq c(n,p) \|\Delta^m u\|_{L^1}^2 \geq c(n,p) \|u\|_{L^p}^2,$$

since we have assumed that $p < n/(n-2m)$. For $\lambda > 0$ sufficiently small, we obtain a contradiction which proves (7.167).

Similarly, the second inequality is a direct consequence of the definition of Σ_a in (7.137) combined with Lemma 7.56 and the nonexistence statement (for $a \leq 4$) in Theorem 7.55. This shows in fact that $\Sigma_4 = S$. $\square$

7.9 Critical Growth Problems in Geometrically Complicated Domains

We consider the semilinear biharmonic equation with purely critical nonlinearity

$$\Delta^2 u = |u|^{8/(n-4)} u \qquad \text{in } \Omega \tag{7.169}$$

either with Navier boundary conditions

$$u = \Delta u = 0 \qquad \text{on } \partial\Omega \tag{7.170}$$

or with Dirichlet boundary conditions

$$u = |\nabla u| = 0 \qquad \text{on } \partial\Omega. \tag{7.171}$$

Here $\Omega \subset \mathbb{R}^n$ ($n > 4$) is a bounded smooth domain.

According to the results in Sections 7.5.1 and 7.6, in geometrically simple domains obstructions to existence of certain solutions (e.g. positive or, in balls, radially symmetric) have to be observed. Here, we address the question whether existence of "relatively simple" nontrivial solutions of (7.169) with one of the above boundary conditions can be proved in sufficiently complicated domains. In the second order analogue the question was intensively discussed, see e.g. [29, 128, 333], references therein and subsequent works. In the next two sections we will specify what is meant by "complicated" domains. First, we consider domains which are noncontractible and, for sake of completeness, we quote without proof results from [33, 157]. Then we consider contractible domains which are however "almost" noncontractible and we prove the corresponding statement in all details.

7.9.1 Existence Results in Domains with Nontrivial Topology

We consider here classes of domains having a nontrivial topology in a suitable sense. If Ω is an annulus and if one restricts the functions to be radial, the problem is no longer critical and both (7.169), (7.170) and (7.169), (7.171) admit a nontrivial solution. Therefore, it appears reasonable to consider domains which are in a topological sense "at least as complicated as an annulus". Although they are slightly beyond the scope of this book, for the reader's convenience we quote two results in this direction without giving proofs.

For the Dirichlet boundary conditions, only the existence of *nontrivial* solutions in domains Ω which are contained in suitable annuli could be shown. More precisely, we quote from Bartsch-Weth-Willem [33] the following statement.

Theorem 7.61. *Let $\Omega \subset \mathbb{R}^n$ ($n > 4$) be a bounded smooth domain. Then there exists $q > 1$ such that (7.169), (7.171) admits a nontrivial solution provided*

$$\{x \in \mathbb{R}^n; \rho \leq |x - a| \leq \rho q\} \subset \Omega \quad and \quad \{x \in \mathbb{R}^n; |x - a| < \rho\} \not\subset \Omega$$

for some $\rho > 0$ and $a \in \mathbb{R}^n$.

For a given domain $\Omega \subset \mathbb{R}^n$, let $H_k(\Omega, \mathbb{Z}_2)$ denote the homology of dimension k with $\mathbb{Z}_2$-coefficients. Then, as pointed out in [33], the assumption of Theorem 7.61 implies that $H_{n-1}(\Omega, \mathbb{Z}_2) \neq 0$. In fact, the statement in [33] holds for any polyharmonic Dirichlet problem (7.79) with $m \geq 2$ and $\lambda = 0$. The proof of Theorem 7.61 in [33] is obtained as a consequence of a more general result concerning nondegenerate critical manifolds of the associated energy functional. The isolated critical values are related to the Sobolev constant (7.44).

On the other hand, for the Navier boundary value problem Ebobisse-Ould Ahmedou [157] proved the following statement.

Theorem 7.62. *Let $\Omega \subset \mathbb{R}^n$ ($n > 4$) be a bounded smooth domain such that*

$$H_k(\Omega, \mathbb{Z}_2) \neq 0$$

for some positive integer k. Then (7.169), (7.170) admits a positive solution.

The proof of Theorem 7.62 in [157] follows the lines of the celebrated paper by Bahri-Coron [29] which deals with the second order critical growth problem. It is assumed for contradiction that (7.169), (7.170) admits no positive solution. Then it is possible to construct some nontrivial homomorphisms in the k-th homology. A careful expansion of the corresponding energy functional near its critical points at infinity (see (7.46)) enables one to show that whenever the "concentration parameter" ε tends to vanish, the above mentioned homomorphisms become trivial, giving a contradiction. We refer to [157] for the details.

7.9.2 *Existence Results in Contractible Domains*

In this section we are interested in topologically simple but geometrically complicated domains. We show that (7.169), (7.170) admits a *positive* solution in "strange" contractible domains Ω. As before, for the Dirichlet problem we merely show that (7.169), (7.171) admits a *nontrivial* solution in the same kind of contractible domain. On one hand our proof is inspired by strong arguments developed by Passaseo [333], on the other hand we have to face several hard difficulties, especially and somehow unexpectedly under Navier boundary conditions. One of the crucial steps in the approach by Passaseo is to prove that sign changing solutions of (7.169) "double the energy" of the associated functional. For the second order problem this may be shown by the usual technique of testing the equation with the positive and negative parts of the solution. Of course, this technique fails for (7.169) where higher order derivatives are involved. We overcome this difficulty thanks to the decomposition method in dual cones explained in Section 3.1.2. This method enables us to bypass the lack of nonexistence results for nodal solutions of (7.169) in starshaped domains. Moreover, when dealing with Navier boundary conditions, the required generalisation of the Struwe compactness lemma [381] turns out to be very delicate because of the second boundary datum and the lack of a uniform extension operator for $H^2 \cap H_0^1$–functions in families of domains. See Lemma 7.71 and its proof in Section 7.9.5. The same problem arises in Lemma 7.73 where a uniform lower bound for an enlarged optimal Sobolev constant has to be found in a suitable class of domains. An important tool for this lemma is a Sobolev inequality with optimal constant and remainder term, see Theorem 7.60. This inequality is closely related to nonexistence results, which have been discussed above.

For the sake of a precise formulation of the existence results we have to give a number of definitions.

Definition 7.63. Let $K \subset \mathbb{R}^n$ be bounded. We say that $u \in \mathscr{D}^{2,2}(\mathbb{R}^n)$ satisfies $u = 1$ on K in the sense of $\mathscr{D}^{2,2}(\mathbb{R}^n)$ if there exists a sequence (u_k) in $C_c^2(\mathbb{R}^n)$ such that $u_k = 1$ on K for each $k \in \mathbb{N}$ and $u_k \to u$ in $\mathscr{D}^{2,2}(\mathbb{R}^n)$.

Definition 7.64. We define the $(2,2)-capacity$ of K as

$$\operatorname{cap} K = \inf\left\{ \int_{\mathbb{R}^n} |\Delta u|^2\, dx : u = 1 \text{ on } K \text{ in the } \mathscr{D}^{2,2}(\mathbb{R}^n) \text{ sense} \right\}.$$

We set $\operatorname{cap} \emptyset := 0$.

Since the nonempty set

$$\left\{ u \in \mathscr{D}^{2,2}(\mathbb{R}^n) : \ u = 1 \text{ on } K \text{ in the } \mathscr{D}^{2,2}(\mathbb{R}^n) \text{ sense} \right\}$$

is closed and convex, there exists a unique function $z_K \in \mathscr{D}^{2,2}(\mathbb{R}^n)$ such that $z_K = 1$ on K in the sense of $\mathscr{D}^{2,2}(\mathbb{R}^n)$ and

$$\int_{\mathbb{R}^n} |\Delta z_K|^2 \, dx = \operatorname{cap} K.$$

Finally, we specify what we mean by set deformations.

Definition 7.65. Let $\widetilde{\Omega} \subset \mathbb{R}^n$ and let $H, \Omega \subset \widetilde{\Omega}$. We say that H *can be deformed in* $\widetilde{\Omega}$ *into a subset of* Ω if there exists a continuous function

$$\mathscr{H} : H \times [0,1] \to \widetilde{\Omega}$$

such that $\mathscr{H}(x,0) = x$ and $\mathscr{H}(x,1) \in \Omega$ for all $x \in H$.

Our first result states the existence of positive solutions for the critical growth equation (7.169) with Navier boundary conditions. Combined with Theorem 7.52, this shows that the existence of positive solution strongly depends on the geometry of the domain.

Theorem 7.66. *Let* $\widetilde{\Omega}$ *be a smooth bounded domain of* $\mathbb{R}^n$ *($n > 4$) and let H be a closed subset contained in* $\widetilde{\Omega}$. *Then there exists* $\varepsilon > 0$ *such that if* $\Omega \subset \widetilde{\Omega}$ *is a smooth domain with* $\operatorname{cap}(\widetilde{\Omega} \setminus \Omega) < \varepsilon$ *and such that H cannot be deformed in* $\widetilde{\Omega}$ *into a subset of* Ω *then there exists a positive solution of*

$$\begin{cases} \Delta^2 u = |u|^{8/(n-4)} u & \text{in } \Omega, \\ u = \Delta u = 0 & \text{on } \partial\Omega. \end{cases} \tag{7.172}$$

We will prove this result in Section 7.9.4.

Next we turn to Dirichlet boundary conditions. In this case, we merely show the existence of nontrivial solutions to (7.169). The lack of information about the sign of the solution is due to the lack of information about the sign of the corresponding Green function. This issue was discussed in detail in Chapter 6.

Theorem 7.67. *Let* $\widetilde{\Omega}$ *be a smooth bounded domain of* $\mathbb{R}^n$ *($n > 4$) and let H be a closed subset contained in* $\widetilde{\Omega}$. *Then there exists* $\varepsilon > 0$ *such that if* $\Omega \subset \widetilde{\Omega}$ *is a smooth domain with* $\operatorname{cap}(\widetilde{\Omega} \setminus \Omega) < \varepsilon$ *and such that H cannot be deformed in* $\widetilde{\Omega}$ *into a subset of* Ω *then there exists a nontrivial solution of*

$$\begin{cases} \Delta^2 u = |u|^{8/(n-4)} u & \text{in } \Omega, \\ u = |\nabla u| = 0 & \text{on } \partial\Omega. \end{cases} \tag{7.173}$$

The proof of this result, which can be obtained by arguing as in Section 7.9.4 below, is simpler than the one of Theorem 7.66. As we will explain below, for (7.172), one has to study very carefully the behaviour of suitable sequences (u_ℓ) such that $u_\ell \in H^2 \cap H_0^1(\Omega_\ell)$ for varying domains Ω_ℓ. In contrast with the spaces $H_0^2(\Omega_\ell)$, there is no obvious extension operator into $H^2(\mathbb{R}^n)$. Only as far as positivity of solutions is concerned, the problem with Dirichlet boundary conditions is more involved than with Navier boundary conditions. Here we have no positivity statement of the solution because Lemma 7.69 below does not seem to hold.

The solutions given by Theorems 7.61 and 7.62 are not related to the ones in Theorems 7.67 and 7.66 respectively. To explain this, assume that $\widetilde{\Omega}$ satisfies the assumptions of Theorems 7.61 or 7.62. If u_Ω is the solution of (7.172) or (7.173), then by exploiting the proof of Theorem 7.66 one has that u_Ω converges weakly to zero as $\mathrm{cap}(\widetilde{\Omega} \setminus \Omega) \to 0$. Hence, any nontrivial solution in $\widetilde{\Omega}$ cannot be obtained as limit of the solutions u_Ω in Ω. On the other hand, one expects the nontrivial solutions in $\widetilde{\Omega}$ to be stable and to remain under "small" perturbations $\Omega \subset \widetilde{\Omega}$. These solutions in Ω will be different from ours. The latter situation was studied in [128] for the second order problem.

Finally, let us observe that there are also *contractible* domains Ω which satisfy the assumptions of the preceding theorems. In the following example we describe precisely such a situation. Further examples may be adapted to the biharmonic setting from [333, pp. 39-41].

Example 7.68. Let $\widetilde{\Omega} \subset \mathbb{R}^n$, with $n > 5$, be an annular shaped domain and let us drill a sufficiently "thin" cylindrical hole along a segment in order to obtain the smooth contractible subdomain Ω. To be more precise, we assume that for ε small enough, $\widetilde{\Omega} \setminus \Omega_\varepsilon$ is contained in a cylinder with basis $B_\varepsilon \subset \mathbb{R}^{n-1}$ and fixed height. Then by simple scaling arguments one finds that $\mathrm{cap}\left(\widetilde{\Omega} \setminus \Omega_\varepsilon\right) = O(\varepsilon^{n-5}) \to 0$ as $\varepsilon \to 0$, provided that the dimension satisfies $n > 5$.

We choose H to be a spherical hypersurface in $\widetilde{\Omega}$, which cannot be deformed into a subset of Ω_ε. This can be seen by looking at the degree of mapping $d(\mathcal{H}(.,t),H,0)$ for $t \in [0,1]$, where $\mathcal{H}$ is assumed to exist according to Definition 7.65. Instead of the above segment of fixed length, one may consider any bounded piece of a fixed generalised plane, provided that its codimension is at least 5.

7.9.3 *Energy of Nodal Solutions*

One may extend the arguments used for Lemma 7.22 to prove a similar statement in any bounded domain in case of Navier boundary conditions. For Dirichlet boundary conditions this seems to be impossible due to the lack of information about the positivity of the corresponding Green's function.

Lemma 7.69. *Let Ω be a bounded domain of $\mathbb{R}^n$ and assume that u solves (7.172) and changes sign. Then* $\|\Delta u\|_{L^2}^2 \geq 2^{4/n} S \|u\|_{L^{2n/(n-4)}}^2$.

In the case of the half space, by exploiting nonexistence results for positive solutions, we have a stronger result. In fact, we believe that no *nontrivial* solution exists in the half space (either with Navier (7.170) or Dirichlet (7.171) boundary conditions), but the following result is enough for our purposes.

Lemma 7.70. *Let $\Omega = \{x_n > 0\}$ be the half space and assume that $u \in \mathscr{D}^{2,2}(\Omega) \cap C^2(\overline{\Omega})$ solves the equation*

$$\Delta^2 u = |u|^{8/(n-4)} u \quad \text{in } \{x_n > 0\} \tag{7.174}$$

with boundary data either (7.170) or (7.171). Then $\|\Delta u\|_{L^2}^2 \geq 2^{4/n} S \|u\|_{L^{2n/(n-4)}}^2$.

Proof. Notice first that by Theorem 7.52, which also holds in unbounded domains, problem (7.174), (7.170) does not admit positive solutions. A similar nonexistence result holds for boundary conditions (7.171), see Theorem 7.33. This result may be extended to unbounded domains. We refer to [353, Theorem 4].

Therefore, any nontrivial solution of (7.174), with boundary conditions (7.170) or (7.171), necessarily changes sign. We obtain the result by repeating the argument of the proof of Lemma 7.22. With Navier boundary conditions this is straightforward, while with Dirichlet boundary data one may invoke Boggio's principle in the half space, see (2.66). $\qquad\square$

From now on we restrict our attention to the $H^2 \cap H_0^1$–framework. We proceed again by constrained minimisation over the critical unit ball M defined in (7.130) but with some additional topological constraints. We put

$$M = \left\{ u \in H^2 \cap H_0^1(\Omega) : \int_\Omega |u|^{2n/(n-4)} \, dx = 1 \right\}$$

and we minimise the functional

$$F(u) := \int_\Omega |\Delta u|^2 \, dx$$

constrained on the manifold M. This functional was previously represented by F_0 in (7.41).

In the spirit of Definition 7.45, we say that $(u_\ell) \subset M$ is a Palais-Smale sequence for F at level c if

$$\lim_{\ell \to \infty} F(u_\ell) = c, \quad \lim_{\ell \to \infty} \|dF(u_\ell)\|_{(T_{u_\ell} M)'} \to 0$$

where $T_{u_\ell} M$ denotes the tangent space to the manifold M at u_ℓ.

We now state a global compactness result for the biharmonic operator. Even if the proof is similar to the one by Struwe [381], some difficulties arise. We postpone the proof, where we emphasise the main differences with respect to second order equations, until Section 7.9.5

Lemma 7.71. *Let $(u_\ell) \subset M$ be a Palais-Smale sequence for F at level $c \in \mathbb{R}$. Then for a suitable subsequence, one has the following alternative. Either (u_ℓ) has a strongly convergent subsequence in $H^2 \cap H_0^1(\Omega)$ or there exist k nonzero functions $\widehat{u}_1, \ldots, \widehat{u}_k \in \mathscr{D}^{2,2}$, solving either (7.45) or (7.174) with boundary conditions (7.170), and a solution $\widehat{u}_0 \in H^2 \cap H_0^1(\Omega)$ of (7.172) such that*

$$u_\ell \rightharpoonup \widehat{u}_0 \left(\sum_{j=0}^{k} \int |\widehat{u}_j|^{2n/(n-4)} \, dx \right)^{-(n-4)/2n} \quad \text{in } H^2 \cap H_0^1(\Omega) \tag{7.175}$$

and

$$\lim_{\ell \to \infty} F(u_\ell) = \left(\sum_{j=0}^{k} \int |\widehat{u}_j|^{2n/(n-4)} \, dx \right)^{4/n}. \tag{7.176}$$

The domain of integration for $\widehat{u}_0$ is just Ω, while for $\widehat{u}_1, \ldots, \widehat{u}_k$, it is either a half space or the whole $\mathbb{R}^n$.

When working in $H_0^2(\Omega)$ a similar result holds true with Dirichlet boundary conditions instead of Navier boundary conditions. Thanks to Lemma 7.71, we may prove the following compactness property in the second critical energy range.

Lemma 7.72. *Let $(u_\ell) \subset M$ be a Palais-Smale sequence for the functional F at level $c \in (S, 2^{4/n}S)$. Then, up to a subsequence, (u_ℓ) strongly converges in $H^2 \cap H_0^1(\Omega)$.*

Proof. Assume by contradiction that (u_ℓ) does not contain a convergent subsequence in $H^2 \cap H_0^1(\Omega)$. Then by Lemma 7.71 one finds functions $\widehat{u}_0 \in H^2 \cap H_0^1(\Omega)$ and $\widehat{u}_1, \ldots, \widehat{u}_k \in \mathscr{D}^{2,2}$ satisfying (7.175) and (7.176). Assume first that all $\widehat{u}_1, \ldots, \widehat{u}_k$ are positive solutions of (7.45). Then each $\widehat{u}_j$ is of type (7.46) and attains the best Sobolev constant, i.e.

$$\int_{\mathbb{R}^n} |\widehat{u}_j|^{2n/(n-4)} \, dx = \int_{\mathbb{R}^n} |\Delta \widehat{u}_j|^2 \, dx = S \left(\int_{\mathbb{R}^n} |\widehat{u}_j|^{2n/(n-4)} \, dx \right)^{(n-4)/n} \qquad j = 1, \ldots k, \tag{7.177}$$

which implies $\int_{\mathbb{R}^n} |\widehat{u}_j|^{2n/(n-4)} \, dx = S^{n/4}$ for $j = 1, \ldots, k$. In turn one obtains

$$\lim_{\ell \to \infty} F(u_\ell) = \left(\int_\Omega |\widehat{u}_0|^{2n/(n-4)} \, dx + k S^{n/4} \right)^{4/n}.$$

If $\widehat{u}_0 \equiv 0$, we get $F(u_\ell) \to k^{4/n} S$, while if $\widehat{u}_0 \not\equiv 0$ for each $k \geq 1$ one has

$$\left(\int_\Omega |\widehat{u}_0|^{2n/(n-4)} \, dx + k S^{n/4} \right)^{4/n} > (k+1)^{4/n} S.$$

In any case we have a contradiction to $S < \lim_{\ell \to \infty} F(u_\ell) < 2^{4/n} S$.

We now consider the case in which at least one $\widehat{u}_j$ is sign changing or a solution in the half space $\{x_n > 0\}$. By Lemmas 7.22 and 7.70 we have for these $\widehat{u}_j$

$$\int_{\mathbb{R}^n} |\widehat{u}_j|^{2n/(n-4)} \, dx = \int_{\mathbb{R}^n} |\Delta \widehat{u}_j|^2 \, dx \geq 2^{4/n} S \left(\int_{\mathbb{R}^n} |\widehat{u}_j|^{2n/(n-4)} \, dx \right)^{(n-4)/n},$$

and hence $\int_{\mathbb{R}^n} |\widehat{u}_j|^{2n/(n-4)} \, dx \geq 2 S^{n/4}$, while for the remaining $\widehat{u}_j$, (7.177) holds true. In any case, we have $\lim_{\ell \to \infty} F(u_\ell) \geq \left(2 S^{n/4} \right)^{4/n} = 2^{4/n} S$, again a contradiction. $\qquad \square$

7.9.4 The Deformation Argument

The proof of Theorem 7.66 is by far more involved than the proof of Theorem 7.67, because the trivial extension of any function $u \in H^2 \cap H_0^1(\Omega)$ by 0 does not yield a function in $H^2(\mathbb{R}^n)$. In particular, for the space $H_0^2(\Omega)$ Lemma 7.73 below easily follows by this extension argument. A further difference is that the positivity conclusion of Lemma 7.69 does not hold in the Dirichlet boundary value case. But in the latter case, one simply has to drop this argument. Therefore we only deal with the proof of Theorem 7.66.

For all smooth $\Omega \subset \widetilde{\Omega}$ let $\beta : M \to \mathbb{R}^n$ be the "barycenter" map

$$\beta(u) = \int_\Omega x |u(x)|^{2n/(n-4)} \, dx.$$

Since $\widetilde{\Omega}$ is smooth one finds $\widetilde{r} > 0$ such that $\widetilde{\Omega}$ is a deformation retract of

$$\widetilde{\Omega}_+ = \left\{ x \in \mathbb{R}^n : \ \mathrm{dist}(x, \widetilde{\Omega}) < \widetilde{r} \right\}.$$

First we show that the energy or, equivalently, the optimal Sobolev constants will remain relatively large if we prevent the functions from concentrating "too close" to their domain of definition.

Lemma 7.73.

$$\gamma := \inf \left\{ \inf \left\{ F(u) : \ u \in M, \beta(u) \notin \widetilde{\Omega}_+ \right\}, \Omega \text{ smooth subset of } \widetilde{\Omega} \right\} > S.$$

Proof. Assume by contradiction that for each $\varepsilon > 0$ there exists a smooth $\Omega_\varepsilon \subset \widetilde{\Omega}$ and $u_\varepsilon \in H^2 \cap H_0^1 \cap C^\infty(\Omega_\varepsilon)$ such that

$$\int_{\Omega_\varepsilon} |u_\varepsilon|^{2n/(n-4)} \, dx = 1 \tag{7.178}$$

$$\int_{\Omega_\varepsilon} |\Delta u_\varepsilon|^2 \, dx \leq S + \varepsilon \tag{7.179}$$

$$\beta(u_\varepsilon) \notin \widetilde{\Omega}_+. \tag{7.180}$$

Let $U_\varepsilon \in H^2(\mathbb{R}^n)$ be any entire extension of u_ε. Since Ω_ε is smooth, the existence of such an extension is well-known. We emphasise that the quantitative properties of U_ε outside Ω_ε (which are expected to blow up for $\varepsilon \searrow 0$) will *not* be used. Furthermore, let 1_{Ω_ε} be the characteristic function of Ω_ε.

Step I. We claim that $1_{\Omega_\varepsilon} U_\varepsilon \to 0$ in $L^2(\mathbb{R}^n)$ as $\varepsilon \to 0$.

By Theorem 7.60 there exists $C > 0$, independent of ε, such that

$$\|\Delta u\|_{L^2}^2 \geq S \|u\|_{L^{2n/(n-4)}}^2 + \frac{1}{C} \|u\|_{L^1}^2 \qquad \text{for all } u \in H^2 \cap H_0^1(\Omega_\varepsilon). \tag{7.181}$$

Putting together (7.178), (7.179) and (7.181) we have for each $\varepsilon > 0$

$$S + \varepsilon \geq \int_{\Omega_\varepsilon} |\Delta u_\varepsilon|^2 \, dx \geq S\|u_\varepsilon\|_{L^{2n/(n-4)}}^2 + \frac{1}{C}\|u_\varepsilon\|_{L^1}^2 = S + \frac{1}{C}\|1_{\Omega_\varepsilon} U_\varepsilon\|_{L^1}^2$$

so that $\|1_{\Omega_\varepsilon} U_\varepsilon\|_{L^1} \to 0$ as $\varepsilon \to 0$. By (7.178) and classical L^p–interpolation we also get $\|1_{\Omega_\varepsilon} U_\varepsilon\|_{L^2} \to 0$ as $\varepsilon \to 0$.

Step II. We show that the weak-$*$-limit (in the sense of measures) of $1_{\Omega_\varepsilon} |U_\varepsilon|^{2n/(n-4)}$ is a Dirac mass.

An integration by parts shows that $\|\nabla u_\varepsilon\|_{L^2(\Omega_\varepsilon)}^2 \leq \|\Delta u_\varepsilon\|_{L^2(\Omega_\varepsilon)} \|u_\varepsilon\|_{L^2(\Omega_\varepsilon)}$. Then by Step I and (7.179) we infer that

$$\|1_{\Omega_\varepsilon} \nabla U_\varepsilon\|_{L^2(\mathbb{R}^n)} = \|\nabla u_\varepsilon\|_{L^2(\Omega_\varepsilon)} \to 0.$$

Therefore, taking any $\varphi \in C_c^\infty(\mathbb{R}^n)$, we find as $\varepsilon \to 0$

$$\int_{\mathbb{R}^n} 1_{\Omega_\varepsilon} |\Delta(\varphi U_\varepsilon)|^2 \, dx = \int_{\mathbb{R}^n} 1_{\Omega_\varepsilon} |\varphi|^2 |\Delta U_\varepsilon|^2 \, dx + o(1). \tag{7.182}$$

By the $L^1(\mathbb{R}^n)$-boundedness of the sequences $(1_{\Omega_\varepsilon} |U_\varepsilon|^{2n/(n-4)})$ and $(1_{\Omega_\varepsilon} |\Delta U_\varepsilon|^2)$ there exist two bounded nonnegative measures ν, μ on $\mathbb{R}^n$ such that

$$1_{\Omega_\varepsilon} |U_\varepsilon|^{2n/(n-4)} \rightharpoonup^* \nu, \quad 1_{\Omega_\varepsilon} |\Delta U_\varepsilon|^2 \rightharpoonup^* \mu \tag{7.183}$$

in the sense of measures. By the Sobolev inequality in Ω_ε we have

$$S\left(\int_{\mathbb{R}^n} 1_{\Omega_\varepsilon} |U_\varepsilon|^{2n/(n-4)} |\varphi|^{2n/(n-4)} \, dx\right)^{(n-4)/n} \leq \int_{\mathbb{R}^n} 1_{\Omega_\varepsilon} |\Delta(\varphi U_\varepsilon)|^2 \, dx$$

for all $\varepsilon > 0$ and therefore, by (7.182) and (7.183), letting $\varepsilon \to 0$ yields

$$S\left(\int_{\mathbb{R}^n} |\varphi|^{2n/(n-4)} \, d\nu\right)^{(n-4)/n} \leq \int_{\mathbb{R}^n} |\varphi|^2 \, d\mu.$$

By (7.178) and (7.179) we also know that

$$S\left(\int_{\mathbb{R}^n} d\nu\right)^{(n-4)/n} = \int_{\mathbb{R}^n} d\mu.$$

Hence by [278, Lemma I.2] there exist $\bar{x} \in \overline{\widetilde{\Omega}}$ and $\sigma > 0$ such that $\nu = \sigma \delta_{\bar{x}}$. From (7.178) we see that $\sigma = 1$ and the claim follows.

    The contradiction to (7.180) is now obtained by means of Step II, since $\beta(u_\varepsilon) \to \bar{x}$. This completes the proof. $\qquad\square$

Proof of Theorem 7.66. According to Lemma 7.73, we may choose $\overline{\mu}$ such that $S < \overline{\mu} < \min\{2^{4/n}S, \gamma\}$. Let $\varphi \in C_c^\infty(B_1(0))$ be such that

$$\int_{B_1(0)} |\varphi|^{2n/(n-4)}\, dx = 1, \qquad \int_{B_1(0)} |\Delta\varphi|^2\, dx < \overline{\mu}. \tag{7.184}$$

Define for each $\sigma > 0$ and $y \in \mathbb{R}^n$ the function $\varphi_{\sigma,y} : \mathbb{R}^n \to \mathbb{R}$ by setting

$$\varphi_{\sigma,y}(x) = \varphi\left(\frac{x-y}{\sigma}\right)$$

where φ is set equal to zero outside $B_1(0)$. It is readily seen that there exists $\overline{\sigma} > 0$ such that $B_{\overline{\sigma}}(y) \subset \widetilde{\Omega}$ and hence $\varphi_{\overline{\sigma},y} \in C_c^\infty(\widetilde{\Omega})$ for each $y \in H$. For all $\Omega \subsetneqq \widetilde{\Omega}$ let $z_\Omega \in \mathscr{D}^{2,2}(\mathbb{R}^n)$ be such that $z_\Omega = 1$ in $\widetilde{\Omega}\setminus\Omega$ in the sense of $\mathscr{D}^{2,2}(\mathbb{R}^n)$ and

$$\int_{\mathbb{R}^n} |\Delta z_\Omega|^2\, dx = \operatorname{cap}(\widetilde{\Omega}\setminus\Omega),$$

see what follows Definition 7.64. Note that one has

$$(1-z_\Omega)T_y(\varphi) \in H^2 \cap H_0^1(\Omega) \quad \text{for all } y \in H,$$

where $T_y(\varphi) := \dfrac{\varphi_{\overline{\sigma},y}}{\|\varphi_{\overline{\sigma},y}\|_{L^{2n/(n-4)}}}$. Moreover, for each $\delta > 0$ there exists $\varepsilon > 0$ with

$$\sup_{y\in H} \|\Delta(z_\Omega T_y(\varphi))\|_{L^2} < \delta \tag{7.185}$$

whenever $\operatorname{cap}(\widetilde{\Omega}\setminus\Omega) < \varepsilon$. Then one gets

$$\|(1-z_\Omega)T_y(\varphi)\|_{L^{2n/(n-4)}} \neq 0 \qquad \text{for all } y \in H,$$

if ε is sufficiently small. So, we define the map $\Phi_\Omega : H \to M$ by

$$\Phi_\Omega(y) = \frac{(1-z_\Omega)\,T_y(\varphi)}{\|(1-z_\Omega)\,T_y(\varphi)\|_{L^{2n/(n-4)}}}.$$

Taking (7.184) and (7.185) into account, we find $\varepsilon > 0$ such that

$$\sup\{F(\Phi_\Omega(y)) : y \in H\} < \overline{\mu} \tag{7.186}$$

provided that $\operatorname{cap}(\widetilde{\Omega}\setminus\Omega) < \varepsilon$. From now on ε is fixed subject to the previous restrictions. Let $\Omega \subset \widetilde{\Omega}$ be such that $\operatorname{cap}(\widetilde{\Omega}\setminus\Omega) < \varepsilon$ and $r \in (0,\widetilde{r})$ be such that Ω is a deformation retract of

$$\Omega_+ = \{x \in \mathbb{R}^n : \operatorname{dist}(x,\Omega) < r\}.$$

As in Lemma 7.73 one obtains

$$\inf\{F(u):\ u\in M,\ \beta(u)\notin\Omega_+\}>S. \tag{7.187}$$

Notice that

$$\inf\{F(u):\ u\in M,\ \beta(u)\notin\Omega_+\}\le\sup\{F(\Phi_\Omega(y)):\ y\in H\}. \tag{7.188}$$

Otherwise, the map $\mathscr{R}:H\times[0,1]\to\widetilde{\Omega}$ given by

$$\mathscr{R}(y,t)=(1-t)y+t\beta(\Phi_\Omega(y))\qquad\text{for }y\in H\text{ and }t\in[0,1]$$

would deform H in $\widetilde{\Omega}$ into a subset of Ω_+ and then into a subset of Ω (Ω is a deformation retract of Ω_+) contradicting the assumptions. Here, in order to see that $\mathscr{R}(y,t)\in\widetilde{\Omega}$, we used the fact that $\text{support}(\Phi_\Omega(y))\subset B_{\overline{\sigma}}(y)$ and that, since $\varphi\ge 0$, $\beta(\Phi_\Omega(y))\in B_{\overline{\sigma}}(y)\subset\widetilde{\Omega}$.

Therefore, by combining (7.186), (7.187) and (7.188) one gets

$$S<\inf\{F(u):\ u\in M,\ \beta(u)\notin\Omega_+\}\le\sup\{F(\Phi_\Omega(y)):\ y\in H\}$$
$$<\overline{\mu}<\gamma\le\inf\left\{F(u):\ u\in M,\ \beta(u)\notin\widetilde{\Omega}_+\right\}.$$

In what follows we need to find two appropriate levels, such that the corresponding sublevel sets

$$F^c=\{u\in M:\ F(u)\le c\}$$

cannot be deformed into each other. For this purpose let $c_1,c_2>S$ be such that

$$c_1<\inf\{F(u):\ u\in M,\ \beta(u)\notin\Omega_+\},$$

$$c_2=\sup\{F(\Phi_\Omega(y)):\ y\in H\}.$$

Assume by contradiction that there exists a deformation ϑ of F^{c_2} into F^{c_1}, i.e.

$$\vartheta:F^{c_2}\times[0,1]\to F^{c_2},\quad \vartheta(.,0)=Id_{F^{c_2}}\ \text{ and }\ \vartheta(F^{c_2},1)\subset F^{c_1}.$$

We define $\mathscr{H}:H\times[0,1]\to\widetilde{\Omega}$ by setting for each $x\in H$

$$\mathscr{H}(x,t)=\begin{cases}(1-3t)x+3t\beta(\Phi_\Omega(x)) & \text{if }t\in[0,1/3],\\ \widetilde{\rho}(\beta(\vartheta(\Phi_\Omega(x),3t-1))) & \text{if }t\in[1/3,2/3],\\ \widetilde{\rho}\,(\rho\,(\beta\,(\vartheta(\Phi_\Omega(x),1)),3t-2)) & \text{if }t\in[2/3,1],\end{cases}$$

where $\widetilde{\rho}:\widetilde{\Omega}_+\to\widetilde{\Omega}$ is a retraction and $\rho:\Omega_+\times[0,1]\to\Omega_+$ is a continuous map with $\rho(x,0)=x$ and $\rho(x,1)\in\Omega$ for all $x\in\Omega_+$. In order to see that $\mathscr{H}(x,t)\in\widetilde{\Omega}$, one should observe that $c_2 < \gamma$ and $\vartheta(\Phi_\Omega(x), 3t - 1) \in F^{c_2}$, hence we find $\beta(\vartheta(\Phi_\Omega(x), 3t - 1)) \in \widetilde{\Omega}_+$.

As $\vartheta(\Phi_\Omega(x), 1) \in F^{c_1}$ for each $x \in H$ and

$$c_1 < \inf\{F(u) : u \in M, \beta(u) \notin \Omega_+\},$$

then for each $x \in H$

$$\beta(\vartheta(\Phi_\Omega(x), 1)) \in \Omega_+$$

and $\mathscr{H}$ is a deformation of H in $\widetilde{\Omega}$ into a subset of Ω, a contradiction to our assumptions. Then the sublevel set

$$F^{c_2} = \{u \in M : F(u) \le c_2\}$$

cannot be deformed into

$$F^{c_1} = \{u \in M : F(u) \le c_1\}.$$

Hence, by combining Lemma 7.72 with the standard deformation lemma, see [329, Theorem 4.6] and also [140, Lemma 27.2], [382, Ch. II, Theorem 3.11], one obtains a constrained critical point u_Ω such that

$$S < F(u_\Omega) \le \sup\{F(\Phi_\Omega(y)) : y \in H\} < \overline{\mu} < 2^{4/n}S.$$

Finally, u_Ω does not change sign by Lemma 7.69. $\qquad\square$

7.9.5 A Struwe-Type Compactness Result

In this section we give the proof of Lemma 7.71. In fact, we prove the corresponding statement for the "free" functional

$$E_\Omega(u) = \frac{1}{2} \int_\Omega |\Delta u|^2 \, dx - \frac{n-4}{2n} \int_\Omega |u|^{2n/(n-4)} \, dx$$

which is defined on the whole space $H^2 \cap H^1_0(\Omega)$. More precisely, we have

Lemma 7.74. *Let $(u_\ell) \subset H^2 \cap H^1_0(\Omega)$ be a Palais-Smale sequence for E_Ω at level $c \in \mathbb{R}$. Then either (u_ℓ) has a strongly convergent subsequence in $H^2 \cap H^1_0(\Omega)$ or there exist $k > 0$ nonzero functions $\widehat{u}_j \in \mathscr{D}^{2,2}(\Omega_{0,j})$, $j = 1, \ldots, k$, with $\Omega_{0,j}$ either the whole $\mathbb{R}^n$ or a half space, solving either (7.45) or (7.174) with boundary conditions (7.170) and a solution $\widehat{u}_0 \in H^2 \cap H^1_0(\Omega)$ of (7.172) such that, up to a subsequence, as $\ell \to \infty$ we have*

$$u_\ell \rightharpoonup \widehat{u}_0 \text{ in } H^2 \cap H_0^1(\Omega),$$

$$\|\Delta u_\ell\|_{L^2}^2 \to \|\Delta \widehat{u}_0\|_{L^2}^2 + \sum_{j=1}^{k} \|\Delta \widehat{u}_j\|_{L^2}^2,$$

$$E_\Omega(u_\ell) \to E_\Omega(\widehat{u}_0) + \sum_{j=1}^{k} E_{\Omega_{0,j}}(\widehat{u}_j).$$

Proof. The proof is lengthy and delicate, so we divide it into seven steps.
Step I. Reduction to the case $u_\ell \rightharpoonup 0$.

It follows as in Lemma 7.46, see (7.99), that there exists $u \in H^2 \cap H_0^1(\Omega)$ such that $dE_\Omega(u) = 0$ and, up to a subsequence, one has

$$u_\ell \rightharpoonup u \text{ in } H^2 \cap H_0^1(\Omega), \qquad u_\ell \rightharpoonup u \text{ in } L^{s+1}(\Omega),$$

$$u_\ell \to u \text{ in } L^p(\Omega), \ p < s+1, \quad u_\ell \to u \text{ a.e. in } \Omega.$$

As in (7.102) one has

$$E_\Omega(u_\ell) - E_\Omega(u_\ell - u) = E_\Omega(u) + o(1)$$

and, like there, Vitali's convergence theorem proves that

$$\int_\Omega |u_\ell|^{s+1} \, dx - \int_\Omega |u_\ell - u|^{s+1} \, dx = \int_\Omega |u|^{s+1} \, dx + o(1).$$

Together with $dE_\Omega(u) = 0$, this yields

$$dE_\Omega(u_\ell) - dE_\Omega(u_\ell - u) \to 0 \qquad \text{in } (H^2 \cap H_0^1(\Omega))'$$

so that also $(u_\ell - u)$ is a Palais-Smale sequence for E_Ω. Therefore, we may assume that

$$u_\ell \rightharpoonup 0, \qquad E_\Omega(u_\ell) \to c \geq \frac{2}{n} S^{n/4}. \tag{7.189}$$

Indeed, if $c < \frac{2}{n} S^{n/4}$, then we are in the compactness range of E_Ω and by arguing as in Lemma 7.46, we infer that $u_\ell \to 0$ up to a subsequence. In this case the statement follows with $k = 0$.

As in (7.105) the Palais-Smale properties of (u_ℓ) give

$$E_\Omega(u_\ell) = \frac{2}{n} \int_\Omega |\Delta u_\ell|^2 \, dx + o(1).$$

Together with (7.189) this yields

$$\int_\Omega |\Delta u_\ell|^2 \, dx \geq S^{n/4} + o(1). \tag{7.190}$$

Let $L \in \mathbb{N}$ be such that $B_2(0)$ is covered by L balls of radius 1. By continuity of the maps

$$y \mapsto \int_{B_{1/R}(y) \cap \Omega} |\Delta u_\ell|^2 \, dx, \qquad R \mapsto \sup_{y \in \Omega} \int_{B_{1/R}(y) \cap \Omega} |\Delta u_\ell|^2 \, dx$$

and (7.190), for ℓ large enough one finds $R_\ell > 1/\mathrm{diam}(\Omega)$ and $x^\ell \in \overline{\Omega}$ such that

$$\int_{B_{1/R_\ell}(x^\ell) \cap \Omega} |\Delta u_\ell|^2 \, dx = \sup_{y \in \Omega} \int_{B_{1/R_\ell}(y) \cap \Omega} |\Delta u_\ell|^2 \, dx = \frac{S^{n/4}}{2L}. \tag{7.191}$$

When passing to a suitable subsequence, three cases may occur.

Case I. $R_\ell \to +\infty$ and $(R_\ell \, \mathrm{dist}(x^\ell, \partial\Omega))$ is bounded;

Case II. $R_\ell \, \mathrm{dist}(x^\ell, \partial\Omega) \to +\infty$;

Case III. (R_ℓ) is bounded.

Step II. Preliminaries for Case I.

For every $x \in \mathbb{R}^n$ let us write x' for its projection onto $\mathbb{R}^{n-1}$, so that $x = (x', x_n)$. Since $\mathrm{dist}(x^\ell, \partial\Omega) \to 0$ we find that, up to a subsequence $x^\ell \to x_0 \in \partial\Omega$ and $\rho_\ell := R_\ell \, \mathrm{dist}(x^\ell, \partial\Omega) \to \rho$. Moreover, for sufficiently large ℓ there exists a unique $y^\ell \in \partial\Omega$ such that $\mathrm{dist}(x^\ell, \partial\Omega) = |y^\ell - x^\ell|$.

Up to a rotation and a translation, we may assume that $x_0 = 0$ and that $-\eta_n = (0', -1)$ is the exterior unit normal to $\partial\Omega$ in 0. Then $y^\ell \to 0$ and the exterior unit normal to $\partial\Omega$ in y^ℓ converges to $-\eta_n$. We now translate and scale the domain Ω by setting

$$\Omega_\ell := R_\ell(\Omega - x^\ell),$$

so that x^ℓ is mapped into the origin 0 while y^ℓ is mapped into a point at distance ρ_ℓ from the origin and converges to the point $(0', -\rho)$ as $\ell \to \infty$. In view of the smoothness of $\partial\Omega$ we have that Ω_ℓ converges locally uniformly to

$$\Omega_0 = \left\{ (x', x_n) \in \mathbb{R}^n : x_n > -\rho \right\}.$$

In particular, for every $\varphi \in C_c^\infty(\Omega_0)$ we also have that $\varphi \in C_c^\infty(\Omega_\ell)$ for sufficiently large ℓ. This will be used below. Let us now set

$$v_\ell(x) = R_\ell^{\frac{4-n}{2}} u_\ell\left(x^\ell + \frac{x}{R_\ell}\right), \tag{7.192}$$

so that $v_\ell \in H^2 \cap H_0^1(\Omega_\ell)$ and, by (7.191),

$$\sup_{y \in \mathbb{R}^n} \int_{\Omega_\ell \cap B_1(y)} |\Delta v_\ell|^2 \, dx = \int_{\Omega_\ell \cap B_1(0)} |\Delta v_\ell|^2 \, dx = \frac{S^{n/4}}{2L}. \tag{7.193}$$

By boundedness of (u_ℓ) we infer that there exists $C > 0$ such that $\|\Delta v_\ell\|_{L^2(\Omega_\ell)} \le C$. Let 1_{Ω_ℓ} denote the characteristic function of Ω_ℓ. Then the sequence $(1_{\Omega_\ell} v_\ell)$ is bounded in $\mathscr{D}^{1,2n/(n-2)}(\mathbb{R}^n)$, and in $L^{2n/(n-4)}(\mathbb{R}^n)$, so that, up to a subsequence, we have

$$1_{\Omega_\ell} v_\ell \rightharpoonup v_0 \quad \text{in } \mathscr{D}^{1,2n/(n-2)} \cap L^{2n/(n-4)}(\mathbb{R}^n) \tag{7.194}$$

where $\mathrm{support}(v_0) \subset \overline{\Omega_0}$ and $v_0|_{x_n=-\rho} = 0$. Moreover, since $(1_{\Omega_\ell} D^2 v_\ell)$ is bounded in $L^2(\mathbb{R}^n)$, by weak continuity of distributional derivatives, up to a further subsequence we deduce

$$\int_{\Omega_\ell} D^2_{ij} v_\ell \, \varphi \, dx \to \int_{\Omega_0} D^2_{ij} v_0 \, \varphi \, dx$$

for all $\varphi \in C_c^\infty(\Omega_0)$ and $i, j = 1, \ldots, n$. In particular, v_0 has in Ω_0 second order weak derivatives.

Step III. The limiting function v_0 in (7.194) solves (7.172) in Ω_0.

Fix $\varphi \in C_c^\infty(\Omega_0)$. Then for ℓ large enough we have $\mathrm{support}(\varphi) \subset \Omega_\ell$. Define $\varphi_\ell \in C_c^\infty(\Omega)$ by setting

$$\varphi_\ell(x) = R_\ell^{-\frac{n-4}{2}} \varphi\left(R_\ell(x - x^\ell)\right).$$

Therefore, $(D^2 \varphi_\ell)$ being bounded in $L^2(\Omega)$ and taking into account (7.192), one obtains

$$o(1) = \langle dE_\Omega(u_\ell), \varphi_\ell \rangle \tag{7.195}$$

$$= R_\ell^{\frac{n}{2}} \int_\Omega \Delta u_\ell \Delta \varphi\left(R_\ell(x - x^\ell)\right) dx - R_\ell^{-\frac{n-4}{2}} \int_\Omega |u_\ell|^{8/(n-4)} u_\ell \varphi\left(R_\ell(x - x^\ell)\right) dx$$

$$= \int_{\mathbb{R}^n} \Delta v_\ell \Delta \varphi \, dx - \int_{\mathbb{R}^n} |v_\ell|^{8/(n-4)} v_\ell \varphi \, dx \tag{7.196}$$

$$= \int_{\mathbb{R}^n} \Delta v_0 \Delta \varphi \, dx - \int_{\mathbb{R}^n} |v_0|^{8/(n-4)} v_0 \varphi \, dx + o(1).$$

Then $v_0 \in \mathscr{D}^{2,2}(\overline{\Omega_0})$ solves (7.172) in distributional sense. The delicate point is to see that $\Delta v_0 = 0$ on $\partial\Omega_0$. To this end, let $\varphi \in C_c^2(\mathbb{R}^n)$ with $\varphi = 0$ on $\partial\Omega_0$ and let $\mathscr{U}$ be a neighbourhood of the support of φ. Then for ℓ large enough we find smooth diffeomorphisms χ_ℓ such that

$$\chi_\ell(\mathscr{U} \cap \Omega_0) = \chi_\ell(\mathscr{U}) \cap \Omega_\ell, \qquad \chi_\ell(\mathscr{U} \cap \partial\Omega_0) = \chi_\ell(\mathscr{U}) \cap \partial\Omega_\ell$$

and (χ_ℓ) converges locally uniformly to the identity as $\ell \to \infty$. For $x \in \chi_\ell(\mathscr{U})$ we define $\varphi_\ell(x) = \varphi(\chi_\ell^{-1}(x))$ and extend φ_ℓ by 0 to a $C_c^2(\mathbb{R}^n)$-function such that $\varphi_\ell = 0$ on $\partial\Omega_\ell$. Since (χ_ℓ^{-1}) also tends to the identity, we obtain $\varphi_\ell \to \varphi$ in $C^2(\mathbb{R}^n)$ and

$$\int_{\Omega_0} (\Delta v_0 \Delta \varphi - |v_0|^{8/(n-4)} v_0 \varphi) \, dx = \lim_{\ell \to \infty} \int_{\Omega_\ell} (\Delta v_\ell \Delta \varphi_\ell - |v_\ell|^{8/(n-4)} v_\ell \varphi_\ell) \, dx$$

$$= \lim_{\ell \to \infty} \int_{\Omega} \left[\Delta u_\ell \Delta \left(R_\ell^{\frac{n-4}{2}} \varphi_\ell(R_\ell(x - x^\ell)) \right) - |u_\ell|^{8/(n-4)} u_\ell R_\ell^{\frac{n-4}{2}} \varphi_\ell(R_\ell(x - x^\ell)) \right] dx$$

$$= \lim_{\ell \to \infty} \left\langle dE_\Omega(u_\ell), R_\ell^{\frac{n-4}{2}} \varphi_\ell(R_\ell(x - x^\ell)) \right\rangle = 0.$$

Therefore, by extending any given $\varphi \in H^2 \cap H_0^1(\Omega_0)$ with bounded support oddly with respect to $\partial \Omega_0$ as a function of $H^2(\mathbb{R}^n)$ and then by approximating it by a sequence of C^2 functions (φ_k) with $\varphi_k = 0$ on $\partial \Omega_0$, we get

$$\int_{\Omega_0} \Delta v_0 \Delta \varphi \, dx = \int_{\Omega_0} |v_0|^{8/(n-4)} v_0 \varphi \, dx \quad \text{for all } \varphi \in H_0^1 \cap H^2(\Omega_0),$$

which implies that v_0 is also a solution of (7.172) in Ω_0.

Step IV. The limiting function v_0 in (7.194) is nontrivial.

Let $\varphi \in C_c^\infty(\mathbb{R}^n)$ such that $\Omega_0 \cap \text{support}(\varphi) \neq \emptyset$; then for ℓ large enough, we may define

$$\widetilde{v}_{0,\ell} : \overline{\Omega_\ell} \cap \text{support}(\varphi) \to \mathbb{R}, \qquad \widetilde{v}_{0,\ell}(x) = v_0 \left(\chi_\ell^{-1}(x) \right). \tag{7.197}$$

Since $v_0 \in C^2(\overline{\Omega_0})$, $v_0 = \Delta v_0 = 0$ on $\partial \Omega_0$ and (χ_ℓ) converges to the identity, we get as $\ell \to +\infty$

$$\| 1_{\Omega_0} v_0 - 1_{\Omega_\ell} \widetilde{v}_{0,\ell} \|_{L^2(\text{support}(\varphi))} = o(1),$$

$$\| 1_{\Omega_0} v_0 - 1_{\Omega_\ell} \widetilde{v}_{0,\ell} \|_{L^{2n/(n-4)}(\text{support}(\varphi))} = o(1),$$

$$\| 1_{\Omega_0} \nabla v_0 - 1_{\Omega_\ell} \nabla \widetilde{v}_{0,\ell} \|_{L^2(\text{support}(\varphi))} = o(1), \tag{7.198}$$

$$\| 1_{\Omega_0} \Delta v_0 - 1_{\Omega_\ell} \Delta \widetilde{v}_{0,\ell} \|_{L^2(\text{support}(\varphi))} = o(1).$$

To simplify the notations we omit in what follows the 1_{Ω_ℓ} in front of $v_\ell, \nabla v_\ell, \Delta v_\ell$ and the 1_{Ω_0} in front of $v_0, \nabla v_0, \Delta v_0$. Then by (7.198) and some computations one obtains, as $\ell \to +\infty$

$$\int_{\mathbb{R}^n} |\Delta(\varphi v_0 - \varphi v_\ell)|^2 \, dx = \int_{\Omega_\ell \cap \text{support}(\varphi)} |\Delta(\varphi \widetilde{v}_{0,\ell} - \varphi v_\ell)|^2 \, dx + o(1)$$

$$\geq S \left(\int_{\Omega_\ell \cap \text{support}(\varphi)} |\varphi(\widetilde{v}_{0,\ell} - v_0)|^{2n/(n-4)} \, dx \right)^{(n-4)/n} + o(1)$$

$$\geq S \left(\int_{\mathbb{R}^n} |\varphi(v_0 - v_\ell)|^{2n/(n-4)} \, dx \right)^{(n-4)/n} + o(1).$$

By compact embedding one has $v_\ell \to v_0$ in $L^2_{loc}(\mathbb{R}^n)$ and thanks to an integration by parts one gets

$$\|\nabla(\varphi(v_\ell - v_0))\|_{L^2} \leq \|\varphi(v_\ell - v_0)\|_{L^2}\|\Delta(\varphi(v_\ell - v_0))\|_{L^2} \tag{7.199}$$

and therefore $\nabla v_\ell \to \nabla v_0$ in $L^2_{loc}(\mathbb{R}^n)$. Hence the previous inequality yields

$$\int_{\mathbb{R}^n} \varphi^2 |\Delta(v_0 - v_\ell)|^2\, dx \geq S \left(\int_{\mathbb{R}^n} |\varphi|^{2n/(n-4)} |v_0 - v_\ell|^{2n/(n-4)}\, dx \right)^{(n-4)/n} + o(1),$$

which implies

$$\int_{\mathbb{R}^n} \varphi^2\, d\mu \geq S \left(\int_{\mathbb{R}^n} |\varphi|^{2n/(n-4)}\, dv \right)^{(n-4)/n}, \tag{7.200}$$

where $d\mu$ and dv denote respectively the weak-$*$-limits of $|\Delta(v_0 - v_\ell)|^2$ and $|v_0 - v_\ell|^{2n/(n-4)}$ in the sense of measures. By the concentration-compactness principle of Lions [278, Lemma I.2] we know that there exists an at most countable set J and two families $(y_j)_{j\in J} \subset \mathbb{R}^n$ and $(v_j)_{j\in J} \subset (0, +\infty)$ such that $dv = \sum_{j\in J} v_j \delta_{y_j}$, where δ_y is the Dirac distribution supported at y. In fact, since dv is a finite measure satisfying (7.200), also the set J is finite. See [178, Formula (2.8)] for the details. Therefore, there exists a finite number $N \in \mathbb{N}$ such that

$$|v_0 - v_\ell|^{2n/(n-4)} \rightharpoonup^* dv = \sum_{j=1}^{N} v_j \delta_{y_j}, \tag{7.201}$$

the convergence $\rightharpoonup^*$ being in the weak-$*$-sense of measures.

In order to show that $v_0 \not\equiv 0$ we assume by contradiction that $v_0 \equiv 0$. Since the number of Dirac masses in (7.201) is finite, we may choose $\tau \in (0, \frac{1}{2})$ such that $y_j \notin (\overline{B}_{1+3\tau}(0) \setminus \overline{B}_1(0))$ for all $j = 1, \ldots, N$. For simplicity we write from now on $B_r = B_r(0)$. For any $\varphi \in C_c^\infty(\mathbb{R}^n)$ such that $\text{support}(\varphi) \subset (\overline{B}_{1+3\tau} \setminus \overline{B}_1)$ we know by (7.201), recalling $v_0 \equiv 0$, that

$$\int_{\Omega_\ell} \varphi |v_\ell|^{2n/(n-4)}\, dx \to 0 \qquad \text{as } \ell \to +\infty.$$

Hence, using φv_ℓ as test function in (7.196) and observing (7.199), we see that also

$$\int_{\Omega_\ell} \varphi |\Delta v_\ell|^2\, dx \to 0 \qquad \text{as } \ell \to +\infty.$$

In particular, this shows that

$$\int_{\Omega_\ell \cap (B_{1+2\tau} \setminus B_{1+\tau})} |\Delta v_\ell|^2\, dx \to 0 \qquad \text{as } \ell \to +\infty. \tag{7.202}$$

Take now $\psi \in C_c^\infty(\mathbb{R}^n; [0,1])$ such that $\psi = 1$ in $B_{1+\tau}$ and $\psi = 0$ outside $B_{1+2\tau}$. Since (ψv_ℓ) is bounded, we may use it as test function in (7.196) and obtain

$$o(1) = \int_{\Omega_\ell \cap B_{1+2\tau}} |\Delta(\psi v_\ell)|^2 \, dx - \int_{\Omega_\ell \cap B_{1+2\tau}} |\psi v_\ell|^{2n/(n-4)} \, dx,$$

where we used (7.199) and (7.202). The latter allows us to handle the region $\overline{B}_{1+2\tau} \setminus B_{1+\tau}$ where different powers of ψ and its derivatives play a role. Therefore, by applying to $\psi v_\ell \in H^2 \cap H_0^1(\Omega_\ell \cap B_{1+2\tau})$ the Sobolev inequality derived from (7.61), we infer

$$\left(1 - S^{-\frac{n}{n-4}} \left[\int_{\Omega_\ell \cap B_{1+2\tau}} |\Delta(\psi v_\ell)|^2 \, dx\right]^{\frac{4}{n-4}}\right) \int_{\Omega_\ell \cap B_{1+2\tau}} |\Delta(\psi v_\ell)|^2 \, dx \le o(1).$$

$$(7.203)$$

On the other hand, from (7.193), (7.199), and (7.202) we deduce

$$\int_{\Omega_\ell \cap B_{1+2\tau}} |\Delta(\psi v_\ell)|^2 \, dx \le \int_{\Omega_\ell \cap B_2} |\Delta v_\ell|^2 \, dx + o(1)$$

$$\le L \int_{\Omega_\ell \cap B_1} |\Delta v_\ell|^2 \, dx + o(1) \le \frac{S^{n/4}}{2} + o(1).$$

This, combined with (7.203), yields

$$\left(1 - \frac{1}{2^{4/(n-4)}}\right) \int_{\Omega_\ell \cap B_{1+2\tau}} |\Delta(\psi v_\ell)|^2 \, dx \le o(1),$$

so that

$$\int_{\Omega_\ell \cap B_{1+2\tau}} |\Delta(\psi v_\ell)|^2 \, dx \to 0 \quad \text{as } \ell \to +\infty.$$

In turn, recalling the shape of ψ, this yields

$$\lim_{\ell \to \infty} \int_{\Omega_\ell \cap B_1} |\Delta v_\ell|^2 \, dx = 0$$

which contradicts (7.193) and shows that $v_0 \not\equiv 0$.

Step V. In Case II, a solution of (7.45) appears.

One can proceed similarly as in steps II-IV. However, since in this case we have $\rho_\ell = R_\ell \mathrm{dist}(x_\ell, \partial\Omega) \to \infty$, the boundary $\partial\Omega$ disappears at infinity. Therefore, the previous arguments simplify because one does not have to consider the boundary $\partial\Omega_0$. Defining v_ℓ as in (7.192), one finds that (v_ℓ) converges to a solution $v_0 \in \mathscr{D}^{2,2}(\mathbb{R}^n)$ of (7.45).

Step VI. Case III cannot occur.

By contradiction, assume that (R_ℓ) is bounded. Since $R_\ell > 1/\mathrm{diam}(\Omega)$ we may assume that, up to a subsequence,

$$x^\ell \to x_0 \in \overline{\Omega}, \qquad R_\ell \to R_0 > 0.$$

Let us set $\Omega_0 = R_0(\Omega - x_0)$ and

$$v_\ell = R_0^{\frac{4-n}{2}} u_\ell \left(x_0 + \frac{x}{R_0} \right).$$

As in Case I (with the obvious simplifications), one gets for a subsequence $v_\ell \rightharpoonup v_0 \in H^2 \cap H_0^1(\Omega_0)$ and that $v_0 \neq 0$ solves (7.172) in Ω_0. But this is absurd since $u_\ell \rightharpoonup 0$.

Step VII. Conclusion.

If (u_ℓ) is a Palais-Smale sequence for E_Ω, then by Step I its weak limit $\widehat{u}_0$ solves (7.172). By Steps III, IV and V the "remaining part" $(u_\ell - \widehat{u}_0)$, suitably scaled, gives rise to a nontrivial solution v_0 of (7.45) (if Case II occurs) or (7.174) (if Case I occurs). With the help of this solution, we construct from $(u_\ell - \widehat{u}_0)$ a new Palais-Smale sequence (w_ℓ) for E_Ω in $H^2 \cap H_0^1(\Omega)$ at a strictly lower energy level. In the case where v_0 is a solution of (7.174) in a half space Ω_0 we again have to use the locally deformed versions $\widetilde{v}_{0,\ell}$ in Ω_ℓ of v_0. These have been defined in (7.197). Let $\varphi \in C_c^\infty(\mathbb{R}^n)$ be any cut–off function with $0 \leq \varphi \leq 1$, $\varphi = 1$ in $B_1(0)$ and $\varphi = 0$ outside $B_2(0)$. We put

$$w_\ell(x) := (u_\ell - \widehat{u}_0)(x) - R_\ell^{(n-4)/2} \widetilde{v}_{0,\ell} \left(R_\ell(x - x^\ell) \right) \varphi \left(\sqrt{R_\ell}(x - x^\ell) \right).$$

First we remark that w_ℓ is well-defined since $\varphi = 0$ for $R_\ell|x - x^\ell| \geq 2\sqrt{R_\ell}$ and since the domain of definition of $\widetilde{v}_{0,\ell}$ grows at rate R_ℓ. Further we notice that χ_ℓ and χ_ℓ^{-1} converge uniformly to the identity even on $B_{2\sqrt{R_\ell}}(0)$. For this reason we have

$$\widetilde{v}_{0,\ell}(\,\cdot\,) \varphi \left(\frac{\cdot}{\sqrt{R_\ell}} \right) \to v_0 \quad \text{in } \mathscr{D}^{2,2}(\overline{\Omega_0})$$

and also in $\mathscr{D}^{1,2n/(n-2)}(\overline{\Omega_0})$ and in $L^{2n/(n-4)}(\Omega_0)$. Hence, we have

$$E_\Omega(w_\ell) = E_\Omega(u_\ell) - E_\Omega(\widehat{u}_0) - E_{\Omega_0}(v_0) + o(1)$$

and $dE_\Omega(w_\ell) \to 0$ strongly in $\left(H^2 \cap H_0^1(\Omega) \right)'$. In the case where we find the solution v_0 in the whole space, we argue similarly with some obvious simplifications.

Now, the same procedure from Steps II to VI is applied to (w_ℓ) instead of $(u_\ell - \widehat{u}_0)$. As $v_0 \neq 0$ and hence

$$E_{\Omega_0}(v_0) \geq \frac{2}{n} S^{n/4},$$

this procedure stops after finitely many iterations. $\qquad\square$

7.10 The Conformally Covariant Paneitz Equation in Hyperbolic Space

For a brief review of the geometrical background we refer to Section 1.5. Here we do not aim at giving an overview of the huge field of Paneitz equations, where a lot of work has been done, see the bibliographical notes in Section 7.12. Such a task is beyond the scope of this monograph. We want to discuss a special noncompact manifold and to explain a result for the Paneitz equation in the hyperbolic space. Details concerning the related differential equations problems are explained below and in Section 7.10.1.

The manifold $(\mathcal{M}, g)$ to be considered here is the hyperbolic space $\mathbb{H}^n$ with its standard metric. We focus on finding a complete metric $h = U^{\frac{4}{n-4}} g$ on $\mathbb{H}^n$ such that h has prescribed Q-curvature. We give conditions on Q, which include the case $Q \equiv$ constant, such that an entire continuum of mutually distinct complete radially symmetric conformal metrics exist all having the same prescribed Q-curvature. In the case where $Q \equiv \frac{1}{8}n(n^2 - 4)$ this family contains in its "center" the explicitly known standard hyperbolic Poincaré metric, and at least a sub-continuum of these metrics has negative scalar curvature.

We point out that it is surprising to find such highly non-unique solutions. In previous work on the second order Yamabe problem, uniqueness of metrics with constant scalar curvature was found in the case of $\mathbb{H}^n$ by Loewner-Nirenberg [280]. In the case of $\mathbb{S}^n$ uniqueness, up to isometries, was proved by Obata [326] and later by Caffarelli, Gidas, Spruck [82] and Chen, Li [94]. In the fourth order Paneitz problem uniqueness, again up to isometries, of metrics with constant Q-curvature on $\mathbb{S}^n$ was found by Chang, Yang [93] for $n = 4$, by Wei, Xu [411] and C.S. Lin [275] for $n > 4$ and by Choi, Xu [96] in the exceptional case $n = 3$.

7.10.1 Infinitely Many Complete Radial Conformal Metrics with the Same Q-Curvature

As a model for hyperbolic space $\mathbb{H}^n$ we use the Poincaré ball, i.e. $\mathbb{H}^n$ is represented by the unit ball $B = B_1(0) \subset \mathbb{R}^n$ with standard coordinates $x_1, \dots, x_n$ and the Poincaré metric $g_{ij} = 4\delta_{ij}/(1 - |x|^2)^2$. Since $\mathbb{H}^n$ is conformally flat we may seek the metric h of the form $h_{ij} = U^{\frac{4}{n-4}} g_{ij} = u^{\frac{4}{n-4}} \delta_{ij}$ and the corresponding Paneitz equation (1.28) for u reduces to

$$\Delta^2 u = \frac{n-4}{2} Q u^{(n+4)/(n-4)}, \quad u > 0 \quad \text{in } B, \qquad u|_{\partial B} = \infty. \tag{7.204}$$

The condition $u|_{\partial B} = \infty$ is necessary, and as we shall show also sufficient, for completeness of the metric h. For $U = 1$ we are at the Poincaré metric. In this case the conformal factor is given explicitly by

$$u_0(x) = \left(\frac{2}{1 - |x|^2} \right)^{(n-4)/2}. \tag{7.205}$$

The Poincaré metric $\left(u_0^{4/(n-4)} \delta_{ij} \right)_{ij}$ with u_0 as above has constant Q-curvature $Q \equiv \frac{1}{8}n(n^2 - 4)$.

Theorem 7.75. *For every $\alpha > 0$, there exists a radial solution of the prescribed Q-curvature equation (7.204) in the unit ball B with $Q \equiv \frac{1}{8}n(n^2 - 4)$, infinite boundary values at ∂B and with $u(0) = \alpha$. Moreover,*

1. *the conformal metric $\left(u^{4/(n-4)} \delta_{ij} \right)_{ij}$ on B is complete;*
2. *if $u(0) > 0$ is sufficiently small, then the corresponding solution generates a metric with negative scalar curvature.*

The existence part and Item 2 are proved in Section 7.10.2, see Proposition 7.83, whereas Item 1 is proved in Section 7.10.3.

The equation (7.204) is invariant under Möbius transformations of the unit ball. But the only solution which is invariant under all Möbius transformations of the unit ball is the explicit solution (7.205). Hence, we also have infinitely many distinct nonradial solutions, which is again in striking contrast to the second order analogue of (7.204). The following interesting open problem remains to be studied.

> Find a geometric criterion, which singles out the explicit solution (7.205) among all other solutions of (7.204).

One might guess that among all radially symmetric metrics h the explicit Poincaré metric is uniquely characterised by a condition on the scalar curvature R_h of the kind

$$-C \le R_h \le -\frac{1}{C} < 0$$

with a suitable constant C. This is apparently wrong, since it will follow from a forthcoming result in [142] that for every radial solution u of (7.204) one has that the scalar curvature of the generated metric satisfies $\lim_{r \to 1} R_h = -n(n-1)$. It is however trivially true that the Poincaré metric is the only one with $R_h \equiv -n(n-1)$.

Remark 7.76. A similar result as Theorem 7.75 can be proved for radial nonconstant $Q \in C^1[0,1]$ for which there are two positive constants $Q_0, Q_1 > 0$ such that $Q_0 < Q(r) < Q_1$ on $[0,1]$ and for which $r \mapsto r^q Q(r)$ is monotonically increasing for some $q \in [0,1)$. For details see [206].

7.10.2 Existence and Negative Scalar Curvature

Here we look for radial solutions of (7.204). By means of a shooting method we shall construct infinitely many distinct solutions. Applying the special Möbius transforms

$$\varphi_a : B \to B, \qquad \varphi_a(x) = \frac{1}{|a|^2}\left(a - (|a|^2 - 1)\frac{x - |a|^{-2} a}{\left|x - |a|^{-2} a\right|^2}\right),$$

we even find nonradial solutions by setting

$$\tilde{u} := J_{\varphi_a}^{(n-4)/(2n)}\, u \circ \varphi_a.$$

Here J_{φ_a} is the modulus of the Jacobian determinant of φ_a. All these conformal metrics have constant Q-curvature $\frac{1}{8}n(n^2 - 4)$ and a continuum of them has negative scalar curvature.

Solutions of (7.204) with $Q \equiv \frac{1}{8}n(n^2 - 4)$ are multiples of solutions for the simplified problem

$$\Delta^2 u = u^{(n+4)/(n-4)}, \quad u > 0 \quad \text{in } B, \qquad u|_{\partial B} = \infty.$$

For radial solutions we study the initial value problem

$$\begin{cases} \Delta^2 u(r) = \left(r^{1-n}\dfrac{d}{dr}\left(r^{n-1}\dfrac{d}{dr}\right)\right)^2 u(r) = u(r)^{(n+4)/(n-4)}, & r > 0, \\ u(0) = \alpha, \quad u'(0) = 0, \quad \Delta u(0) = \beta, \quad (\Delta u)'(0) = 0, \end{cases} \tag{7.206}$$

where $\alpha \geq 0$, $\beta \in \mathbb{R}$ are given. If necessary, $u^{(n+4)/(n-4)}$ will denote also the odd extension to the negative reals; however, we mainly focus on positive solutions. It is a routine application or modification of the Banach fixed point theorem or the Picard-Lindelöf result to show that (7.206) always has a unique local C^4–solution.

It is a simple but very useful observation that the initial value problem enjoys a comparison principle by McKenna-Reichel [297].

Lemma 7.77. *Let $u, v \in C^4([0, R))$ and $\tilde{Q} \geq 0$ be such that*

$$\begin{cases} \Delta^2 u(r) - \tilde{Q} u(r)^{(n+4)/(n-4)} \geq \Delta^2 v(r) - \tilde{Q} v(r)^{(n+4)/(n-4)} & \textit{for all } r \in [0, R) \\ u(0) \geq v(0), u'(0) = v'(0) = 0, \Delta u(0) \geq \Delta v(0), (\Delta u)'(0) = (\Delta v)'(0) = 0. \end{cases}$$

Then we have

$$u(r) \geq v(r), u'(r) \geq v'(r), \Delta u(r) \geq \Delta v(r), (\Delta u)'(r) \geq (\Delta v)'(r) \textit{ for all } r \in [0, R).$$

Moreover,

1. *the initial point 0 can be replaced by any initial point $\rho > 0$ if all four initial data at ρ are weakly ordered,*
2. *a strict inequality in one of the initial data at $\rho \geq 0$ or in the differential inequality on (ρ, R) implies a strict ordering of $u, u', \Delta u, \Delta u'$ and $v, v', \Delta v, \Delta v'$ on (ρ, R).*

According to Theorem 7.21, problem (7.206) has the following entire solutions

$$U_\alpha(r) = \alpha \frac{\left(n(n^2-4)(n-4)\right)^{\frac{n-4}{4}}}{\left(\sqrt{n(n^2-4)(n-4)} + \left(\alpha^{2/(n-4)}r\right)^2\right)^{\frac{n-4}{2}}}$$

with $\alpha > 0$. Moreover, these functions are the only positive entire solutions of (7.206), provided β is suitably chosen, say $\beta = \beta_0 = \beta_0(\alpha)$. The metric

$$h = U_\alpha^{\frac{4}{n-4}} \delta_{ij} \tag{7.207}$$

arises as the pullback of the standard metric of the sphere $\mathbb{S}^n$ under a stereographic projection to $\mathbb{R}^n$.

For our purposes it is enough to show that the solution U_α is a separatrix in the r-u-plane, i.e., if we fix $\alpha > 0$ and consider β as a varying parameter then U_α separates the blow-up solutions from the solutions with one sign-change, which lie below U_α.

Lemma 7.78. *Let $\alpha > 0$ be fixed. Then for $\beta > \beta_0$, the solution $u = u_{\alpha,\beta}$ blows up on a finite interval $[0, R(\alpha,\beta))$. The blow-up radius $R(\alpha,\beta)$ is monotonically decreasing in β.*

Proof. It is useful to have the explicit solutions

$$V_\alpha(r) = \alpha \left(1 - \left(\frac{r}{\lambda_\alpha}\right)^2\right)^{-(n-4)/2} \tag{7.208}$$

of (7.206) at hand, where $\lambda_\alpha = \alpha^{-2/(n-4)} \left(n(n^2-4)(n-4)\right)^{1/4}$. We fix any $\alpha > 0$, some $\beta > \beta_0(\alpha)$ and look at the corresponding solution $u = u_{\alpha,\beta}$ of (7.206). In order to see that $u'(r) - U_\alpha'(r)$ is strictly increasing, note first by Lemma 7.77 that $\Delta u(r) - \Delta U_\alpha(r)$ is positive and strictly increasing. Since $u'(r) - U_\alpha'(r) = \int_0^1 r t^{n-1}(\Delta u - \Delta U_\alpha)(rt)\,dt$ it follows that $u'(r) - U_\alpha'(r)$ is also strictly increasing. So $u(r)$ cannot converge to 0 and hence has to become unbounded as $r \to \infty$. By integrating successively the differential equation of u we find R large enough such that

$$u(R) > 0, \qquad u'(R) > 0, \qquad \Delta u(R) > 0, \qquad (\Delta u)'(R) > 0.$$

Since $\lim_{\tilde\alpha \to 0} V_{\tilde\alpha}(r) = 0$ locally uniformly in C^4, we can find a sufficiently small $\tilde\alpha > 0$ such that

$$u(R) > V_{\tilde\alpha}(R), \quad u'(R) > V_{\tilde\alpha}'(R), \quad \Delta u(R) > \Delta V_{\tilde\alpha}(R), \quad (\Delta u)'(R) > (\Delta V_{\tilde\alpha})'(R).$$

But then the comparison principle Lemma 7.77 shows that for all $r > R$ we have $u(r) > V_{\tilde\alpha}(r)$ and hence, blow-up of u at some finite radius $R(\alpha,\beta)$. The monotonicity of $R(\alpha,\beta)$ is also a direct consequence of Lemma 7.77. $\qquad\square$

Lemma 7.79. *Let $\alpha > 0$ be fixed. The blow-up radius $R(\alpha,\beta)$ is a continuous function of $\beta \in (\beta_0,\infty)$.*

Proof. Let $\beta > \beta_0$ be arbitrary but fixed and let $u = u_{\alpha,\beta}$ denote the corresponding solution of (7.206). The continuity from the right

$$\beta_k \searrow \beta \quad \Rightarrow \quad R(\alpha,\beta_k) \to R(\alpha,\beta)$$

follows directly from monotonicity and lower semicontinuity of $\beta \mapsto R(\alpha,\beta)$. Hence, only continuity from the left has to be proved.

First we show that for r close enough to $R = R(\alpha,\beta)$ the functions u, u', Δu and $(\Delta u)'$ are eventually strictly increasing. For u, $r^{n-1} u'$, Δu and $r^{n-1}(\Delta u)'$, this follows from successive integration of the differential equation, since the relevant quantities become – at least eventually – positive.

It remains to consider $u'(R)$ and $(\Delta u)'(R)$. We observe that

$$\infty = R^{n-1} u'(R) = \int_0^R r^{n-1} \Delta u \, dr;$$
$$\infty = R^{n-1}(\Delta u)'(R) = \int_0^R r^{n-1} \Delta^2 u \, dr = \int_0^R r^{n-1} u^{(n+4)/(n-4)} \, dr. \tag{7.209}$$

From this we conclude for $r \nearrow R$ that

$$(\Delta u)'(r) = \int_0^r \left(\frac{s}{r}\right)^{n-1} u^{(n+4)/(n-4)}(s) \, ds = r \int_0^1 (u(rt))^{(n+4)/(n-4)} t^{n-1} \, dt;$$
$$(\Delta u)''(r) = \int_0^1 (u(rt))^{(n+4)/(n-4)} t^{n-1} \, dt + r \frac{n+4}{n-4} \int_0^1 (u(rt))^{8/(n-4)} u'(rt) t^n \, dt$$
$$\to +\infty \text{ by } (7.209);$$

$$u'(r) = \int_0^r \left(\frac{s}{r}\right)^{n-1} \Delta u(s) \, ds = r \int_0^1 t^{n-1} \Delta u(rt) \, dt;$$
$$u''(r) = \int_0^1 t^{n-1} \Delta u(rt) \, dt + r \int_0^1 t^n (\Delta u)'(rt) \, dt$$
$$\to +\infty \text{ by } (7.209).$$

Moreover, for later purposes we note that for $r \nearrow R$

$$u'''(r) = 2 \int_0^1 t^n (\Delta u)'(rt) \, dt + r \int_0^1 t^{n+1} (\Delta u)''(rt) \, dt$$
$$\geq \frac{2}{r^{n+1}} \int_0^r s^n (\Delta u)'(s) \, ds - C \geq \frac{1}{C} \Delta u(r) - C \to +\infty.$$

Here, C denotes a constant which depends on the solution u.

Let us consider a sequence $\beta_k \nearrow \beta$. By monotonicity we have $R(\alpha,\beta_k) \geq R(\alpha,\beta)$. For $t_k > 1$, which will be adequately chosen below, we define the function

$$v_k(r) := t_k^{(4-n)/2} u_{\alpha,\beta}\left(\frac{r}{t_k}\right),$$

which solves the same differential equation as $u_{\alpha,\beta}$. We find values $r_0 - \delta < r_0 < R(\alpha,\beta)$ such that

$$u_{\alpha,\beta}(r_0) > 0, \quad u'_{\alpha,\beta}(r_0) > 0, \quad \Delta u_{\alpha,\beta}(r_0) > 0, \quad (\Delta u_{\alpha,\beta})'(r_0) > 0,$$

and all these quantities are strictly increasing on $(r_0 - \delta, R(\alpha,\beta))$. By continuous dependence on data, for β_k close enough to β we also have

$$u_{\alpha,\beta_k}(r_0) > 0, \quad u'_{\alpha,\beta_k}(r_0) > 0, \quad \Delta u_{\alpha,\beta_k}(r_0) > 0, \quad (\Delta u_{\alpha,\beta_k})'(r_0) > 0.$$

For suitably chosen t_k we conclude that

$$v_k(r_0) = t_k^{(4-n)/2} u_{\alpha,\beta}\left(\frac{r_0}{t_k}\right) \leq t_k^{(4-n)/2} u_{\alpha,\beta}(r_0) < u_{\alpha,\beta_k}(r_0),$$

$$v'_k(r_0) = t_k^{(2-n)/2} u'_{\alpha,\beta}\left(\frac{r_0}{t_k}\right) \leq t_k^{(2-n)/2} u'_{\alpha,\beta}(r_0) < u'_{\alpha,\beta_k}(r_0),$$

$$\Delta v_k(r_0) = t_k^{-n/2} \Delta u_{\alpha,\beta}\left(\frac{r_0}{t_k}\right) \leq t_k^{-n/2} \Delta u_{\alpha,\beta}(r_0) < \Delta u_{\alpha,\beta_k}(r_0),$$

$$(\Delta v_k)'(r_0) = t_k^{(-n-2)/2}(\Delta u_{\alpha,\beta})'\left(\frac{r_0}{t_k}\right) \leq t_k^{(-n-2)/2}(\Delta u_{\alpha,\beta})'(r_0) < (\Delta u_{\alpha,\beta_k})'(r_0).$$

By continuous dependence on data, we may achieve

$$t_k \searrow 1 \qquad (k \to \infty).$$

The comparison result of Lemma 7.77 yields for $r \geq r_0$,

$$u_{\alpha,\beta_k}(r) \geq v_k(r).$$

This gives finally

$$R(\alpha,\beta) \leq R(\alpha,\beta_k) \leq R(v_k) = R(\alpha,\beta)\, t_k \to R(\alpha,\beta) \quad \text{as } k \to \infty,$$

where $R(v_k)$ denotes the blow-up radius of v_k. The proof is complete. $\qquad\square$

Lemma 7.80. *Let $\alpha > 0$ be fixed. Then for the limits of the blow-up radius $R(\alpha,\beta)$, one has*

$$\lim_{\beta \searrow \beta_0} R(\alpha,\beta) = \infty, \qquad \lim_{\beta \nearrow \infty} R(\alpha,\beta) = 0.$$

Proof. The first claim is just a consequence of the global existence of the solution for $\beta = \beta_0$ and continuous dependence of solutions on the initial data. The proof of the second statement bases upon some scaling arguments. First we note that the same argument as in the proof of Lemma 7.78 shows that $R(0,1) < \infty$. By the comparison result from Lemma 7.77 we conclude that

$$R(\alpha',1) \leq R(0,1) < \infty \qquad \text{for all } \alpha' > 0. \tag{7.210}$$

For $\beta > 0$ we find the relation

$$u_{\alpha,\beta}(r) = \left(\frac{\alpha}{\alpha'}\right) u_{\alpha',1}\left(\left(\frac{\alpha}{\alpha'}\right)^{2/(n-4)} r\right), \tag{7.211}$$

where α' is chosen such that

$$\beta = \left(\frac{\alpha}{\alpha'}\right)^{n/(n-4)}, \quad \text{i.e. } \alpha' = \alpha \, \beta^{(4-n)/n}.$$

Obviously, $\alpha' \searrow 0$ for $\beta \nearrow \infty$. We read from (7.211) and (7.210) that

$$R(\alpha,\beta) = R(\alpha',1)\left(\frac{\alpha'}{\alpha}\right)^{2/(n-4)} \leq R(0,1)\left(\frac{\alpha'}{\alpha}\right)^{2/(n-4)} = R(0,1)\beta^{-2/n},$$

which tends to 0 as $\beta \to \infty$. $\qquad\square$

Lemma 7.81. *Let $u : B \to (0,\infty)$ be a C^4-function such that $-\Delta u \leq 0$ in B. Then the conformal metric h given by (7.207), namely*

$$h_{ij} = u^{4/(n-4)}\delta_{ij},$$

satisfies $R_u \leq 0$ in B.

Proof. In order to compute the scalar curvature it is more convenient to write the conformal factor as

$$h_{ij} = v^{4/(n-2)}\delta_{ij},$$

i.e. we set $v := u^{(n-2)/(n-4)}$, $u = v^{(n-4)/(n-2)}$. The scalar curvature R_u of the metric $(h_{ij})_{ij}$ is then given by

$$\begin{aligned}
R_u &= -\frac{4(n-1)}{(n-2)}v^{-(n+2)/(n-2)}\Delta v = -\frac{4(n-1)}{(n-2)}u^{-(n+2)/(n-4)}\Delta u^{(n-2)/(n-4)} \\
&= -\frac{4(n-1)}{(n-4)}u^{-(n+2)/(n-4)}\left(u^{2/(n-4)}\Delta u + \frac{2}{(n-4)}u^{(6-n)/(n-4)}|\nabla u|^2\right)
\end{aligned}$$

so that $R_u \leq 0$. $\qquad\square$

Remark 7.82. The subharmonicity assumption in Lemma 7.81 is justified by the fact that, for radially symmetric solutions, it is also a necessary condition in order to have negative scalar curvature, see [206, Proposition 1].

We now prove the existence part and Item 2 of Theorem 7.75 which is summarised in

Proposition 7.83. *For every $\alpha > 0$ there exists a radial solution of (7.206) with $u(0) = \alpha$ which blows up at $r = 1$. Moreover,*

1. *if $u, \tilde{u}$ are two such solutions with $u(0) < \tilde{u}(0)$, then $\Delta u(0) > \Delta \tilde{u}(0)$,*
2. *if $0 < u(0) \leq (n(n^2 - 4)(n - 4))^{\frac{n-4}{8}}$, then the corresponding solution generates a metric with negative scalar curvature.*

Proof. Let $\alpha > 0$ be fixed and let $u_{\alpha,\beta}$ denote the solution of (7.206). According to Lemmas 7.79 and 7.80, we find a suitable $\beta > \beta_0(\alpha)$ such that for the blow-up radius we have precisely $R(\alpha,\beta) = 1$. Item 1 is a consequence of Lemma 7.77. To prove Item 2, we consider V_α as defined in (7.208) and note that under the hypothesis $0 < u(0) < V_{\alpha_0}(0)$ with $\alpha_0 = (n(n^2 - 4)(n - 4))^{\frac{n-4}{8}}$ we find by Item 1 that $\Delta u(0) > \Delta V_{\alpha_0}(0) > 0$ and hence $\Delta u > 0$ on $[0, 1)$. Thus by Lemma 7.81 the solution u generates a metric with negative scalar curvature. $\square$

In order to complete the proof of Theorem 7.75, it remains to show the completeness of the induced metrics.

7.10.3 Completeness of the Conformal Metric

Completeness of the metric $h = u^{\frac{4}{n-4}} \delta_{ij}$ on B means that every maximally extended geodesic curve has infinite length. However, the following lemma reduces this to a property which is simpler to check.

Lemma 7.84. *Let u be a radial solution of (7.204). The induced metric $u^{\frac{4}{n-4}} \delta_{ij}$ on $\mathbb{H}^n$ is complete if and only if*

$$\int_0^1 u(r)^{2/(n-4)} \, dr = \infty.$$

Proof. To see the necessity of the above condition note that for fixed $z \in \mathbb{R}^n \setminus \{0\}$ the curve $\gamma(r) = rz/|z|$ for $r \in (-1, 1)$ is a maximally extended geodesic and its length is given by

$$2 \int_0^1 (\gamma'(r), \gamma'(r))_h^{1/2} \, dr = 2 \int_0^1 u^{\frac{2}{n-4}} \, dr.$$

Next we prove the sufficiency. Let γ be a maximally extended geodesic in (B, h) parameterised over $\mathbb{R}$. Then $\lim_{t \to \pm\infty} |\gamma(t)| = 1$. Clearly γ has infinite length if $\delta(t) = \operatorname{dist}_h(\gamma(t), 0)$ becomes unbounded for $t \to \pm\infty$. Since

$$\delta(t) = \int_0^{|\gamma(t)|} u^{\frac{2}{n-4}}(r)\, dr$$

the claim follows.   □

Diaz-Lazzo-Schmidt [142, 143] announced that for the unbounded solutions with constant Q-curvature constructed in Theorem 7.75, one has asymptotically for $r \nearrow 1$

$$u(r) \sim C(1 - r^2)^{(4-n)/2}, \tag{7.212}$$

where $C = C(n)$ does not depend on the solution. Furthermore, the derivatives of u exhibit a corresponding uniform behaviour. This is an even more precise information than just completeness of the conformal metric since it shows in particular that

$$\int^1 u(r)^{2/(n-4)}\, dr = \infty$$

and so, the completeness of the conformal metric in view of Lemma 7.84. Their work is announced to cover a very general situation, will be quite involved and rely on a deep result of Mallet-Paret and Smith [283] on Poincaré-Bendixson results for monotone cyclic feedback systems.

In what follows we give an independent and relatively simple proof of the statement of completeness by means of a suitable transformation and energy considerations. The proof applies in the same way both to the case of constant and non-constant Q-curvature functions, which is useful in view of Remark 7.76. The final statement concerning completeness is given in Proposition 7.88 below.

In what follows we consider for simplicity $(\frac{n-4}{2}Q)^{(n-4)/8}u$ instead of u, so that in radial coordinates the Paneitz equation (7.204) reads

$$u^{iv}(r) + \frac{2(n-1)}{r}u'''(r) + \frac{(n-1)(n-3)}{r^2}u''(r) - \frac{(n-1)(n-3)}{r^3}u'(r) = u^{\frac{n+4}{n-4}}(r).$$

$$\tag{7.213}$$

We estimate the maximal blow-up rate for solutions of (7.213). The following statement proves "half" of the optimal asymptotic behaviour (7.212) and is therefore of independent interest.

Proposition 7.85. *Let* $u : [0,1) \to [0,\infty)$ *be an unbounded smooth solution to* (7.213). *Then there exists a constant* $C = C(u)$ *such that*

$$u(r) \le C\left(\frac{1}{1 - r^2}\right)^{\frac{n-4}{2}}.$$

Proof. As was shown in the proof of Lemma 7.79, we may choose $\rho \in (0,1)$ such that

$$u, u', u'', \Delta u, (\Delta u)' > 0 \ \text{ are increasing in } \ (\rho, 1).$$

Rewriting the Pohožaev-Rellich-type identity (7.77) for the solution u on B_r for any $r \in (0,1)$, the volume integrals (over B_r) vanish and therefore

$$0 = r^{n-1}(\Delta u)'\left(ru' + \frac{n-4}{2}u\right) + \frac{n}{2}r^{n-1}u'\Delta u - r^n\left(\frac{1}{2}(\Delta u)^2 + \frac{n-4}{2n}u^{\frac{2n}{n-4}}\right). \quad (7.214)$$

In the sequel, C denotes a constant depending on u. By using the analogue of (7.214) on the interval $[\rho, r]$ we obtain for all $r \in (\rho, 1)$

$$\frac{n-4}{2n}r^n u^{\frac{2n}{n-4}}(r) + \frac{r^n}{2}(\Delta u(r))^2 = r^{n-1}(\Delta u)'\left(ru' + \frac{n-4}{2}u\right) + \frac{n}{2}r^{n-1}u'\Delta u + C.$$

$$(7.215)$$

We estimate the two sides of (7.215) separately.

Right-hand side. The following estimates for $r > \rho$ are obtained by integration

$$u(r) = u(\rho) + \int_\rho^r u'(s)\,ds \leq u'(r) + C,$$

$$\Delta u(r) \leq (\Delta u)'(r) + C.$$

Hence the entire right-hand side of (7.215) can be estimated by $C_1 u'(r)(\Delta u)'(r) + C_2$ and since $u'(r), (\Delta u)'(r) \to \infty$ for $r \to 1$ we find that $Cu'(r)(\Delta u)'(r)$ for $\rho < r < 1$ is an upper estimate for the right-hand side of (7.215).

Left-hand side. After dropping the last term in the left-hand side of (7.215) a lower bound is given by

$$\frac{n-4}{2n}r^n u^{\frac{2n}{n-4}}(r).$$

Hence, (7.215) yields the existence of a constant $C = C(u, \rho, \varepsilon)$ such that

$$u^{\frac{2n}{n-4}} \leq Cu'(\Delta u)' \text{ on } [\rho, 1].$$

The multiplication by u' leads to

$$\left(u^{\frac{3n-4}{n-4}}\right)' \leq Cu'^2(\Delta u)' = C\left(u'^2\Delta u\right)' - 2Cu'u''\Delta u \leq C\left(u'^2\Delta u\right)' \text{ on } [\rho, 1],$$

and an integration shows

$$u^{\frac{3n-4}{n-4}} \leq C_1 u'^2\Delta u + C_2 \leq Cu'^2\Delta u \text{ on } [\rho, 1].$$

Now, as above, we can estimate

$$\Delta u(r) = u''(r) + \frac{n-1}{r}u'(\rho) + \frac{n-1}{r}\int_\rho^r u''(s)\,ds \leq Cu''(r) + C \leq Cu''(r)$$

and we may proceed to the inequality

$$u^{\frac{3n-4}{n-4}} \leq C(u')^2 \Delta u \leq C(u')^2 u''.$$

In a similar way the multiplication by u' and an integration leads to

$$u^{\frac{4n-8}{n-4}} \leq Cu'^4 \quad \text{on } [\rho,1), \qquad u^{\frac{n-2}{n-4}} \leq Cu' \quad \text{on } [\rho,1).$$

Solutions of $Cv' = v^{\frac{n-2}{n-4}}$ on some interval $[\rho,\delta)$ with $0 < \delta \leq 1$ are given by

$$v_\delta(r) = \left(\frac{n-4}{2}C\right)^{\frac{n-4}{2}} (\delta - r)^{\frac{4-n}{2}}.$$

If for some value of $r_0 \in [\rho,1)$ we would have $u(r_0) > v_1(r_0)$, then $u(r_0) > v_\delta(r_0)$ for some $\delta \in (0,1)$. Then u stays strictly above v_δ and hence u blows up somewhere in the interval (ρ,δ), i.e., strictly before the point 1. This contradiction shows that $u(r) \leq v_1(r)$ for all $r \in [\rho,1)$ which in turn proves the claim. $\qquad\square$

Next, we establish a first lower bound for the blow-up rate of radial solutions to the Paneitz equation (7.204). This bound is far from being optimal, see (7.212).

Lemma 7.86. *Let $u : [0,1) \to [0,\infty)$ be an unbounded smooth solution to (7.213). Then there exists a constant $C = C(u)$ such that*

$$u(r) \geq C\left(\frac{1}{1-r^2}\right)^{\frac{n-4}{4}} \quad \text{on } [1/2,1).$$

Proof. Let $u = u(r)$ solve (7.213) on $[0,1)$ with $u(1) = \infty$. Then for some $r_0 \in (0,1)$ we may assume that $u'(r) \geq 0$ and $(\Delta u)'(r) \geq 0$ whenever $r \geq r_0$. Thus

$$(\Delta u)'(r) = \left(\frac{r_0}{r}\right)^{n-1} (\Delta u)'(r_0) + \int_{r_0}^r \left(\frac{s}{r}\right)^{n-1} u^{\frac{n+4}{n-4}}(s)\,ds \leq (\Delta u)'(r_0) + u^{\frac{n+4}{n-4}}(r),$$

and hence

$$\Delta u(r) \leq \Delta u(r_0) + (\Delta u)'(r_0) + u^{\frac{n+4}{n-4}}(r) = K + u^{\frac{n+4}{n-4}}(r)$$

with suitably chosen $K = K(u) > 0$. Now let v be the unique radial solution of

$$\Delta v = K + v^{\frac{n+4}{n-4}} \quad \text{for } r_0 < r < 1, \quad v(r_0) = u(r_0), \quad v(1) = \infty.$$

Then v is a subsolution for u and

$$u(r) \geq v(r) \geq C\left(\frac{1}{1-r^2}\right)^{\frac{n-4}{4}} \quad \text{on } [r_0,1),$$

where $C = C(r_0;u)$. $\qquad\square$

Instead of proving the expected asymptotic behaviour (7.212), we show the weaker result that $\int^1 u^{2n/(n-4)} = +\infty$. This is enough for our purposes and the proof appears to be much simpler. To this end we employ some dynamical systems techniques. We apply a sort of Emden-Fowler transformation to (7.213) in order to obtain an "asymptotically autonomous" differential equation. The stability analysis performed for the limit autonomous equation can be carried over to this asymptotically autonomous equation. Energy considerations then allow to reach a contradiction if $\int^1 u^{2n/(n-4)} < +\infty$.

With the transformation

$$u(r) = (1-r^2)^{\frac{4-n}{2}} v(-\log(1-r^2)), \quad v(t) = e^{(4-n)t/2} u(\sqrt{1-e^{-t}}), \quad t \in (0,\infty),$$

(7.213) becomes

$$K_4(t)v^{iv}(t) + K_3(t)v'''(t) + K_2(t)v''(t) + K_1(t)v'(t) + K_0 v(t) = \frac{1}{16} v^{\frac{n+4}{n-4}}(t), \quad (7.216)$$

where the boundary $r = 1$ is transformed into $t = +\infty$ and

$$K_0 = \frac{1}{16}(n+2)n(n-2)(n-4),$$

$$K_1(t) = \frac{1}{16}\Big((1-e^{-t})^2(-4n^2+24n-32)+(1-e^{-t})(4n^3-16n^2-16n+64)$$
$$+ 4n^3-4n^2-24n\Big),$$

$$K_2(t) = \frac{1}{16}\Big((1-e^{-t})^2(4n^2-40n+80)+(1-e^{-t})(16n^2-16n-96)$$
$$+ 4n^2+8n\Big),$$

$$K_3(t) = (1-e^{-t})^2(n-4)+(1-e^{-t})(n+2),$$
$$K_4(t) = (1-e^{-t})^2.$$

Note that (7.216) is asymptotically autonomous, since all the $K_i(t)$ have finite limits as $t \to \infty$, and that (7.216) has the constant solutions $v_0 \equiv 0$ and $v_1 \equiv (16K_0)^{\frac{n-4}{8}}$.

Motivated by the observation that

$$u'(r) = 0 \Leftrightarrow v'(t) + \frac{n-4}{2} v(t) = 0$$

we transform (7.216) into a system for $w(t) = (w_1(t), w_2(t), w_3(t), w_4(t))^T$ by setting

$$w_1(t) = v(t), \qquad w_2(t) = v'(t) + \frac{n-4}{2} v(t),$$

$$w_3(t) = v''(t) + \frac{n-4}{2} v'(t), \qquad w_4(t) = v'''(t) + \frac{n-4}{2} v''(t).$$

The resulting system is

$$\begin{cases} w_1'(t) = -\frac{n-4}{2}w_1(t) + w_2(t) \\ w_2'(t) = w_3(t) \\ w_3'(t) = w_4(t) \\ K_4(t)w_4'(t) = C_2(t)w_2(t) + C_3(t)w_3(t) + C_4(t)w_4(t) + \frac{1}{16}w_1(t)^{\frac{n+4}{n-4}}, \end{cases} \tag{7.217}$$

where

$$C_m(t) = -\sum_{k=m-1}^{4} K_k(t)\left(\frac{4-n}{2}\right)^{k+1-m}.$$

By explicit calculations we get $C_1(t) \equiv 0$ and

$$C_2(t) = -\frac{1}{8}n^3 + \frac{1}{2}n,$$

$$C_3(t) = 1 - \frac{3}{4}n^2 + \frac{n(n-2)}{2}e^{-t} + \frac{n-2}{2}e^{-2t},$$

$$C_4(t) = -\frac{3}{2}n + e^{-t}(2n-2) + e^{-2t}\left(2 - \frac{1}{2}n\right).$$

To get an idea of the behaviour of the asymptotically autonomous system (7.217) we replace the functions $C_i(t)$ by their limit $C_i^\infty = \lim_{t\to\infty} C_i(t), i = 2,3,4$. We study the autonomous limit system

$$\begin{cases} w_1'(t) = -\frac{n-4}{2}w_1(t) + w_2(t) \\ w_2'(t) = w_3(t) \\ w_3'(t) = w_4(t) \\ w_4'(t) = C_2^\infty w_2(t) + C_3^\infty w_3(t) + C_4^\infty w_4(t) + \frac{1}{16}w_1(t)^{\frac{n+4}{n-4}}, \end{cases} \tag{7.218}$$

where

$$C_2^\infty = -\frac{1}{8}n^3 + \frac{1}{2}n, \quad C_3^\infty = 1 - \frac{3}{4}n^2, \quad C_4^\infty = -\frac{3}{2}n.$$

The autonomous system has the steady-states

$$O = (0,0,0,0) \text{ and } P = \left((16K_0)^{\frac{n-4}{8}}, \frac{n-4}{2}(16K_0)^{\frac{n-4}{8}}, 0, 0\right).$$

Note that O and P are also steady states for the asymptotically autonomous system (7.218). At the point O the system (7.218) admits the linearised matrix

$$M_O = \begin{pmatrix} \frac{4-n}{2} & 1 & 0 & 0 \\ 0 & 0 & 1 & 0 \\ 0 & 0 & 0 & 1 \\ 0 & C_2^\infty & C_3^\infty & C_4^\infty \end{pmatrix}$$

with four negative eigenvalues

$$\lambda_1 = 2 - \frac{n}{2} > \lambda_2 = 1 - \frac{n}{2} > \lambda_3 = -\frac{n}{2} > \lambda_4 = -1 - \frac{n}{2}$$

and corresponding eigenvectors

$$\phi_1 = (1,0,0,0), \qquad \phi_2 = \left(1,-1,-1+\frac{n}{2},-\frac{(n-2)^2}{4}\right),$$

$$\phi_3 = \left(1,-2,n,-\frac{n^2}{2}\right), \qquad \phi_4 = \left(1,-3,3+\frac{3n}{2},-\frac{3(n+2)^2}{4}\right).$$

Thus O is asymptotically stable for (7.218). At the point P the linearised matrix is

$$M_P = \begin{pmatrix} \frac{4-n}{2} & 1 & 0 & 0 \\ 0 & 0 & 1 & 0 \\ 0 & 0 & 0 & 1 \\ \frac{n+4}{n-4}K_0 & C_2^\infty & C_3^\infty & C_4^\infty \end{pmatrix}$$

with the eigenvalues

$$\mu_1 = 1, \qquad\qquad \mu_2 = -n,$$
$$\mu_3 = \tfrac{1-n}{2} - \tfrac{i}{2}\sqrt{n^2+2n-9}, \qquad \mu_4 = \tfrac{1-n}{2} + \tfrac{i}{2}\sqrt{n^2+2n-9}.$$

Thus P has a three-dimensional stable manifold and a one-dimensional unstable manifold.

Lemma 7.87. *The origin O is an asymptotically stable steady state of system (7.217). Moreover, the following facts hold.*

1. *If w is a solution to system (7.217) such that $w(t_k) \to O$ for a sequence $t_k \to \infty$, then for any $\varepsilon > 0$ one has that eventually*

$$|w(t)| \le \exp\left(\left(\frac{4-n}{2}+\varepsilon\right)t\right).$$

2. *The corresponding solution $u(r) = (1-r^2)^{\frac{4-n}{2}} w_1(-\log(1-r^2))$ of the original equation (7.204) is bounded near $r = 1$.*

Proof. System (7.217) has the form

$$w'(t) = M_O w(t) + G(t,w(t)),$$

$$G(t,w) = \left(\frac{1}{16}+O(e^{-t})\right)\left(0,0,0,w_1^{(n+4)/(n-4)}\right)^T + e^{-t}Bw + e^{-2t}Cw,$$

with constant 4×4–matrices B and C. In particular

$$\lim_{t \to \infty, w \to O} \frac{G(t,w)}{|w|} = 0,$$

i.e. condition [224, Ch. X, (8.11)] is satisfied. Since all eigenvalues of M_O are smaller than $(4-n)/2$, the corollary of [224, Ch. X, Theorem 8.1] shows asymptotic stability of the origin O. Moreover, for a solution w with $w(t_k) \to 0$, it follows from this corollary that

$$\limsup_{t \to \infty} \frac{\log|w(t)|}{t} \le \frac{4-n}{2}.$$

Hence, for any $\varepsilon > 0$, one has that eventually

$$|w(t)| \le \exp\left(\left(\frac{4-n}{2} + \varepsilon\right)t\right).$$

For the solution u of the original equation (7.204) this means that for $r < 1$ close enough to 1

$$u(r) \le \left(1 - r^2\right)^{-\varepsilon}.$$

In view of the minimal blow-up rate for unbounded solutions proved in Lemma 7.86, this shows that $r \mapsto u(r)$ has to remain bounded near $r = 1$. $\square$

We are now ready to prove Item 1 in Theorem 7.75, namely the completeness of the conformal metric. To this end, thanks to Lemma 7.84, it suffices to prove the following statement.

Proposition 7.88. *Let $u : B \to [0,\infty)$ be an unbounded smooth radial solution of the Paneitz equation (7.204). Then*

$$\int^1 u(r)^{2/(n-4)}\,dr = \infty.$$

Proof. From Proposition 7.85 we infer that v is bounded as long as it solves (7.216). Then $t \mapsto K_4(t)v^{iv}(t) + K_3(t)v'''(t) + K_2(t)v''(t) + K_1(t)v'(t)$ is also bounded for $t \in (0,\infty)$. Hence, by local L^q–estimates for fourth order elliptic equations (see Section 2.5.2), we infer that for any $q > 1$ there exists a constant $C_q > 0$ such that for any $t > 1$ we have

$$\|v\|_{W^{4,q}(t-1,t+2)} \le C_q \|v\|_{L^\infty(0,\infty)}.$$

By combining Sobolev embeddings and local Schauder estimates we conclude that there exists a positive constant independent of t, still called C_q, such that

$$\|v\|_{C^{4,\gamma}(t,t+1)} \le C_q \|v\|_{L^\infty(0,\infty)}.$$

Therefore, all the derivatives $v' \dots v^{iv}$ are bounded in $(0, \infty)$ and may be used as test functions. Let us assume by contradiction that

$$\int^1 u(r)^{2/(n-4)} dr < \infty,$$

which gives that

$$\int_0^\infty v(s)^2 \, ds \le C \int_0^\infty v(s)^{2/(n-4)} \, ds < \infty.$$

Consider the values $K_i^\infty = \lim_{t \to \infty} K_i(t)$, i.e,

$$K_0^\infty = K_0 = \frac{1}{16}(n+2)n(n-2)(n-4), \qquad K_1^\infty = \frac{1}{2}(n-1)(n^2 - 2n - 4),$$

$$K_2^\infty = \frac{3}{2}n^2 - 3n - 1, \qquad K_3^\infty = 2n - 2, \qquad K_4^\infty = 1.$$

Testing the differential equation (7.216) once with v and once with v' gives that for $t \to \infty$

$$\int_0^t v''(s)^2 \, ds - K_2^\infty \int_0^t v'(s)^2 \, ds = O(1),$$

$$K_3^\infty \int_0^t v''(s)^2 \, ds - K_1^\infty \int_0^t v'(s)^2 \, ds = O(1).$$

Observe that only the terms with constant coefficients are relevant since all other terms contain a factor e^{-t} and produce finite integrals.

Since $K_2^\infty K_3^\infty \ne K_1^\infty$, the two above estimates show that

$$\int_0^\infty v'(s)^2 \, ds < \infty, \qquad \int_0^\infty v''(s)^2 \, ds < \infty.$$

Testing the differential equation (7.216) with v''' finally gives

$$\int_0^\infty v'''(s)^2 \, ds < \infty$$

so that

$$\int_0^\infty \left(w_1(s)^2 + w_2(s)^2 + w_3(s)^2 + w_4(s)^2 \right) ds < \infty.$$

Consequently there is a sequence $t_k \nearrow \infty$ such that

$$\lim_{k \to \infty} \left(w_1, w_2, w_3, w_4 \right)(t_k) = 0.$$

Since $O = (0,0,0,0)$ is stable, this shows that

$$\lim_{t \to \infty} \left(w_1, w_2, w_3, w_4 \right)(t) = 0.$$

From Lemma 7.87 we conclude that $u(r)$ remains bounded near $r = 1$, contradicting the assumption on u. $\qquad \square$

7.11 Fourth Order Equations with Supercritical Terms

It has become clear that the growth of the nonlinear term in semilinear elliptic equations has a crucial influence on its properties. After a brief overview of subcritical problems, see Section 7.2, we have so far focussed on problems with critical growth. Since phenomena and proofs are quite different beyond critical growth – any variational approach seems to break down completely – we finally study here a fourth order model Dirichlet problem

$$\begin{cases} \Delta^2 u = \lambda (1+u)^p & \text{in } B, \\ u > 0 & \text{in } B, \\ u = |\nabla u| = 0 & \text{on } \partial B, \end{cases} \tag{7.219}$$

of supercritical growth, i.e. we assume that $n > 4$ and $p > \frac{n+4}{n-4}$. As before, $B \subset \mathbb{R}^n$ is the unit ball and $\lambda > 0$ a nonlinear eigenvalue parameter. By regularity theory for the biharmonic operator, see Chapter 2, any bounded solution u of (7.219) satisfies $u \in C^\infty(\overline{B})$ and, by Theorem 7.1, it is also radially symmetric. Moreover, by Boggio's principle Lemma 2.27, the sub-supersolution method applies in B while it may fail in general domains. For all these reasons, we restrict ourselves to balls. As already shown for (7.38), when p is supercritical unbounded weak solutions to (7.219) may exist. Similar problems for higher polyharmonic operators $(-\Delta)^m$ $(m \geq 2)$ could also be considered but the proofs appear significantly more technical.

According to related work on second order equations, see e.g. [70, 73, 188, 196, 240, 303, 409], we address questions concerning existence/nonexistence, smoothness and stability of positive minimal solutions, regularity of the extremal solution within a certain regime for p and existence of singular solutions for a suitable value of the parameter λ. Moreover, we characterise radial singular solutions in terms of critical points of associated dynamical systems and give some further qualitative properties of singular solutions. Here, a singular solution is always understood to be singular at the origin $x = 0$.

Since the problem is of supercritical growth, we cannot work within a variational framework and there is no canonical space for weak solutions to (7.219).

Definition 7.89. We say that $u \in L^p(B)$ is a *solution* of (7.219) if $u \geq 0$ and if for all $\varphi \in C^4(\overline{B}) \cap H_0^2(B)$ one has

$$\int_B u \Delta^2 \varphi \, dx = \lambda \int_B (1+u)^p \varphi \, dx.$$

We call u *singular* if $u \notin L^\infty(B)$, and *regular* if $u \in L^\infty(B)$. Finally, we call a solution u of (7.219) *minimal* if $u \leq v$ a.e. in B for any further solution v of (7.219).

In order to state the results we recall that $\Lambda_{2,1} > 0$ denotes the first eigenvalue for the biharmonic operator with Dirichlet boundary conditions in B. By Theorem 3.7

we know that $\Lambda_{2,1}$ is and simple and that the corresponding eigenfunction φ_1 does not change sign. Define

$$\Lambda := \{\lambda > 0 : (7.219) \text{ admits a solution}\} ; \qquad \lambda^* := \sup \Lambda . \tag{7.220}$$

A first result concerns the existence of minimal solutions. It shows that one has existence of stable regular minimal solutions to (7.219) for $\lambda \in (0, \lambda^*)$, while for $\lambda > \lambda^*$, not even singular solutions exist. In second order problems this immediate switch from existence of regular to nonexistence of singular solutions is established by using suitable functions of possibly existing singular solutions as bounded supersolutions, see [70]. Such techniques fail completely for fourth and higher order problems. Here, we employ dynamical systems arguments, and this is one further reason why we have to formulate our results for positive – hence radial – solutions in the ball.

Theorem 7.90. *Let $n > 4$ and $p > (n+4)/(n-4)$. Then the following holds true.*

1. *For $\lambda \in (0, \lambda^*)$ problem (7.219) admits a minimal regular solution u_λ. This solution is radially symmetric and strictly decreasing in $r = |x|$.*
2. *For $\lambda = \lambda^*$ problem (7.219) admits at least one solution $u^* \in H^2(B)$ which is the a.e. pointwise monotone limit of u_λ as $\lambda \nearrow \lambda^*$.*
3. *For $\lambda > \lambda^*$ problem (7.219) admits no (not even singular) solutions.*

Moreover

$$\lambda^* \in \left[K_0, \frac{\Lambda_{2,1}}{p} \right) , \tag{7.221}$$

where

$$K_0 = \frac{4}{p-1} \left(\frac{4}{p-1} + 2 \right) \left(n - 2 - \frac{4}{p-1} \right) \left(n - 4 - \frac{4}{p-1} \right) . \tag{7.222}$$

For the proof see Section 7.11.2. Next, we establish that the regular minimal solution is stable.

Theorem 7.91. *Let $n > 4$ and $p > (n+4)/(n-4)$. Assume that $\lambda \in (0, \lambda^*)$ and let u_λ be the corresponding minimal solution of (7.219). Let $\mu_1(\lambda)$ be the first eigenvalue of the linearised operator $\Delta^2 - \lambda p(1+u_\lambda)^{p-1}$. Then $\mu_1(\lambda) > 0$.*

For the proof see Section 7.11.4. We remark that Theorems 7.90 and 7.91 also hold in the subcritical and critical range, i.e. for any $p > 1$. For $1 < p \le \frac{n+4}{n-4}$, we may define the action functional J_λ associated with the Euler-Lagrange equation (7.219)

$$J_\lambda(u) = \frac{1}{2} \int_B |\Delta u|^2 \, dx - \frac{\lambda}{p+1} \int_B |1+u|^{p+1} \, dx \qquad \text{for all } u \in H_0^2(B).$$

Then variational methods enable us to show that for any $\lambda \in (0, \lambda^*)$ the corresponding minimal solution u_λ is a strict local minimum for the functional J_λ. Moreover, according to the related result [40, Theorem 2.2] we expect for any $\lambda \in (0, \lambda^*)$ the existence of a second positive solution, a mountain-pass critical point for J_λ.

For the corresponding second order problem an explicit singular solution for a suitable value of the parameter λ turned out to play a fundamental role for the description of the shape of the corresponding bifurcation diagram, see in particular [73] and (7.38). When turning to the biharmonic problem (7.219) the second boundary condition $|\nabla u| = 0$ prevents to find an explicit singular solution directly from the entire singular solution to the differential equation $\Delta^2 u = |u|^{p-1} u$ in $\mathbb{R}^n \setminus \{0\}$, namely

$$\overline{u}(r) = K_0^{1/(p-1)} r^{-4/(p-1)}, \tag{7.223}$$

where K_0 is as in (7.222). Nevertheless, the existence of a singular (i.e. unbounded) solution u_s for a suitable eigenvalue parameter λ_s can be proved.

Theorem 7.92. *Let $n > 4$ and $p > (n+4)/(n-4)$. Then there exists a unique parameter $\lambda_s > 0$ such that for $\lambda = \lambda_s$, problem (7.219) admits a radial singular solution and this radial solution is unique.*

This result is proved in Section 7.11.5, where supercriticality is intensively exploited.

By means of energy considerations we can give bounds for radial singular solutions and the corresponding singular parameter. The precise blow-up rate $\sim C|x|^{-4/(p-1)}$ at $x = 0$ is determined and an explicit estimate from below is deduced. For this purpose we transform in Section 7.11.1 the differential equation in (7.219) into an autonomous system of ordinary differential equations and apply subtle energy estimates. This technique has proved to be very powerful for studying the precise asymptotic behaviour of entire solutions in $\mathbb{R}^n$ in [182]. Moreover, a characterisation of singular (respectively regular) radial solutions to (7.219) in terms of the corresponding dynamical system is given. This system is shown to have two critical points, the unstable manifolds of which are related to singular (respectively regular) radial solutions.

Theorem 7.93. *Let $n > 4$ and $p > (n+4)/(n-4)$. Assume that u_s is a singular radial solution of (7.219) with parameter λ_s. Then $\lambda_s > K_0$ and*

$$u_s(x) > \left(\frac{K_0}{\lambda_s}\right)^{1/(p-1)} |x|^{-4/(p-1)} - 1,$$

$$u_s(x) \sim \left(\frac{K_0}{\lambda_s}\right)^{1/(p-1)} |x|^{-4/(p-1)} \quad as\ x \to 0.$$

In particular, any radial solution to (7.219) for $\lambda \leq K_0$ is regular.

For the proof we refer to Section 7.11.3. Again supercriticality is crucial here.

Furthermore, we address the question whether the extremal solution is regular or not. This question seems to be quite difficult. Inspired by stability techniques developed in [131], we give here a partial result. In this respect a further critical exponent p_c arises for $n > 12$, see Proposition 7.97 below, which is defined as the unique solution $p_c \in \left(\frac{n+4}{n-4}, \infty\right)$ of

$$\frac{4p_c}{p_c - 1}\left(\frac{4}{p_c - 1} + 2\right)\left(n - 2 - \frac{4}{p_c - 1}\right)\left(n - 4 - \frac{4}{p_c - 1}\right) = \frac{n^2(n-4)^2}{16}. \quad (7.224)$$

Equivalently, p_c is the unique value of $p > \frac{n+4}{n-4}$ such that

$$-(n-4)(n^3 - 4n^2 - 128n + 256)(p-1)^4 + 128(3n-8)(n-6)(p-1)^3$$
$$+ 256(n^2 - 18n + 52)(p-1)^2 - 2048(n-6)(p-1) + 4096 = 0$$

and is given by

$$p_c = \frac{n+2 - \sqrt{n^2 + 4 - n\sqrt{n^2 - 8n + 32}}}{n-6 - \sqrt{n^2 + 4 - n\sqrt{n^2 - 8n + 32}}}.$$

In Section 7.11.4 we prove the following result.

Theorem 7.94. *Let $p_c \in ((n+4)/(n-4), \infty)$ be the number which is defined by (7.224) for $n \geq 13$. We assume that*

$$\frac{n+4}{n-4} < p < p_c \text{ if } n \geq 13, \qquad \frac{n+4}{n-4} < p < \infty \text{ if } 5 \leq n \leq 12.$$

Let $u^* \in H_0^2 \cap L^p(B)$ be the extremal radial solution of (7.219) corresponding to the extremal parameter λ^*, which is obtained as monotone limit of the minimal regular solutions u_λ for $\lambda \nearrow \lambda^*$. Then u^* is regular.

This result suggests that, in the parameter regime between $(n+4)/(n-4)$ and p_c, (λ^*, u^*) is a turning point on the branch of solutions to (7.219). This and much more has been recently proved by Dávila-Flores-Guerra [132]. Concerning the asymptotic behaviour of the branch of regular radial solutions they have the following result.

Theorem 7.95. *Let $p_c \in ((n+4)/(n-4), \infty)$ be the number which is defined by (7.224) for $n \geq 13$. We assume that*

$$\frac{n+4}{n-4} < p < p_c \text{ if } n \geq 13, \qquad \frac{n+4}{n-4} < p < \infty \text{ if } 5 \leq n \leq 12.$$

Then (7.219) has infinitely many regular radial solutions for $\lambda = \lambda_s$. For $\lambda \neq \lambda_s$, there are finitely many regular radial solutions the number of which becomes unbounded when $\lambda \to \lambda_s$.

We sketch the basic ideas of the proof at the end of Section 7.11.5.

7.11.1 An Autonomous System

In radial coordinates $r = |x| \in [0,1]$, the differential equation in (7.219) reads

$$u^{iv}(r) + \frac{2(n-1)}{r}u'''(r) + \frac{(n-1)(n-3)}{r^2}u''(r) - \frac{(n-1)(n-3)}{r^3}u'(r) = \lambda\left(1+u(r)\right)^p.$$

$$(7.225)$$

We put first

$$U(x) := 1 + u(x/\sqrt[4]{\lambda}) \text{ for } x \in B_{\sqrt[4]{\lambda}}(0), \quad u(x) = U(\sqrt[4]{\lambda}x) - 1 \text{ for } x \in B.$$

For $x \in B_{\sqrt[4]{\lambda}}(0)$ one has

$$\Delta^2 U(x) = U(x)^p. \tag{7.226}$$

Our purpose here is to transform (7.226) first into an autonomous equation and, subsequently, into an autonomous system. For some of the estimates which follow, it is convenient to rewrite the original assumption $p > \frac{n+4}{n-4}$ as

$$(n-4)(p-1) > 8. \tag{7.227}$$

Next, for $t < \frac{\log\lambda}{4}$ we set

$$U(r) = r^{-4/(p-1)}v(\log r), r \in (0, \sqrt[4]{\lambda}), \quad v(t) = e^{4t/(p-1)}U\left(e^t\right). \tag{7.228}$$

After the change (7.228) equation (7.226) takes the form

$$v^{iv}(t) + K_3 v'''(t) + K_2 v''(t) + K_1 v'(t) + K_0 v(t) = v^p(t), \qquad t < \frac{\log\lambda}{4}, \tag{7.229}$$

where the constants $K_i = K_i(n,p)$, with $i = 1,\ldots,3$, are given by

$$K_1 = -2\frac{(n-2)(n-4)(p-1)^3 + 4(n^2-10n+20)(p-1)^2 - 48(n-4)(p-1) + 128}{(p-1)^3},$$

$$K_2 = \frac{(n^2-10n+20)(p-1)^2 - 24(n-4)(p-1) + 96}{(p-1)^2},$$

$$K_3 = 2\frac{(n-4)(p-1)-8}{p-1}.$$

For K_0, we refer to (7.222). By using the supercriticality assumption (7.227), it is not difficult to show that

$$K_0 > 0, \qquad K_1 < 0, \qquad K_3 > 0.$$

On the other hand, the sign of K_2 depends on n and p.

Finally, we put

$$z(t) := v(-t), \qquad t > -\frac{\log \lambda}{4}.$$

For z, we have the differential equation analogous to (7.229), namely

$$z^{iv}(t) - K_3 z'''(t) + K_2 z''(t) - K_1 z'(t) + K_0 z(t) = z^p(t) \ \text{ for } t > -\frac{\log \lambda}{4}. \qquad (7.230)$$

In order to study the possibly singular behaviour of u near $r = 0$, we have to investigate the behaviour of z for $t \to \infty$. Equation (7.230) has two equilibrium points, namely 0 and $K_0^{1/(p-1)}$. First we show that once the solution converges to an equilibrium point, then all derivatives converge to 0 as $t \to \infty$.

Proposition 7.96. *Assume that $z : [T_0, \infty) \to \mathbb{R}$ exists for some T_0 and solves a constant coefficient fourth order equation*

$$z^{iv}(t) - K_3 z'''(t) + K_2 z''(t) - K_1 z'(t) = f(z(t)) \ \text{ for } t > -T_0, \qquad (7.231)$$

where $f \in C^1(\mathbb{R})$ and where the coefficients may be considered as arbitrary real numbers $K_j \in \mathbb{R}$. Moreover, let z_0 be such that $f(z_0) = 0$ and assume that z satisfies $\lim_{t\to\infty} z(t) = z_0$. Then for $k = 1,\ldots,4$, one has that

$$\lim_{t \to \infty} z^{(k)}(t) = 0. \qquad (7.232)$$

Proof. By assumption we have for any $q > 1$ that

$$\lim_{t\to\infty} \int_{t-2}^{t+3} |f(z(\tau))|^q \, d\tau = 0, \qquad \lim_{t\to\infty} \|z(\,.\,) - z_0\|_{C^0([t-2,t+3])} = 0.$$

We consider (7.231) as a fourth order "elliptic" equation and apply local L^q-estimates, which could of course be directly obtained in a much easier way for the ordinary differential equation (7.231), and conclude

$$\lim_{t\to\infty} \|z(\,.\,) - z_0\|_{W^{4,q}(t-1,t+2)} = 0.$$

By combining Sobolev embedding, see Theorem 2.6, and classical local Schauder estimates (Theorem 2.19) we then have that

$$\lim_{t\to\infty} \|z(\,.\,) - z_0\|_{C^{4,\gamma}([t,t+1])} = 0$$

which proves (7.232). $\qquad\qquad\square$

We now write (7.229) as a system in $\mathbb{R}^4$. We obtain from (7.228)

$$\frac{U'(r)}{r^3} = r^{-4p/(p-1)} \left(v'(t) - \frac{4}{p-1} v(t) \right) \qquad (7.233)$$

so that

$$U'(r) = 0 \iff v'(t) = \frac{4}{p-1}v(t) .$$

This fact suggests the definition

$$w_1(t) = v(t), \quad w_2(t) = v'(t) - \frac{4}{p-1}v(t),$$

$$w_3(t) = v''(t) - \frac{4}{p-1}v'(t), \quad w_4(t) = v'''(t) - \frac{4}{p-1}v''(t)$$

so that (7.229) becomes

$$\begin{cases} w_1'(t) = \frac{4}{p-1}w_1(t) + w_2(t) \\ w_2'(t) = w_3(t) \\ w_3'(t) = w_4(t) \\ w_4'(t) = C_2 w_2(t) + C_3 w_3(t) + C_4 w_4(t) + w_1^p(t), \end{cases} \tag{7.234}$$

where for $w_1 < 0$ we interpret $w_1^p := |w_1|^{p-1}w_1$ as its odd extension and

$$C_m = - \sum_{k=m-1}^{4} \frac{K_k 4^{k+1-m}}{(p-1)^{k+1-m}} \qquad \text{for } m = 1,2,3,4 \quad \text{with } K_4 = 1 .$$

This gives first that $C_1 = 0$ so that the term $C_1 w_1(t)$ does not appear in the last equation of (7.234). Moreover, we have the explicit formulae

$$C_2 = \frac{p-1}{4}K_0,$$
$$C_3 = -\frac{1}{(p-1)^2}\left((n^2 - 10n + 20)(p-1)^2 - 16(n-4)(p-1) + 48\right),$$
$$C_4 = -\frac{2}{p-1}\left((n-4)(p-1) - 6\right).$$

System (7.234) has the two stationary points (corresponding to $v_0 := 0$ and $v_s := K_0^{1/(p-1)}$)

$$O\left(0,0,0,0\right) \qquad \text{and} \qquad P\left(K_0^{1/(p-1)}, -\frac{4}{p-1}K_0^{1/(p-1)}, 0, 0\right).$$

Let us study first the "regular point" O. The linearised matrix at O is

$$M_O = \begin{pmatrix} \frac{4}{p-1} & 1 & 0 & 0 \\ 0 & 0 & 1 & 0 \\ 0 & 0 & 0 & 1 \\ 0 & C_2 & C_3 & C_4 \end{pmatrix}$$

and the characteristic polynomial is

$$\mu \mapsto \mu^4 + K_3\mu^3 + K_2\mu^2 + K_1\mu + K_0.$$

Then the eigenvalues are given by

$$\mu_1 = 2\frac{p+1}{p-1}, \quad \mu_2 = \frac{4}{p-1}, \quad \mu_3 = \frac{4p}{p-1} - n, \quad \mu_4 = 2\frac{p+1}{p-1} - n.$$

Since we assume that $p > \frac{n+4}{n-4} > \frac{n}{n-4} > \frac{n+2}{n-2}$, we have

$$\mu_1 > \mu_2 > 0 > \mu_3 > \mu_4.$$

This means that O is a hyperbolic point and that both the stable and the unstable manifolds are two-dimensional.

At the "singular point" P, the linearised matrix of the system (7.234) is given by

$$M_P = \begin{pmatrix} \frac{4}{p-1} & 1 & 0 & 0 \\ 0 & 0 & 1 & 0 \\ 0 & 0 & 0 & 1 \\ pK_0 & C_2 & C_3 & C_4 \end{pmatrix}.$$

The corresponding characteristic polynomial is

$$v \mapsto v^4 + K_3 v^3 + K_2 v^2 + K_1 v + (1-p)K_0$$

and the eigenvalues are given by

$$v_1 = \frac{N_1 + \sqrt{N_2 + 4\sqrt{N_3}}}{2(p-1)}, \qquad v_2 = \frac{N_1 - \sqrt{N_2 + 4\sqrt{N_3}}}{2(p-1)},$$

$$v_3 = \frac{N_1 + \sqrt{N_2 - 4\sqrt{N_3}}}{2(p-1)}, \qquad v_4 = \frac{N_1 - \sqrt{N_2 - 4\sqrt{N_3}}}{2(p-1)},$$

where

$$N_1 := -(n-4)(p-1) + 8, \qquad N_2 := (n^2 - 4n + 8)(p-1)^2,$$
$$N_3 := (9n - 34)(n-2)(p-1)^4 + 8(3n-8)(n-6)(p-1)^3$$
$$+ (16n^2 - 288n + 832)(p-1)^2 - 128(n-6)(p-1) + 256$$
$$= (p-1)^4 \left((n-2)^2 + pK_0\right).$$

The stability of the stationary point P is described by the following

Proposition 7.97. *Assume that $n > 4$ and $p > \frac{n+4}{n-4}$.*

1. We have $v_1, v_2 \in \mathbb{R}$ and $v_2 < 0 < v_1$.
2. For $5 \leq n \leq 12$ we have $v_3, v_4 \notin \mathbb{R}$ and $\mathrm{Re}\, v_3 = \mathrm{Re}\, v_4 < 0$.

3. *For $n \geq 13$ there exists a unique $p_c > \frac{n+4}{n-4}$ satisfying (7.224) and:*
   if $p < p_c$, then $v_3, v_4 \notin \mathbb{R}$ and $\operatorname{Re} v_3 = \operatorname{Re} v_4 < 0$,
   if $p = p_c$, then $v_3, v_4 \in \mathbb{R}$ and $v_4 = v_3 < 0$,
   if $p > p_c$, then $v_3, v_4 \in \mathbb{R}$ and $v_4 < v_3 < 0$.

Proof. We first observe that (7.227) is equivalent to

$$N_1 < 0 \tag{7.235}$$

and that (7.227) implies

$$N_2 - N_1^2 = 4(n-2)(p-1)^2 + 16(n-4)(p-1) - 64 > 4(n-2)(p-1)^2 + 64 > 0.$$

Next, we show that

$$N_3 > \frac{(N_2 - N_1^2)^2}{16}. \tag{7.236}$$

Indeed, by exploiting again (7.227), we have:

$$
\begin{aligned}
N_3 &- \frac{(N_2 - N_1^2)^2}{16} \\
&= 8(n-2)(n-4)(p-1)^4 + 16(n^2 - 10n + 20)(p-1)^3 - 128(n-4)(p-1)^2 \\
&\quad + 256(p-1) \\
&> 16(n^2 - 6n + 12)(p-1)^3 - 128(n-4)(p-1)^2 + 256(p-1) \\
&= 64(p-1)^3 + 16(n-2)(n-4)(p-1)^3 - 128(n-4)(p-1)^2 + 256(p-1) \\
&> 64(p-1)^3 + 128(n-2)(p-1)^2 - 128(n-4)(p-1)^2 + 256(p-1) \\
&= 64(p-1)^3 + 256(p-1)^2 + 256(p-1) = 64(p+1)^2(p-1) > 0.
\end{aligned}
$$

In particular, (7.236) implies that $N_3 > 0$. In turn, together with the fact that $N_2 > N_1^2$, this shows that $\sqrt{N_2 + 4\sqrt{N_3}} > |N_1|$ which proves Item 1.

In order to discuss the stability properties of the eigenvalues v_3 and v_4 we introduce the function

$$
\begin{aligned}
N_4 &:= 16N_3 - N_2^2 \\
&= -(n-4)(n^3 - 4n^2 - 128n + 256)(p-1)^4 + 128(3n-8)(n-6)(p-1)^3 \\
&\quad + 256(n^2 - 18n + 52)(p-1)^2 - 2048(n-6)(p-1) + 4096 \\
&= -(n-4)^2 n^2 (p-1)^4 \\
&\quad + 64p(p-1)^3 \left(2 + \frac{4}{p-1}\right)\left(n - 2 - \frac{4}{p-1}\right)\left(n - 4 - \frac{4}{p-1}\right). \tag{7.237}
\end{aligned}
$$

For $1.939\ldots < n < 12.565\ldots$, the first coefficient in (7.237) is positive, so that assuming

$$5 \leq n \leq 12,$$

we obtain with the help of (7.227):

$$
\begin{aligned}
N_4 &= -(n-4)(n^3-4n^2-128n+256)(p-1)^4+128(3n-8)(n-6)(p-1)^3 \\
&\quad +256(n^2-18n+52)(p-1)^2-2048(n-6)(p-1)+4096 \\
&> -8(n^3-4n^2-128n+256)(p-1)^3+128(3n-8)(n-6)(p-1)^3 \\
&\quad +256(n^2-18n+52)(p-1)^2-2048(n-6)(p-1)+4096 \\
&= 64n^2(p-1)^3-8(n-4)(n^2-40n+128)(p-1)^3 \\
&\quad +256(n^2-18n+52)(p-1)^2-2048(n-6)(p-1)+4096 \\
&> 64n(n-4)(p-1)^3-64(n^2-40n+128)(p-1)^2 \\
&\quad +256(n^2-18n+52)(p-1)^2-2048(n-6)(p-1)+4096 \\
&> 512n(p-1)^2+64(n-4)(3n-20)(p-1)^2-2048(n-6)(p-1)+4096 \\
&= 2048(p-1)^2+192(n-4)^2(p-1)^2-2048(n-6)(p-1)+4096 \\
&> 2048(p-1)^2+1536(n-4)(p-1)-2048(n-6)(p-1)+4096 \\
&= 2048(p-1)^2-512(n-12)(p-1)+4096>0,
\end{aligned}
$$

since $n \le 12$. This, together with (7.235), proves Item 2.

In order to prove Item 3, we assume that $n \ge 13$ and we study $N_4 = N_4(n,p)$ as a function of p. We compute its second derivative (with respect to p):

$$
-\frac{\partial^2 N_4}{\partial p^2} = 12(n-4)(n^3-4n^2-128n+256)(p-1)^2
$$
$$
-768(3n-8)(n-6)(p-1)-512(n^2-18n+52).
$$

This is a quadratic function of p which tends to $+\infty$ as $p \to +\infty$. Its minimum is smaller than the Sobolev exponent $(n+4)/(n-4)$ if and only if

$$
0 < (n^3-4n^2-128n+256)-4(3n-8)(n-6) = (n-18)(n^2+2n+12)+280.
$$

This is certainly true for $n \ge 18$, while for $n = 13,\ldots,17$, we have $\frac{\partial^2 N_4}{\partial p^2}\left(n,\frac{n+4}{n-4}\right) < 0$. Summarising, for $p > (n+4)/(n-4)$, $\frac{\partial^2 N_4}{\partial p^2}$ has at most one zero. Therefore, for $p > \frac{n+4}{n-4}$,

$$
p \mapsto N_4(n,p) \quad \text{is either always concave or it is first convex and then concave.}
$$
$$
\tag{7.238}
$$

Moreover, since the first coefficient in (7.237) is negative (recall $n \ge 13$), we have

$$
\lim_{p\to\infty} N_4(n,p) = -\infty \qquad \text{for all } n \ge 13. \tag{7.239}
$$

Finally, note that

$$N_4\left(n,\frac{n+4}{n-4}\right)=\frac{32768\,n^2}{(n-4)^3}>0\quad\text{and}\quad\frac{\partial N_4}{\partial p}\left(n,\frac{n+4}{n-4}\right)=\frac{20480\,n^2}{(n-4)^2}>0.\quad(7.240)$$

By (7.238)-(7.239)-(7.240) there exists a unique $p_c>(n+4)/(n-4)$ such that

$$N_4(n,p)>0\text{ for all }p<p_c\,,\qquad N_4(n,p_c)=0\,,\qquad N_4(n,p)<0\text{ for all }p>p_c.$$

This proves Item 3 and completes the proof of the proposition. $\qquad\qquad\square$

According to Proposition 7.97 we have in all cases

$$v_1>0,\qquad v_2<0,\qquad\operatorname{Re}v_4\le\operatorname{Re}v_3<0.$$

This means that P has a three-dimensional stable manifold and a one-dimensional unstable manifold.

We now show that there are only a few possible values for $\lim_{t\to-\infty}v(t)$, provided the limit exists. Here, v is as in (7.228). The following proposition holds independently of the signs of the coefficients K_i $(i=1,2,3)$.

Proposition 7.98. *Let v be a positive solution of (7.229) on $\left(-\infty,\frac{1}{4}\log\lambda\right)$ and assume that there exists $L\in[0,+\infty]$ such that*

$$\lim_{t\to-\infty}v(t)=L\,.$$

Then $L\in\{0,K_0^{1/(p-1)}\}$.

Proof. In order to avoid confusion with respect to the time direction we switch to the solution z of (7.230):

$$z^{iv}(t)-K_3z'''(t)+K_2z''(t)-K_1z'(t)+K_0z(t)=z^p(t)\text{ for }t>-\frac{1}{4}\log\lambda\,.$$

For contradiction, assume first that L is finite and $L\notin\{0,K_0^{1/(p-1)}\}$. Then $z^p(t)-K_0z(t)\to\alpha:=L^p-K_0L\neq0$ and for all $\varepsilon>0$ there exists $T>0$ such that

$$\alpha-\varepsilon\le z^{iv}(t)-K_3z'''(t)+K_2z''(t)-K_1z'(t)\le\alpha+\varepsilon\qquad\text{for all }t\ge T.\quad(7.241)$$

Take $\varepsilon\in(0,|\alpha|)$ such that $\alpha-\varepsilon$ and $\alpha+\varepsilon$ have the same sign and let

$$\delta:=\sup_{t\ge T}|z(t)-z(T)|<\infty.$$

Integrating (7.241) over $[T,t]$ for any $t\ge T$ yields

$$(\alpha-\varepsilon)(t-T)+C-|K_1|\delta\le z'''(t)-K_3z''(t)+K_2z'(t)\le(\alpha+\varepsilon)(t-T)+C+|K_1|\delta,$$

where $C = C(T)$ is a constant containing all the terms $z(T)$, $z'(T)$, $z''(T)$ and $z'''(T)$. Repeating this procedure twice more gives

$$\frac{\alpha - \varepsilon}{6}(t - T)^3 + O(t^2) \leq z'(t) \leq \frac{\alpha + \varepsilon}{6}(t - T)^3 + O(t^2) \qquad \text{as } t \to \infty .$$

This contradicts the assumption that z admits a finite limit as $t \to +\infty$.

Next, we exclude the case $L = +\infty$. Assume by contradiction that

$$\lim_{t \to +\infty} z(t) = +\infty. \tag{7.242}$$

Then there exists $T \in \mathbb{R}$ such that

$$z^{iv}(t) - K_3 z'''(t) + K_2 z''(t) - K_1 z'(t) \geq \frac{z^p(t)}{2} \qquad \text{for all } t \geq T.$$

Moreover, by integrating this inequality over $[T,t]$ (for $t \geq T$), we get

$$z'''(t) - K_3 z''(t) + K_2 z'(t) - K_1 z(t) \geq \frac{1}{2}\int_T^t z^p(s)ds + C \qquad \text{for all } t \geq T, \tag{7.243}$$

where $C = C(T)$ is a constant containing all the terms $z(T)$, $z'(T)$, $z''(T)$ and $z'''(T)$. From (7.242) and (7.243) we deduce that there exists $T' \geq T$ such that $\alpha := z'''(T') - K_3 z''(T') + K_2 z'(T') - K_1 z(T') > 0$. Since (7.229) is autonomous, we may assume that $T' = 0$. Therefore, we have

$$z^{iv}(t) - K_3 z'''(t) + K_2 z''(t) - K_1 z'(t) \geq \frac{z^p(t)}{2} \qquad \text{for all } t \geq 0 , \tag{7.244}$$

$$z'''(0) - K_3 z''(0) + K_2 z'(0) - K_1 z(0) = \alpha > 0 . \tag{7.245}$$

We may now apply the test function method developed by Mitidieri-Pohožaev [309]. More precisely, fix $T_1 > T > 0$ and a nonnegative function $\phi \in C_c^4[0,\infty)$ such that

$$\phi(t) = \begin{cases} 1 & \text{for } t \in [0,T] \\ 0 & \text{for } t \geq T_1 . \end{cases}$$

In particular, these properties imply that $\phi(T_1) = \phi'(T_1) = \phi''(T_1) = \phi'''(T_1) = 0$. Hence, multiplying inequality (7.244) by $\phi(t)$, integrating by parts, and recalling (7.245) yields

$$\int_0^{T_1} [\phi^{iv}(t) + K_3 \phi'''(t) + K_2 \phi''(t) + K_1 \phi'(t)]z(t)dt \geq \frac{1}{2}\int_0^{T_1} z^p(t)\phi(t)dt + \alpha . \tag{7.246}$$

We now apply Young's inequality. For any $\varepsilon > 0$ there exists $C(\varepsilon) > 0$ such that

$$z\phi^{(i)} = z\phi^{1/p}\frac{\phi^{(i)}}{\phi^{1/p}} \leq \varepsilon z^p \phi + C(\varepsilon)\frac{|\phi^{(i)}|^{p/(p-1)}}{\phi^{1/(p-1)}}, \qquad \phi^{(i)} = \frac{d^i \phi}{dt^i} \text{ for } i = 1,2,3,4.$$

Then, provided ε is chosen sufficiently small, (7.246) becomes

$$C\sum_{i=1}^{4}\int_{0}^{T_1}\frac{|\phi^{(i)}(t)|^{p/(p-1)}}{\phi^{1/(p-1)}(t)}\,dt \geq \frac{1}{4}\int_{0}^{T} z^p(t)\,dt + \alpha \qquad (7.247)$$

where $C = C(\varepsilon, K_i) > 0$. We now choose $\phi(t) = \phi_0(\frac{t}{T})$, where $\phi_0 \in C_c^4[0,\infty)$, $\phi_0 \geq 0$ and

$$\phi_0(\tau) = \begin{cases} 1 & \text{for } \tau \in [0,1] \\ 0 & \text{for } \tau \geq \tau_1 > 1 \,. \end{cases}$$

As noticed in [309], there exists a function ϕ_0 in such class satisfying moreover

$$\int_{0}^{\tau_1}\frac{|\phi_0^{(i)}(\tau)|^{p/(p-1)}}{\phi_0^{1/(p-1)}(\tau)}\,d\tau =: A_i < \infty \qquad (i = 1,2,3,4).$$

To see this, it suffices to fix any nonnegative nontrivial $\varphi \in C_c^4[0,\infty)$ and take $\phi_0 = \varphi^k$ for a sufficiently large integer power k. Thanks to a change of variables in the integrals, (7.247) becomes

$$C\sum_{i=1}^{4} A_i T^{1-ip/(p-1)} \geq \frac{1}{4}\int_{0}^{T} z^p(t)\,dt + \alpha \qquad \text{for all } T > 0.$$

Letting $T \to \infty$, the previous inequality contradicts (7.242). $\qquad\square$

7.11.2 Regular Minimal Solutions

The main goal here is to prove Theorem 7.90. We first prove the estimate (7.221) for λ^*. Let K_0 be as in (7.222). Then the function

$$u(x) := |x|^{-4/(p-1)} - 1$$

is an explicit singular solution of the differential equation in (7.219) with $\lambda = K_0$ and also a weak supersolution for (7.219) with parameter $\lambda = K_0$. To see this, one observes that $u \in H^2(B)$ in view of (7.227) and that also for biharmonic equations one has a kind of Hopf lemma for the boundary data, see Theorem 5.7 and Lemma 7.9. This shows $\lambda^* \geq K_0$.

In order to show that $\lambda^* < \Lambda_{2,1}/p$, we proceed as for (7.43), that is, we multiply (7.219) by a positive first eigenfunction φ_1 (see Theorem 3.7) of the Dirichlet problem and obtain

$$\Lambda_{2,1}\int_{B} u\varphi_1\,dx = \int_{B} u\Delta^2\varphi_1\,dx = \lambda\int_{B}(1+u)^p\varphi_1\,dx > p\lambda\int_{B} u\varphi_1\,dx, \qquad (7.248)$$

thereby proving the desired inequality.

We now turn to the most difficult part of Theorem 7.90, namely the immediate switch from existence of regular minimal solutions to nonexistence even of singular solutions. We start by proving boundedness of the transformed function v.

Lemma 7.99. *Let u be a radial solution of the Dirichlet problem (7.219), and define the corresponding functions $U = U(r)$ and $v = v(t)$ according to (7.226) and (7.228) respectively for $r \in (0, \sqrt[4]{\lambda})$ and $t \in (-\infty, \frac{1}{4}\log\lambda)$. Then v is bounded.*

Proof. By contradiction, assume that v is not bounded. In view of Proposition 7.98 we may exclude that the limit as $t \to -\infty$ exists. Hence, we assume that

$$0 \le \liminf_{t \to -\infty} v(t) < \limsup_{t \to -\infty} v(t) = +\infty.$$

Then there exists a sequence $t_k \to -\infty$ of local maxima for v such that for all k

$$\lim_{k \to +\infty} v(t_k) = +\infty, \quad v'(t_k) = 0. \tag{7.249}$$

Define

$$\lambda_k = v^{p-1}(t_k) \tag{7.250}$$

so that

$$\lim_{k \to +\infty} \lambda_k = +\infty.$$

Since (7.229) is an autonomous equation, the translated function

$$\tilde{v}_k(t) = v\left(t + t_k - \frac{1}{4}\log\lambda_k\right), \qquad t \in \left(-\infty, \frac{1}{4}\log\lambda - t_k + \frac{1}{4}\log\lambda_k\right)$$

also solves (7.229). In particular, the function

$$\tilde{U}_k(r) = r^{-4/(p-1)}\tilde{v}_k(\log r)$$

is a radial solution of equation (7.226) which satisfies the conditions

$$\tilde{U}_k(\sqrt[4]{\lambda_k}) = \lambda_k^{-1/(p-1)}\tilde{v}_k\left(\frac{1}{4}\log\lambda_k\right) = \lambda_k^{-1/(p-1)}v(t_k) = 1 \tag{7.251}$$

and by (7.233), (7.249), (7.250)

$$\tilde{U}_k'(\sqrt[4]{\lambda_k}) = -\frac{4}{p-1}\lambda_k^{-1/4} < 0. \tag{7.252}$$

Next, we define the radial function

$$u_k(r) = \tilde{U}_k(\sqrt[4]{\lambda_k}r) - 1 = \lambda_k^{-1/(p-1)}e^{4t_k/(p-1)}U\left(re^{t_k}\right) - 1$$

so that by (7.251) and (7.252) we have

$$\begin{cases} \Delta^2 u_k = \lambda_k(1+u_k)^p, \ u_k > 0 \text{ in } B, \\ u_k = 0 & \text{on } \partial B, \\ -\dfrac{\partial u_k}{\partial \nu} = \dfrac{4}{p-1} > 0 & \text{on } \partial B. \end{cases}$$

This boundary value problem is solved in a weak sense, since U is a weak solution of (7.226). In view of the comparison principle in B with respect to the boundary datum $\frac{\partial u}{\partial \nu}$, see Theorem 5.6, this shows that u_k is a weak supersolution for the problem

$$\begin{cases} \Delta^2 u = \lambda_k(1+u)^p, \ u > 0 \text{ in } B, \\ u = |\nabla u| = 0 & \text{on } \partial B. \end{cases} \tag{7.253}$$

By standard arguments, see for example [40, Lemma 3.3], we infer that for any λ_k problem (7.253) admits a weak solution. Since $\lambda_k \to +\infty$ this contradicts the nonexistence of solutions of (7.219) for large λ, see (7.248). This completes the proof of the lemma. $\qquad\square$

In Section 7.11.1 we transformed the differential equation in (7.219) for radial functions into the autonomous system (7.234) which has two critical points O and P. With the help of the former we give a precise characterisation of regular solutions of (7.219).

Proposition 7.100. *Let $u = u(r)$ be a radial solution of (7.219) and let*

$$W(t) = (w_1(t), w_2(t), w_3(t), w_4(t))$$

be the corresponding trajectory relative to (7.234). Then u is regular (i.e. $u \in L^\infty(B)$) if and only if

$$\lim_{t \to -\infty} W(t) = O.$$

Proof. If W corresponds to a regular solution, it is obvious that $\lim_{t\to-\infty} W(t) = O$. Let us now assume conversely that $\lim_{t\to-\infty} W(t) = O$; we have to prove that the corresponding solution u of (7.219) is regular. We calculate the eigenvectors of M_O corresponding to the positive eigenvalues i.e. spanning the unstable manifold. These are

$$W_1 = \left(1, 2, 4\frac{p+1}{p-1}, 8\left(\frac{p+1}{p-1}\right)^2\right) \text{ for } \mu_1 = 2\frac{p+1}{p-1};$$

$$W_2 = (1,0,0,0) \text{ for } \mu_2 = \frac{4}{p-1}.$$

Since $\mu_1 > \mu_2$, all trajectories approaching O as $t \to -\infty$ are tangent to W_2 except one, which is tangent to W_1, see [36] for $p \neq 3$ and [220] for $p = 3$.

But for a solution to (7.219), the latter case cannot occur, since one always has $u(r) > 0, u'(r) \le 0$, i.e. $w_1 > 0, w_2 \le 0$. So, for any solution of (7.219), we may conclude that $ru'(r) = o(u(r))$ for $r \searrow 0$. That means that for any $\varepsilon > 0$ and $r > 0$ close enough to 0 we have

$$-\varepsilon < \frac{ru'(r)}{u(r)} \le 0.$$

Integration yields that for $r \searrow 0$

$$0 \le u(r) \le Cr^{-\varepsilon}.$$

Using this information, the differential equation and $\lim_{t \to \infty} W(t) = 0$, i.e.

$$r^{4/(p-1)}u(r) \to 0, \qquad r^{1+4/(p-1)}u'(r) \to 0,$$
$$r^{2+4/(p-1)}\Delta u(r) \to 0, \quad r^{3+4/(p-1)}(\Delta u)'(r) \to 0,$$

successive integrations of (7.219) show that

$$(\Delta u)'(r) = O(1), \quad \Delta u(r) = O(1), \quad u'(r) = O(1), \quad u(r) = O(1)$$

for $r \searrow 0$. This proves that u is regular. $\square$

We may now prove

Proposition 7.101. *Assume that u_s is a solution of (7.219) with parameter λ_s. Then the Dirichlet problem (7.219) has for any $\lambda \in (0, \lambda_s)$ a regular radially decreasing minimal solution.*

Proof. Suppose that u_s is a solution of (7.219) corresponding to $\lambda = \lambda_s$. After possibly replacing u_s by the minimal solution of (7.219) corresponding to $\lambda = \lambda_s$, we may assume that u_s is radial. We look for a regular solution of (7.219) for a fixed $\lambda \in (0, \lambda_s)$. Put $u_0 = u_s$ and define $u_1 = \frac{\lambda}{\lambda_s} u_0$ so that u_1 solves

$$\int_B u_1 \Delta^2 \varphi \, dx = \lambda \int_B (1 + u_0)^p \varphi \, dx \qquad \text{for all } \varphi \in C^4(\overline{B}) \cap H_0^2(B).$$

We define by iteration u_k as the unique solution of

$$\int_B u_k \Delta^2 \varphi \, dx = \lambda \int_B (1 + u_{k-1})^p \varphi \, dx \qquad \text{for all } \varphi \in C^4(\overline{B}) \cap H_0^2(B). \tag{7.254}$$

By Proposition 3.6 we deduce that

$$0 < u_{\min} \le u_k \le u_{k-1} \qquad \text{for all } k \ge 1, \tag{7.255}$$

where $u_{\min}$ denotes the minimal solution of (7.219) with respect to the parameter λ. By monotone convergence it follows that there exists $u \in L^p(B)$ such that $u_k \to u$ in

$L^p(B)$ as $k \to \infty$, $u \geq u_{\min}$. Moreover, passing to the limit in (7.254) we have

$$\int_B u \Delta^2 \varphi \, dx = \lambda \int_B (1+u)^p \varphi \, dx \qquad \text{for all } \varphi \in C^4(\overline{B}) \cap H_0^2(B).$$

Fix $\overline{\vartheta} \in (\frac{\lambda}{\lambda_s}, 1)$ and introduce a strictly increasing sequence (ϑ_k) with $\frac{\lambda}{\lambda_s} < \vartheta_k < \overline{\vartheta}$ for any $k \geq 1$. Note that for any $\alpha > 0$ and for any $\beta > \alpha$ there exists $\gamma > 0$ such that for all $s \geq 0$

$$(1 + \alpha s)^p \leq \beta^p (1+s)^p + \gamma. \tag{7.256}$$

By (7.256) there exists $C_1 > 0$ such that for all $\varphi \in C^4(\overline{B})$, $\varphi \geq 0$, $\varphi = |\nabla \varphi| = 0$ on ∂B

$$\int_B u_2 \Delta^2 \varphi \, dx = \lambda \int_B (1+u_1)^p \varphi \, dx = \lambda \int_B \left(1 + \frac{\lambda}{\lambda_s} u_0\right)^p \varphi \, dx$$

$$\leq \lambda \int_B \left(\vartheta_1^p (1+u_0)^p + C_1\right) \varphi \, dx = \int_B (\vartheta_1^p u_1 + \lambda C_1 \psi) \Delta^2 \varphi \, dx,$$

where ψ is the unique solution of the Dirichlet problem

$$\begin{cases} \Delta^2 \psi = 1 & \text{in } B, \\ \psi = |\nabla \psi| = 0 & \text{on } \partial B. \end{cases}$$

The weak comparison principle (Proposition 3.6) yields

$$u_2 \leq \vartheta_1^p u_1 + \lambda C_1 \psi \leq \vartheta_1 u_1 + \lambda C_1 \psi.$$

Iterating this procedure we prove that for any $k \geq 1$ there exists $C_k > 0$ such that

$$u_{k+1} \leq \vartheta_k u_k + \lambda C_k \psi. \tag{7.257}$$

Since we chose $\frac{\lambda}{\lambda_s} < \vartheta_k < \overline{\vartheta} < 1$ it follows by (7.257) that for any $k \geq 1$

$$u_k \leq (\overline{\vartheta})^k u_0 + D_k \qquad \text{for all } k \geq 1 \tag{7.258}$$

for suitable $D_k > 0$. Therefore, for any $\varepsilon > 0$ there exists $\overline{k}$ such that $(\overline{\vartheta})^{\overline{k}} < \varepsilon$ and hence, by (7.255) and (7.258), we have

$$0 \leq u \leq u_{\overline{k}} \leq \varepsilon u_0 + D_{\overline{k}}.$$

From this we deduce that for any $\varepsilon > 0$

$$0 \leq \limsup_{r \searrow 0} r^{4/(p-1)} u(r) \leq \limsup_{r \searrow 0} \left(\varepsilon r^{4/(p-1)} u_0(r) + r^{4/(p-1)} D_{\overline{k}}\right) = \varepsilon L,$$

where $L = \limsup_{r \searrow 0} r^{4/(p-1)} u_0(r)$ is finite according to Lemma 7.99. This proves that

$$\lim_{r \searrow 0} r^{4/(p-1)} u(r) = 0.$$

Finally by (7.228), Propositions 7.96 and 7.100 we conclude that $u \in L^\infty(B)$.

The minimal solution $u_{\min}$ may now be obtained by means of an iterative procedure starting with 0. Radial symmetry is hence obvious, see also Theorem 7.1. Furthermore, the equation written in radial coordinates shows that u is radially decreasing. $\qquad\square$

Proof of Theorem 7.90. First, we remark that Items 1 and 3 are proved by Proposition 7.101.

As for Item 2, i.e. existence of a possibly singular solution for the extremal parameter λ^*, we obtain by means of a generalised Pohožaev identity uniform bounds for minimal regular solutions to (7.219) (for $\lambda \in (0, \lambda^*)$) in $H_0^2 \cap L^{p+1}(B)$, which allows to take a monotone limit as $\lambda \nearrow \lambda^*$.

The estimates (7.221) are proved at the beginning of this section. $\qquad\square$

7.11.3 Characterisation of Singular Solutions

Let u be a radial singular solution of (7.219) and let $v = v(t)$ be the corresponding function defined in (7.228). Let $z(t) = v(-t)$ so that $z(t)$ solves the equation (7.230) for $t > -\frac{1}{4} \log \lambda$.

We introduce here the energy function

$$E(t) := \frac{1}{p+1} z(t)^{p+1} - \frac{K_0}{2} z(t)^2 - \frac{K_2}{2} |z'(t)|^2 + \frac{1}{2} |z''(t)|^2. \tag{7.259}$$

This energy function will prove to be a useful tool to establish integrability properties of z. These are essential in order to characterise singular solutions in terms of critical points of the dynamical system (7.234). See Proposition 7.107 below which is the main result of this section.

The first result is analogous to Proposition 7.96.

Lemma 7.102. *Let $z : \left(-\frac{1}{4} \log \lambda, \infty\right) \to \mathbb{R}$ be the solution of (7.230) corresponding to a radial singular solution of (7.219). Then for $k = 1, \ldots, 4$ the functions z and $z^{(k)}$ are bounded in $\left(-\frac{1}{4} \log \lambda, \infty\right)$.*

Proof. By Lemma 7.99 it follows that $z(t) = v(-t)$ is bounded in $\left(-\frac{1}{4} \log \lambda, \infty\right)$. Put $I = \left(-\frac{1}{4} \log \lambda, \infty\right)$ and $t_0 = -\frac{1}{4} \log \lambda$. Then $z^p(t) - K_0 z(t)$ is bounded in I and hence, by local L^q–estimates for fourth order elliptic equations (see Section 2.5.2), we infer that for any $q > 1$ there exists a constant $C_q > 0$ such that for any $t > t_0 + 1$ we have

$$\|z(\,.\,)\|_{W^{4,q}(t-1,t+2)} \le C_q \|z\|_{L^\infty(I)}.$$

By combining Sobolev embeddings and local Schauder estimates we conclude that there exists a positive constant independent of t, still called C_q, such that

$$\|z(\cdot)\|_{C^{4,\gamma}(t,t+1)} \leq C_q \|z\|_{L^\infty(I)}.$$

$\square$

In the next four lemmas we prove some summability properties for the function z and its derivatives.

Lemma 7.103. *Let $t_0 = -\frac{1}{4}\log\lambda$. Then*

$$\int_{t_0}^{\infty} |z'(s)|^2 ds + \int_{t_0}^{\infty} |z''(s)|^2 ds < \infty.$$

Proof. Let $E(t)$ be the function defined in (7.259). For any $t > t_0$ we obtain by integration by parts and exploiting (7.230)

$$E(t) - E(t_0) = \int_{t_0}^{t} E'(s)\,ds = \int_{t_0}^{t} \left(z^p z' - K_0 z z' - K_2 z' z'' + z'' z'''\right)\,ds$$

$$= z'(t)z'''(t) - z'(t_0)z'''(t_0) + \int_{t_0}^{t} z'\left(z^p - K_0 z - K_2 z'' - z^{iv}\right)\,ds$$

$$= z'(t)z'''(t) - z'(t_0)z'''(t_0) + \int_{t_0}^{t} z'\left(-K_3 z''' - K_1 z'\right)\,ds$$

$$= z'(t)z'''(t) - z'(t_0)z'''(t_0) - K_3 z'(t)z''(t) + K_3 z'(t_0)z''(t_0)$$

$$+ \int_{t_0}^{t} \left(K_3 z''(s)^2 - K_1 z'(s)^2\right)\,ds. \tag{7.260}$$

By Lemma 7.102 it follows that $E(t)$ and the functions $z'(t), z''(t), z'''(t)$ are bounded in $I = (t_0,\infty)$, while around t_0 they are obviously smooth. This together with (7.260) and the fact that $K_3 > 0, K_1 < 0$ proves the claim. $\square$

Lemma 7.104. *We have*

$$\int_{t_0}^{\infty} |z'''(s)|^2 ds < \infty.$$

Proof. We multiply the equation (7.230) by z'' and integrate over (t_0,t) to obtain

$$\int_{t_0}^{t} \left(z^{iv}(s) - K_3 z'''(s) + K_2 z''(s) - K_1 z'(s) + K_0 z(s)\right) z''(s)\,ds = \int_{t_0}^{t} z^p(s)z''(s)\,ds. \tag{7.261}$$

First, we prove that all the lower order terms in the integral identity (7.261) are bounded. By Lemmas 7.102, 7.103, and integrating by parts we have

$$\left|\int_{t_0}^{t} z(s)z''(s)\,ds\right| \leq |z(t)z'(t)| + |z(t_0)z'(t_0)| + \int_{t_0}^{t} |z'(s)|^2 ds = O(1) \quad \text{as } t \to \infty.$$

By Lemma 7.103 and Hölder's inequality we have

$$\left| \int_{t_0}^{t} z'(s)z''(s)\, ds \right| \leq \left(\int_{t_0}^{t} |z'(s)|^2 ds \right)^{1/2} \left(\int_{t_0}^{t} |z''(s)|^2 ds \right)^{1/2} = O(1) \quad \text{as } t \to \infty.$$

By Lemma 7.102, integrating by parts, and Lemma 7.103, we obtain

$$\left| \int_{t_0}^{t} z^p(s)z''(s)\, ds \right| \leq |z^p(t)z'(t)| + |z^p(t_0)z'(t_0)| + \left| \int_{t_0}^{t} p z^{p-1}(s)z'(s)^2 \, ds \right| = O(1)$$

as $t \to \infty$. Using again Lemma 7.102 we conclude that

$$\left| \int_{t_0}^{t} z'''(s)z''(s)\, ds \right| \leq \frac{1}{2}|z''(t)|^2 + \frac{1}{2}|z''(t_0)|^2 = O(1) \quad \text{as } t \to \infty.$$

Finally, after integrating by parts we infer that

$$\int_{t_0}^{t} |z'''(s)|^2 ds = z'''(t)z''(t) - z'''(t_0)z'(t_0) - \int_{t_0}^{t} z^{iv}(s)z''(s)\, ds = O(1) \quad \text{as } t \to \infty$$

in view of Lemmas 7.102, 7.103, and by inserting the above estimates into (7.261). This completes the proof of the lemma.   □

Lemma 7.105. *We have*

$$\int_{t_0}^{\infty} |z^{iv}(s)|^2 ds < \infty.$$

Proof. We multiply the equation (7.230) by z^{iv} and integrate over (t_0, t) to obtain

$$\int_{t_0}^{t} |z^{iv}(s)|^2 ds = \int_{t_0}^{t} \left(z^p(s) - K_0 z(s) + K_1 z'(s) - K_2 z''(s) + K_3 z'''(s) \right) z^{iv}(s)\, ds.$$

$$\tag{7.262}$$

Arguing as in the proof of Lemma 7.104 one can easily prove that the right hand side of (7.262) remains bounded as $t \to \infty$. This completes the proof of the lemma.   □

Lemma 7.106. *We have*

$$\int_{t_0}^{\infty} z^2(s)|z^{p-1}(s) - K_0|^2 ds < \infty.$$

Proof. Using the differential equation (7.230) we obtain

$$\left(z^{iv}(s) - K_3 z'''(s) + K_2 z''(s) - K_1 z'(s) \right)^2 = z^2(s)|z^{p-1}(s) - K_0|^2.$$

The proof of the lemma follows immediately from Lemmas 7.103 and 7.105.   □

By considering the autonomous system (7.234) we can now characterise singular solutions of (7.219). The next result is the complement to Proposition 7.100

Proposition 7.107. *Let $u = u(r)$ be a radial solution of (7.219) and let*

$$W(t) = (w_1(t), w_2(t), w_3(t), w_4(t))$$

be the corresponding trajectory relative to (7.234). Then u is singular if and only if

$$\lim_{t \to -\infty} W(t) = P.$$

Proof. Let $W = (w_1, w_2, w_3, w_4)$ be the solution of the dynamical system (7.234) corresponding to a radial singular solution u of (7.219). In view of Lemmas 7.103 and 7.106, we infer that

there exists $\{\sigma_k\}$ such that $\sigma_{k+1} < \sigma_k$, $\displaystyle\lim_{k \to \infty} \sigma_k = -\infty$, $\displaystyle\lim_{k \to +\infty} (\sigma_{k+1} - \sigma_k) = 0$

and such that
$$\text{either} \quad \lim_{k \to \infty} W(\sigma_k) = P \quad \text{or} \quad \lim_{k \to \infty} W(\sigma_k) = O. \tag{7.263}$$

Then we claim that (7.263) holds on the whole real line, without extracting a subsequence, that is

$$\text{either} \quad \lim_{t \to -\infty} W(t) = P \quad \text{or} \quad \lim_{t \to -\infty} W(t) = O \tag{7.264}$$

respectively in the two cases of (7.263). We prove that the first alternative in (7.263) implies the first alternative in (7.264), the implication with the corresponding second alternative (obtained by replacing P with O) being similar. So, assume for contradiction that $\lim_{k \to \infty} W(\sigma_k) = P$ and that $\lim_{t \to -\infty} W(t)$ does not exist. Then there would exist a subsequence $(k_\ell)_{\ell \in \mathbb{N}}$ with the following properties: for any small enough $\varepsilon > 0$ there exists ℓ_ε such that for all $\ell \geq \ell_\varepsilon$ one has that

$$\left| W(\sigma_{k_\ell}) - P \right| < \varepsilon, \qquad \sigma_{k_\ell} - \sigma_{k_\ell+1} < \varepsilon^2$$

and moreover there exists $\theta_\ell \in (\sigma_{k_\ell+1}, \sigma_{k_\ell})$ with

$$|W(s) - P| < 2\varepsilon \quad \text{for all } s \in (\theta_\ell, \sigma_{k_\ell}) \quad \text{and} \quad |W(\theta_\ell) - P| = 2\varepsilon.$$

The triangle inequality shows that $|W(\theta_\ell) - W(\sigma_{k_\ell})| > \varepsilon$ and hence

$$\frac{1}{\sigma_{k_\ell} - \theta_\ell} \left| W(\theta_\ell) - W(\sigma_{k_\ell}) \right| > \frac{1}{\varepsilon}.$$

By the mean value theorem we conclude that

$$\frac{1}{\varepsilon} < \frac{1}{\sigma_{k_\ell} - \theta_\ell} \left| \int_{\theta_\ell}^{\sigma_{k_\ell}} W'(s)\, ds \right| \leq \frac{1}{\sigma_{k_\ell} - \theta_\ell} \int_{\theta_\ell}^{\sigma_{k_\ell}} \left| W'(s) \right| ds$$

so that there exists $\tau_\ell \in [\theta_\ell, \sigma_{k_\ell}]$ with

$$\left| W'(\tau_\ell) \right| > \frac{1}{\varepsilon}.$$

Since ε is arbitrarily small, $\left| W(\sigma_{k_\ell}) - P \right| < \varepsilon$, $\left| W(\tau_\ell) - P \right| \leq 2\varepsilon$ and since W solves system (7.234), this is impossible for large enough ℓ. This contradiction shows that (7.264) holds.

In view of Proposition 7.100 we may exclude the second case in (7.264) since it would imply that u is a regular solution. Therefore, only the first case may occur and this proves that if W corresponds to a radial singular solution u of (7.219), then $\lim_{t \to -\infty} W(t) = P$. The converse conclusion is obvious. $\qquad\square$

The energy function defined in (7.259) will help us to specify further the behaviour of singular solutions of (7.219) near $r = 0$. To this end we may assume by Proposition 7.107 that

$$\lim_{t \to \infty} z(t) = K_0^{1/(p-1)}.$$

Next we prove the following.

Lemma 7.108. *Let u_s be a singular solution of (7.219) with parameter λ_s and let $z(t) : (-\frac{1}{4} \log \lambda_s, \infty) \to (0, \infty)$ be the corresponding solution of (7.230). Then it cannot happen that $z'(t_0) = 0$ for some t_0.*

Proof. Assume by contradiction that $z'(t_0) = 0$ for some t_0. Then by (7.233), we have that $z'(-\frac{1}{4} \log \lambda_s) \neq 0$ and hence, z is not a constant. We infer from (7.260) that for any $t > t_0$

$$E(t) - E(t_0) = z'(t)z'''(t) - K_3 z'(t)z''(t) + \int_{t_0}^{t} \left(K_3 z''(s)^2 - K_1 z'(s)^2 \right) ds.$$

Letting $t \to \infty$ and observing Proposition 7.96 yields

$$E(\infty) - E(t_0) = \int_{t_0}^{\infty} \left(K_3 z''(s)^2 - K_1 z'(s)^2 \right) ds > 0$$

$$\Rightarrow -\frac{p-1}{2(p+1)} K_0^{(p+1)/(p-1)} > \frac{1}{p+1} z(t_0)^{p+1} - \frac{K_0}{2} z(t_0)^2 + \frac{1}{2} |z''(t_0)|^2$$

$$\geq \min_{\zeta \geq 0} \left(\frac{\zeta^{p+1}}{p+1} - \frac{K_0}{2} \zeta^2 \right) = -\frac{p-1}{2(p+1)} K_0^{(p+1)/(p-1)},$$

a contradiction. $\qquad\square$

Proof of Theorem 7.93. We know from (7.233) that

$$v'\left(\frac{1}{4}\log \lambda_s\right) = \frac{4}{p-1}v\left(\frac{1}{4}\log \lambda_s\right) > 0$$

so that

$$z'\left(-\frac{1}{4}\log \lambda_s\right) < 0.$$

Lemma 7.108 then shows that for all $t \geq -\frac{1}{4}\log \lambda_s$ we have

$$z'(t) < 0 \quad \Rightarrow \quad z(t) > K_0^{1/(p-1)} \quad \Rightarrow \quad U(x) > K_0^{1/(p-1)}|x|^{-4/(p-1)}$$

so that

$$u_s(x) > \left(\frac{K_0}{\lambda_s}\right)^{1/(p-1)}|x|^{-4/(p-1)} - 1.$$

In particular $0 = u_s(1)$, which entails $\lambda_s > K_0$.

Finally, as a straightforward consequence of Proposition 7.107, we obtain the asymptotic behaviour of singular solutions at the origin. If u_s is a singular radial solution of (7.219) with parameter λ_s, then

$$u_s(r) \sim \left(\frac{K_0}{\lambda_s}\right)^{1/(p-1)} r^{-4/(p-1)} \qquad \text{as } r \to 0.$$

This completes the proof. $\qquad\qquad\qquad\qquad\qquad\qquad\qquad\qquad\qquad\qquad\square$

Remark 7.109. With a completely analogous proof one can show a similar result for different nonlinearities such as $f(u) = e^u$ or $f(u) = (1-u)^{-k}$ for $k > 1$. See [39]. Here one can also find related results under Steklov boundary conditions.

7.11.4 Stability of the Minimal Regular Solution

In this section we shall give the proof of Theorem 7.91.

Let $\lambda \in (0, \lambda^*)$, and let u_λ be the corresponding minimal solution. By Proposition 7.101 we know that u_λ is a regular solution. Consider the following weighted η-eigenvalue problem

$$\begin{cases} \Delta^2 \psi = \eta \lambda p(1+u_\lambda)^{p-1}\psi & \text{in } B, \\ \psi = |\nabla \psi| = 0 & \text{on } \partial B, \end{cases} \qquad (7.265)$$

and let

$$\eta_1(\lambda) = \inf_{\psi \in H_0^2(B) \setminus \{0\}} \frac{\int_B |\Delta \psi|^2 \, dx}{\lambda p \int_B (1 + u_\lambda)^{p-1} \psi^2 \, dx} \tag{7.266}$$

be the corresponding first eigenvalue. Since $u_\lambda \in L^\infty(B)$, we infer by compactness of the embedding $H_0^2(B) \subset L^2(B)$ that the minimum in (7.266) is achieved. Note that by the Lagrange multiplier method any minimiser ψ_1 of $\eta_1(\lambda)$ solves (7.265) with $\eta = \eta_1(\lambda)$.

Since u_λ is a regular solution of (7.219), we infer by L^q–estimates for fourth order elliptic equations and Schauder estimates that both u_λ and ψ_1 are classical solutions of (7.219) and (7.265), respectively. In the next lemma we show that ψ_1 does not change sign in B.

Lemma 7.110. *Let ψ_1 be a minimiser for $\eta_1(\lambda)$. Then $\psi_1 > 0$ in B up to a constant multiple.*

Proof. This can be obtained by using the dual cones decomposition, exactly in the same way as in the proof of Theorem 3.7. $\qquad\square$

Lemma 7.111. *Let $\eta_1(\lambda)$ be the first eigenvalue of (7.265). Then $\eta_1(\lambda) > 1$.*

Proof. Fix $\overline{\lambda} \in (\lambda, \lambda^*)$ and consider the corresponding minimal solution $u_{\overline{\lambda}}$ of (7.219). Since $u_\lambda, u_{\overline{\lambda}}$ are minimal solutions for the respective problems, we have that $u_\lambda \leq u_{\overline{\lambda}}$ in B. Boggio's maximum principle yields $u_\lambda < u_{\overline{\lambda}}$ in B. By Lemma 7.110 we may fix a positive minimiser ψ_1 of (7.266). Convexity of $s \mapsto (1+s)^p$ yields

$$\lambda p \eta_1(\lambda) \int_B (u_{\overline{\lambda}} - u_\lambda)(1 + u_\lambda)^{p-1} \psi_1 \, dx$$

$$= \int_B (u_{\overline{\lambda}} - u_\lambda) \Delta^2 \psi_1 \, dx$$

$$= \int_B \left[\overline{\lambda}(1 + u_{\overline{\lambda}})^p - \lambda(1 + u_\lambda)^p \right] \psi_1 \, dx > \lambda \int_B \left[(1 + u_{\overline{\lambda}})^p - (1 + u_\lambda)^p \right] \psi_1 \, dx$$

$$\geq \lambda p \int_B (1 + u_\lambda)^{p-1} (u_{\overline{\lambda}} - u_\lambda) \psi_1 \, dx.$$

This proves that $\eta_1(\lambda) > 1$. $\qquad\square$

Proof of Theorem 7.91. Consider now the first eigenvalue $\mu_1(\lambda)$ for the linearised operator $\Delta^2 - \lambda p(1 + u_\lambda)^{p-1}$. We have

$$\mu_1(\lambda) = \inf_{w \in H_0^2(B) \setminus \{0\}} \frac{\int_B |\Delta w|^2 \, dx - \lambda p \int_B (1 + u_\lambda)^{p-1} w^2 \, dx}{\int_B w^2 \, dx}.$$

Observing that $\eta_1(\lambda) > 1$ by Lemma 7.111, we have for any $w \in H_0^2(B)$

$$\int_B |\Delta w|^2 dx - \lambda p \int_B (1+u_\lambda)^{p-1} w^2 \, dx \geq \left(1 - \frac{1}{\eta_1(\lambda)}\right) \int_B |\Delta w|^2 \, dx$$
$$\geq \Lambda_{2,1} \left(1 - \frac{1}{\eta_1(\lambda)}\right) \int_B w^2 \, dx,$$

where $\Lambda_{2,1}$ denotes the first eigenvalue of the linear Dirichlet problem. It follows that

$$\mu_1(\lambda) \geq \Lambda_{2,1} \left(1 - \frac{1}{\eta_1(\lambda)}\right) > 0.$$

This completes the proof of Theorem 7.91. $\qquad\qquad\square$

Proof of Theorem 7.94. We make use of an idea from [131]. Let u_λ denote the positive minimal regular solution of (7.219) for $0 \leq \lambda < \lambda^*$. According to Theorem 7.91 these are stable so that one has in particular

$$\int_B (\Delta\varphi(x))^2 \, dx - p\lambda \int_B (1+u_\lambda(x))^{p-1}\varphi(x)^2 \, dx \geq 0 \qquad \text{for all } \varphi \in C_c^\infty(B).$$

By taking the monotone limit we obtain that

$$\int_B (\Delta\varphi(x))^2 \, dx - p\lambda^* \int_B (1+u^*(x))^{p-1}\varphi(x)^2 \, dx \geq 0 \qquad \text{for all } \varphi \in C_c^\infty(B).$$
$$(7.267)$$

We assume now for contradiction that u^* is singular. Then according to Theorem 7.93 we have the following estimate from below:

$$u^*(x) > \left(\frac{K_0}{\lambda^*}\right)^{1/(p-1)} |x|^{-4/(p-1)} - 1.$$

Combining this with (7.267) yields

$$\int_B (\Delta\varphi(x))^2 \, dx \geq pK_0 \int_B \frac{\varphi(x)^2}{|x|^4} \, dx \qquad \text{for all } \varphi \in C_c^\infty(B).$$

However under the "subcriticality" assumptions that we made it follows that we have $pK_0 > n^2(n-4)^2/16$. This contradicts the optimality of the constant in Hardy's inequality

$$\int_B (\Delta\varphi(x))^2 \, dx \geq \frac{n^2(n-4)^2}{16} \int_B \frac{\varphi(x)^2}{|x|^4} \, dx \qquad \text{for all } \varphi \in C_c^\infty(B),$$

see [354] and the more recent paper [130]. Therefore, u^* is regular. $\qquad\square$

7.11.5 Existence and Uniqueness of a Singular Solution

Existence of a singular solution for a parameter λ to be specified will be shown by passing to the limit in a sequence of suitably rescaled regular solutions u to (7.219). We recall the transformation used throughout Section 7.11.1

$$U(x) = 1 + u(x/\sqrt[4]{\lambda}) \qquad \text{for } x \in B_{\sqrt[4]{\lambda}}(0)$$

so that we shall study the equation

$$\Delta^2 U = U^p \qquad \text{in } B_{\sqrt[4]{\lambda}}(0). \tag{7.268}$$

Since the equation (7.268) is invariant under the scaling

$$U_a(x) = aU(a^{\frac{p-1}{4}} x)$$

i.e. U is a solution of (7.268) if and only if U_a is a solution of (7.268), it is not restrictive to concentrate our attention on solutions U of the equation (7.268) which satisfy the condition $U(0) = 1$.

Next we define $U_\gamma = U_\gamma(r)$ as the unique solution of the initial value problem (see (7.225))

$$U_\gamma^{iv} + \frac{2(n-1)}{r}U_\gamma''' + \frac{(n-1)(n-3)}{r^2}U_\gamma'' - \frac{(n-1)(n-3)}{r^3}U_\gamma' = |U_\gamma|^{p-1}U_\gamma, \tag{7.269}$$

$$U_\gamma(0) = 1, \qquad U_\gamma'(0) = U_\gamma'''(0) = 0, \qquad U_\gamma''(0) = \gamma < 0.$$

We report here a crucial result from [182].

Lemma 7.112. *Let $n > 4$ and $p > (n+4)/(n-4)$.*

1. *There exists a unique $\overline{\gamma} < 0$ such that the solution $U_{\overline{\gamma}}$ of (7.269) exists on the whole interval $[0,\infty)$, it is positive everywhere, it vanishes at infinity and it satisfies $U_{\overline{\gamma}}'(r) < 0$ for any $r \in (0,\infty)$.*
2. *If $\gamma < \overline{\gamma}$, there exist $0 < R_1 < R_2 < \infty$ such that the solution U_γ of (7.269) satisfies $U_\gamma(R_1) = 0$, $\lim_{r \nearrow R_2} U_\gamma(r) = -\infty$ and $U_\gamma'(r) < 0$ for any $r \in (0, R_2)$.*
3. *If $\gamma > \overline{\gamma}$, there exist $0 < R_1 < R_2 < \infty$ such that the solution U_γ of (7.269) satisfies $U_\gamma'(r) < 0$ for $r \in (0,R_1)$, $U_\gamma'(R_1) = 0$, $U_\gamma'(r) > 0$ for $r \in (R_1,R_2)$ and $\lim_{r \nearrow R_2} U_\gamma(r) = +\infty$.*
4. *If $\gamma_1 < \gamma_2 < 0$, then the corresponding solutions $U_{\gamma_1}, U_{\gamma_2}$ of (7.269) satisfy $U_{\gamma_1} < U_{\gamma_2}$ and $U_{\gamma_1}' < U_{\gamma_2}'$ as long as they both exist.*
5. $r^{4/(p-1)}U_{\overline{\gamma}}(r) \to K_0^{1/(p-1)}$ *as $r \to \infty$.*

Proof. See the statements and proofs of [182, Theorem 2, Theorem 3, Lemma 2]. $\square$

For any $\gamma < 0$ let U_γ be the unique local solution of (7.269). Thanks to Item 3 in Lemma 7.112, we may define R_γ for $\gamma > \overline{\gamma}$ as the unique value of $r > 0$ for which we have $U_\gamma'(R_\gamma) = 0$.

The idea in constructing a singular solution to (7.219) consists in suitably scaling $U_\gamma(\,.\,) - U_\gamma(R_\gamma)|_{B_{R_\gamma}}$ to B and in finding a suitable subsequence for $\gamma \searrow \overline{\gamma}$, which locally converges in $B \setminus \{0\}$ to a singular solution. A first step is proving $R_\gamma \to \infty$ for $\gamma \searrow \overline{\gamma}$.

Lemma 7.113. *Let $n > 4$, $p > (n+4)/(n-4)$ and $\overline{\gamma}$ be as in the statement of Lemma 7.112. Then the map $\gamma \mapsto R_\gamma$ is non-increasing on the interval $(\overline{\gamma}, 0)$ and*

$$\lim_{\gamma \searrow \overline{\gamma}} R_\gamma = +\infty.$$

Proof. The fact that the map $\gamma \mapsto R_\gamma$ is non-increasing on the interval $(\overline{\gamma}, 0)$ follows immediately by Items 3 and 4 of Lemma 7.112 and the definition of R_γ. This shows that the function $\gamma \mapsto R_\gamma$ admits a limit as $\gamma \to \overline{\gamma}$. Suppose by contradiction that

$$\overline{R} := \lim_{\gamma \searrow \overline{\gamma}} R_\gamma < +\infty.$$

Then by Lemma 7.112, Items 1 and 4, we have for all $\gamma \in (\overline{\gamma}, 0)$ that

$$U_\gamma(R_\gamma) > U_{\overline{\gamma}}(R_\gamma) \geq U_{\overline{\gamma}}(\overline{R}) > 0. \tag{7.270}$$

Define for any $\gamma \in (\overline{\gamma}, 0)$, $r \in [0,1]$ the function

$$u_\gamma(r) = \frac{U_\gamma(R_\gamma r)}{U_\gamma(R_\gamma)} - 1. \tag{7.271}$$

Then u_γ solves the Dirichlet problem

$$\begin{cases} \Delta^2 u_\gamma = R_\gamma^4 U_\gamma(R_\gamma)^{p-1}(1+u_\gamma)^p & \text{in } B, \\ u_\gamma = |\nabla u_\gamma| = 0 & \text{on } \partial B. \end{cases} \tag{7.272}$$

Since the function U_γ is decreasing on the interval $(0, R_\gamma)$ we find that

$$U_\gamma(R_\gamma) \leq U_\gamma(r) \leq U_\gamma(0) = 1 \qquad \text{for all } r \in [0, R_\gamma]. \tag{7.273}$$

Then by (7.270) and (7.273) we obtain for all $\gamma \in (\overline{\gamma}, 0)$ and all $r \in [0, 1)$ that

$$0 \leq u_\gamma(r) \leq \frac{1}{U_\gamma(R_\gamma)} - 1 \leq \frac{1}{U_{\overline{\gamma}}(\overline{R})} - 1.$$

This shows that the set $\{u_\gamma : \gamma \in (\overline{\gamma}, 0)\}$ is bounded in $L^\infty(B)$ and hence by a bootstrap argument, from (7.272) and the fact that $R_\gamma^4 U_\gamma(R_\gamma)^{p-1} \leq \lambda^*$ (see (7.220) for the definition of λ^*), we deduce that there exists a sequence $\gamma_k \searrow \overline{\gamma}$ and a function $\overline{u} \in H_0^2(B) \cap C^\infty(\overline{B})$ such that

$$u_{\gamma_k} \to \overline{u} \qquad \text{in } C^4(\overline{B}) \tag{7.274}$$

as $k \to \infty$. Take any $r \in [0, \overline{R})$. Since $R_{\gamma_k} \nearrow \overline{R}$, there exists $\overline{k} = \overline{k}(r)$ such that $r < R_{\gamma_k}$ for any $k > \overline{k}$. Hence, for $k > \overline{k}$, we may take r/R_{γ_k} instead of r in (7.271) and obtain

$$U_{\gamma_k}(r) = U_{\gamma_k}(R_{\gamma_k})\left(u_{\gamma_k}(r/R_{\gamma_k}) + 1\right). \tag{7.275}$$

Since the sequence (γ_k) is decreasing, we infer by Items 3 and 4 of Lemma 7.112 that $U_{\gamma_k}(R_{\gamma_k})$ is non-increasing. By (7.270), $U_{\gamma_k}(R_{\gamma_k})$ is also bounded from below and hence admits a strictly positive limit. Thanks to (7.274) we also have $u_{\gamma_k}(r/R_{\gamma_k}) \to \overline{u}(r/\overline{R})$ as $k \to \infty$. Therefore by (7.275) we deduce that for any $r \in [0, \overline{R})$

$$\overline{U}(r) := \lim_{k \to \infty} U_{\gamma_k}(r) = \left(\lim_{k \to \infty} U_{\gamma_k}(R_{\gamma_k})\right)\left(\overline{u}(r/\overline{R}) + 1\right). \tag{7.276}$$

In fact, we deduce from (7.274) and (7.275) that $U_{\gamma_k} \to \overline{U}$ in $C^4([0, R])$ for any $0 < R < \overline{R}$. Since $\overline{u} \in H_0^2(B)$, (7.276) shows that

$$\lim_{r \nearrow \overline{R}} \overline{U}'(r) = 0. \tag{7.277}$$

On the other hand, by continuous dependence on the initial conditions it follows

$$\lim_{k \to \infty} U_{\gamma_k}(r) = U_{\overline{\gamma}}(r) \quad \text{for all } r \in [0, \overline{R})$$

and hence $\overline{U}(r) = U_{\overline{\gamma}}(r)$ for any $r \in [0, \overline{R})$. This with (7.277) implies

$$\lim_{r \nearrow \overline{R}} U_{\overline{\gamma}}'(r) = 0,$$

which is absurd since $U_{\overline{\gamma}}'(\overline{R}) < 0$, see Item 1 in Lemma 7.112. This completes the proof of the lemma. $\qquad\square$

Lemma 7.114. *Let $n > 4$ and $p > (n+4)/(n-4)$ and let u be a regular solution of (7.219). Then*

$$u(x) \le \left(\frac{\lambda^*}{\lambda}\right)^{1/(p-1)} |x|^{-4/(p-1)} - 1 \quad \text{for all } x \in B \setminus \{0\}.$$

Proof. Let u be a regular solution of (7.219) for some $\lambda > 0$ and define the scaled function

$$U(x) = \frac{1}{1 + u(0)}\left(1 + u\left(\frac{x}{\sqrt[4]{\lambda}(1 + u(0))^{\frac{p-1}{4}}}\right)\right) \tag{7.278}$$

so that U satisfies

$$\Delta^2 U = U^p \quad \text{in } B_R(0) \qquad \text{and} \qquad U(0) = 1$$

where we put $R = \sqrt[4]{\lambda}(1 + u(0))^{\frac{p-1}{4}}$. Let

$$M = \max_{r \in [0,R]} r^{4/(p-1)} U(r)$$

and let $\overline{R} \in (0,R]$ be such that $\overline{R}^{4/(p-1)} U(\overline{R}) = M$. Defining

$$w(r) = \frac{U(\overline{R}r)}{U(\overline{R})} - 1,$$

we obtain a solution of

$$\begin{cases} \Delta^2 w = \overline{R}^4 U(\overline{R})^{p-1}(1+w)^p & \text{in } B, \\ w = 0 & \text{on } \partial B, \\ w' \leq 0 & \text{on } \partial B. \end{cases}$$

This proves that $M^{p-1} = \overline{R}^4 U(\overline{R})^{p-1} \leq \lambda^*$ since otherwise by the super-subsolution method, see [40, Lemma 3.3] for the details, we would obtain a solution of (7.219) for $\lambda = \overline{R}^4 U(\overline{R})^{p-1} > \lambda^*$. This yields for all $r \in [0,R]$ that

$$U(r) \leq M r^{-4/(p-1)} \leq (\lambda^*)^{1/(p-1)} r^{-4/(p-1)}. \tag{7.279}$$

Then reversing the identity (7.278), by (7.279) we obtain

$$u(r) = \lambda^{-1/(p-1)} R^{4/(p-1)} U(Rr) - 1 \leq \left(\frac{\lambda^*}{\lambda}\right)^{1/(p-1)} r^{-4/(p-1)} - 1$$

which completes the proof of the lemma. $\qquad\square$

Proof of Theorem 7.92. For $\gamma \in (\overline{\gamma},0)$ consider the corresponding solution U_γ of the Cauchy problem (7.269) and the function u_γ introduced in (7.271). If we put $\lambda_\gamma = R_\gamma^4 U_\gamma(R_\gamma)^{p-1}$, then by (7.272) we have that u_γ solves

$$\begin{cases} \Delta^2 u_\gamma = \lambda_\gamma(1+u_\gamma)^p & \text{in } B, \\ u_\gamma = |\nabla u_\gamma| = 0 & \text{on } \partial B. \end{cases} \tag{7.280}$$

We show that λ_γ remains bounded away from zero for $\gamma > \overline{\gamma}$ sufficiently close to $\overline{\gamma}$, which is defined in Lemma 7.112. By Item 5 of Lemma 7.112 we infer that for a fixed $\varepsilon \in (0, K_0^{1/(p-1)})$ there exists a corresponding $r_\varepsilon > 0$ such that

$$U_{\overline{\gamma}}(r) > (K_0^{1/(p-1)} - \varepsilon) r^{-4/(p-1)} \quad \text{for all } r > r_\varepsilon. \tag{7.281}$$

On the other hand, by Lemma 7.113, we deduce that there exists $\gamma_0 \in (\bar{\gamma}, 0)$ such that for any $\gamma \in (\bar{\gamma}, \gamma_0)$ then $R_\gamma > r_\varepsilon$. Therefore by Item 4 of Lemma 7.112 we obtain for all $\gamma \in (\bar{\gamma}, \gamma_0)$

$$U_\gamma(R_\gamma) > U_{\bar{\gamma}}(R_\gamma) > (K_0^{1/(p-1)} - \varepsilon)R_\gamma^{-4/(p-1)}$$

and this yields for all $\gamma \in (\bar{\gamma}, \gamma_0)$

$$\lambda_\gamma > (K_0^{1/(p-1)} - \varepsilon)^{p-1} =: C. \tag{7.282}$$

Combining (7.282) and Lemma 7.114 we obtain for all $\gamma \in (\bar{\gamma}, \gamma_0)$, $x \in B\backslash\{0\}$

$$u_\gamma(x) \le \left(\frac{\lambda^*}{C}\right)^{1/(p-1)} |x|^{-4/(p-1)} - 1. \tag{7.283}$$

Since $p > (n+4)/(n-4)$ and u_γ solves (7.280), we obtain by (7.283)

$$\int_B |\Delta u_\gamma|^2 dx = \lambda_\gamma \int_B (1 + u_\gamma)^p u_\gamma dx \le \lambda^* \int_B (1 + u_\gamma)^{p+1} dx$$

$$\le \frac{(\lambda^*)^{\frac{2p}{p-1}}}{C^{\frac{p+1}{p-1}}} \int_B |x|^{-\frac{4(p+1)}{p-1}} dx < +\infty.$$

This proves that the set $\{u_\gamma : \gamma \in (\bar{\gamma}, \gamma_0)\}$ is bounded in $H_0^2(B)$ and hence there exists a sequence $\gamma_k \searrow \bar{\gamma}$ and a function $u \in H_0^2(B)$ such that $u_{\gamma_k} \rightharpoonup u$ in $H_0^2(B)$. Moreover, by (7.283) and applying Lebesgue's theorem, u weakly solves (7.219) for a suitable $\tilde{\lambda} \ge C$.

It remains to prove that the function u is unbounded. For simplicity, in the rest of the proof $u_{\gamma_k}, U_{\gamma_k}, R_{\gamma_k}, \lambda_{\gamma_k}$ will be called respectively u_k, U_k, R_k, λ_k.

By compact embedding we have that $u_k \to u$ in $L^1(B)$ and hence we have

$$\lim_{r \searrow 0} \frac{1}{|B_r(0)|} \int_{B_r(0)} u(x)dx = \lim_{r \searrow 0} \left(\frac{1}{e_n r^n} \lim_{k \to \infty} \int_{B_r(0)} u_k(x)dx\right)$$

and passing to radial coordinates, we obtain by (7.271) and Item 4 of Lemma 7.112

$$\lim_{r \searrow 0} \frac{1}{e_n r^n} \int_{B_r(0)} u(x)dx \tag{7.284}$$

$$= \lim_{r \searrow 0} \left(-1 + \frac{n}{r^n} \lim_{k \to \infty} \int_0^r \frac{U_k(R_k \rho)}{U_k(R_k)} \rho^{n-1} d\rho\right)$$

$$= \lim_{r \searrow 0} \left(-1 + \frac{n}{r^n} \lim_{k \to \infty} \frac{1}{R_k^n U_k(R_k)} \int_0^{R_k r} U_k(\rho)\rho^{n-1} d\rho\right)$$

$$\ge \lim_{r \searrow 0} \left(-1 + \frac{n}{r^n} \lim_{k \to \infty} \frac{1}{R_k^n U_k(R_k)} \int_0^{R_k r} U_{\bar{\gamma}}(\rho)\rho^{n-1} d\rho\right). \tag{7.285}$$

By (7.281) we have that there exist $C, R_0 > 0$ such that

$$U_{\bar{\gamma}}(\rho) > C\rho^{-4/(p-1)} \qquad \text{for all } \rho \in (R_0, \infty).$$

Hence, we have for $k > \bar{k} = \bar{k}(r)$

$$\int_0^{R_k r} U_{\bar{\gamma}}(\rho)\rho^{n-1}d\rho \geq \int_0^{R_0} U_{\bar{\gamma}}(\rho)\rho^{n-1}d\rho + \frac{C}{n - \frac{4}{p-1}}\left(R_k^{n-\frac{4}{p-1}} r^{n-\frac{4}{p-1}} - R_0^{n-\frac{4}{p-1}}\right).$$
$$(7.286)$$

Since $p > (n+4)/(n-4) > (n+4)/n$, since λ_k is bounded away from zero by (7.282), and since $R_k \to \infty$ by Lemma 7.113 as $k \to \infty$, we find

$$\lim_{k \to \infty} R_k^n U_k(R_k) = \lim_{k \to \infty} R_k^{n-\frac{4}{p-1}} \lambda_k^{\frac{1}{p-1}} = +\infty.$$

Hence, we obtain by (7.286)

$$\lim_{k \to \infty} \frac{1}{R_k^n U_k(R_k)} \int_0^{R_k r} U_{\bar{\gamma}}(\rho)\rho^{n-1}d\rho$$

$$\geq \liminf_{k \to \infty} \frac{C}{\left(n - \frac{4}{p-1}\right) R_k^n U_k(R_k)}\left(R_k^{n-\frac{4}{p-1}} r^{n-\frac{4}{p-1}} - R_0^{n-\frac{4}{p-1}}\right)$$

$$= \liminf_{k \to \infty} \frac{Cr^{n-\frac{4}{p-1}}}{\left(n - \frac{4}{p-1}\right) \lambda_k^{1/(p-1)}} \geq \frac{Cr^{n-\frac{4}{p-1}}}{\left(n - \frac{4}{p-1}\right)(\lambda^*)^{1/(p-1)}} =: \tilde{C}r^{n-\frac{4}{p-1}}. \quad (7.287)$$

Inserting (7.287) in (7.285) we obtain

$$\lim_{r \searrow 0} \frac{1}{e_n r^n} \int_{B_r(0)} u(x)dx \geq \lim_{r \searrow 0}(-1 + n\tilde{C}r^{-4/(p-1)}) = +\infty.$$

This proves that $u \notin L^\infty(B)$. The existence of a parameter $\lambda_s > 0$ and of the corresponding singular solution u_s is so proved.

Concerning uniqueness of λ_s and u_s, we refer to Propositions 3.1 and 3.2 by Dávila-Flores-Guerra [132] which are based on a slightly modified version of (7.234) for $W = (w_1, w_2, w_3, w_4)$, see (2.12) in their paper. $\qquad \square$

Sketch of the basic ideas of the proof of Theorem 7.95. The analysis by Dávila-Flores-Guerra [132] is based on the Emden-Fowler transform (7.228) of (7.219) and on the just mentioned modified version of (7.234) for $W = (w_1, w_2, w_3, w_4)$. The stability analysis of the stationary points O and P, which is performed in Section 7.11.1, plays a crucial role. According to Proposition 7.100, regular solutions are characterised by $\lim_{t \to -\infty} W(t) = O$ while, according to Proposition 7.107, singular solutions correspond to $\lim_{t \to -\infty} W(t) = P$. Dirichlet boundary conditions mean that the corresponding trajectories intersect the hyperplane $\mathscr{H} = \{w_2 = 0\}$. The goal is to understand all trajectories $W(.)$ solving (7.234) and intersecting $\mathscr{H}$.

As a first observation, Ferrero-Gazzola-Grunau-Karageorgis [168, 182] proved that in the parameter regime below p_c one has a heteroclinic orbit W_e connecting O to P which, for $t \to \infty$, spirals infinitely many times around P in its stable manifold. This corresponds to an entire solution of $\Delta^2 U = U^p$ oscillating around the singular solution $\bar{u}$, see (7.223).

Secondly, one has – up to a shift in time – precisely one singular orbit W_s – i.e. $\lim_{t \to -\infty} W_s(t) = P$ – intersecting $\mathscr{H}$ in a point Q_0. This trajectory corresponds to the uniquely determined singular solution of (7.219), see Theorem 7.92.

One then studies trajectories $W(.)$ "very close" to W_e in the unstable manifold of O, which is shown to intersect the stable manifold of P transversally. This means that $W(.)$ eventually hits a local three-dimensional manifold close to W_s and to P and "almost parallel" to the stable manifold of P. If this intersection point is close enough to W_s the singular orbit takes it for increasing t until $W(.)$ finally intersects $\mathscr{H}$ in a point Q close Q_0. Such a trajectory $W(.)$ gives rise to a regular solution to (7.219). The spiraling behaviour of W_e around P yields that that the intersection points Q spiral around and converge to Q_0 in $\mathscr{H}$. Transforming this result back into the λ-u-picture proves Theorem 7.95. $\square$

7.12 Bibliographical Notes

The seminal paper by Serrin [369] which is based on Alexandrov's technique of moving planes [7] initiated a huge series of work on symmetry properties of second order elliptic equations and systems. As prototypes we only mention the work by Gidas-Ni-Nirenberg [195] and Troy [397], the latter referring to systems and therefore to Navier boundary conditions, see Theorem 7.3. Concerning the Dirichlet problem, we refer to preliminary work by Bennett [38] and Dalmasso [122] who extended with different proofs Serrin's result to the biharmonic problem. Subsequently, Lazer-McKenna [270] and Dalmasso [124] extended the result by Gidas-Ni-Nirenberg under quite restrictive assumptions on the nonlinearity f. Moreover, first symmetry results concerning minimisers of Sobolev embeddings were obtained in by Ferrero-Gazzola-Weth [166]. Theorem 7.1 is due to Berchio-Gazzola-Weth [45] and takes advantage of refined monotonicity properties of the Green function with respect to reflections at hyperplanes, which were first observed in [166]. The moving plane method has been previously applied to some integral equations in papers by Chang-Yang [93], Li [274], Birkner, López-Mimbela, Wakolbinger [56], and Chen-Li-Ou [95]. The proof of Theorem 7.1 has common points with some of these papers but also contains new features since, in particular, it deals with very general nonlinearities f and reduces the problem to Green's function inequalities. Finally, the counterexample 7.4 is due to Sweers [387].

Proposition 7.15, namely the regularity result for at most critical problems, is due to Luckhaus [282]. Theorems 7.17 and 7.18 are quite standard and cannot be precisely located in literature. Let us just mention that, among many others, some work on polyharmonic problems of subcritical growth is contained in papers by von Wahl [406], Oswald [328], Dalmasso [121], Clément-de Figueiredo-Mitidieri [106], Soranzo [376]. Theorem 7.19 is an application of the decomposition method in dual cones developed in Section 3.1.2.

The best Sobolev constant S in (7.50) was determined by Lions [278] and Swanson [383], see also previous work in [160, 391]. Theorem 7.21 is obtained as a combination of results in [95, 275, 278, 383, 411] and uses Lemma 7.22 which is the generalisation of [186, Lemma 2] from the case $m = 2$ to the general case $m \geq 2$. This is a further application of the decomposition method in dual cones. Concerning the Sobolev inequalities in the critical case on bounded domains, Theorem 7.23 is due to Lions [278] whereas Theorem 7.24 is due to van der Vorst [400] when $m = 2$ and to Ge [192] for general $m \geq 2$. Here, we give a relatively simple proof for bounded smooth domains which directly extends to the non-Hilbertian spaces $W_{\vartheta}^{m,p}$ $(1 < p < \infty)$ and is taken from [187]. For $p = 1$, such a result was proved to become false, see [86].

It appears to be impossible to survey all the huge literature concerning second order semilinear problems of critical growth, namely (7.72) when $m = 1$. However, the starting point may be identified with the investigation of the Yamabe problem [418]. The so-called "positive case" was solved 1976 by Aubin [26] (see also [27]) and the general case 1984 by Schoen [363]. At the same time, further critical growth problems from geometry and physics like the system for parametric surfaces of prescribed mean curvature and the Yang-Mills-functional were under investigation. In 1983, Brezis-Nirenberg studied in their seminal paper [72] a semilinear model problem of critical growth and opened the way for a systematic investigation of relevant compactness phenomena. Struwe [381] made clear in which way the problem lacks compactness, we come back to this issue below. Since then, many contributions on (7.72) when $m = 1$ have appeared. With no hope of being complete, let us just mention the papers which are more closely related to the spirit of the present book, namely [15, 20, 25, 29, 37, 84, 87, 88, 104, 128, 190, 333, 340, 341, 419]. We also refer to the monograph [382] for a survey of results and further references.

The critical semilinear polyharmonic model problem (7.72) for $m \geq 2$ was studied e.g. in [236, 307, 328, 348, 349, 376], as far as nonexistence is concerned, and e.g. in [84, 160, 179, 205, 383] as for existence. The Navier problem was treated in [52, 106, 193, 308, 399–401] and many other papers. For a recent contribution on both boundary conditions see [273]. We also refer to [31] for related results on *systems* of equations and to [44, 189] for results concerning the Steklov problem (7.75). The material here is related to [44, 53, 179, 180, 185, 186, 189, 202–205]. In [181] this subject was viewed from a Sobolev inequality point of view.

Theorem 7.27 was proved by Pohožaev in his celebrated paper [340] for $m = 1$, and later extended to any $m \geq 1$ in [341, Lemma 3], see also [348, 349]. The identity in Theorem 7.29 was found independently and at the same time by Mitidieri [308] and van der Vorst [399]. The Pohožaev identity of Theorem 7.30 is taken from [44].

The application of Pohožaev's identity in order to prove Theorem 7.31 is due to Brezis-Nirenberg [72] when $m = 1$ and to Pucci-Serrin [348] when $m \geq 2$. Theorem

7.33 was proved by Oswald [328]. Theorem 7.34 is due to Lazzo-Schmidt [273]. In the particular cases when $m = 2$ or $m = 3$, Theorem 7.34 was previously proved using the identities in [349], which are based on refined choices of testing functions, see [205, Theorem 3.11]. Moreover, for the biharmonic case $m = 2$, an even simpler proof may be found in [186, Theorem 4]. As already mentioned, Theorem 7.34 excludes, in particular, the existence of positive solutions to (7.81), a fact that was already observed by Soranzo [376].

The existence part of Theorem 7.38 is taken from [203], although the biharmonic case $m = 2$ was treated earlier in [160]. Critical dimensions were defined by Pucci-Serrin [349] and subsequently emphasised from different points of view in [53, 181, 202, 236, 307]. The nonexistence part of Theorem 7.38 was proved in [204] for radial solutions since Theorem 7.1 was not known at that time. Theorem 7.40 is taken from [179] and may be proved essentially as Theorem 7.38. Theorem 7.44 was proved independently by Gazzola [179] and Grunau [205]. The proof suggested here is taken from [179] and uses the orthogonalisation method developed in [190]. The proof in [205] follows ideas of Capozzi-Fortunato-Palmieri [84]. The proof of Item 1 is based on the work by Cerami-Fortunato-Struwe [87], see again [179, 205]. Finally, let us also mention that the proof given here allows to treat more general subcritical perturbations $g(x, u)$ instead of λu, see [179]. For the second order special case $m = 1$, we also refer to [15, 84, 419].

Theorem 7.52 is taken from [399, Theorem 3.10] and also [308, Theorem 3.3]. Theorem 7.53 was first proved in [186] when $m = 2$ but the proof presented here is taken from Lazzo-Schmidt [273]. Theorem 7.54 is due to Ge [192], see also previous work when $m = 2$ by Gazzola-Grunau-Squassina [186] (Item 1) and van der Vorst [401, Theorem 3] (Item 2). The nonexistence part of Theorem 7.55 is proved in [44] where one can also find results about (7.131) in general bounded domains Ω. The existence and uniqueness parts are taken from [189], see also [44] for preliminary partial results. Finally, we refer to [43] for existence and nonexistence results for sign-changing solutions of (7.131) in general domains.

The proof of (7.159) for positive functions is obtained by adjusting the arguments by Brezis-Lieb [71] in the case $m = 1$ when there is no need to distinguish between positive and sign-changing functions because it suffices to replace u with $|u|$. In higher order spaces, this trick is no longer available and one has to treat sign-changing functions differently. For this case, our proof is taken from Gazzola-Grunau [180], and once more, the decomposition method with respect to dual cones (see Section 3.1.2) proves to be quite helpful. The two inequalities in Theorem 7.60 are taken respectively from [192] (see also [186] for indepedent work when $m = 2$) and [44]. Theorem 7.61 is due to Bartsch-Weth-Willem [33] whereas Theorem 7.62 is due to Ebobisse-Ould Ahmedou [157]. We also refer to Ge [193] for related results under Navier boundary conditions and superlinear subcritical perturbations of u^s. All the other material from Section 7.9 (including the Struwe-type compactness result in Lemma 7.74) is taken from [186]. We emphasise that the proof of Lemma 7.74 is similar to the original second order proof ($m = 1$) by Struwe [381, Proposition 2.1], see also the simplified proof in [382, Ch. III, Theorem 3.1]. It also takes advantage of the work by Alves-do Ó [12, Theorem 1] in the biharmonic Dirichlet case. The first occurrence of a Struwe-type compactness result in the context of higher order equations is due to Hebey-Robert [225], while related generalisations can also be found in [33, 157]. However, none of these applies directly to our situation since a particular difficulty arises from the existence of the boundary $\partial\Omega$ in combination with Navier boundary conditions.

Concerning the field of fourth order Paneitz-type equations, numerous important papers are devoted to this subject, too many to be recalled here in detail. We only give a brief and by far non-exhaustive survey on some of these results. In Chang-Yang [92], Wei-Xu [410] and Gursky [219] existence results for the constant Q-curvature problem in compact 4-manifolds are given. Recent work of Djadli-Malchiodi [145] provides further extensions and completions of these works. On compact manifolds of dimension greater than 4 existence results were given for Einstein manifolds by Djadli-Hebey-Ledoux [144] and Esposito-Robert [161] and in the case of invariance of both the manifold and Q-curvature function under a group of isometries by Robert [357]. On the sphere $\mathbb{S}^n$ we refer to results of Djadli-Malchiodi-Ould Ahmedou [146, 147] and Felli [163]. Finally, we refer to the monograph [89] and the references therein. Section 7.10 is based on the work by Grunau-Ould Ahmedou-Reichel [206]. Results closely related to the proof of completeness are announced by Diaz-Lazzo-Schmidt [142, 143]. The procedure in [206] takes advantage of some techniques developed for radial solutions to fourth order semilinear equations in [167, 182].

For semilinear elliptic equations of second order with supercritical growth of the kind of (7.219), it is almost impossible to review the existing literature; we only refer to the fundamental contributions [70, 73, 196, 240, 409] and references therein. Entire solutions to higher order supercritical problems were studied in [18, 123, 131–133, 168, 182, 217, 243, 370, 407], while for Dirichlet problems we refer to Arioli-Gazzola-Grunau-Mitidieri [19], Berchio-Gazzola [40], Cassani-do Ó-Ghoussoub [85], Cowan-Esposito-Ghoussoub-Moradifam [113] Dávila-Dupaigne-Guerra-Montenegro [131], Dávila-Flores-Guerra [132, 133], Ferrero-Grunau [167], Ferrero-Grunau-Karageorgis [168], Ferrero-Warnault [169], Guo-Wei [218] and references therein. Uniqueness issues were addressed by Reichel [352]. The material presented in Section 7.11 is mostly taken from Gazzola-Grunau [182], Ferrero-Grunau [167], Ferrero-Grunau-Karageorgis [168], and takes advantage of some tools introduced in [18, 19], see also [196, 240, 409] for preliminary work on second order problems. Theorem 7.95 and the uniqueness part in Theorem 7.92 were proved by Dávila-Flores-Guerra [132].

Here, we do not cover "coercive" problems like

$$\begin{cases} (-\Delta)^m u + |u|^{p-1}u = f & \text{in } \Omega \subset \mathbb{R}^n, \ n > 2m, \\ D^\alpha u = 0 & \text{on } \partial\Omega \text{ for } |\alpha| < m. \end{cases}$$

If $m = 1$, thanks to the maximum principle, one has a classical solution for any exponent $p \geq 1$ and any $f \in C^{0,\gamma}(\overline{\Omega})$ in any bounded $C^{2,\gamma}$-smooth domain. Due to the lack of maximum principles, the situation is still completely different for higher order equations $m \geq 2$ and, in general, one imposes additional growth conditions on p. Assuming at most critical growth, i.e. $p \le s = (n+2m)/(n-2m)$, allows for proving existence of classical solutions, see e.g. [395, 405, 406]. This is achieved by means of a continuity method, where a priori bounds are found in dependence and in the course of this continuity procedure. Also variants of growth conditions were studied, see e.g. [201, 210, 396] and references therein. The biharmonic case $m = 2$ appears to be somehow special and intermediate between the second order and the general polyharmonic case, see e.g. [396]. While weak solutions exist for any m and any p (see e.g. [75]), we are not aware neither of any result concerning existence of classical solutions where, for general m, growth restrictions on p could have been completely removed nor of any nonexistence result.

Chapter 8
Willmore Surfaces of Revolution

This last chapter serves to give a first existence result for a priori bounded classical solutions of the Dirichlet problem for Willmore surfaces and thereby to outline possible directions of further research. In order to see which kind of phenomena and results concerning compact embedded solutions in $\mathbb{R}^3$ of boundary value problems for the corresponding equation might be expected, we investigate Dirichlet problems in a particularly symmetric situation.

8.1 An Existence Result

We look at surfaces of revolution, which are obtained by rotating in $\mathbb{R}^3$ a graph over the x-axis around the x-axis. These are described by sufficiently smooth functions

$$u : [-1,1] \to (0,\infty)$$

which themselves are supposed to be symmetric, meaning that $u(x) = u(-x)$, and are parametrised as follows:

$$(x,\varphi) \mapsto (x, u(x)\cos\varphi, u(x)\sin\varphi), \quad x \in [-1,1], \quad \varphi \in [0,2\pi].$$

We consider the Willmore problem under Dirichlet boundary conditions, where the height $u(\pm 1) = \alpha > 0$ and a horizontal angle $u'(\pm 1) = 0$ are prescribed at the boundary:

Theorem 8.1. *For each $\alpha > 0$, there exists a smooth function $u \in C^\infty([-1,1],(0,\infty))$ such that the corresponding surface of revolution solves the Dirichlet problem for the quasilinear Willmore equation*

$$\begin{cases} \Delta_g H + 2H(H^2 - K) = 0 & \text{in } (-1,1), \\ u(\pm 1) = \alpha, \quad u'(\pm 1) = 0. \end{cases} \tag{8.1}$$

F. Gazzola et al., *Polyharmonic Boundary Value Problems*,
Lecture Notes in Mathematics 1991, DOI 10.1007/978-3-642-12245-3_8,
© Springer-Verlag Berlin Heidelberg 2010

The solution u may be taken to be even and to have the following additional properties:

$$0 \leq x + u(x)u'(x), \qquad u'(x) \leq 0 \qquad \text{for all } x \in [0,1].$$

$$\alpha \leq u(x) \leq \alpha + 1, \qquad |u'(x)| \leq \frac{1}{\alpha} \qquad \text{for all } x \in [-1,1].$$

When comparing this result with the situation for minimal surfaces of revolution one may be surprised that existence holds true even for $\alpha \searrow 0$. Moreover, with a quite involved proof one can show existence when prescribing any boundary angle $u'(-1) = -u'(1) = \beta$ with $\beta \in \mathbb{R}$ together with the position $u(\pm 1) = \alpha$, see [116]. There also the singular limit $\alpha \searrow 0$ is discussed. The solutions constructed in Theorem 8.1 minimise the Willmore energy in suitable classes and converge locally uniformly in $(-1,1)$ to $x \mapsto \sqrt{1-x^2}$. This means that the corresponding surfaces of revolution converge to the unit sphere when $\alpha \searrow 0$, where the second Dirichlet datum $u'(\pm 1) = 0$ gets lost.

We solve (8.1) by minimising the Willmore functional in the class of surfaces of revolution, which are given by even functions $u : [-1,1] \to (0,\infty)$. A priori, minimising sequences need not be bounded in the Sobolev space $H^2(-1,1)$. This observation reflects the lack of uniformity in the ellipticity of the corresponding Euler-Lagrange equation. The goal is to pass to suitable minimising sequences where strong enough quantitative information is available. In the following section we give a reformulation as a minimisation problem of the elastic energy for curves in the hyperbolic half plane. This point of view opens the possibility for specific geometric constructions. In Section 8.3, taking advantage of using geodesic arcs in the hyperbolic half plane and refined energy reducing constructions, we show that one may construct suitable minimising sequences satisfying quite strong a priori estimates ensuring sufficient compactness. These constructions take advantage of the conformal invariance of the Willmore functional; i.e. applying any Möbius transformation of $\mathbb{R}^3$ leaves this functional unchanged, see e.g. [361]. Further interesting properties of minimising sequences and of the minimal Willmore energy, such as e.g. monotonicity in α of the latter, are also proved in Section 8.3.

Langer and Singer [268] gave explicit expressions for the curvature of elastic curves in the hyperbolic half plane in terms of the arclength of the unknown curve. However, there does not seem to be a direct way to use these results for proving Theorem 8.1. Moreover, we think that the constructions made below in order to improve the properties of minimising sequences are of independent interest and *explain* to a good extent the shape of solutions.

8.2 Geometric Background

8.2.1 Geometric Quantities for Surfaces of Revolution

The calculations below are based on the formulae given in the Notations-Section on pages 393 and forward. For a more profound geometric background one may also see [59]. Let

$$u : [-1,1] \to (0,\infty)$$

be a sufficiently smooth function. We consider the surface generated by the graph of u, the parametrisation of which is given by

$$\mathscr{R}: \quad (x,\varphi) \mapsto (x, u(x)\cos\varphi, u(x)\sin\varphi), \quad x \in [-1,1], \quad \varphi \in [0,2\pi].$$

Here we consider $x = x_1$ as first and $\varphi = x_2$ as second parameter. First and second fundamental forms and the inward pointing normal on the surface of revolution are given as follows:

$$(g_{ij}) = \begin{pmatrix} 1 + u'(x)^2 & 0 \\ 0 & u(x)^2 \end{pmatrix},$$

$$g = \det(g_{ij}) = u(x)^2 \left(1 + u'(x)^2\right),$$

$$(L_{ij}) = \frac{1}{\sqrt{1+u'(x)^2}} \begin{pmatrix} -u''(x) & 0 \\ 0 & u(x) \end{pmatrix},$$

$$v(x,\varphi) = \frac{1}{\sqrt{1+u'(x)^2}} \left(u'(x), -\cos\varphi, -\sin\varphi\right).$$

We use the sign convention that the mean curvature H is positive if the surface is mean convex and negative if it is mean concave with respect to the inward pointing normal v. The mean and Gaussian curvature are then given respectively by

$$H(x) = -\frac{u''(x)}{2\left(1+u'(x)^2\right)^{3/2}} + \frac{1}{2u(x)\sqrt{1+u'(x)^2}}$$

$$= \frac{1}{2u(x)u'(x)} \frac{\partial}{\partial x} \left(\frac{u(x)}{\sqrt{1+u'(x)^2}} \right), \tag{8.2}$$

$$K(x) = -\frac{u''(x)}{u(x)\left(1+u'(x)^2\right)^2}.$$

The Laplace-Beltrami operator on the surface of revolution acts on smooth functions $(x,\varphi) \mapsto h(x,\varphi)$ as follows

$$\Delta_g h = \frac{1}{\sqrt{g}} \sum_{i,j=1}^{2} \partial_i \left(\sqrt{g}\, g^{ij} \partial_j h\right) = \frac{1}{u(x)\sqrt{1+u'(x)^2}}$$

$$\times \left(\frac{\partial}{\partial x} \left(\frac{u(x)}{\sqrt{1+u'(x)^2}} \frac{\partial}{\partial x} h \right) + \frac{\partial}{\partial\varphi} \left(\frac{\sqrt{1+u'(x)^2}}{u(x)} \frac{\partial}{\partial\varphi} h \right) \right),$$

where g^{ij} are the entries of the inverse of $(g_{ij})_{i,j}$. The terms in equation (8.1) for a surface of revolution are then

$$
\Delta_g H = \frac{1}{u(x)\sqrt{1+u'(x)^2}}
$$
$$
\times \frac{\partial}{\partial x}\left(\frac{u(x)}{\sqrt{1+u'(x)^2}}\frac{\partial}{\partial x}\left(\frac{1}{2u(x)\sqrt{1+u'(x)^2}}-\frac{u''(x)}{2\left(1+u'(x)^2\right)^{3/2}}\right)\right),
$$

$$
2H(H^2-K) = \frac{1}{4\left(1+u'(x)^2\right)^{3/2}}\left(\frac{1}{u(x)}-\frac{u''(x)}{1+u'(x)^2}\right)\left(\frac{1}{u(x)}+\frac{u''(x)}{1+u'(x)^2}\right)^2.
$$

For surfaces of revolution $\mathscr{R}$ generated by the graph of u as described above, the Willmore functional reads

$$
W(u) = \int_{\mathscr{R}} H^2 d\omega
$$
$$
= \frac{\pi}{2}\int_{-1}^{1}\left(\frac{1}{u(x)\sqrt{1+u'(x)^2}}-\frac{u''(x)}{\left(1+u'(x)^2\right)^{3/2}}\right)^2 u(x)\sqrt{1+u'(x)^2}\,dx.
$$

We show that the Euler-Lagrange equation for this functional is indeed the differential equation in (8.1).

Lemma 8.2. *Let $u \in C^4([-1,1],(0,\infty))$. Then for all $\varphi \in H^2 \cap H_0^1(-1,1)$ we have*

$$
\frac{-1}{2\pi}\frac{d}{dt}W(u+t\varphi)|_{t=0}
$$
$$
= \left[H(x)\frac{u(x)\varphi'(x)}{1+u'(x)^2}\right]_{-1}^{1} + \int_{-1}^{1} u(x)\varphi(x)\left(\Delta_g H(x)+2H(x)(H(x)^2-K(x))\right)\,dx.
$$

Proof.

$$
\frac{1}{2\pi}\frac{d}{dt}W(u+t\varphi)|_{t=0} = \int_{-1}^{1} H(x)u(x)\sqrt{1+u'(x)^2}
$$
$$
\times \left(\frac{-\varphi(x)}{u(x)^2\sqrt{1+u'(x)^2}}-\frac{u'(x)\varphi'(x)+u(x)\varphi''(x)}{u(x)\left(1+u'(x)^2\right)^{3/2}}+3\frac{u'(x)\varphi'(x)u''(x)}{\left(1+u'(x)^2\right)^{5/2}}\right)\,dx
$$
$$
+ \int_{-1}^{1} H^2(x)\left(\varphi(x)\sqrt{1+u'(x)^2}+\frac{u(x)u'(x)\varphi'(x)}{\sqrt{1+u'(x)^2}}\right)\,dx
$$

$$= -\int_{-1}^{1} \mathrm{H}(x)\frac{\varphi(x)}{u(x)}\,dx - \int_{-1}^{1}\mathrm{H}(x)\frac{u'(x)\varphi'(x)}{1+u'(x)^2}\,dx$$

$$-\int_{-1}^{1}\mathrm{H}(x)\frac{\partial}{\partial x}\left(\frac{\varphi'(x)}{(1+u'(x)^2)^{3/2}}\right)u(x)\sqrt{1+u'(x)^2}\,dx$$

$$+\int_{-1}^{1}\mathrm{H}^2(x)\varphi(x)\sqrt{1+u'(x)^2}\,dx - \int_{-1}^{1}\mathrm{H}^2(x)\varphi(x)\frac{u'(x)^2}{\sqrt{1+u'(x)^2}}\,dx$$

$$-\int_{-1}^{1}\mathrm{H}^2(x)u(x)\varphi(x)\frac{\partial}{\partial x}\left(\frac{u'(x)}{\sqrt{1+u'(x)^2}}\right)dx$$

$$-2\int_{-1}^{1}\mathrm{H}(x)\mathrm{H}'(x)\frac{\varphi(x)u(x)u'(x)}{\sqrt{1+u'(x)^2}}\,dx$$

$$= -\int_{-1}^{1}\mathrm{H}(x)\frac{\varphi(x)}{u(x)}\,dx$$

$$-\left[\mathrm{H}(x)\frac{u(x)\varphi'(x)}{1+u'(x)^2}\right]_{-1}^{1} + \int_{-1}^{1}\mathrm{H}'(x)\frac{\varphi'(x)u(x)}{1+u'(x)^2}\,dx$$

$$+\int_{-1}^{1}\mathrm{H}(x)\frac{\varphi'(x)u(x)u'(x)u''(x)}{(1+u'(x)^2)^2}\,dx + \int_{-1}^{1}\mathrm{H}^2(x)\varphi(x)\frac{1}{\sqrt{1+u'(x)^2}}\,dx$$

$$-\int_{-1}^{1}\mathrm{H}^2(x)u(x)\left(-2\mathrm{H}(x)+\frac{1}{u(x)\sqrt{1+u'(x)^2}}\right)\varphi(x)\,dx$$

$$-2\int_{-1}^{1}\mathrm{H}(x)\mathrm{H}'(x)\frac{u(x)u'(x)\varphi(x)}{\sqrt{1+u'(x)^2}}\,dx$$

$$= -\int_{-1}^{1}\mathrm{H}(x)\frac{\varphi(x)}{u(x)}\,dx - \left[\mathrm{H}(x)\frac{u(x)\varphi'(x)}{1+u'(x)^2}\right]_{-1}^{1}$$

$$-\int_{-1}^{1}\frac{\varphi(x)}{\sqrt{1+u'(x)^2}}\frac{\partial}{\partial x}\left(\frac{u(x)\mathrm{H}'(x)}{\sqrt{1+u'(x)^2}}\right)dx$$

$$+\int_{-1}^{1}\frac{u(x)u'(x)u''(x)}{(1+u'(x)^2)^2}\left(\mathrm{H}'(x)\varphi(x)+\mathrm{H}(x)\varphi'(x)\right)dx$$

$$+2\int_{-1}^{1}\mathrm{H}^3(x)u(x)\varphi(x)\,dx - 2\int_{-1}^{1}\mathrm{H}(x)\mathrm{H}'(x)\frac{u(x)u'(x)\varphi(x)}{\sqrt{1+u'(x)^2}}\,dx$$

$$= -\int_{-1}^{1}\frac{\mathrm{H}(x)\varphi(x)}{u(x)}\,dx - \left[\frac{\mathrm{H}(x)u(x)\varphi'(x)}{1+u'(x)^2}\right]_{-1}^{1}$$

$$-\int_{-1}^{1}u(x)\varphi(x)\left(\Delta_g\mathrm{H}(x)-2\mathrm{H}(x)^3\right)dx$$

$$-\int_{-1}^{1}\left(2\mathrm{H}(x)\frac{u(x)u'(x)}{\sqrt{1+u'(x)^2}}-\frac{u'(x)}{1+u'(x)^2}\right)\frac{\partial}{\partial x}\left(\mathrm{H}(x)\varphi(x)\right)dx$$

$$-2\int_{-1}^{1}\mathrm{H}(x)\mathrm{H}'(x)\frac{u(x)u'(x)\varphi(x)}{\sqrt{1+u'(x)^2}}\,dx$$

$$= -\int_{-1}^{1} \frac{\mathrm{H}(x)\varphi(x)}{u(x)}\, dx - \left[\frac{\mathrm{H}(x)u(x)\varphi'(x)}{1+u'(x)^2}\right]_{-1}^{1}$$

$$- \int_{-1}^{1} u(x)\varphi(x)\left(\Delta_g \mathrm{H}(x) - 2\mathrm{H}(x)^3\right) dx$$

$$+ 2\int_{-1}^{1} \mathrm{H}^2(x)\varphi(x)\left(\frac{u'(x)^2}{\sqrt{1+u'(x)^2}} - 2\mathrm{H}(x)u(x) + \frac{1}{\sqrt{1+u'(x)^2}}\right) dx$$

$$- \int_{-1}^{1} \mathrm{H}(x)\varphi(x)\left(\frac{u''(x)}{1+u'(x)^2} - 2\frac{u'(x)^2 u''(x)}{(1+u'(x)^2)^2}\right) dx$$

$$= -\int_{-1}^{1} \frac{\mathrm{H}(x)\varphi(x)}{u(x)}\, dx - \left[\frac{\mathrm{H}(x)u(x)\varphi'(x)}{1+u'(x)^2}\right]_{-1}^{1}$$

$$- \int_{-1}^{1} u(x)\varphi(x)\left(\Delta_g \mathrm{H}(x) + 2\mathrm{H}(x)^3\right) dx$$

$$+ 2\int_{-1}^{1} \mathrm{H}^2(x)\varphi(x)\sqrt{1+u'(x)^2}\, dx - 2\int_{-1}^{1} \mathrm{H}(x)\varphi(x)\frac{u''(x)}{(1+u'(x)^2)^2}\, dx$$

$$+ \int_{-1}^{1} \mathrm{H}(x)\varphi(x)\frac{u''(x)}{1+u'(x)^2}\, dx$$

$$= -\int_{-1}^{1} \frac{\mathrm{H}(x)\varphi(x)}{u(x)}\, dx - \left[\frac{\mathrm{H}(x)u(x)\varphi'(x)}{1+u'(x)^2}\right]_{-1}^{1}$$

$$- \int_{-1}^{1} u(x)\varphi(x)\left(\Delta_g \mathrm{H}(x) + 2\mathrm{H}(x)^3\right) dx$$

$$+ \int_{-1}^{1} \mathrm{H}(x)\varphi(x)\left(-\frac{u''(x)}{1+u'(x)^2} + \frac{1}{u(x)} - \underbrace{2\frac{u''(x)}{(1+u'(x)^2)^2} + \frac{u''(x)}{1+u'(x)^2}}_{-2\mathrm{K}(x)u(x)}\right) dx$$

$$= -\left[\frac{\mathrm{H}(x)u(x)\varphi'(x)}{1+u'(x)^2}\right]_{-1}^{1} - \int_{-1}^{1} u(x)\varphi(x)\left(\Delta_g \mathrm{H}(x) + 2\mathrm{H}(x)^3 - 2\mathrm{H}(x)\mathrm{K}(x)\right) dx.$$

$$\square$$

8.2.2 Surfaces of Revolution as Elastic Curves in the Hyperbolic Half Plane

We give here a different interpretation and a reformulation of the Willmore functional that is more suitable for our purposes. It turns out that the Willmore energy of surfaces of revolution may equivalently be described by the elastic energy of its generating graphs in the hyperbolic half plane. The following formulae and calculations are mainly based on [267].

The hyperbolic half plane $\mathbb{R}^2_+ := \{(x,y) : y > 0\}$ is equipped with the metric

$$ds^2 = \frac{1}{y^2}(dx^2 + dy^2).$$

In order to introduce the elastic energy for curves in the hyperbolic half plane and to take advantage of the hyperbolic geodesics we first deduce formulae for the hyperbolic curvature.

Lemma 8.3. *Let* $s \mapsto \gamma(s) = (\gamma^1(s), \gamma^2(s))$ *be a curve in* $\mathbb{R}^2_+$ *parametrised with respect to its arclength, i.e.*

$$1 \equiv \frac{(\gamma^{1'}(s))^2 + (\gamma^{2'}(s))^2}{(\gamma^2(s))^2}. \tag{8.3}$$

Then its hyperbolic curvature is given by

$$\kappa(s) = -\frac{(\gamma^2(s))^2}{\gamma^{2'}(s)} \frac{d}{ds}\left(\frac{\gamma^{1'}(s)}{(\gamma^2(s))^2}\right) = \frac{(\gamma^2(s))^2}{\gamma^{1'}(s)}\left(\frac{1}{\gamma^2(s)} + \frac{d}{ds}\left(\frac{\gamma^{2'}(s)}{(\gamma^2(s))^2}\right)\right). \tag{8.4}$$

It seems that this is the most frequently used sign convention. However, the arguments below are not affected by choosing the opposite sign.

Proof. We take any point $(x,y) = (\gamma^1(s), \gamma^2(s))$ and keep it fixed in what follows. In this point the tangent of γ is

$$(\tau^1, \tau^2) = (\gamma^{1'}, \gamma^{2'})$$

and its normal – up to sign –

$$(v^1, v^2) = (-\tau^2, \tau^1) = \left(-\gamma^{2'}, \gamma^{1'}\right).$$

In order to calculate the covariant derivative of τ we need the Christoffel symbols $\Gamma^k_{ij} = \frac{1}{2} g^{k\ell}\left(g_{i\ell,j} + g_{j\ell,i} - g_{ij,\ell}\right)$, where here $g_{ij} = \frac{1}{y^2}\delta_{ij}$ refers to the hyperbolic metric:

$$\Gamma^1_{11} = 0, \quad \Gamma^1_{12} = \Gamma^1_{21} = -\frac{1}{y}, \quad \Gamma^1_{22} = 0,$$
$$\Gamma^2_{11} = \frac{1}{y}, \quad \Gamma^2_{12} = \Gamma^2_{21} = 0, \quad \Gamma^2_{22} = -\frac{1}{y}.$$

The covariant derivative of τ along τ is parallel to the normal v, the proportionality factor defining the curvature of γ. Making also use of (8.3), we calculate

$$\frac{\nabla \tau}{ds} = \frac{d\tau}{ds} + \left(\Gamma^k_{ij}\tau^i\tau^j\right)_{k=1,2} = \frac{d^2}{ds^2}\gamma + \frac{1}{\gamma^2}\left(-2\gamma^{1'}\gamma^{2'}, (\gamma^{1'})^2 - (\gamma^{2'})^2\right)$$

$$\overset{!}{=} \kappa\left(-\gamma^{2'}, \gamma^{1'}\right)$$

$$\Rightarrow \kappa = -\frac{\gamma^{1''}}{\gamma^{2'}} + 2\frac{\gamma^{1'}}{\gamma^2} = \frac{\gamma^{2''}}{\gamma^{1'}} + \frac{\gamma^2}{\gamma^{1'}} - 2\frac{(\gamma^{2'})^2}{\gamma^2\gamma^{1'}}$$

This proves (8.4). $\qquad\square$

Considering graphs $[-1,1] \ni x \mapsto (x, u(x)) \in \mathbb{R}_+^2$ as curves in the hyperbolic half plane, their arclength is given by

$$s(x) = \int_{-1}^{x} \frac{\sqrt{1 + u'(\xi)^2}}{u(\xi)} \, d\xi$$

so that by means of

$$\frac{d}{ds} = \frac{u(x)}{\sqrt{1 + u'(x)^2}} \frac{d}{dx},$$

formula (8.4) yields

$$\kappa(x) = -\frac{u(x)^2}{u'(x)} \frac{d}{dx} \left(\frac{1}{u(x)\sqrt{1 + u'(x)^2}} \right) = \frac{u(x)u''(x)}{(1 + u'(x)^2)^{3/2}} + \frac{1}{\sqrt{1 + u'(x)^2}}. \tag{8.5}$$

From these formulae it is immediate that hyperbolic geodesics are circular arcs centered on the x-axis and lines parallel to the y-axis. The first will play a crucial role in choosing suitable minimising sequences for the modified Willmore functional, which we are going to define now.

We introduce the hyperbolic Willmore energy of graphs $[-1,1] \ni x \mapsto (x, u(x)) \in \mathbb{R}_+^2$ as their elastic energy and relate it to the original Willmore energy of the corresponding surface of revolution.

$$\widehat{W}(u) := \int_{-1}^{1} \kappa(x)^2 \, ds(x) = \int_{-1}^{1} \kappa(x)^2 \frac{\sqrt{1 + u'(x)^2}}{u(x)} \, dx$$

$$= \int_{-1}^{1} \left(\frac{u''(x)}{(1 + u'(x)^2)^{3/2}} - \frac{1}{u(x)\sqrt{1 + u'(x)^2}} \right)^2 u(x)\sqrt{1 + u'(x)^2} \, dx$$

$$+ 4 \int_{-1}^{1} \frac{u''(x)}{(1 + u'(x)^2)^{3/2}} \, dx$$

$$= \frac{2}{\pi} \int_{\mathscr{R}} \mathrm{H}^2 \, d\omega - \frac{2}{\pi} \int_{\mathscr{R}} \mathrm{K} \, d\omega = \frac{2}{\pi} \int_{\mathscr{R}} \mathrm{H}^2 \, d\omega + 4 \left[\frac{u'(x)}{\sqrt{1 + u'(x)^2}} \right]_{-1}^{1},$$

with H and K as given in (8.2). The latter identity for $\int \mathrm{K}$ may be viewed as a kind of Gauss-Bonnet theorem for the surface of revolution $\mathscr{R}$ with boundary. We conclude that

$$W(u) = \frac{\pi}{2} \widehat{W}(u) - 2\pi \left[\frac{u'(x)}{\sqrt{1 + u'(x)^2}} \right]_{-1}^{1},$$

where the Willmore energy $W(u)$ of the surface of revolution $\mathscr{R}$ generated by u is defined in (8.3). In our situation where we assume Dirichlet data

$$u(\pm 1) = \alpha, \qquad u'(\pm 1) = 0,$$

we even have

$$W(u) = \frac{\pi}{2}\widehat{W}(u).\tag{8.6}$$

In proving Theorem 8.1, we benefit a lot from considering $\widehat{W}$ instead of W. We do not only take technical advantage from this point of view, but we think that it is geometrically more suitable as the constructions in Section 8.3 will make clear.

Concerning the Euler-Lagrange equation for critical points of the "hyperbolic Willmore functional" $\widehat{W}$ one has:

Lemma 8.4. *Assume that $u \in C^4([-1,1])$ is positive and such that for all $\varphi \in C_c^\infty(-1,1)$ one has that $0 = \frac{d}{dt}\widehat{W}(u+t\varphi)|_{t=0}$. Then u satisfies the following Euler-Lagrange equation:*

$$\frac{u(x)}{\sqrt{1+u'(x)^2}}\frac{d}{dx}\left(\frac{u(x)}{\sqrt{1+u'(x)^2}}\kappa'(x)\right) - \kappa(x) + \frac{1}{2}\kappa(x)^3 = 0, \qquad x \in (-1,1),$$

$$\tag{8.7}$$

with κ as defined in (8.5).

One may observe that $\frac{u(x)}{\sqrt{1+u'(x)^2}}\frac{d}{dx}$ is the derivative with respect to the hyperbolic arclength of $x \mapsto (x,u(x))$. The following proof of Lemma 8.4 will also be used in proving regularity for our Willmore surfaces of revolution.

Proof. In order to calculate the Euler-Lagrange equation for the functional $\widehat{W}$, we observe first that for arbitrary $\varphi \in C_c^\infty(-1,1)$:

$$\frac{d}{dt}\kappa[u+t\varphi]|_{t=0} = -\frac{d}{dt}\left\{\frac{(u+t\varphi)^2}{u'+t\varphi'}\frac{d}{dx}\left(\frac{1}{(u+t\varphi)\sqrt{1+(u'+t\varphi')^2}}\right)\right\}\bigg|_{t=0}$$

$$= -2\frac{u\varphi}{u'}\frac{d}{dx}\left(\frac{1}{u\sqrt{1+u'^2}}\right) + \frac{u^2\varphi'}{u'^2}\frac{d}{dx}\left(\frac{1}{u\sqrt{1+u'^2}}\right)$$

$$+ \frac{u^2}{u'}\frac{d}{dx}\left(\frac{\varphi}{u^2\sqrt{1+u'^2}}\right) + \frac{u^2}{u'}\frac{d}{dx}\left(\frac{u'\varphi'}{u(1+u'^2)^{3/2}}\right)$$

and writing it in terms of κ

$$\frac{d}{dt}\kappa[u+t\varphi]|_{t=0} = 2\frac{\varphi}{u}\kappa - \frac{\varphi'}{u'}\kappa - \frac{\varphi}{u}\kappa + \frac{u}{u'\sqrt{1+u'^2}}\left(\frac{\varphi}{u}\right)'$$

$$- \frac{u'\varphi'}{1+u'^2}\kappa + \frac{u}{u'\sqrt{1+u'^2}}\left(\frac{u'\varphi'}{1+u'^2}\right)'$$

$$= \frac{\varphi}{u}\kappa - \frac{\varphi'}{u'}\kappa - \frac{u'\varphi'}{1+u'^2}\kappa + \frac{\varphi'}{u'\sqrt{1+u'^2}} - \frac{\varphi}{u\sqrt{1+u'^2}}$$

$$+ \frac{u}{u'\sqrt{1+u'^2}}\left(\frac{\varphi''u'}{1+u'^2} + \frac{\varphi'u''}{1+u'^2} - 2\frac{\varphi'u'^2u''}{(1+u'^2)^2}\right).$$

As for the last large bracket we have

$$\left(\frac{\varphi''u'}{1+u'^2}+\frac{\varphi'u''}{1+u'^2}-2\frac{\varphi'u'^2u''}{(1+u'^2)^2}\right)=\frac{\varphi''u'}{1+u'^2}-\frac{\varphi'u''}{1+u'^2}+2\frac{\varphi'u''}{(1+u'^2)^2}$$

$$=\frac{\varphi''u'}{1+u'^2}+\varphi'\sqrt{1+u'^2}\left(-\frac{\kappa}{u}+\frac{1}{u\sqrt{1+u'^2}}\right)$$

$$-\frac{2\varphi'}{\sqrt{1+u'^2}}\left(-\frac{\kappa}{u}+\frac{1}{u\sqrt{1+u'^2}}\right)$$

$$=\frac{\varphi''u'}{1+u'^2}-\frac{\kappa\varphi'}{u}\sqrt{1+u'^2}+\frac{\varphi'}{u}$$

$$+\frac{2\kappa\varphi'}{u\sqrt{1+u'^2}}-\frac{2\varphi'}{u(1+u'^2)}$$

so that the variation of κ becomes

$$\frac{d}{dt}\kappa[u+t\varphi]|_{t=0}=\frac{\varphi\kappa}{u}-3\frac{u'\varphi'\kappa}{1+u'^2}-\frac{\varphi}{u\sqrt{1+u'^2}}+\frac{2u'\varphi'+u\varphi''}{(1+u'^2)^{3/2}}.$$

So, if $u\in C^4([-1,1],(0,\infty))$ is such that for all $\varphi\in C_c^\infty(-1,1)$ one has that $0=\frac{d}{dt}\widehat{W}(u+t\varphi)|_{t=0}$, it follows:

$$0=\frac{d}{dt}\widehat{W}(u+t\varphi)|_{t=0}=\frac{d}{dt}\int_{-1}^{1}\kappa[u+t\varphi]^2\frac{\sqrt{1+(u'+t\varphi')^2}}{u+t\varphi}\,dx\bigg|_{t=0}$$

$$=\int_{-1}^{1}2\kappa\frac{\sqrt{1+u'^2}}{u}\left(\frac{\varphi\kappa}{u}-3\frac{u'\varphi'\kappa}{1+u'^2}-\frac{\varphi}{u\sqrt{1+u'^2}}+\frac{2u'\varphi'+u\varphi''}{(1+u'^2)^{3/2}}\right)dx$$

$$+\int_{-1}^{1}\kappa^2\left(\frac{u'\varphi'}{u\sqrt{1+u'^2}}-\frac{\varphi\sqrt{1+u'^2}}{u^2}\right)dx$$

$$=\int_{-1}^{1}\kappa^2\frac{\sqrt{1+u'^2}}{u^2}\varphi\,dx-5\int_{-1}^{1}\kappa^2\frac{u'}{u\sqrt{1+u'^2}}\varphi'\,dx-2\int_{-1}^{1}\kappa\frac{1}{u^2}\varphi\,dx$$

$$+4\int_{-1}^{1}\kappa\frac{u'}{u(1+u'^2)}\varphi'\,dx+2\int_{-1}^{1}\kappa\frac{1}{1+u'^2}\varphi''\,dx \qquad (8.8)$$

(integrating by parts first the last integral and then the second one)

$$=\int_{-1}^{1}\kappa^2\frac{\sqrt{1+u'^2}}{u^2}\varphi\,dx-\int_{-1}^{1}\kappa^2\frac{u'}{u\sqrt{1+u'^2}}\varphi'\,dx-2\int_{-1}^{1}\kappa\frac{1}{u^2}\varphi\,dx$$

$$-2\int_{-1}^{1}\kappa'\frac{1}{1+u'^2}\varphi'\,dx$$

$$=\int_{-1}^{1}\kappa^2\frac{\sqrt{1+u'^2}}{u^2}\varphi\,dx+\int_{-1}^{1}\kappa^2u'\left(\frac{1}{u\sqrt{1+u'^2}}\right)'\varphi\,dx$$

$$+2\int_{-1}^{1}\kappa\kappa'\frac{u'}{u\sqrt{1+u'^2}}\varphi\,dx+\int_{-1}^{1}\kappa^2\frac{u''}{u\sqrt{1+u'^2}}\varphi\,dx$$

$$-2\int_{-1}^{1}\kappa\frac{1}{u^2}\varphi\,dx-2\int_{-1}^{1}\kappa'\frac{1}{1+u'^2}\varphi'\,dx$$

$$= \int_{-1}^{1} \kappa^2 \varphi \left(\frac{\sqrt{1+u'^2}}{u^2} - \frac{u'^2}{u^2\sqrt{1+u'^2}} - \frac{u'^2 u''}{u(1+u'^2)^{3/2}} + \frac{u''}{u\sqrt{1+u'^2}} \right) dx$$

$$- 2\int_{-1}^{1} \kappa \frac{1}{u^2} \varphi \, dx + 2\int_{-1}^{1} \kappa \kappa' \frac{u'}{u\sqrt{1+u'^2}} \varphi \, dx$$

$$- 2\int_{-1}^{1} \frac{u}{\sqrt{1+u'^2}} \kappa' \frac{1}{u\sqrt{1+u'^2}} \varphi' \, dx$$

(integrating by parts the last term)

$$= \int_{-1}^{1} \kappa^3 \frac{1}{u^2} \varphi \, dx - 2\int_{-1}^{1} \kappa \frac{1}{u^2} \varphi \, dx + 2\int_{-1}^{1} \frac{u}{\sqrt{1+u'^2}} \frac{d}{dx} \left(\frac{u}{\sqrt{1+u'^2}} \kappa' \right) \frac{1}{u^2} \varphi \, dx.$$

$$\square$$

8.3 Minimisation of the Willmore Functional

For $\alpha \in (0,\infty)$ we define

$$N_\alpha := \{ u \in C^{1,1}([-1,1]), u \text{ is even and positive}, u(1) = \alpha, u'(1) = 0 \}, \qquad (8.9)$$

and

$$M_\alpha := \inf\{ \widehat{W}(u) : u \in N_\alpha \}. \qquad (8.10)$$

In this section, we show that M_α is attained, i.e. that there exists $u_\alpha \in N_\alpha$, which is even in $C^\infty([-1,1])$, such that $\widehat{W}(u_\alpha) = M_\alpha$.

According to (8.6) we have for all $u \in N_\alpha$

$$W(u) = \frac{\pi}{2} \int_{-1}^{1} \kappa(x)^2 \, ds(x) = \frac{\pi}{2} \widehat{W}(u).$$

Hence, the surface of revolution generated by the graph of u_α is a minimiser of the Willmore functional in the class of surfaces of revolution generated by graphs of functions in N_α. According to Lemma 8.2 the corresponding Euler-Lagrange equation is the Dirichlet problem (8.1) for the Willmore equation.

Remark 8.5. We use the following rescaling property which is a special case of the conformal invariance of the Willmore functional. If u is a positive function in $C^{1,1}([-r,r])$ for some $r > 0$, then the function $v \in C^{1,1}([-1,1])$ defined by $v(x) = \frac{1}{r} u(rx)$ satisfies

$$\widehat{W}(v) = \int_{-r}^{r} \kappa^2[u] ds[u].$$

Here and in the following $\kappa[u]$ denotes the curvature of the graph of u in the hyperbolic half plane defined in (8.5) and $ds[u]$ denotes the corresponding line element.

Beside rescaling, a number of geometric constructions will be involved in the minimisation process. All these will be based on gluing geodesic arcs $C^{1,1}$-smoothly to suitable parts of comparison functions for M_α or elements of minimising sequences respectively. In a first step we will give an accurate bound from above for the optimal Willmore energy M_α. The proof will be developed further in order to show that $\alpha \mapsto M_\alpha$ is decreasing. Finally, by further refining these gluing techniques we come up with suitable minimising sequences obeying strong C^1 a priori bounds and related qualitative properties. Basing on these properties a minimiser of the Willmore functional is obtained by using direct methods from the calculus of variation.

8.3.1 An Upper Bound for the Optimal Energy

Lemma 8.6. *Let M_α be defined as in (8.10). Then*

$$M_\alpha \le 8 \tanh\left(\frac{1}{\alpha}\right).$$

In particular, $\lim_{\alpha \to \infty} M_\alpha = 0$.

Proof. For $x \ne 0$ let $v(x) := \alpha \cosh((|x| - 1)/\alpha)$. The corresponding surface of revolution consists of two branches of minimal surfaces with $v(\pm 1) = \alpha$, $v'(\pm 1) = 0$. Moreover, let $x_0 \in (0,1)$ be the uniquely determined point in $(0,1)$ such that $0 = x_0 + v(x_0)v'(x_0)$, and $r = \sqrt{x_0^2 + v(x_0)^2}$. One may observe that $x_0 + v(x_0)v'(x_0)$ is the intersection point of the euclidean normal of the graph of v in $(x_0, v(x_0))$ with the x-axis. We consider the $C^{1,1}$ comparison function

$$u(x) := \begin{cases} v(x) & \text{for } x_0 \le |x| \le 1, \\ \sqrt{r^2 - x^2} & \text{for } 0 \le |x| \le x_0. \end{cases}$$

See Figure 8.1.

For the hyperbolic Willmore energy we compute

$$\widehat{W}(u) = 2 \int_{x_0}^{1} \left(\frac{u''}{(1 + u'^2)^{\frac{3}{2}}} - \frac{1}{u\sqrt{1 + u'^2}} \right)^2 u\sqrt{1 + u'^2}\, dx + 8 \int_{x_0}^{1} \frac{u''}{(1 + u'^2)^{\frac{3}{2}}}\, dx$$

$$= 8 \int_{x_0}^{1} \frac{u''}{(1 + u'^2)^{\frac{3}{2}}}\, dx = -8 \frac{v'(x_0)}{\sqrt{1 + v'(x_0)^2}} = 8 \tanh\left(\frac{1 - x_0}{\alpha}\right)$$

$$\le 8 \tanh\left(\frac{1}{\alpha}\right) < 8 \min\{1, 1/\alpha\},$$

and the claim follows. $\qquad\square$

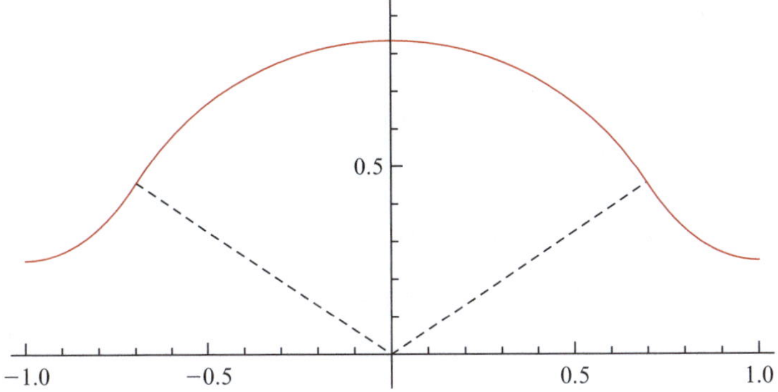

Fig. 8.1 Comparison function to estimate M_α. Here $\alpha = 1/4$.

8.3.2 Monotonicity of the Optimal Energy

In order to show that $\alpha \mapsto M_\alpha$ is decreasing we introduce several geometric constructions which will also be used to suitably modify minimising sequences for M_α. One key observation will be that functions in N_α may be shortened while maintaining the boundary values and decreasing the hyperbolic Willmore energy. As a first step we introduce this procedure for functions in N_α which are decreasing on $[0,1]$. Since this result will also be applied on rescaled intervals we formulate it on arbitrary intervals $[-a,a]$.

Lemma 8.7. *Fix $a > 0$. Assume that $u \in C^{1,1}([-a,a])$ has only finitely many critical points and is positive and symmetric with $u'(a) = 0$ and such that $u'(x) \le 0$ for all $x \in [0,a]$. Then for each $\rho \in (0,a]$ there exists a positive symmetric function $u_\rho \in C^{1,1}([-\rho,\rho])$ such that $u_\rho(\rho) = u(a)$, $u'_\rho(\rho) = 0$, u_ρ has at most as many critical points as u and satisfies for all $x \in [0,\rho]$*

$$u'_\rho(x) \le 0 \ \text{ as well as } \ \int_{-\rho}^{\rho} \kappa[u_\rho]^2\, ds[u_\rho] \le \int_{-a}^{a} \kappa[u]^2\, ds[u].$$

In particular if $a = 1$, then

$$\int_{-\rho}^{\rho} \kappa[u_\rho]^2\, ds[u_\rho] \le \widehat{W}(u).$$

Proof. Let $r \in (0,a)$ be a parameter. The normal to the graph of u in $(r,u(r))$ has direction $(-u'(r),1)$. The straight line generated by the normal intersects the x-axis left of r, since u is decreasing. We take this intersection point $(c(r),0)$ as center for a geodesic circular arc, where the radius is chosen such that the arc is tangential to the graph of u in $(r,u(r))$. This means that the radius is given by the distance between $(c(r),0)$ and $(r,u(r))$. We build a new symmetric function with smaller curvature integral as follows. On $[c(r),r]$ we take this geodesic arc, which has horizontal tangent

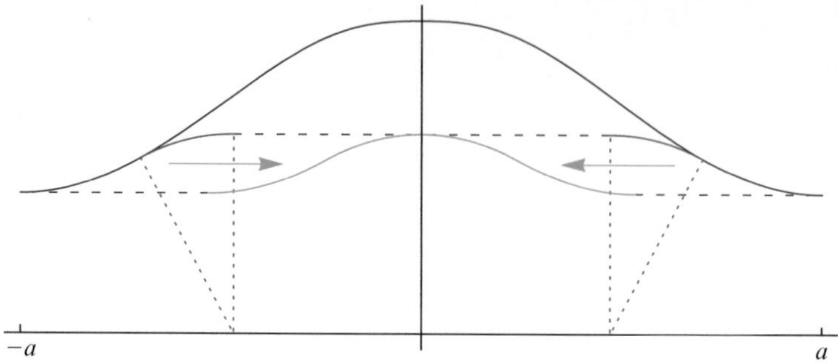

Fig. 8.2 Proof of Lemma 8.7.

in $c(r)$, while on $[r,a]$ we take u. By construction, this function is $C^{1,1}([c(r),a])$ and decreasing. We shift it such that $c(r)$ is moved to 0, and extend this to an even function, which is again $C^{1,1}$, now on a suitable interval $[-\ell(r),\ell(r)]$. This function has the same boundary values as u, at most as many critical points as u and, by construction, a smaller curvature integral. This construction yields the claim since $r \mapsto \ell(r)$ is continuous and $\lim_{r\searrow 0}\ell(r) = a$, $\lim_{r\nearrow a}\ell(r) = 0$. The procedure is illustrated in Figure 8.2. $\qquad\qquad\Box$

Next, it is important to observe that for functions in N_α bending upwards requires less energy than bending downwards. This statement cannot be verified by just reflecting the graph about the straight line through its boundary points. Instead, it requires a refined geometric construction based on the previous observation.

Lemma 8.8. *Fix $a > 0$. Assume that $u \in C^{1,1}([-a,a])$ has only finitely many critical points and is symmetric, positive with $u'(a) = 0$ and such that $u'(x) \geq 0$ for all $x \in [0,a]$. Then there exists a positive symmetric function $v \in C^{1,1}([-a,a])$ with $v(a) = u(a)$, $v'(a) = 0$, v has at most as many critical points as u and*

$$v'(x) \leq 0 \ \text{for all } x \in [0,a], \ \text{as well as} \ \int_{-a}^{a} \kappa[v]^2\,ds[v] \leq \int_{-a}^{a} \kappa[u]^2\,ds[u].$$

In particular if $a = 1$, $\widehat{W}(v) \leq \widehat{W}(u)$.

Proof. We may assume that $u(0) < u(a)$. We consider

$$\tilde{u}(x) := \begin{cases} u(x+a) & \text{if } x \in [-a,0], \\ u(x-a) & \text{if } x \in [0,a]. \end{cases}$$

We apply the procedure of Lemma 8.7 to $\tilde{u}$ and find for all $\rho \in (0,a]$ a symmetric positive function $\tilde{u}_\rho \in C^{1,1}([-\rho,\rho])$ with lower Willmore energy, at most as many critical point as $\tilde{u}$ and such that $\tilde{u}_\rho(\rho) = \tilde{u}(a) = u(0)$, $\tilde{u}_\rho'(\rho) = 0$ and $\tilde{u}_\rho'(x) \leq 0$ for all $x \in [0,\rho]$. Let $\rho_0 \in (0,a]$ be such that $\tilde{u}(a) = u(0) = \frac{\rho_0}{a}u(a)$. Then by rescaling

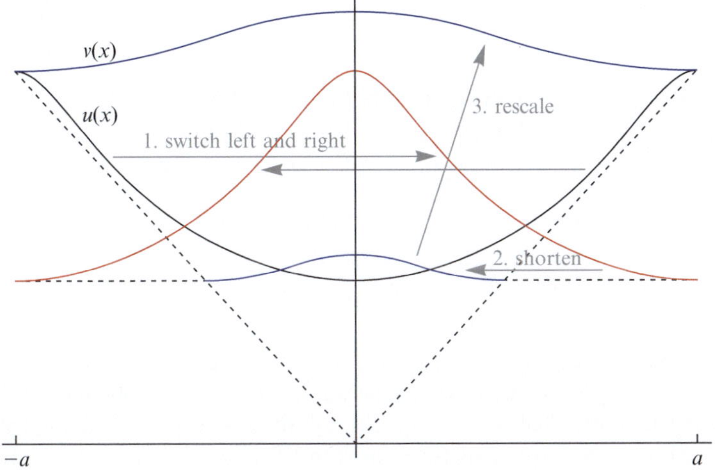

Fig. 8.3 Proof of Lemma 8.8.

(see Remark 8.5) the function $v(x) = \frac{a}{\rho_0}\tilde{u}_{\rho_0}(\frac{\rho_0}{a}x)$ defined on $[-a,a]$ is the desired decreasing function with smaller Willmore energy. The procedure is illustrated in Figure 8.3. $\qquad\square$

In order to extend the shortening procedure of Lemma 8.7 to functions not necessarily decaying on $[0,1]$ we combine both previous constructions and proceed iteratively.

Lemma 8.9. *Fix $a > 0$. Assume that $u \in C^{1,1}([-a,a])$ is a symmetric, positive function having only finitely many critical points and satisfying $u'(a) = 0$. Then for each $\rho \in (0,a]$ there exists a symmetric positive function $u_\rho \in C^{1,1}([-\rho,\rho])$ with $u'_\rho(\rho) = 0$ and $u_\rho(\rho) = u(a)$ with at most as many critical points as u such that*

$$\int_{-\rho}^{\rho} \kappa[u_\rho]^2\, ds[u_\rho] \leq \int_{-a}^{a} \kappa[u]^2\, ds[u].$$

If $u'(x) < 0$ for x close to a, the same may be achieved for $u'_\rho(x)$ for x close to ρ. In particular if $a = 1$

$$\int_{-\rho}^{\rho} \kappa[u_\rho]^2\, ds[u_\rho] \leq \widehat{W}(u).$$

Proof. We may assume that u is not a constant. Let $x_0 > 0$ be such that $[-x_0,x_0]$ is the smallest possible symmetric interval with $u'(x_0) = 0$. In $[0,x_0]$ the derivative of u has a fixed sign. If $u'(x) \geq 0$ in $[0,x_0]$, then by Lemma 8.8 there is a positive symmetric function $v \in C^{1,1}([-x_0,x_0])$ with lower Willmore energy such that $v(x_0) = u(x_0)$, $v'(x_0) = 0$ and $v'(x) \leq 0$ in $[0,x_0]$. Hence we may assume that $u'(x) \leq 0$ in $[0,x_0]$. By Lemma 8.7 for all $r \in (0,x_0]$ there exists a positive symmetric function

$v_r \in C^{1,1}([-r,r])$ such that $v_r(r) = u(x_0)$ and $v_r'(r) = 0$ and $v_r''(x) \leq 0$ in $[0,r]$. Hence the function

$$u_r(x) := \begin{cases} u(x+x_0-r) & \text{if } r < x \leq a+r-x_0, \\ v_r(x) & \text{if } -r \leq x \leq r, \\ u(x-x_0+r) & \text{if } -a-r+x_0 < x \leq -r, \end{cases}$$

is in $C^{1,1}([-a-r+x_0, a+r-x_0])$, is symmetric, $u_r'(a+r-x_0) = 0$, $u_r(a+r-x_0) = u(a)$ and

$$\int_{-(a+r-x_0)}^{a+r-x_0} \kappa[u_r]^2 \, ds[u_r] \leq \int_{-a}^{a} \kappa[u]^2 \, ds[u].$$

With this construction the claim is proved for $\rho \geq a - x_0$.

For $\rho < a - x_0$ we start from the function just constructed obtained at the limit for r going to zero. That is $v(x) = u(x+x_0)$ for $x \in [0, a-x_0]$ and extended by symmetry on $[-a+x_0, 0]$. This function is in $C^{1,1}([-a+x_0, a-x_0])$, positive and symmetric. We can repeat the same construction just done. We continuously decrease the interval of definition and, at the same time, the curvature integral. Since we have only finitely many critical points and at each iteration step we decrease the number of critical points, this procedure is well-defined and terminates after finitely many iterations.

If $u' < 0$ close to a the same may be achieved for u_ρ' since in the construction we do not change the function near the end-points of the interval of definition. $\qquad \square$

Corollary 8.10. *Fix $a > 0$ and $\alpha > 0$. For each positive symmetric $u \in C^{1,1}([-a,a])$ having only finitely many critical points and satisfying*

$$u(\pm a) = \alpha, \qquad u'(\pm a) = 0$$

and for each $\beta \geq \alpha$, there exists a symmetric positive function $v \in C^{1,1}([-a,a])$ having at most as many critical points as u, satisfying

$$v(\pm a) = \beta, \qquad v'(\pm a) = 0$$

and

$$\int_{-a}^{a} \kappa[v]^2 \, ds[v] \leq \int_{-a}^{a} \kappa[u]^2 \, ds[u].$$

If $u'(x) < 0$ for x close to a, the same may be achieved for v'. In particular, if $a = 1$ then $\widehat{W}(v) \leq \widehat{W}(u)$.

Proof. By Lemma 8.9 for each $\rho \in (0, a]$, there exists a symmetric positive function $u_\rho \in C^{1,1}([-\rho, \rho])$ having at most as many critical points as u with $u_\rho'(\rho) = 0$ and $u_\rho(\rho) = u(a) = \alpha$ such that

$$\int_{-\rho}^{\rho} \kappa[u_\rho]^2 \, ds[u_\rho] \leq \int_{-a}^{a} \kappa[u]^2 \, ds[u].$$

Choosing ρ_0 such that $\frac{a}{\rho_0}\alpha = \beta$ the function $v(x) = \frac{a}{\rho_0} u_{\rho_0}(\frac{\rho_0}{a} x)$ for $x \in [-a, a]$ yields the claim. $\qquad \square$

Theorem 8.11. *Let M_α for $\alpha \in (0,\infty)$ be as defined in (8.10). Then for $0 < \alpha < \widehat{\alpha}$ we have that*

$$M_{\widehat{\alpha}} \leq M_\alpha.$$

Proof. Since the polynomials are dense in H^2, a minimising sequence for M_α may be chosen in N_α, which consists of symmetric positive polynomials. Corollary 8.10 yields the claim. $\qquad\square$

8.3.3 Properties of Minimising Sequences

The first main step consists in finding a procedure which does not increase the Willmore energy but allows to restrict oneself to functions v in N_α such that $v'(x) \leq 0$ for all $x \in [0,1]$. We recall that the set N_α is defined in (8.9). Here, the techniques developed in Section 8.3.2 are used essentially.

Theorem 8.12. *Let N_α be as defined in (8.9). For each $u \in N_\alpha$ having only finitely many critical points, we find $v \in N_\alpha$ having at most as many critical points as u, satisfying*

$$v'(x) \leq 0 \text{ for all } x \in [0,1] \text{ and } \widehat{W}(v) \leq \widehat{W}(u).$$

Proof. If u does not have the claimed property then there exist $x_0, x_1 \in [0,1]$, $x_0 < x_1$, with $u'(x) > 0$ in (x_0,x_1), $u'(x_0) = u'(x_1) = 0$ and $u'(x) \leq 0$ in $[x_1,1]$. Using that $u(x_0) < u(x_1)$, we construct a positive symmetric function $v_1 \in C^{1,1}([-x_1,x_1])$ such that v_1 has at most as many critical points as $u|_{[-x_1,x_1]}$, $v_1'(x) \leq 0$ in $[\tilde{x}_0,x_1]$, where $\tilde{x}_0$ is the largest critical point of v_1 below x_1. Moreover,

$$v_1'(x_1) = 0, \quad v_1(x_1) = u(x_1), \quad \int_{-x_1}^{x_1} \kappa[v_1]^2 ds[v_1] \leq \int_{-x_1}^{x_1} \kappa[u]^2 ds[u]. \tag{8.11}$$

The claim will then follow by finitely many iterations proceeding from the boundary points towards the central point 0.

We consider $u|_{[-x_0,x_0]}$ and apply Corollary 8.10 with $\beta = u(x_1)$. If $x_0 = 0$ one simply skips this first step. There exists a symmetric positive function $w_1 \in C^{1,1}([-x_0,x_0])$ with $w_1(x_0) = u(x_1)$, $w_1'(x_0) = 0$, having no more critical points than $u|_{[-x_0,x_0]}$ and satisfying

$$\int_{-x_0}^{x_0} \kappa[w_1]^2 ds[w_1] \leq \int_{-x_0}^{x_0} \kappa[u]^2 ds[u].$$

Interchanging the parts of u over $[-x_1,-x_0]$ and $[x_0,x_1]$, we define on $[-x_1,x_1]$

$$\tilde{v}_1(x) := \begin{cases} u(x+x_1+x_0) & \text{if } x \in [-x_1,-x_0], \\ w_1(x) & \text{if } x \in [-x_0,x_0], \\ u(x-x_1-x_0) & \text{if } x \in [x_0,x_1]. \end{cases}$$

Certainly, $\tilde{v}_1 \in C^{1,1}([-x_1,x_1])$ is positive, symmetric and it does not have more critical points than $u|_{[-x_1,x_1]}$. Moreover, $\tilde{v}'_1(x) \leq 0$ for $x \in [x_0,x_1]$ and

$$\int_{-x_1}^{x_1} \kappa[\tilde{v}_1]^2 ds[\tilde{v}_1] \leq \int_{-x_1}^{x_1} \kappa[u]^2 ds[u], \quad \tilde{v}_1(x_1) = u(x_0), \quad \tilde{v}'_1(x_1) = 0.$$

Corollary 8.10 now yields a positive symmetric function $v_1 \in C^{1,1}([-x_1,x_1])$, having no more critical points than $u|_{[-x_1,x_1]}$ and satisfying (8.11), with $v'_1(x) \leq 0$ in $[\tilde{x}_0,x_1]$, where $\tilde{x}_0$ is the largest critical point of v_1 below x_1. The last property is verified first close to x_1; it holds on the whole interval since no further critical points arise. $\square$

Moreover, in choosing a minimising sequence for M_α we may restrict ourselves to functions in N_α satisfying

$$0 \leq x + v(x)v'(x) \quad \text{for all } x \in [0,1]. \tag{8.12}$$

For $x = 0$ and $x = 1$, this inequality is trivially satisfied and for $x = 1$, it is even strict. If for some $x_0 \in (0,1)$ we have that $0 = x_0 + v(x_0)v'(x_0)$, then the euclidean normal in $(x_0, v(x_0))$ to the graph of v goes through the origin. Hence, with the same construction as in Lemma 8.6 we could substitute over $[-x_0,x_0]$ the original graph by a geodesic circular arc lowering the Willmore energy. Observe that this procedure, applied to a positive symmetric $C^{1,1}$-function with $v'(x) \leq 0$ for all $x \in [0,1]$ preserves all these properties. See Figure 8.4.

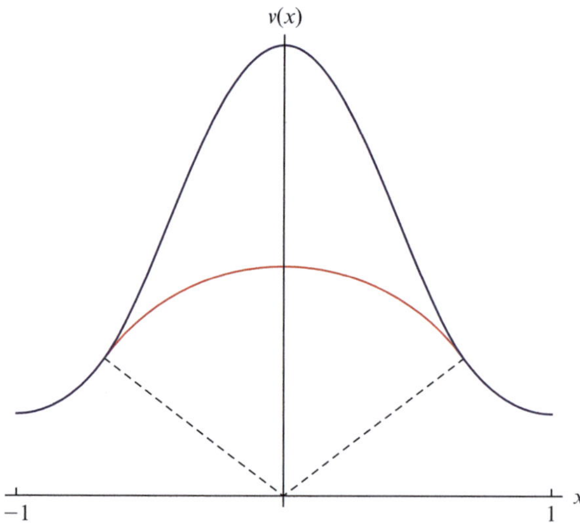

Fig. 8.4 The idea how to achieve the condition $0 \leq x + v(x)v'(x)$.

Combining (8.12) with Theorem 8.12 we may restrict ourselves to minimising sequences (v_k) for the optimal Willmore energy M_α defined in (8.10), which have the following properties:

$$v_k \in C^{1,1}([-1,1]) \text{ are positive, symmetric and satisfy}$$
$$\text{for all } x \in [0,1]: \ 0 \le x + v_k(x)v_k'(x), \ v_k'(x) \le 0. \tag{8.13}$$

This implies immediately a priori estimates for this suitably chosen minimising sequence. For all $x \in [-1,1]$ we have

$$\alpha \le v_k(x) \le \sqrt{\alpha^2 + 1 - x^2} \le \alpha + 1, \qquad |v_k'(x)| \le \frac{|x|}{\alpha}. \tag{8.14}$$

8.3.4 Attainment of the Minimal Energy

We are now able to state and to prove a more precise result than Theorem 8.1.

Theorem 8.13. *For arbitrary $\alpha > 0$, there exists a positive symmetric function $u \in H^2(-1,1) \cap C^{1,1/2}([-1,1])$ satisfying*

$$u(\pm 1) = \alpha, \qquad u'(\pm 1) = 0,$$

such that

$$\widehat{W}(u) = M_\alpha \overset{def}{=} \inf\{\widehat{W}(v) : v \in C^{1,1}([-1,1]), v \text{ is even}, v(\pm 1) = \alpha, v'(\pm 1) = 0\}.$$

This minimum is a weak solution to the Dirichlet problem (8.1) satisfying

$$0 \le x + u(x)u'(x), \qquad u'(x) \le 0 \text{ for all } x \in [0,1]. \tag{8.15}$$

$$\alpha \le u(x) \le \sqrt{\alpha^2 + 1 - x^2} \le \alpha + 1, \qquad |u'(x)| \le \frac{|x|}{\alpha} \text{ for } x \in [-1,1]. \tag{8.16}$$

Moreover, u is a classical solution, i.e. $u \in C^\infty([-1,1])$.

Proof. Step 1. Existence and quantitative properties of a minimiser.
Let $(v_k) \subset N_\alpha$ be a minimising sequence for M_α satisfying (8.13)-(8.14). By the uniform bounds in (8.14) we find

$$\widehat{W}(v_k) = \int_{-1}^1 \frac{v_k''(x)^2 v_k(x)}{(1 + v_k'(x)^2)^{5/2}} \, dx + \int_{-1}^1 \frac{1}{v_k(x)\sqrt{1 + v_k'(x)^2}} \, dx$$

$$\ge \frac{\alpha}{\left(1 + \frac{1}{\alpha^2}\right)^{5/2}} \int_{-1}^1 v_k''(x)^2 \, dx + \frac{2}{(\alpha + 1)\sqrt{1 + \frac{1}{\alpha^2}}}.$$

This shows uniform boundedness of (v_k) in $H^2(-1,1)$. After passing to a subsequence, we find a positive symmetric function $u \in H^2(-1,1)$ such that

$$v_k \rightharpoonup u \text{ in } H^2(-1,1), \quad v_k \to u \in C^1([-1,1]),$$

and satisfying (8.15)–(8.16). Since

$$M_\alpha + o(1) = \widehat{W}(v_k) = \int_{-1}^{1} \frac{v_k''(x)^2 u(x)}{(1+u'(x)^2)^{5/2}}\,dx + \int_{-1}^{1} \frac{1}{u(x)\sqrt{1+u'(x)^2}}\,dx + o(1)$$

$$\geq \int_{-1}^{1} \frac{u''(x)^2 u(x)}{(1+u'(x)^2)^{5/2}}\,dx + \int_{-1}^{1} \frac{1}{u(x)\sqrt{1+u'(x)^2}}\,dx + o(1),$$

it follows that u minimises $\widehat{W}$ in the class of all positive symmetric $H^2(-1,1)$-functions v, satisfying $v(\pm 1) = \alpha$, $v'(\pm 1) = 0$. So, u weakly solves (8.1), see Lemma 8.2. At a first instance only symmetric testing functions are admissible. But since the integral of odd functions vanishes and each function may be decomposed into an odd and an even part this does not give rise to any restriction. Analogously, (8.7) is solved in the sense of (8.17) below. See also (8.8).

Step 2. Regularity of the minimiser.
From the calculations in the proof of Lemma 8.4 we see that for any *even* $\varphi \in C^2([-1,1])$ with $\varphi(1) = 0$, $\varphi'(1) = 0$ one has that

$$-2\int_{-1}^{1} \kappa \frac{1}{1+u'^2}\varphi''\,dx = \int_{-1}^{1} \kappa^2 \frac{\sqrt{1+u'^2}}{u^2}\varphi\,dx - 5\int_{-1}^{1} \kappa^2 \frac{u'}{u\sqrt{1+u'^2}}\varphi'\,dx$$

$$-2\int_{-1}^{1} \kappa \frac{1}{u^2}\varphi\,dx + 4\int_{-1}^{1} \kappa \frac{u'}{u(1+u'^2)}\varphi'\,dx, \quad (8.17)$$

see (8.8). First, we observe that (8.17) is still true for any $\varphi \in C^2([-1,1])$ with $\varphi(\pm 1) = 0$ and $\varphi'(\pm 1) = 0$. This follows by decomposing φ in its even and odd part and using that these satisfy the same boundary conditions. We use further that integrals over odd functions vanish. For arbitrary $\eta \in C_c^\infty(-1,1)$ we take

$$\varphi(x) := \int_{-1}^{x}\int_{-1}^{y} \eta(s)\,ds\,dy - \beta(x+1)^2 - \gamma(x+1)^3,$$

where

$$\beta = -\frac{1}{2}\int_{-1}^{1}\eta(s)\,ds + \frac{3}{4}\int_{-1}^{1}\int_{-1}^{y}\eta(s)\,ds\,dy,$$

$$\gamma = \frac{1}{4}\int_{-1}^{1}\eta(s)\,ds - \frac{1}{4}\int_{-1}^{1}\int_{-1}^{y}\eta(s)\,ds\,dy$$

are chosen such that $\varphi(\pm 1) = 0$ and $\varphi'(\pm 1) = 0$. Since $\widehat{W}(u)$ is finite, u obeys (8.16) and since

$$\beta, \gamma, \|\varphi\|_{C^1} \leq C\|\eta\|_{L^1},$$

we can conclude from (8.17) that for each $\eta \in C_c^\infty(-1,1)$,

$$\left| \int_{-1}^{1} \kappa \frac{1}{1+u'^2} \eta \, dx \right| \leq C(u)\|\eta\|_{L^1}.$$

By the bounds on u in (8.16), the above inequality shows that κ is bounded and so,

$$u \in W^{2,\infty}(-1,1).$$

Next, for arbitrary $\eta \in C_c^\infty(-1,1)$ we choose

$$\varphi(x) = \int_{-1}^{x} \eta(s)\,ds - \frac{3}{4}\left(\int_{-1}^{1}\eta(s)\,ds\right)(x+1)^2 + \frac{1}{4}\left(\int_{-1}^{1}\eta(s)\,ds\right)(x+1)^3$$

so that

$$\varphi(\pm 1) = 0, \qquad \varphi'(\pm 1) = 0, \qquad \|\varphi\|_{C^0} \leq C\|\eta\|_{L^1}, \qquad \|\varphi'\|_{L^1} \leq C\|\eta\|_{L^1}.$$

Since we already know that κ is bounded, we conclude from (8.17) that for each $\eta \in C_c^\infty(-1,1)$,

$$\left| \int_{-1}^{1} \kappa \frac{1}{1+u'^2} \eta'(x)\,dx \right| \leq C(u)\|\eta\|_{L^1}.$$

This proves that

$$\kappa \frac{1}{1+u'^2} \in W^{1,\infty}(-1,1), \quad \kappa \in W^{1,\infty}([-1,1]) = C^{0,1}([-1,1]),$$

$$u \in W^{3,\infty}([-1,1]) = C^{2,1}([-1,1]).$$

Finally, rewriting (8.7) as follows

$$\frac{d}{dx}\left(\frac{u(x)}{\sqrt{1+u'(x)^2}} \kappa'(x) \right) = \frac{\sqrt{1+u'(x)^2}}{u(x)}\left(\kappa(x) - \frac{1}{2}\kappa(x)^3 \right) \quad \text{in } (-1,1),$$

we get an equation for κ with $W^{1,\infty}$-coefficients and right hand side. Hence, $\kappa \in W^{3,\infty}([-1,1]) = C^{2,1}([-1,1])$, $u \in C^{4,1}([-1,1])$ and finally, by straightforward bootstrapping, $u \in C^\infty([-1,1])$. $\qquad\square$

8.4 Bibliographical Notes

A historical survey over the Willmore functional and a profound discussion of modeling aspects is provided by Nitsche [325]. There, also Euler-Lagrange equations are deduced and sets of natural boundary conditions are discussed. The Willmore functional was already considered as a model for the elastic energy of thin plates in the 19th century, see e.g. [342], but a mathematical treatment seemed to have been out of reach for more than 100 years. It was popularised again in the second half of the 20th century by Willmore's work, see e.g. [414, 415].

Existence of closed Willmore surfaces of prescribed genus was proved by Simon and Bauer-Kuwert [35, 372]. Rivière [356] showed a far reaching regularity result. Also, local and global existence results for the Willmore flow of closed surfaces are available, see e.g. [263–265, 373]. On the other hand, Mayer and Simonett [288] gave a numerical example providing evidence that for particular initial data the Willmore flow may develop geometric singularities – change of topology – in finite time. An analytic proof for occurrence of such a singularity in finite or infinite time for the same initial data was given by Blatt [60]. The Willmore flow for one-dimensional closed curves was studied by Dziuk-Kuwert-Schätzle [156] and Polden [343].

Numerical experiments concerning Willmore surfaces of revolutions were performed by Fröhlich [175], where as boundary data, the position $\alpha > 0$ and the mean curvature $\mathrm{H} = 0$ were prescribed. Bryant, Griffiths [77, 78] and Hertrich-Jeromin, Pinkall (see e.g. [230]) observed that Willmore surfaces of revolution can be considered as elastic curves in the hyperbolic half space. This observation was exploited in [267, 268], where also many properties of elastic curves in the hyperbolic half space were deduced. This reformulation helped a lot in proving our main result from [115], which is a joint work of Deckelnick, Dall'Acqua, and Grunau. The underlying geometric constructions benefit from previous works on related one-dimensional problems [137, 139]. More general Dirichlet problems for Willmore surfaces of revolution are studied in [116], while Navier-type boundary value problems are subject of [50, 138].

Quite recently, Schätzle [361] proved an important general result concerning existence of branched Willmore immersions in $\mathbb{S}^n$ satisfying Dirichlet boundary conditions. Assuming the boundary data to obey some explicit geometrically motivated smallness condition these immersions can even be shown to be embedded. By working in $\mathbb{S}^n$, some compactness problems could be overcome. On the other hand, when pulling pack these immersions to $\mathbb{R}^n$ it cannot be excluded that they contain ∞. Moreover, in general, the existence of branch points cannot be ruled out, and due to the generality of the approach, it seems to us that only little topological information about the solutions can be extracted from the existence proof.

Notations

C	positive constants in estimates, which may change their value from term to term.
t^+	$\max\{t,0\}$ for $t \in \mathbb{R}$.
t^-	$\max\{-t,0\}$ for $t \in \mathbb{R}$.
$a_k \sim b_k$	$\displaystyle\lim_{k \to \infty} \frac{a_k}{b_k} = 1$.
$\mathbb{N}$	$= \{0,1,2,3\ldots\}$.
$\mathbb{N}^+$	$= \{1,2,3\ldots\}$.
n	space dimension.
$\mathbb{R}^n_+$	$= \{x \in \mathbb{R}^n : x_1 > 0\}$, half space.
Ω	domain, an open and connected subset of $\mathbb{R}^n$.
$\partial\Omega$	$\overline{\Omega} \cap \overline{\mathbb{R}^n \setminus \Omega}$, the boundary of Ω.
$\Omega_0 \subset\subset \Omega$	$\overline{\Omega_0}$ is compact and $\overline{\Omega_0} \subset \Omega$.
A^c	for $A \subset \Omega$: complement of A in Ω, $\Omega \setminus A$.
$d\omega$	surface element for $\partial\Omega$.
ν	exterior unit normal at $\partial\Omega$.
$B_r(x)$	open ball with radius r and centre x.
B	$= B_1(0)$, open unit ball in $\mathbb{R}^n$.
$\mathbb{S}^{n-1}$	$= \partial B \subset \mathbb{R}^n$, unit sphere.
e_n	$= \dfrac{\pi^{n/2}}{\Gamma(1+n/2)}$, volume of the n-dimensional unit ball $B \subset \mathbb{R}^n$. So $n e_n$ is the $(n-1)$-dimensional measure of the unit sphere.

$d(x)$	$= \mathrm{dist}(x, \partial\Omega)$, for $x \in \Omega$.
$d(x)$	$= 1 - |x|$, for $x \in B$.
$[XY]$	$= \left| |x|y - \dfrac{x}{|x|} \right|$, for $x, y \in B$.
r	$= |x|$, $x \in \mathbb{R}^n$.
$D_{i_1,\ldots,i_k}$	$= \dfrac{\partial^k}{\partial x_{i_1} \cdots \partial x_{i_k}}$.
α, β	multiindices $\in \mathbb{N}_0^n$, $|\alpha| = \sum_{i=1}^{n} \alpha_i$.
D^α	$= \prod_{i=1}^{n} \left(\dfrac{\partial}{\partial x_i} \right)^{\alpha_i}$ with $|\alpha| = \sum_{i=1}^{n} \alpha_i$.
$C_c^\infty(\Omega)$	space of $C^\infty(\Omega)$-functions having compact support in Ω.
$W^{m,p}(\Omega)$	Sobolev space of the m-times weakly differentiable functions in Ω with L^p-derivatives.
$D^k u \cdot D^k v$	$= \sum_{i_1,\ldots,i_k=1}^{n} \dfrac{\partial^k u}{\partial x_{i_1} \ldots \partial x_{i_k}} \cdot \dfrac{\partial^k v}{\partial x_{i_1} \ldots \partial x_{i_k}}$.
$|D^k u|$	$= \left(D^k u \cdot D^k u \right)^{1/2}$.
$\|u\|_{W^{m,p}}$	$= \left(\|u\|_{L^p(\Omega)}^p + \|D^m u\|_{L^p(\Omega)}^p \right)^{1/p}$.
$\|u\|_{W_0^{m,p}}$	$= \|D^m u\|_{L^p(\Omega)}$.
$W_0^{m,p}(\Omega)$	in bounded domains Ω, closure of $C_c^\infty(\Omega)$ with respect to the norm $\|\cdot\|_{W_0^{m,p}}$.
$W_0^{m,p}(\Omega)$	in any domain Ω, closure of $C_c^\infty(\Omega)$ with respect to the norm $\|\cdot\|_{W^{m,p}}$.
$\mathscr{D}^{m,p}(\Omega)$	in unbounded domains Ω, closure of $C_c^\infty(\Omega)$ with respect to the norm $\|\cdot\|_{W_0^{m,p}}$.
$H^m(\Omega)$	$= W^{m,2}(\Omega)$.
$H_0^m(\Omega)$	$= W_0^{m,2}(\Omega)$.
$H_\vartheta^m(\Omega)$	$= \left\{ v \in H^m(\Omega); \Delta^j v = 0 \text{ on } \partial\Omega \text{ for } j < \dfrac{m}{2} \right\}$.
$H^{-m}(\Omega)$	dual space $(H_0^m(\Omega))'$.

$\|u\|_{H_0^m}^2$ $\qquad = \displaystyle\sum_{i_1,\dots,i_m=1}^{n} \int_\Omega |D_{i_1,\dots,i_m}u|^2\,dx = (\text{see } (2.12))$

$$= \begin{cases} \displaystyle\int_\Omega \left(\Delta^{m/2}u\right)^2 dx & \text{if } m \text{ is even,} \\[2mm] \displaystyle\int_\Omega \left|\nabla\Delta^{(m-1)/2}u\right|^2 dx & \text{if } m \text{ is odd.} \end{cases}$$

$(\,.\,,.\,)_{H_0^m}$ $\qquad$ corresponding scalar product in $H_0^m(\Omega)$.

$\|u\|_{\mathscr{D}^{m,2}}^2$ $\qquad = \begin{cases} \displaystyle\int_\Omega \left(\Delta^{m/2}u\right)^2 dx & \text{if } m \text{ is even,} \\[2mm] \displaystyle\int_\Omega \left|\nabla\Delta^{(m-1)/2}u\right|^2 dx & \text{if } m \text{ is odd.} \end{cases}$

$(\,.\,,.\,)_{\mathscr{D}^{m,2}}$ $\qquad$ corresponding scalar product in $\mathscr{D}^{m,2}$.

$\langle f,u\rangle$ $\qquad$ dual pairing: $u \in$ Banach space, $f \in$ its dual.

$\Lambda_{m,j}$ $\qquad$ j-th Dirichlet-eigenvalue of $(-\Delta)^m$, according to its multiplicity.

φ_j $\qquad$ corresponding eigenfunctions, orthonormal in $H_0^m(\Omega)$.

$G_{m,n}, \mathscr{G}_{m,n}$ $\qquad$ Green's function, Green's operator, resp. for $(-\Delta)^m$ under Dirichlet boundary conditions in $B \subset \mathbb{R}^n$.

$G_{(-\Delta)^m,\Omega}, \mathscr{G}_{(-\Delta)^m,\Omega}$ $\qquad$ the same for $\Omega \subset \mathbb{R}^n$.

$G_{m,n,\mathscr{A}}, \mathscr{G}_{m,n,\mathscr{A}}$ $\qquad$ the same for $(-\Delta)^m + \mathscr{A}$ in $B \subset \mathbb{R}^n$, where $\mathscr{A}u = \displaystyle\sum_{|\alpha|\le 2m-1} a_\alpha D^\alpha u$.

$G_{(-\Delta)^m,\Omega,\mathscr{A}}, \mathscr{G}_{(-\Delta)^m,\Omega,\mathscr{A}}$ $\qquad$ the same in $\Omega \subset \mathbb{R}^n$.

s $\qquad = \dfrac{n+2m}{n-2m}$, for $n > 2m$; critical Sobolev exponent.

For measurable functions f, g:

$f > 0$ $\qquad$ $f(x) > 0$ for almost all x.

$f > g$ $\qquad$ $f - g > 0$.

$f \not\geq 0$ $\qquad$ $f(x) < 0$ for x in a set of positive measure.

$f \not\geq g$ $\qquad$ $f - g \not\geq 0$.

$f \gneqq 0$ $\qquad$ $f(x) \neq 0$ for x in a set of positive measure and $f(x) \geq 0$ for almost all x.

$f \gneqq g$ $\qquad$ $f - g \gneqq 0$.

$f(t) \simeq g(t)$ $\qquad$ $\exists C > 0\ \forall t: \quad \dfrac{1}{C}f(t) \le g(t) \le Cf(t);\ \text{ for } f,g \ge 0.$

$f(t) \preceq g(t)$ $\qquad$ $\exists C > 0\ \forall t: \quad f(t) \le Cg(t),\ \text{for } f,g \ge 0.$

Ω^* $\qquad$ a ball centered at the origin such that $|\Omega| = |\Omega^*|$.

u^*

spherical rearrangement of a measurable function u, see Definition 3.10.

$(g_{ij})_{i,j=1,\ldots,n}$

Riemannian metric, positive definite tensor. In case of parametrisations X in $\mathbb{R}^3$ over a two dimensional parameter domain: $g_{ij} = \partial_i X \cdot \partial_j X$.

g

$= \det\left((g_{ij})_{i,j=1,\ldots,n}\right)$, Gram's determinant.

$(g^{ij})_{i,j=1,\ldots,n}$

inverse of the metric tensor.

$(L_{ij})_{i,j=1,\ldots,n}$

second fundamental form. In case of parametrisations X in $\mathbb{R}^3$ over a two dimensional parameter domain:

$$L_{ij} = \frac{1}{\sqrt{g}} \det\left(\partial_i \partial_j X, \partial_1 X, \partial_2 X\right).$$

H

mean curvature. In case of parametrisations X in $\mathbb{R}^3$ over a two dimensional parameter domain:

$$\mathrm{H} = \frac{1}{2} \sum_{i,j=1}^{2} g^{ij} L_{ij} = \frac{1}{2g}\left(g_{22} L_{11} - 2 g_{12} L_{12} + g_{11} L_{22}\right).$$

K

Gaussian curvature. In case of parametrisations X in $\mathbb{R}^3$. over a two dimensional parameter domain:

$$\mathrm{K} = \frac{\det\left((L_{ij})_{i,j=1,2}\right)}{g}.$$

Γ_{ij}^k

$$= \sum_{\ell=1}^{n} \frac{1}{2} g^{k\ell}\left(\partial_j g_{i\ell} + \partial_i g_{j\ell} - \partial_\ell g_{ij}\right), \text{Christoffel symbols.}$$

$R_{kij}^\ell = -R_{kji}^\ell$

$$= \partial_i \Gamma_{jk}^\ell - \partial_j \Gamma_{ik}^\ell + \sum_{m=1}^{n}\left(\Gamma_{im}^\ell \Gamma_{jk}^m - \Gamma_{jm}^\ell \Gamma_{ik}^m\right),$$

Riemannian curvature tensor.

$R_{k\ell ij} = -R_{\ell k ij} = R_{ijk\ell}$

$$= \sum_{m=1}^{n} g_{km} R_{\ell ij}^m.$$

R_{ij}

$$= \sum_{k,\ell=1}^{n} g^{k\ell} R_{ikj\ell} = \sum_{k=1}^{n} R_{ikj}^k, \text{Ricci tensor.}$$

R

$$= \sum_{i,j=1}^{n} g^{ij} R_{ij}, \text{scalar curvature.}$$

S_{ij}

$$= \frac{1}{n-2}\left(2 R_{ij} - \frac{R}{(n-1)} g_{ij}\right), \text{Schouten tensor.}$$

$W_{ijk\ell}$

$$= R_{ijk\ell} - \frac{1}{n-2}\left(R_{ik} g_{j\ell} - R_{i\ell} g_{jk} + R_{j\ell} g_{ik} - R_{jk} g_{i\ell}\right)$$
$$+ \frac{R}{(n-1)(n-2)}\left(g_{j\ell} g_{ik} - g_{jk} g_{i\ell}\right), \text{Weyl tensor.}$$

References

1. R.A. Adams. *Sobolev Spaces*, volume 65 of *Pure and Applied Mathematics*. Academic Press, New York, London, 1975.
2. V. Adolfsson. L^2-integrability of second-order derivatives for Poisson's equation in nonsmooth domains. *Math. Scand.*, 70:146–160, 1992.
3. S. Agmon. The coerciveness problem for integro-differential forms. *J. Analyse Math.*, 6:183–223, 1958.
4. S. Agmon. Maximum theorems for solutions of higher order elliptic equations. *Bull. Amer. Math. Soc.*, 66:77–80, 1960.
5. S. Agmon, A. Douglis, and L. Nirenberg. Estimates near the boundary for solutions of elliptic partial differential equations satisfying general boundary conditions. I. *Comm. Pure Appl. Math.*, 12:623–727, 1959.
6. S. Agmon, A. Douglis, and L. Nirenberg. Estimates near the boundary for solutions of elliptic partial differential equations satisfying general boundary conditions. II. *Comm. Pure Appl. Math.*, 17:35–92, 1964.
7. A.D. Alexandrov. A characteristic property of the spheres. *Ann. Mat. Pura Appl.*, (4) 58:303–354, 1962.
8. E. Almansi. Sull'integrazione dell'equazione differenziale $\Delta^2\Delta^2 = 0$. *Rom. Acc. L. Rend.*, (5) 8_1:104–107, 1899.
9. E. Almansi. Sull'integrazione dell'equazione differenziale $\Delta^{2n} = 0$. *Annali di Mat.*, (3) 2:1–51, 1899.
10. F.J. Almgren and E.H. Lieb. Symmetric decreasing rearrangement is sometimes continuous. *J. Amer. Math. Soc.*, 2:683–773, 1989.
11. H.W. Alt. *Lineare Funktionalanalysis, eine anwendungsorientierte Einführung.* 4., überarbeitete und erweiterte Aufl. Springer-Verlag, Berlin etc., 2002.
12. C.O. Alves and J.M. do Ó. Positive solutions of a fourth-order semilinear problem involving critical growth. *Adv. Nonlinear Stud.*, 2:437–458, 2002.
13. H. Amann. Fixed point equations and nonlinear eigenvalue problems in ordered Banach spaces. *SIAM Rev.*, 18:620–709, 1976. (Erratum in: *SIAM Rev.* 19: vii, 1977.).
14. A. Ambrosetti and P.H. Rabinowitz. Dual variational methods in critical point theory and applications. *J. Funct. Anal.*, 14:349–381, 1973.
15. A. Ambrosetti and M. Struwe. A note on the problem $-\Delta u = \lambda u + u|u|^{2^*-2}$. *Manuscr. Math.*, 54:373–379, 1986.
16. O.H. Ammann, Th. von Kármán, and G.B. Woodruff. The failure of the Tacoma Narrows Bridge. *Federal Works Agency*, 1941.
17. G.E. Andrews, R. Askey, and R. Roy. *Special Functions*, volume 71 of *Encyclopedia of Mathematics and its applications*. Cambridge University Press, Cambridge, 1999.
18. G. Arioli, F. Gazzola, and H.-Ch. Grunau. Entire solutions for a semilinear fourth order elliptic problem with exponential nonlinearity. *J. Differential Equations*, 230:743–770, 2006.
19. G. Arioli, F. Gazzola, H.-Ch. Grunau, and E. Mitidieri. A semilinear fourth order elliptic problem with exponential nonlinearity. *SIAM J. Math. Anal.*, 36:1226–1258, 2005.

20. G. Arioli, F. Gazzola, H.-Ch. Grunau, and E. Sassone. The second bifurcation branch for radial solutions of the Brezis-Nirenberg problem in dimension four. *NoDEA Nonlinear Differential Equations Appl.*, 15:69–90, 2008.

21. N. Aronszajn, T.M. Creese, and L.J. Lipkin. *Polyharmonic functions*. Oxford Mathematical Monographs. Oxford Science Publications. The Clarendon Press, Oxford University Press, New York, 1983.

22. M.S. Ashbaugh and R.D. Benguria. On Rayleigh's conjecture for the clamped plate and its generalization to three dimensions. *Duke Math. J.*, 78:1–17, 1995.

23. M.S. Ashbaugh and D. Bucur. On the isoperimetric inequality for the buckling of a clamped plate. *Z. Angew. Math. Phys.*, 54:756–770, 2003.

24. M.S. Ashbaugh and R.S. Laugesen. Fundamental tones and buckling loads of clamped plates. *Ann. Scuola Norm. Sup. Pisa Cl. Sci.*, (4) 23:383–402, 1996.

25. F.V. Atkinson, H. Brezis, and L.A. Peletier. Nodal solutions of elliptic equations with critical Sobolev exponents. *J. Differential Equations*, 85:151–170, 1990.

26. Th. Aubin. Équations différentielles non linéaires et problème de Yamabe concernant la courbure scalaire. *J. Math. Pures Appl.*, (9) 55:269–296, 1976.

27. Th. Aubin. *Some nonlinear problems in Riemannian geometry*. Springer Monographs in Mathematics. Springer-Verlag, Berlin etc., 1998.

28. I. Babuška. Stabilität des Definitionsgebietes mit Rücksicht auf grundlegende Probleme der Theorie der partiellen Differentialgleichungen auch im Zusammenhang mit der Elastizitätstheorie. I, II (Russian, German summary). *Czechoslovak Math. J.*, 11 (86):76–105, 165–203, 1961.

29. A. Bahri and J.M. Coron. On a nonlinear elliptic equation involving the critical Sobolev exponent: the effect of the topology of the domain. *Comm. Pure Appl. Math.*, 41:253–294, 1988.

30. P. Bartolo, V. Benci, and D. Fortunato. Abstract critical point theorems and applications to some nonlinear problems with "strong" resonance at infinity. *Nonlinear Anal., Theory Methods Appl.*, 7:981–1012, 1983.

31. Th. Bartsch and Y. Guo. Existence and nonexistence results for critical growth polyharmonic elliptic systems. *J. Differential Equations*, 220:531–543, 2006.

32. Th. Bartsch, M. Schneider, and T. Weth. Multiple solutions of a critical polyharmonic equation. *J. Reine Angew. Math.*, 571:131–143, 2004.

33. Th. Bartsch, T. Weth, and M. Willem. A Sobolev inequality with remainder term and critical equations on domains with topology for the polyharmonic operator. *Calc. Var. Partial Differential Equations*, 18:253–268, 2003.

34. L. Bauer and E.L. Reiss. Block five diagonal matrices and the fast numerical solution of the biharmonic equation. *Math. Comp.*, 26:311–326, 1972.

35. M. Bauer and E. Kuwert. Existence of minimizing Willmore surfaces of prescribed genus. *Int. Math. Res. Not.*, 2003 (10):553–576, 2003.

36. G.R. Belickiĭ. Functional equations and conjugacy of local diffeomorphisms of a finite smoothness class. *Funct. Anal. Appl.*, 7:268–277, 1973. Russian original: *Funkcional. Anal. i Priložen.* 7: 17–28, 1973.

37. V. Benci and G. Cerami. The effect of the domain topology on the number of positive solutions of nonlinear elliptic problems. *Arch. Rational Mech. Anal.*, 114:79–93, 1991.

38. A. Bennett. Symmetry in an overdetermined fourth order elliptic boundary value problem. *SIAM J. Math. Anal.*, 17:1354–1358, 1986.

39. E. Berchio, D. Cassani, and F. Gazzola. Hardy-Rellich inequalities with boundary remainder terms and applications. *Manuscr. Math.*, 131:427–458, 2010.

40. E. Berchio and F. Gazzola. Some remarks on biharmonic elliptic problems with positive, increasing and convex nonlinearities. *Electron. J. Differential Equations*, 2005 (34):1–20, 2005.

41. E. Berchio and F. Gazzola. Best constants and minimizers for embeddings of second order Sobolev spaces. *J. Math. Anal. Appl.*, 320:718–735, 2006.

42. E. Berchio, F. Gazzola, and E. Mitidieri. Positivity preserving property for a class of biharmonic elliptic problems. *J. Differential Equations*, 229:1–23, 2006.

43. E. Berchio, F. Gazzola, and D. Pierotti. Nodal solutions to critical growth elliptic problems under Steklov boundary conditions. *Comm. Pure Appl. Anal.*, 8:533–557, 2009.

44. E. Berchio, F. Gazzola, and T. Weth. Critical growth biharmonic elliptic problems under Steklov-type boundary conditions. *Adv. Differential Equations*, 12:381–406, 2007.
45. E. Berchio, F. Gazzola, and T. Weth. Radial symmetry of positive solutions to nonlinear polyharmonic Dirichlet problems. *J. Reine Angew. Math.*, 620:165–183, 2008.
46. H. Berestycki, L. Nirenberg, and S.R.S. Varadhan. The principal eigenvalue and maximum principle for second-order elliptic operators in general domains. *Comm. Pure Appl. Math.*, 47:47–92, 1994.
47. M. Berger, P. Gauduchon, and E. Mazet. *Le spectre d'une variété Riemannienne.* Lecture Notes in Mathematics, Vol. 194. Springer-Verlag, Berlin, 1971.
48. M.S. Berger and P.C. Fife. On von Karman's equations and the buckling of a thin elastic plate. *Bull. Amer. Math. Soc.*, 72:1006–1011, 1966.
49. M.S. Berger and P.C. Fife. Von Kármán's equations and the buckling of a thin elastic plate. II. Plate with general edge conditions. *Comm. Pure Appl. Math.*, 21:227–241, 1968.
50. M. Bergner, A. Dall'Acqua, and S. Fröhlich. Symmetric Willmore surfaces of revolution satisfying natural boundary conditions. *Calc. Var. Partial Differential Equations*, to appear.
51. F. Bernis. Change of sign of the solutions to some parabolic problems. In V. Lakshmikantham, editor, *Nonlinear Analysis and Applications (Arlington, Texas, 1986)*, volume 109 of *Lecture Notes in Pure and Appl. Math.*, pages 75–82. Marcel Dekker, New York, 1987.
52. F. Bernis, J. García Azorero, and I. Peral Alonso. Existence and multiplicity of nontrivial solutions in semilinear critical problems of fourth order. *Adv. Differential Equations*, 1:219–240, 1996.
53. F. Bernis and H.-Ch. Grunau. Critical exponents and multiple critical dimensions for polyharmonic operators. *J. Differential Equations*, 117:469–486, 1995.
54. J. Bernoulli II. Essay théoretique sur les vibrations des plaques élastiques, rectangulaires et libres. *Nova Acta Acad. Sci. Imperialis Petropolitanae*, 5:197–219, 1789.
55. I. Birindelli, E. Mitidieri, and G. Svirs. Existence of the principal eigenvalue for cooperative elliptic systems in a general domain (Russian). *Differ. Uravn.*, 35:325–333, 429, 1999.
56. M. Birkner, J.A. López-Mimbela, and A. Wakolbinger. Comparison results and steady states for the Fujita equation with fractional Laplacian. *Ann. Inst. H. Poincaré Anal. Non Linéaire*, 22:83–97, 2005.
57. M.Š. Birman. On Trefftz's variational method for the equation $\Delta^2 u = f$ (Russian). *Dokl. Akad. Nauk SSSR (N.S.)*, 101:201–204, 1955.
58. M.Š. Birman. Variational methods of solution of boundary problems analogous to the method of Trefftz (Russian). *Vestnik Leningrad. Univ.*, 11:69–89, 1956.
59. W. Blaschke and K. Leichtweiss. *Elementare Differentialgeometrie,* Fünfte Auflage. Die Grundlehren der mathematischen Wissenschaften, Band 1. Springer-Verlag, Berlin etc., 1973.
60. S. Blatt. A singular example for the Willmore flow. *Analysis (Munich)*, 29:407–430, 2009.
61. F. Bleich, C.B. McCullough, R. Rosecrans, and G.S. Vincent. *The Mathematical Theory of Suspension Bridges.* U.S. Dept. of Commerce, Bureau of Public Roads, Washington, D.C., 1950.
62. T. Boggio. Sull'equilibrio delle piastre elastiche incastrate. *Rend. Acc. Lincei*, 10:197–205, 1901.
63. T. Boggio. Sulle funzioni di Green d'ordine *m. Rend. Circ. Mat. Palermo*, 20:97–135, 1905.
64. R. Böhme. Die Lösung der Verzweigungsgleichungen für nichtlineare Eigenwertprobleme. *Math. Z.*, 127:105–126, 1972.
65. M. Borsuk and V. Kondratiev. *Elliptic Boundary Value Problems of Second Order in Piecewise Smooth Domains*, volume 69 of *North-Holland Mathematical Library*. Elsevier Science B.V., Amsterdam, 2006.
66. T.P. Branson. Differential operators canonically associated to a conformal structure. *Math. Scand.*, 57:293–345, 1985.
67. T.P. Branson. Group representations arising from Lorentz conformal geometry. *J. Funct. Anal.*, 74:199–291, 1987.
68. T. Bräu. A decomposition method with respect to dual cones and its application to higher order Sobolev spaces. Studienarbeit, Universität Magdeburg, 2005, downloadable at http://www-ian.math.uni-magdeburg.de/home/grunau/papers/BraeuStudEnglisch.pdf.

69. H. Brezis. *Analyse fonctionnelle.* Théorie et applications. Collection Mathématiques Appliquées pour la Maîtrise. Masson, Paris, 1983.

70. H. Brezis, Th. Cazenave, Y. Martel, and A. Ramiandrisoa. Blow up for $u_t - \Delta u = g(u)$ revisited. *Adv. Differential Equations*, 1:73–90, 1996.

71. H. Brezis and E.H. Lieb. Sobolev inequalities with remainder terms. *J. Funct. Anal.*, 62:73–86, 1985.

72. H. Brezis and L. Nirenberg. Positive solutions of nonlinear elliptic equations involving critical Sobolev exponents. *Comm. Pure Appl. Math.*, 36:437–477, 1983.

73. H. Brezis and J.L. Vazquez. Blow-up solutions of some nonlinear elliptic problems. *Rev. Mat. Univ. Complutense Madrid*, 10:443–469, 1997.

74. J. Brothers and W. Ziemer. Minimal rearrangements of Sobolev functions. *J. Reine Angew. Math.*, 384:153–179, 1988.

75. F.E. Browder. Existence theory for boundary value problems for quasilinear elliptic systems with strongly nonlinear lower order terms. In *Partial differential equations (Proc. Sympos. Pure Math., Vol. XXIII, Univ. California, Berkeley, Calif., 1971)*, pages 269–286. Amer. Math. Soc., Providence, R.I., 1973.

76. B.M. Brown, E.B. Davies, P.K. Jimack, and M.D. Mihajlovic. A numerical investigation of the solution of a class of fourth-order eigenvalue problems. *R. Soc. Lond. Proc. Ser. A. Math. Phys. Eng. Sci.*, 456:1505–1521, 1998.

77. R. Bryant. A duality theorem for Willmore surfaces. *J. Differential Geom.*, 20:23–53, 1984.

78. R. Bryant and P. Griffiths. Reduction for constrained variational problems and $\int \frac{1}{2}k^2 \, ds$. *Amer. J. Math.*, 108:525–570, 1986.

79. L.L. Bucciarelli and N. Dworsky. *Sophie Germain,* An essay in the history of the theory of elasticity, volume 6 of *Studies in the History of Modern Science.* D. Reidel Publishing Co., Dordrecht, 1980.

80. D. Bucur, A. Ferrero, and F. Gazzola. On the first eigenvalue of a fourth order Steklov problem. *Calc. Var. Partial Differential Equations*, 35:103–131, 2009.

81. D. Bucur and N. Varchon. Global minimizing domains for the first eigenvalue of an elliptic operator with non-constant coefficients. *Electron. J. Differential Equations*, 2000 (36):1–10, 2000.

82. L. Caffarelli, B. Gidas, and J. Spruck. Asymptotic symmetry and local behavior of semilinear equations with critical Sobolev growth. *Comm. Pure Appl. Math.*, 42:271–297, 1989.

83. A.P. Calderón and A. Zygmund. On the existence of certain singular integrals. *Acta Math.*, 88:85–139, 1952.

84. A. Capozzi, D. Fortunato, and G. Palmieri. An existence result for nonlinear elliptic problems involving critical Sobolev exponent. *Ann. Inst. H. Poincaré Anal. Non Linéaire*, 2:463–470, 1985.

85. D. Cassani, J.M. do Ó, and N. Ghoussoub. On a fourth order elliptic problem with a singular nonlinearity. *Adv. Nonlinear Stud.*, 9:177–197, 2009.

86. D. Cassani, B. Ruf, and C. Tarsi. Best constants in a borderline case of second order Moser type inequalities. *Anal. Institut H. Poincaré, Anal. Non Linéaire*, 2009.

87. G. Cerami, D. Fortunato, and M. Struwe. Bifurcation and multiplicity results for nonlinear elliptic problems involving critical Sobolev exponents. *Anal. Institut H. Poincaré, Anal. Non Linéaire*, 1:341–350, 1984.

88. G. Cerami, S. Solimini, and M. Struwe. Some existence results for superlinear elliptic boundary value problems involving critical exponents. *J. Funct. Anal.*, 69:289–306, 1986.

89. S.-Y.A. Chang. *Non-linear Elliptic Equations in Conformal Geometry.* Zurich Lectures in Advanced Mathematics. European Mathematical Society (EMS), Zürich, 2004.

90. S.-Y.A. Chang, M.J. Gursky, and P.C. Yang. An equation of Monge-Ampère type in conformal geometry, and four-manifolds of positive Ricci curvature. *Ann. of Math.*, (2) 155:709–787, 2002.

91. S.-Y.A. Chang, M.J. Gursky, and P.C. Yang. A conformally invariant sphere theorem in four dimensions. *Publ. Math. Inst. Hautes Études Sci.*, 98:105–143, 2003.

92. S.-Y.A. Chang and P.C. Yang. Extremal metrics of zeta function determinants on 4-manifolds. *Ann. of Math.*, (2) 142:171–212, 1995.

93. S.-Y.A. Chang and P.C. Yang. On uniqueness of solutions of n-th order differential equations in conformal geometry. *Math. Res. Lett.*, 4:91–102, 1997.

94. W. Chen and C. Li. Classification of solutions of some nonlinear elliptic equations. *Duke Math. J.*, 63:615–622, 1991.

95. W. Chen, C. Li, and B. Ou. Classification of solutions for an integral equation. *Comm. Pure Appl. Math.*, 59:330–343, 2006. (Corrigendum in: *Comm. Pure Appl. Math.*, 59: 1064, 2006.).

96. Y.S. Choi and X. Xu. Nonlinear biharmonic equations with negative exponents. *J. Differential Equations*, 246:216–234, 2009.

97. K. Chung and Z. Zhao. *From Brownian motion to Schrödinger's equation*. Grundlehren der Mathematischen Wissenschaften, Band 312. Springer-Verlag, Berlin etc., 1995.

98. A. Cianchi. Second-order derivatives and rearrangements. *Duke Math. J.*, 105:355–385, 2000.

99. A. Cianchi. Symmetrization and second-order Sobolev inequalities. *Ann. Mat. Pura Appl.*, (4) 183:45–77, 2004.

100. A. Cianchi, L. Esposito, N. Fusco, and C. Trombetti. A quantitative Pólya-Szegö principle. *J. Reine Angew. Math.*, 614:153–189, 2008.

101. P.G. Ciarlet. A justification of the von Kármán equations. *Arch. Rational Mech. Anal.*, 73:349–389, 1980.

102. P.G. Ciarlet. *Mathematical Elasticity. Vol. II.* Theory of plates, volume 27 of *Studies in Mathematics and its Applications*. North-Holland Publishing Co., Amsterdam, 1997.

103. P.G. Ciarlet and P. Destuynder. A justification of the two-dimensional linear plate model. *J. Mécanique*, 18:315–344, 1979.

104. M. Clapp and T. Weth. Multiple solutions for the Brezis-Nirenberg problem. *Adv. Differential Equations*, 10:463–480, 2005.

105. U. Clarenz, U. Diewald, G. Dziuk, M. Rumpf, and R. Rusu. A finite element method for surface restoration with smooth boundary conditions. *Comput. Aided Geom. Design*, 21:427–445, 2004.

106. Ph. Clément, D.G. de Figueiredo, and E. Mitidieri. Positive solutions of semilinear elliptic systems. *Comm. Partial Differential Equations*, 17:923–940, 1992.

107. Ch.V. Coffman. On the structure of solutions to $\Delta^2 u = \lambda u$ which satisfy the clamped plate conditions on a right angle. *SIAM J. Math. Anal.*, 13:746–757, 1982.

108. Ch.V. Coffman and R.J. Duffin. On the structure of biharmonic functions satisfying the clamped plate conditions on a right angle. *Adv. in Appl. Math.*, 1:373–389, 1980.

109. Ch.V. Coffman and R.J. Duffin. On the fundamental eigenfunctions of a clamped punctured disk. *Adv. in Appl. Math.*, 13:142–151, 1992.

110. Ch.V. Coffman, R.J. Duffin, and D.H. Shaffer. The fundamental mode of vibration of a clamped annular plate is not of one sign. In *Constructive approaches to mathematical models (Proc. Conf. in honor of R.J. Duffin, Pittsburgh, Pa., 1978)*, pages 267–277. Academic Press, New York, London, 1979.

111. Ch.V. Coffman and C.L. Grover. Obtuse cones in Hilbert spaces and applications to partial differential equations. *J. Funct. Anal.*, 35:369–396, 1980.

112. R. Courant. *Dirichlet's Principle, Conformal Mapping, and Minimal Surfaces*. Interscience, New York, 1950.

113. C. Cowan, P. Esposito, N. Ghoussoub, and A. Moradifam. The critical dimension for a fourth order elliptic problem with singular nonlinearity. *Arch. Rational Mech. Anal.*, to appear.

114. M. Cranston, E. Fabes, and Z. Zhao. Conditional gauge and potential theory for the Schrödinger operator. *Trans. Amer. Math. Soc*, 307:171–194, 1988.

115. A. Dall'Acqua, K. Deckelnick, and H.-Ch. Grunau. Classical solutions to the Dirichlet problem for Willmore surfaces of revolution. *Adv. Calc. Var.*, 1:379–397, 2008.

116. A. Dall'Acqua, S. Fröhlich, H.-Ch. Grunau, and F. Schieweck. Symmetric Willmore surfaces of revolution satisfying arbitrary Dirichlet boundary data. *Preprint*, 2008.

117. A. Dall'Acqua, Ch. Meister, and G. Sweers. Separating positivity and regularity for fourth order Dirichlet problems in 2d-domains. *Analysis (Munich)*, 25:205–261, 2005.

118. A. Dall'Acqua and G. Sweers. Estimates for Green function and Poisson kernels of higher-order Dirichlet boundary value problems. *J. Differential Equations*, 205:466–487, 2004.

119. A. Dall'Acqua and G. Sweers. On domains for which the clamped plate system is positivity preserving. In *Partial differential equations and inverse problems*, volume 362 of *Contemp. Math.*, pages 133–144. Amer. Math. Soc., Providence, 2004.

120. A. Dall'Acqua and G. Sweers. The clamped-plate equation for the limaçon. *Ann. Mat. Pura Appl.*, (4) 184:361–374, 2005.

121. R. Dalmasso. Problème de Dirichlet homogène pour une équation biharmonique semi-linéaire dans une boule. *Bull. Sc. Math. 2ᵉ Série*, 114:123–137, 1990.

122. R. Dalmasso. Un problème de symétrie pour une équation biharmonique. *Ann. Fac. Sci. Toulouse*, (5) 11:45–53, 1990.

123. R. Dalmasso. Positive entire solutions of superlinear biharmonic equations. *Funkcial. Ekvac.*, 34:403–422, 1991.

124. R. Dalmasso. Symmetry properties in higher order semilinear elliptic equations. *Nonlinear Anal., Theory Methods Appl.*, 24:1–7, 1995.

125. R. Dalmasso. Uniqueness theorems for some fourth-order elliptic equations. *Proc. Amer. Math. Soc.*, 123:1177–1183, 1995.

126. R. Dalmasso. A priori estimates for some semilinear elliptic equations of order $2m$. *Nonlinear Anal., Theory Methods Appl.*, 29:1433–1452, 1997.

127. R. Dalmasso. Existence and uniqueness results for polyharmonic equations. *Nonlinear Anal., Ser. A*, 36:131–137, 1999.

128. E.N. Dancer. A note on an equation with critical exponent. *Bull. London Math. Soc.*, 20:600–602, 1988.

129. E.B. Davies. L^p spectral theory of higher-order elliptic differential operators. *Bull. London Math. Soc.*, 29:513–546, 1997.

130. E.B. Davies and A.M. Hinz. Explicit constants for Rellich inequalities in $L_p(\Omega)$. *Math. Z.*, 227:511–523, 1998.

131. J. Dávila, L. Dupaigne, I. Guerra, and M. Montenegro. Stable solutions for the bilaplacian with exponential nonlinearity. *SIAM J. Math. Anal.*, 39:565–592, 2007.

132. J. Dávila, I. Flores, and I. Guerra. Multiplicity of solutions for a fourth order equation with power-type nonlinearity. *Math. Ann.*, to appear.

133. J. Dávila, I. Flores, and I. Guerra. Multiplicity of solutions for a fourth order problem with exponential nonlinearity. *J. Differential Equations*, 247:3136–3162, 2009.

134. C. Davini. Γ-convergence of external approximations in boundary value problems involving the bi-Laplacian. *Proceedings of the 9th International Congress on Computational and Applied Mathematics (Leuven, 2000), J. Comput. Appl. Math.*, 140:185–208, 2002.

135. C. Davini. Gaussian curvature and Babuška's paradox in the theory of plates. In *Rational Continua, Classical and New*, pages 67–87. Springer Italia, Milan, 2003.

136. B. de Pagter. Irreducible compact operators. *Math. Z.*, 192:149–153, 1986.

137. K. Deckelnick and H.-Ch. Grunau. Boundary value problems for the one-dimensional Willmore equation. *Calc. Var. Partial Differential Equations*, 30:293–314, 2007.

138. K. Deckelnick and H.-Ch. Grunau. A Navier boundary value problem for Willmore surfaces of revolution. *Analysis (Munich)*, 29:229–258, 2009.

139. K. Deckelnick and H.-Ch. Grunau. Stability and symmetry in the Navier problem for the one-dimensional Willmore equation. *SIAM J. Math. Anal.*, 40:2055 – 2076, 2009.

140. K. Deimling. *Nonlinear Functional Analysis*. Springer-Verlag, Berlin etc., 1985.

141. Ph. Destuynder and M. Salaun. *Mathematical Analysis of Thin Plate Models*, volume 24 of *Mathématiques & Applications (Berlin)*. Springer-Verlag, Berlin, 1996.

142. J.I. Diaz, M. Lazzo, and P.G. Schmidt. Asymptotic behavior of large radial solutions of a polyharmonic equation with superlinear growth. In preparation.

143. J.I. Diaz, M. Lazzo, and P.G. Schmidt. Large radial solutions of a polyharmonic equation with superlinear growth. *Electron. J. Differential Equations,* Proceedings of the 2006 International Conference in honor of J. Fleckinger, Conference 16:103–128, 2007.

144. Z. Djadli, E. Hebey, and M. Ledoux. Paneitz-type operators and applications. *Duke Math. J.*, 104:129–169, 2000.

145. Z. Djadli and A. Malchiodi. Existence of conformal metrics with constant Q-curvature. *Ann. Math.*, (2) 168:813–858, 2008.

146. Z. Djadli, A. Malchiodi, and M. Ould Ahmedou. Prescribing a fourth order conformal invariant on the standard sphere. I. A perturbation result. *Commun. Contemp. Math.*, 4:375–408, 2002.

147. Z. Djadli, A. Malchiodi, and M. Ould Ahmedou. Prescribing a fourth order conformal invariant on the standard sphere. II. Blow up analysis and applications. *Ann. Sc. Norm. Super. Pisa Cl. Sci.*, (5) 1:387–434, 2002.

148. A. Douglis and L. Nirenberg. Interior estimates for elliptic systems of partial differential equations. *Comm. Pure Appl. Math.*, 8:503–538, 1955.

149. O. Druet, E. Hebey, and F. Robert. *Blow-up Theory for Elliptic PDEs in Riemannian Geometry*, volume 45 of *Mathematical Notes*. Princeton University Press, Princeton, 2004.

150. R.J. Duffin. On a question of Hadamard concerning super-biharmonic functions. *J. Math. Phys.*, 27:253–258, 1949.

151. R.J. Duffin. Continuation of biharmonic functions by reflection. *Duke Math. J.*, 22:313–324, 1955.

152. R.J. Duffin. Some problems of mathematics and science. *Bull. Amer. Math. Soc.*, 80:1053–1070, 1974.

153. R.J. Duffin and A. Schild. The effect of small constraints on natural vibrations. In *Proceedings of Symposia in Applied Mathematics, Vol. V, Wave motion and vibration theory*, pages 155–163. McGraw-Hill, New York-Toronto-London, 1954.

154. R.J. Duffin and A. Schild. On the change of natural frequencies induced by small constraints. *J. Math. Mech.*, 6:731–758, 1957.

155. R.J. Duffin and D.H. Shaffer. On the modes of vibration of a ring-shaped plate. *Bull. Amer. Math. Soc.*, 58:652, 1952.

156. G. Dziuk, E. Kuwert, and R. Schätzle. Evolution of elastic curves in $\mathbb{R}^n$: Existence and computation. *SIAM J. Math. Anal.*, 33:1228–1245, 2002.

157. F. Ebobisse and M. Ould Ahmedou. On a nonlinear fourth-order elliptic equation involving the critical Sobolev exponent. *Nonlinear Anal., Theory Methods Appl.*, 52:1535–1552, 2003.

158. J. Edenhofer. Eine Integraldarstellung der Lösung der Dirichletschen Aufgabe bei der Polypotentialgleichung im Falle einer Hyperkugel. *Math. Nachr.*, 69:149–162, 1975.

159. J. Edenhofer. Integraldarstellung einer m-polyharmonischen Funktion, deren Funktionswerte und erste $m - 1$ Normalableitungen auf einer Hypersphäre gegeben sind. *Math. Nachr.*, 68:105–113, 1975.

160. D.E. Edmunds, D. Fortunato, and E. Jannelli. Critical exponents, critical dimensions and the biharmonic operator. *Arch. Rational Mech. Anal.*, 112:269–289, 1990.

161. P. Esposito and F. Robert. Mountain pass critical points for Paneitz-Branson operators. *Calc. Var. Partial Differential Equations*, 15:493–517, 2002.

162. G. Faber. Beweis, dass unter allen homogenen Membranen von gleicher Fläche und gleicher Spannung die kreisförmige den tiefsten Grundton gibt. *Sitz. Ber. Bayer. Akad. Wiss.*, 1923:169–172, 1923.

163. V. Felli. Existence of conformal metrics on S^n with prescribed fourth-order invariant. *Adv. Differential Equations*, 7:47–76, 2002.

164. A. Ferrero, F. Gazzola, and H.-Ch. Grunau. Decay and eventual local positivity for biharmonic parabolic equations. *Discrete Cont. Dynam. Systems*, 21:1129–1157, 2008.

165. A. Ferrero, F. Gazzola, and T. Weth. On a fourth order Steklov eigenvalue problem. *Analysis (Munich)*, 25:315–332, 2005.

166. A. Ferrero, F. Gazzola, and T. Weth. Positivity, symmetry and uniqueness for minimizers of second-order Sobolev inequalities. *Ann. Mat. Pura Appl.*, (4) 186:565–578, 2007.

167. A. Ferrero and H.-Ch. Grunau. The Dirichlet problem for supercritical biharmonic equations with power-type nonlinearity. *J. Differential Equations*, 234:582–606, 2007.

168. A. Ferrero, H.-Ch. Grunau, and P. Karageorgis. Supercritical biharmonic equations with power-type nonlinearity. *Ann. Mat. Pura Appl.*, (4) 188:171–185, 2009.

169. A. Ferrero and G. Warnault. On solutions of second and fourth order elliptic equations with power-type nonlinearities. *Nonlinear Anal., Theory Methods Appl.*, 70:2889–2902, 2009.

170. G. Fichera. Su un principio di dualità per talune formole di maggiorazione relative alle equazioni differenziali. *Atti Accad. Naz. Lincei*, (8) 19:411–418, 1955.

171. D. Fortunato and E. Jannelli. Infinitely many solutions for some nonlinear elliptic problems in symmetrical domains. *Proc. Royal. Soc. Edinburgh*, A 105:205–213, 1987.

172. L. Friedlander. Remarks on the membrane and buckling eigenvalues for planar domains. *Mosc. Mat. J.*, 4:369–375, 2004.

173. K. Friedrichs. Die Randwert- und Eigenwertprobleme aus der Theorie der elastischen Platten (Anwendung der direkten Methoden der Variationsrechnung). *Math. Ann.*, 98:205–247, 1927.

174. G. Friesecke, R.D. James, and S. Müller. A hierarchy of plate models derived from nonlinear elasticity by gamma-convergence. *Arch. Ration. Mech. Anal.*, 180, 2006.

175. S. Fröhlich. Katenoidähnliche Lösungen geometrischer Variationsprobleme. Preprint 2322, FB Mathematik, TU Darmstadt (2004), 2004.

176. P.R. Garabedian. A partial differential equation arising in conformal mapping. *Pacific J. Math.*, 1:485–524, 1951.

177. P.R. Garabedian. *Partial Differential Equations*. Chelsea Publishing Co., New York, second edition, 1986.

178. J. García Azorero and I. Peral Alonso. Multiplicity of solutions for elliptic problems with critical exponent or with a nonsymmetric term. *Trans. Amer. Math. Soc.*, 323:877–895, 1991.

179. F. Gazzola. Critical growth problems for polyharmonic operators. *Proc. Roy. Soc. Edinburgh*, A 128:251–263, 1998.

180. F. Gazzola and H.-Ch. Grunau. On the role of space dimension $n = 2 + 2\sqrt{2}$ in the semilinear Brezis-Nirenberg eigenvalue problem. *Analysis (Munich)*, 20:395–399, 2000.

181. F. Gazzola and H.-Ch. Grunau. Critical dimensions and higher order Sobolev inequalities with remainder terms. *NoDEA Nonlinear Differential Equations Appl.*, 8:35–44, 2001.

182. F. Gazzola and H.-Ch. Grunau. Radial entire solutions for supercritical biharmonic equations. *Math. Ann.*, 334:905–936, 2006.

183. F. Gazzola and H.-Ch. Grunau. Global solutions for superlinear parabolic equations involving the biharmonic operator for initial data with optimal slow decay. *Calc. Var. Partial Differential Equations*, 30:389–415, 2007.

184. F. Gazzola and H.-Ch. Grunau. Eventual local positivity for a biharmonic heat equation in $\mathbb{R}^n$. *Discrete Cont. Dyn. Syst., Ser. S*, 1:83–87, 2008.

185. F. Gazzola, H.-Ch. Grunau, and E. Mitidieri. Hardy inequalities with optimal constants and remainder terms. *Trans. Amer. Math. Soc.*, 356:2149–2168, 2004.

186. F. Gazzola, H.-Ch. Grunau, and M. Squassina. Existence and nonexistence results for critical growth biharmonic elliptic equations. *Calc. Var. Partial Differential Equations*, 18:117–143, 2003.

187. F. Gazzola, H.-Ch. Grunau, and G. Sweers. Optimal Sobolev and Hardy-Rellich constants under Navier boundary conditions. *Ann. Mat. Pura Appl.*, to appear.

188. F. Gazzola and A. Malchiodi. Some remarks on the equation $-\Delta u = \lambda (1 + u)^p$ for varying λ, p and varying domains. *Comm. Partial Differential Equations*, 27:809–845, 2002.

189. F. Gazzola and D. Pierotti. Positive solutions to critical growth biharmonic elliptic problems under Steklov boundary conditions. *Nonlinear Anal., Theory Methods Appl.*, 71:232–238, 2009.

190. F. Gazzola and B. Ruf. Lower-order perturbations of critical growth nonlinearities in semilinear elliptic equations. *Adv. Differential Equations*, 2:555–572, 1997.

191. F. Gazzola and G. Sweers. On positivity for the biharmonic operator under Steklov boundary conditions. *Arch. Rational Mech. Anal*, 188:399–427, 2008.

192. Y. Ge. Sharp Sobolev inequalities in critical dimensions. *Michigan Math. J.*, 51:27–45, 2003.

193. Y. Ge. Positive solutions in semilinear critical problems for polyharmonic operators. *J. Math. Pures Appl.*, (9) 84:199–245, 2005.

194. I.M. Gel'fand. Some problems in the theory of quasilinear equations. *Amer. Math. Soc. Transl.*, (2) 29:295–381, 1963. Due to G.I. Barenblatt. Russian original: *Uspekhi Mat. Nauk* (86) 14:87–158, 1959.

195. B. Gidas, W.M. Ni, and L. Nirenberg. Symmetry and related properties via the maximum principle. *Comm. Math. Phys.*, 68:209–243, 1979.

196. B. Gidas and J. Spruck. Global and local behavior of positive solutions of nonlinear elliptic equations. *Comm. Pure Appl. Math.*, 34:525–598, 1981.

197. B. Gidas and J. Spruck. A priori bounds for positive solutions of nonlinear elliptic equations. *Comm. Partial Differential Equations*, 6:883–901, 1981.

198. D. Gilbarg and N.S. Trudinger. *Elliptic Partial Differential Equations of Second Order.* Grundlehren der Mathematischen Wissenschaften, Band 224. Springer-Verlag, Berlin etc., second edition, 1983.

199. S. Giuffrè. Strong solvability of boundary value contact problems. *Appl. Math. Optim.*, 51:361–372, 2005.

200. P. Grisvard. *Singularities in Boundary Value Problems*, volume 22 of *Recherches en Mathématiques Appliquées*. Masson and Springer-Verlag, Paris and Berlin etc., 1992.

201. H.-Ch. Grunau. The Dirichlet problem for some semilinear elliptic differential equations of arbitrary order. *Analysis (Munich)*, 11:83–90, 1991.

202. H.-Ch. Grunau. Critical exponents and multiple critical dimensions for polyharmonic operators. II. *Boll. Un. Mat. Ital*, B (7) 9:815–847, 1995.

203. H.-Ch. Grunau. Positive solutions to semilinear polyharmonic Dirichlet problems involving critical Sobolev exponents. *Calc. Var. Partial Differential Equations*, 3:243–252, 1995.

204. H.-Ch. Grunau. On a conjecture of P. Pucci and J. Serrin. *Analysis (Munich)*, 16:399–403, 1996.

205. H.-Ch. Grunau. Polyharmonische Dirichletprobleme: Positivität, kritische Exponenten und kritische Dimensionen, Habilitationsschrift, Universität Bayreuth, 1996.

206. H.-Ch. Grunau, M. Ould Ahmedou, and W. Reichel. The Paneitz equation in hyperbolic space. *Anal. Institut H. Poincaré, Anal. Non Linéaire*, 25:847–864, 2008.

207. H.-Ch. Grunau and F. Robert. Boundedness of the negative part of biharmonic Green's functions under Dirichlet boundary conditions in general domains. *C. R. Math. Acad. Sci. Paris*, Ser. I 347:163–166, 2009.

208. H.-Ch. Grunau and F. Robert. Positivity and almost positivity of biharmonic Green's functions under Dirichlet boundary conditions. *Arch. Rational Mech. Anal.*, 195:865–898, 2010.

209. H.-Ch. Grunau and G. Sweers. Positivity for perturbations of polyharmonic operators with Dirichlet boundary conditions in two dimensions. *Math. Nachr.*, 179:89–102, 1996.

210. H.-Ch. Grunau and G. Sweers. Classical solutions for some higher order semilinear elliptic equations under weak growth conditions. *Nonlinear Anal., Theory Methods Appl.*, 28:799–807, 1997.

211. H.-Ch. Grunau and G. Sweers. Positivity for equations involving polyharmonic operators with Dirichlet boundary conditions. *Math. Ann.*, 307:589–626, 1997.

212. H.-Ch. Grunau and G. Sweers. The role of positive boundary data in generalized clamped plate equations. *Z. Angew. Math. Phys.*, 49:420–435, 1998.

213. H.-Ch. Grunau and G. Sweers. Sign change for the Green function and for the first eigenfunction of equations of clamped-plate type. *Arch. Rational Mech. Anal.*, 150:179–190, 1999.

214. H.-Ch. Grunau and G. Sweers. Sharp estimates for iterated Green functions. *Proc. Roy. Soc. Edinburgh*, A 132:91–120, 2002.

215. H.-Ch. Grunau and G. Sweers. Regions of positivity for polyharmonic Green functions in arbitrary domains. *Proc. Amer. Math. Soc.*, 135:3537–3546, 2007.

216. M. Grüter and K.-O. Widman. The Green function for uniformly elliptic equations. *Manuscr. Math.*, 37:303–342, 1982.

217. Z. Guo and J. Wei. Entire solutions and global bifurcations for a biharmonic equation with singular non-linearity in $\mathbb{R}^3$. *Adv. Differential Equations*, 13:753–780, 2008.

218. Z. Guo and J. Wei. On a fourth order nonlinear elliptic equation with negative exponent. *SIAM J. Math. Anal.*, 40:2034–2054, 2008/09.

219. M.J. Gursky. The principal eigenvalue of a conformally invariant differential operator, with an application to semilinear elliptic PDE. *Comm. Math. Phys.*, 207:131–143, 1999.

220. M. Guysinsky, B. Hasselblatt, and V. Rayskin. Differentiability of the Hartman-Grobman linearization. *Discrete Contin. Dyn. Syst. A*, 9:979–984, 2003.

221. W. Hackbusch and G. Hofmann. Results of the eigenvalue problem for the plate equation. *Z. Angew. Math. Phys.*, 31:730–739, 1980.

222. J. Hadamard. *Mémoire sur le problème d'analyse relatif à l'équilibre des plaques élastiques encastrées.* In: Œuvres de Jacques Hadamard, Tome II, pages 515–641. CNRS Paris, 1968. Reprint of: *Mémoires présentés par divers savants a l'Académie des Sciences* (2) **33**:1–128, 1908.

223. J. Hadamard. *Sur certains cas intéressants du problème biharmonique.* In: Œuvres de Jacques Hadamard, Tome III, pages 1297–1299. CNRS Paris, 1968. Reprint of: *Atti IV Congr. Intern. Mat. Rome* 12-14, 1908.

224. P. Hartman. *Ordinary Differential Equations.* John Wiley and Sons, New York etc., 1964.

225. E. Hebey and F. Robert. Coercivity and Struwe's compactness for Paneitz type operators with constant coefficients. *Calc. Var. Partial Differential Equations*, 13:491–517, 2001.

226. P.J.H. Hedenmalm. A computation of Green functions for the weighted biharmonic operators $\Delta|z|^{-2\alpha}\Delta$, with $\alpha > -1$. *Duke Math. J.*, 75:51–78, 1994.

227. P.J.H. Hedenmalm, S. Jakobsson, and S. Shimorin. A biharmonic maximum principle for hyperbolic surfaces. *J. Reine Angew. Math.*, 550:25–75, 2002.

228. W. Helfrich. Elastic properties of lipid bilayers: Theory and possible experiments. *Z. Naturforsch. C*, 28:693–703, 1973.

229. A. Henrot. *Extremum Problems for Eigenvalues of Elliptic Operators.* Frontiers in Mathematics. Birkhäuser-Verlag, Basel, 2006.

230. U. Hertrich-Jeromin and U. Pinkall. Ein Beweis der Willmoreschen Vermutung für Kanaltori. *J. Reine Angew. Math.*, 430:21–34, 1992.

231. L. Hörmander. On the regularity of the solutions of boundary problems. *Acta Math.*, 99:225–264, 1958.

232. L. Hörmander. *The Analysis of Linear Partial Differential Operators. I.* Distribution theory and Fourier analysis. Classics in Mathematics, Reprint of the second (1990) edition. Springer-Verlag, Berlin etc., 2003.

233. A. Huber. On the reflection principle for polyharmonic functions. *Comm. Pure Appl. Math.*, 9:471–478, 1956.

234. H. Hueber and M. Sieveking. Uniform bounds for quotients of Green functions on $C^{1,1}$-domains. *Ann. Inst. Fourier (Grenoble)*, 32:105–117, 1982.

235. R.A. Hunt. On $L(p,q)$ spaces. *Enseignement Math.*, (2) 12:249–276, 1966.

236. E. Jannelli. The role played by space dimension in elliptic critical problems. *J. Differential Equations*, 156:407–426, 1999.

237. R. Jentzsch. Über Integralgleichungen mit positivem Kern. *J. Reine Angew. Math.*, 141:235–244, 1912.

238. D.S. Jerison and C.E. Kenig. The Neumann problem on Lipschitz domains. *Bull. Amer. Math. Soc.*, 4:203–207, 1981.

239. D.S. Jerison and C.E. Kenig. Boundary value problems on Lipschitz domains. In *Studies in partial differential equations*, volume 23 of *MAA Stud. Math.*, pages 1–68. Math. Assoc. America, Washington, DC, 1982.

240. D.D. Joseph and T.S. Lundgren. Quasilinear Dirichlet problems driven by positive sources. *Arch. Rational Mech. Anal.*, 49:241–269, 1972/73.

241. J. Kadlec. The regularity of the solution of the Poisson problem in a domain whose boundary is similar to that of a convex domain (Russian). *Czechoslovak Math. J.*, 14 (89):386–393, 1964.

242. A. Kameswara Rao and K. Rajaiah. Polygon-circle paradox of simply supported thin plates under uniform pressure. *AIAA Journal*, 6:155–156, 1968.

243. P. Karageorgis. Stability and intersection properties of solutions to the nonlinear biharmonic equation. *Nonlinearity*, 22:1653–1661, 2009.

244. B. Kawohl. On the isoperimetric nature of a rearrangement inequality and its consequences for some variational problems. *Arch. Rational Mech. Anal.*, 94:227–243, 1986.

245. B. Kawohl. Some nonconvex shape optimization problems. In *Optimal shape design (Tróia, 1998)*, volume 1740 of *Lecture Notes in Math.*, pages 7–46. Springer-Verlag, Berlin etc., 2000.

246. B. Kawohl, H.A. Levine, and W. Velte. Buckling eigenvalues for a clamped plate embedded in an elastic medium and related questions. *SIAM J. Math. Anal.*, 24:327–340, 1993.

247. B. Kawohl and G. Sweers. On 'anti'-eigenvalues for elliptic systems and a question of McKenna and Walter. *Indiana Univ. Math. J.*, 51:1023–1040, 2002.

248. C.E. Kenig. Restriction theorems, Carleman estimates, uniform Sobolev inequalities and unique continuation. In *Harmonic analysis and partial differential equations (El Escorial, 1987)*, volume 1384 of *Lecture Notes in Math.*, pages 69–90. Springer-Verlag, Berlin etc., 1989.

249. D. Kinderlehrer and G. Stampacchia. *An Introduction to Variational Inequalities and Their Applications*. Academic Press, New York, London, 1980.

250. G.R. Kirchhoff. Über das Gleichgewicht und die Bewegung einer elastischen Scheibe. *J. Reine Angew. Math.*, 40:51–88, 1850.

251. Ch. Köckritz. Lokale Positivität der polyharmonischen Greenschen Funktion. Studienarbeit, Universität Magdeburg, 2007, downloadable at http://www-ian.math.uni-magdeburg.de/home/grunau/papers/StudienarbeitKoeckritz.pdf.

252. V.A. Kondrat'ev. Boundary value problems for elliptic equations in domains with conical or angular points (Russian). *Trudy Moskov. Mat. Obšč.*, 16:209–292, 1967.

253. V.A. Kozlov, V.A. Kondratiev, and V.G. Maz'ya. On sign variation and the absence of strong zeros of solutions of elliptic equations. *Math. USSR Izvestiya*, 34:337–353, 1990. Russian original: *Izv. Akad. Nauk SSSR Ser. Mat.* 53: 328–344, 1989.

254. E. Krahn. Über eine von Rayleigh formulierte Minimaleigenschaft des Kreises. *Math. Ann.*, 94:97–100, 1925.

255. E. Krahn. Über Minimaleigenschaften der Kugel in drei und mehr Dimensionen. *Acta Comm. Univ. Dorpat.*, A 9:1–44, 1926.

256. Ju.P. Krasovskiĭ. Investigation of potentials associated with boundary problems for elliptic equations. *Math. USSR Izv.*, 1:569–622, 1967. Russian original: *Izv. Akad. Nauk SSSR Ser. Mat.* 31: 587–640, 1967.

257. Ju.P. Krasovskiĭ. Isolation of singularities of the Green's function. *Math. USSR Izv.*, 1:935–966, 1967. Russian original: *Izv. Akad. Nauk SSSR Ser. Mat.* 31: 977–1010, 1967.

258. M.G. Kreĭn and M.A. Rutman. Linear operators leaving invariant a cone in a Banach space. *Amer. Math. Soc. Translation*, 1950(26), 1950. Russian original: *Uspehi Matem. Nauk (N.S.)* 23: 3–95, 1948.

259. J.R. Kuttler. Remarks on a Stekloff eigenvalue problem. *SIAM J. Numer. Anal.*, 9:1–5, 1972.

260. J.R. Kuttler. Dirichlet eigenvalues. *SIAM J. Numer. Anal.*, 16:332–338, 1979.

261. J.R. Kuttler and V.G. Sigillito. Inequalities for membrane and Stekloff eigenvalues. *J. Math. Anal. Appl.*, 23:148–160, 1968.

262. J.R. Kuttler and V.G. Sigillito. *Estimating Eigenvalues with A Posteriori/A Priori Inequalities*, volume 135 of *Research Notes in Mathematics*. Pitman (Advanced Publishing Program), Boston, 1985.

263. E. Kuwert and R. Schätzle. The Willmore flow with small initial energy. *J. Differential Geom.*, 57:409–441, 2001.

264. E. Kuwert and R. Schätzle. Gradient flow for the Willmore functional. *Comm. Anal. Geom.*, 10:307–339, 2002.

265. E. Kuwert and R. Schätzle. Removability of point singularities of Willmore surfaces. *Annals of Math.*, (2) 160:315–357, 2004.

266. R.S. Lakes. Foam structures with a negative Poisson's ratio. *Science*, 235:1038–1040, 1987.

267. J.C. Langer and D.A. Singer. Curves in the hyperbolic plane and mean curvature of tori in 3-space. *Bull. London Math. Soc.*, 16:531–534, 1984.

268. J.C. Langer and D.A. Singer. The total squared curvature of closed curves. *J. Differential Geom.*, 20:1–22, 1984.

269. G. Lauricella. Integrazione dell'equazione $\Delta^2(\Delta^2 u) = 0$ in un campo di forma circolare. *Torino Atti*, 31:1010–1018, 1896.

270. A.C. Lazer and P.J. McKenna. A symmetry theorem and applications to nonlinear partial differential equations. *J. Differential Equations*, 72:95–106, 1988.

271. A.C. Lazer and P.J. McKenna. Large-amplitude periodic oscillations in suspension bridges: some new connections with nonlinear analysis. *SIAM Rev.*, 32:537–578, 1990.

272. A.C. Lazer and P.J. McKenna. Nonlinear periodic flexing in a floating beam. *J. Comput. Appl. Math., Special issue: Oscillations in nonlinear systems: applications and numerical aspects*, 52:287–303, 1994.

273. M. Lazzo and P.G. Schmidt. Nonexistence criteria for polyharmonic boundary-value problems. *Analysis (Munich)*, 28:449–460, 2008.

274. Y.Y. Li. Remark on some conformally invariant integral equations: the method of moving spheres. *J. Eur. Math. Soc.*, 6:153–180, 2004.

275. C.S. Lin. A classification of solutions of a conformally invariant fourth order equation in $\mathbb{R}^n$. *Comment. Math. Helv.*, 73:206–231, 1998.

276. J.-L. Lions and E. Magenes. *Problèmes aux limites non homogènes et applications. Vol. 1.* Travaux et Recherches Mathématiques, No. 17. Dunod, Paris, 1968.

277. P.-L. Lions. The concentration-compactness principle in the calculus of variations. The locally compact case. I. *Ann. Inst. H. Poincaré Anal. Non Linéaire*, 1:109–145, 1984.

278. P.-L. Lions. The concentration-compactness principle in the calculus of variations. The limit case. I. *Rev. Mat. Iberoamericana*, 1:145–201, 1985.

279. Ch. Loewner. On generation of solutions of the biharmonic equation in the plane by conformal mappings. *Pacific J. Math.*, 3:417–436, 1953.

280. Ch. Loewner and L. Nirenberg. Partial differential equations invariant under conformal or projective transformations. In *L. Ahlfors (ed.), Contributions to analysis*, pages 245–272. Academic Press, New York, London, 1974.

281. A.E.H. Love. *A Treatise on the Mathematical Theory of Elasticity.* Fourth edition. Cambridge Univ. Press, Cambridge, 1927.

282. S. Luckhaus. Existence and regularity of weak solutions to the Dirichlet problem for semilinear elliptic systems of higher order. *J. Reine Angew. Math.*, 306:192–207, 1979.

283. J. Mallet-Paret and H.L. Smith. The Poincaré-Bendixson theorem for monotone cyclic feedback systems. *J. Dynam. Differential Equations*, 2:367–421, 1990.

284. V.A. Malyshev. Hadamard's conjecture and estimates of the Green function. *St. Petersburg Math. J.*, 4:633–666, 1993.

285. S. Mayboroda and V. Maz'ya. Boundedness of the Hessian of a biharmonic function in a convex domain. *Comm. Partial Differential Equations*, 33:1439–1454, 2008.

286. S. Mayboroda and V. Maz'ya. Boundedness of the gradient of a solution and Wiener test of order one for the biharmonic equation. *Invent. Math.*, 175:287–334, 2009.

287. S. Mayboroda and V. Maz'ya. Pointwise estimates for the polyharmonic Green function in general domains. In V. Adamyan, Y.M. Berezansky, I. Gohberg, M.L. Gorbachuk, V. Gorbachuk, A.N. Kochubei, H. Langer, and G. Popov, editors, *Analysis, Partial Differential Equations and Applications, The Vladimir Maz'ya Anniversary Volume*, volume 193 of *Operator Theory: Advances and Applications*, pages 143–158. Birkhäuser-Verlag, Basel etc., 2009.

288. U.F. Mayer and G. Simonett. A numerical scheme for axisymmetric solutions of curvature-driven free boundary problems, with applications to the Willmore flow. *Interfaces Free Bound.*, 4:89–109, 2002.

289. W.G. Mazja, S.A. Nasarow, and B.A. Plamenewski. *Asymptotische Theorie elliptischer Randwertaufgaben in singulär gestörten Gebieten. I.* Störungen isolierter Randsingularitäten, volume 82 of *Mathematische Lehrbücher und Monographien, II. Abteilung: Mathematische Monographien.* Akademie-Verlag, Berlin, 1991.

290. W.G. Mazja, S.A. Nasarow, and B.A. Plamenewski. *Asymptotische Theorie elliptischer Randwertaufgaben in singulär gestörten Gebieten. II.* Nichtlokale Störungen, volume 83 of *Mathematische Lehrbücher und Monographien, II. Abteilung: Mathematische Monographien.* Akademie-Verlag, Berlin, 1991.

291. V. Maz'ya and J. Rossmann. On the Agmon-Miranda maximum principle for solutions of strongly elliptic equations in domains of $\mathbb{R}^n$ with conical points. *Ann. Global Anal. Geom.*, 10:125–150, 1992.

292. V.G. Maz'ya. *Sobolev Spaces,* Translated from the Russian by T.O. Shaposhnikova. Springer Series in Soviet Mathematics. Springer-Verlag, Berlin etc., 1985.

293. V.G. Maz'ya and S.A. Nazarov. On the Sapondzhyan-Babushka paradox in problems of the theory of thin plates (Russian). *Dokl. Akad. Nauk Arm. SSR*, 78:127–130, 1984.

294. V.G. Maz'ya and S.A. Nazarov. Paradoxes of the passage to the limit in solutions of boundary value problems for the approximation of smooth domains by polygons (Russian). *Izv. Akad. Nauk SSSR Ser. Mat.*, 50:1156–1177 and 1343, 1986.

295. V.G. Maz'ya and T.O. Shaposhnikova. *Jaques Hadamard, A Universal Mathematician*. History of Mathematics Vol. 14. American Mathematical Society, London Mathematical Society, Rhode Island and London, 1998.

296. P.J. McKenna. Large-amplitude periodic oscillations in simple and complex mechanical systems: outgrowths from nonlinear analysis. *Milan J. Math.*, 74:79–115, 2006.

297. P.J. McKenna and W. Reichel. Radial solutions of singular nonlinear biharmonic equations and applications to conformal geometry. *Electronic J. Differential Equations*, 2003 (37):1–13, 2003.

298. P.J. McKenna and W. Walter. Nonlinear oscillations in a suspension bridge. *Arch. Rational Mech. Anal.*, 98:167–177, 1987.

299. P.J. McKenna and W. Walter. Travelling waves in a suspension bridge. *SIAM J. Appl. Math.*, 50:703–715, 1990.

300. V.V. Meleshko. Selected topics in the history of the two-dimensional biharmonic problem. *Applied Mechanics Reviews*, 56:33–85, 2003.

301. N.G. Meyers and J. Serrin. $H = W$. *Proc. Nat. Acad. Sci. USA*, 51:1055–1056, 1964.

302. E. Miersemann. Über positive Lösungen von Eigenwertgleichungen mit Anwendungen auf elliptische Gleichungen zweiter Ordnung und auf ein Beulproblem für die Platte. *Z. Angew. Math. Mech.*, 59:189–194, 1979.

303. F. Mignot and J.-P. Puel. Sur une classe de problèmes non linéaires avec non linéarité positive, croissante, convexe. *Comm. Partial Diff. Equations*, 5:791–836, 1980.

304. S.G. Mikhlin. *Variational Methods in Mathematical Physics*. Translated by T. Boddington; A Pergamon Press Book. The Macmillan Co., New York, 1964.

305. C. Miranda. Formule di maggiorazione e teorema di esistenza per le funzioni biarmoniche di due variabili. *Giorn. Mat. Battaglini*, (4) 2 (78):97–118, 1948.

306. C. Miranda. Teorema del massimo modulo e teorema di esistenza e di unicità per il problema di Dirichlet relativo alle equazioni ellittiche in due variabili. *Ann. Mat. Pura Appl.*, (4) 46:265–311, 1958.

307. E. Mitidieri. On the definition of critical dimension. unpublished manuscript, 1993.

308. E. Mitidieri. A Rellich type identity and applications. *Comm. Partial Diff. Equations*, 18:125–151, 1993.

309. E. Mitidieri and S.I. Pohožaev. Apriori estimates and blow-up of solutions to nonlinear partial differential equations and inequalities. *Proc. Steklov Inst. Math.*, 234:1–362, 2001. Russian original: *Tr. Mat. Inst. Steklova* 234: 1–384, 2001.

310. E. Mitidieri and G. Sweers. Weakly coupled elliptic systems and positivity. *Math. Nachr.*, 173:259–286, 1995.

311. E. Mohr. Über die Rayleighsche Vermutung: Unter allen Platten von gegebener Fläche und konstanter Dichte und Elastizität hat die kreisförmige den tiefsten Grundton. *Ann. Mat. Pura Appl.*, (4) 104:85–122, 1975.

312. J.-J. Moreau. Décomposition orthogonale d'un espace hilbertien selon deux cônes mutuellement polaires. *C. R. Acad. Sci. Paris*, 255:238–240, 1962.

313. Ch.B. Morrey. On the analyticity of the solutions of analytic non-linear elliptic systems of partial differential equations. I. Analyticity in the interior. *Amer. J. Math.*, 80:198–218, 1958.

314. Ch.B. Morrey. On the analyticity of the solutions of analytic non-linear elliptic systems of partial differential equations. II. Analyticity at the boundary. *Amer. J. Math.*, 80:219–237, 1958.

315. N.W. Murray. The polygon-circle paradox and convergence in thin plate theory. *Proc. Camb. Philos. Soc.*, 73:279–282, 1973.

316. N.S. Nadirashvili. Rayleigh's conjecture on the principal frequency of the clamped plate. *Arch. Rational Mech. Anal.*, 129:1–10, 1995.

317. M. Nakai and L. Sario. Green's function of the clamped punctured disk. *J. Austral. Math. Soc.*, B 20:175–181, 1977.

318. M. Nakai and L. Sario. On Hadamard's problem for higher dimensions. *J. Reine Angew. Math.*, 291:145–148, 1977.

319. S.A. Nazarov and B.A. Plamenevsky. *Elliptic Problems in Domains with Piecewise Smooth Boundaries*, volume 13 of *de Gruyter Expositions in Mathematics*. Walter de Gruyter & Co., Berlin, 1994.

320. S.A. Nazarov and G.Kh. Svirs. Boundary value problems for the biharmonic equation and the iterated Laplacian in a three-dimensional domain with an edge. *Journal of Mathematical Sciences*, 143:2936–2960, 2007. Russian original: *Zap. Nauchn. Sem. S.-Peterburg. Otdel. Mat. Inst. Steklov. (POMI)*, 336: 153–198 and 276–277, 2006.

321. S.A. Nazarov and G. Sweers. A hinged plate equation and iterated Dirichlet Laplace operator on domains with concave corners. *J. Differential Equations*, 233:151–180, 2007.

322. J. Nečas. *Les méthodes directes en théorie des équations elliptiques*. Masson and Academia, Paris and Prague, 1967.

323. Z. Nehari. On the biharmonic Green's function. In *Studies in Mathematics and Mechanics, presented to Richard von Mises*, pages 111–117. Academic Press, New York, London, 1954.

324. M. Nicolesco. *Les fonctions polyharmoniques*. Hermann, Paris, 1936.

325. J.C.C. Nitsche. Boundary value problems for variational integrals involving surface curvatures. *Quarterly Appl. Math.*, 51:363–387, 1993.

326. M. Obata. Certain conditions for a Riemannian manifold to be isometric with a sphere. *J. Math. Soc. Japan*, 14:333–340, 1962.

327. S. Osher. On Green's function for the biharmonic equation in a right angle wedge. *J. Math. Anal. Appl.*, 43:705–716, 1973.

328. P. Oswald. On a priori estimates for positive solutions of a semilinear biharmonic equation in a ball. *Comment. Math. Univ. Carolinae*, 26:565–577, 1985.

329. R.S. Palais. Critical point theory and the minimax principle. In *Global Analysis (Proc. Sympos. Pure Math., Vol. XV, Berkeley, Calif, 1968)*, pages 185–212. Amer. Math. Soc., Providence, 1970.

330. S.M. Paneitz. Essential unitarization of symplectics and application to field quantization. *J. Funct. Anal.*, 48:310–359, 1982.

331. S.M. Paneitz. A quartic conformally covariant differential operator for arbitrary pseudo-Riemannian manifolds. *Symmetry, Integrability and Geometry: Methods and Applications (SIGMA)*, 4:036, 3 pages, 2008. Posthumous printing of the 1983 preprint.

332. E. Parini and A. Stylianou. On the positivity preserving property of hinged plates. *SIAM J. Math. Anal.*, 41:2031–2037, 2009.

333. D. Passaseo. The effect of the domain shape on the existence of positive solutions of the equation $\Delta u + u^{2^*-1} = 0$. *Topol. Methods Nonlinear Anal.*, 3:27–54, 1994.

334. L.E. Payne. Isoperimetric inequalities and their applications. *SIAM Rev.*, 9:453–488, 1967.

335. L.E. Payne. Some isoperimetric inequalities for harmonic functions. *SIAM J. Math. Anal.*, 1:354–359, 1970. Erratum in: *SIAM J. Math. Anal.*, 5: 847, 1974.

336. L.E. Payne, G. Pólya, and H.F. Weinberger. On the ratio of consecutive eigenvalues. *J. Math. Phys.*, 35:289–298, 1956.

337. R.N. Pederson. On the unique continuation theorem for certain second and fourth order elliptic equations. *Comm. Pure Appl. Math.*, 11:67–80, 1958.

338. J. Pipher and G. Verchota. A maximum principle for biharmonic functions in Lipschitz and C^1 domains. *Comment. Math. Helv.*, 68:385–414, 1993.

339. J. Pipher and G. Verchota. Maximum principles for the polyharmonic equation on Lipschitz domains. *Potential Anal.*, 4:615–636, 1995.

340. S.I. Pohožaev. Eigenfunctions of the equation $\Delta u + \lambda f(u) = 0$. *Soviet. Math. Dokl.*, 6:1408–1411, 1965. Russian original: *Dokl. Akad. Nauk SSSR*, 165: 36–39, 1965.

341. S.I. Pohožaev. On the eigenfunctions of quasilinear elliptic problems. *Math. USSR Sbornik*, 11:171–188, 1970. Russian original: *Mat. Sb. (N.S.)*, 82 (124): 192–212, 1970.

342. S.D. Poisson. Mémoire sur les surfaces élastiques. *Cl. Sci. Mathém. Phys. Inst. de France, 2nd printing*, pages 167–225, 1812.

343. A. Polden. *Curves and surfaces of least total curvature and fourth-order flows*. PhD thesis, University of Tübingen, 1996.

344. G. Pólya and G. Szegö. *Isoperimetric Inequalities in Mathematical Physics*. Annals of Mathematics Studies, no. 27. Princeton University Press, Princeton, N. J., 1951.

345. C. Pommerenke. *Boundary Behaviour of Conformal Maps*. Grundlehren der Mathematischen Wissenschaften, Band 299. Springer-Verlag, Berlin etc., 1992.

346. M.H. Protter. Unique continuation for elliptic equations. *Trans. Amer. Math. Soc.*, 95:81–91, 1960.

347. M.H. Protter and H.F. Weinberger. *Maximum principles in differential equations*. Prentice-Hall Inc., Englewood Cliffs, N.J., 1967.

348. P. Pucci and J. Serrin. A general variational identity. *Indiana Univ. Math. J.*, 35:681–703, 1986.

349. P. Pucci and J. Serrin. Critical exponents and critical dimensions for polyharmonic operators. *J. Math. Pures Appl.*, (9) 69:55–83, 1990.

350. P.H. Rabinowitz. Some global results for nonlinear eigenvalue problems. *J. Funct. Analysis*, 7:487–513, 1971.

351. J.W.S. Rayleigh. *The Theory of Sound. I,II*. 2nd edition, revised and enlarged. Mc Millan - London, London, 1926. Reprint of the 1894 and 1896 edition.

352. W. Reichel. Uniqueness results for semilinear polyharmonic boundary value problems on conformally contractible domains, I & II. *J. Math. Anal. Appl.*, 287:61–74 & 75–89, 2003.

353. W. Reichel and T. Weth. A priori bounds and a Liouville theorem on a half-space for higher-order elliptic Dirichlet problems. *Math. Z.*, 261:805–827, 2009.

354. F. Rellich. Halbbeschränkte Differentialoperatoren höherer Ordnung. In *Proceedings of the International Congress of Mathematicians, 1954, Amsterdam, vol. III*, pages 243–250. Erven P. Noordhoff N.V., Groningen, 1956.

355. G. Rieder. On the plate paradox of Sapondzhyan and Babuška. *Mechanics Research Communications*, 1:51–53, 1974.

356. T. Rivière. Analysis aspects of Willmore surfaces. *Invent. Math.*, 174:1–45, 2008.

357. F. Robert. Positive solutions for a fourth order equation invariant under isometries. *Proc. Amer. Math. Soc.*, 131:1423–1431, 2003.

358. O.M. Sapondžyan. Bending of a simply supported polygonal plate (Russian). *Akad. Nauk Armyan. SSR. Izv. Fiz.-Mat. Estest. Tehn. Nauki*, 5:29–46, 1952.

359. E. Sassone. Positivity for polyharmonic problems on domains close to a disk. *Ann. Mat. Pura Appl.*, (4) 186:419–432, 2007.

360. H.H. Schaefer. *Banach Lattices and Positive Operators,* Grundlehren der mathematischen Wissenschaften, Band 215. Springer-Verlag, Berlin etc., 1974.

361. R. Schätzle. The Willmore boundary value problem. *Calc. Var. Partial Differential Equations*, 37:275–302, 2010.

362. M.U. Schmidt. A proof of the Willmore conjecture. Preprint, available at http://arxiv.org/abs/math/0203224v2, 2002.

363. R. Schoen. Conformal deformation of a Riemannian metric to constant scalar curvature. *J. Differential Geom.*, 20:479–495, 1984.

364. J. Schröder. Randwertaufgaben vierter Ordnung mit positiver Greenscher Funktion. *Math. Z.*, 90:429–440, 1965.

365. J. Schröder. On linear differential inequalities. *J. Math. Anal. Appl.*, 22:188–216, 1968.

366. J. Schröder. *Operator Inequalities*, volume 147 of *Mathematics in Science and Engineering*. Academic Press, New York, London, 1980.

367. H.A. Schwarz. Notizia sulla rappresentazione conforme di un'area ellittica sopra un'area circolare. *Ann. Mat. Pura Appl.*, (2) 3:166–170, 1869.

368. J.B. Seif. On the Green's function for the biharmonic equation in an infinite wedge. *Trans. Amer. Math. Soc.*, 182:241–260, 1973.

369. J. Serrin. A symmetry problem in potential theory. *Arch. Rational Mech. Anal.*, 43:304–318, 1971.

370. J. Serrin and H. Zou. Existence of positive solutions of the Lane-Emden system. *Atti Semin. Mat. Fis. Univ. Modena*, 46 (Suppl.):369–380, 1998.

371. H.S. Shapiro and M. Tegmark. An elementary proof that the biharmonic Green function of an eccentric ellipse changes sign. *SIAM Rev.*, 36:99–101, 1994.
372. L. Simon. Existence of surfaces minimizing the Willmore functional. *Comm. Anal. Geom.*, 1:281–326, 1993.
373. G. Simonett. The Willmore flow near spheres. *Differential Integral Equations*, 14:1005–1014, 2001.
374. J. Smith. The coupled equation approach to the numerical solution of the biharmonic equation by finite differences, I. *SIAM J. Numer. Anal.*, 5:323–339, 1968.
375. J. Smith. The coupled equation approach to the numerical solution of the biharmonic equation by finite differences, II. *SIAM J. Numer. Anal.*, 7:104–111, 1970.
376. R. Soranzo. A priori estimates and existence of positive solutions of a superlinear polyharmonic equation. *Dynam. Systems Appl.*, 3:465–487, 1994.
377. R. Souam. Schiffer's problem and an isoperimetric inequality for the first buckling eigenvalue of domains on $\mathbb{R}^2$. *Ann. Global Anal. Geom.*, 27:341–354, 2005.
378. S. Spanne. Sur le principe de maximum et le théorème de Fatou pour les solutions des équatons elliptiques d'ordre quelconque. *C. R. Acad. Sci. Paris, Sér. A*, 262:1407–1410, 1966.
379. E. Sperner Jr. Symmetrisierung für Funktionen mehrerer reeller Variablen. *Manuscripta Math.*, 11:159–170, 1974.
380. W. Stekloff. Sur les problèmes fondamentaux de la physique mathématique. *Ann. Sci. École Norm. Sup.*, (3) 19:191–259 and 455–490, 1902.
381. M. Struwe. A global compactness result for elliptic boundary value problems involving limiting nonlinearities. *Math. Z.*, 187:511–517, 1984.
382. M. Struwe. *Variational Methods.* Applications to nonlinear partial differential equations and Hamiltonian systems, volume 34 of *Ergebnisse der Mathematik und ihrer Grenzgebiete, 3. Folge/A Series of Modern Surveys in Mathematics.* Springer-Verlag, Berlin etc., fourth edition, 2008.
383. Ch.A. Swanson. The best Sobolev constant. *Appl. Anal.*, 47:227–239, 1992.
384. G. Sweers. Strong positivity in $C(\overline{\Omega})$ for elliptic systems. *Math. Z.*, 209:251–271, 1992.
385. G. Sweers. Positivity for a strongly coupled elliptic system by Green function estimates. *J. Geom. Anal.*, 4:121–142, 1994.
386. G. Sweers. When is the first eigenfunction for the clamped plate equation of fixed sign? *Electron. J. Differ. Equ. Conf.*, 6:285–296, 2001.
387. G. Sweers. No Gidas-Ni-Nirenberg type result for semilinear biharmonic problems. *Math. Nachr.*, 246/247:202–206, 2002.
388. G. Sweers. A survey on boundary conditions for the biharmonic. *Complex Variables and Elliptic Equations*, 54:79–93, 2009.
389. G. Szegö. On membranes and plates. *Proc. Nat. Acad. Sci. U.S.A.*, 36:210–216, 1950.
390. G. Szegö. Remark on the preceding paper of Charles Loewner. *Pacific J. Math.*, 3:437–446, 1953.
391. G. Talenti. Best constant in Sobolev inequality. *Ann. Mat. Pura Appl.*, (4) 110:353–372, 1976.
392. G. Talenti. Elliptic equations and rearrangements. *Ann. Scuola Norm. Sup. Pisa Cl. Sci.*, (4) 3:697–718, 1976.
393. G. Talenti. On the first eigenvalue of the clamped plate. *Ann. Mat. Pura Appl.*, (4) 129:265–280 (1982), 1981.
394. G. Talenti. Inequalities in rearrangement invariant function spaces. In *Nonlinear analysis, function spaces and applications, Vol. 5 (Prague, 1994)*, pages 177–230. Prometheus, Prague, 1994.
395. F. Tomi. Über semilineare elliptische Differentialgleichungen zweiter Ordnung. *Math. Z.*, 111:350–366, 1969.
396. F. Tomi. Über elliptische Differentialgleichungen 4. Ordnung mit einer starken Nichtlinearität. *Nachr. Akad. Wiss. Göttingen, II. Math.-Phys. Klasse*, pages 33–42, 1976.
397. W.C. Troy. Symmetry properties in systems of semilinear elliptic equations. *J. Differential Equations*, 42:400–413, 1981.
398. C. Truesdell. Sophie Germain: Fame earned by stubborn error. *Boll. Storia Sci. Mat.*, 11:3–24, 1991.

399. R.C.A.M. van der Vorst. Variational identities and applications to differential systems. *Arch. Rational Mech. Anal.*, 116:375–398, 1991.

400. R.C.A.M. van der Vorst. Best constant for the embedding of the space $H^2 \cap H_0^1(\Omega)$ into $L^{2N/(N-4)}(\Omega)$. *Differential Integral Equations*, 6:259–276, 1993.

401. R.C.A.M. van der Vorst. Fourth-order elliptic equations with critical growth. *C.R. Acad. Sci. Paris Sér. I Math.*, 320:295–299, 1995.

402. P. Villaggio. *Mathematical Models for Elastic Structures.* Cambridge University Press, Cambridge, 1997.

403. V. Volterra. Osservazioni sulla nota precedente del prof. Lauricella e sopra una nota di analogo argomento dell'ing. Almansi. *Torino Atti*, 31:1018–1021, 1896.

404. Th. von Kármán. Festigkeitsprobleme im Maschinenbau. *Encycl. der Mathematischen Wissenschaften, Leipzig*, IV/4 C:348–352, 1910.

405. W. von Wahl. Über semilineare elliptische Gleichungen mit monotoner Nichtlinearität. *Nachr. Akad. Wiss. Göttingen, II. Math.-Phys. Klasse*, pages 27–33, 1975.

406. W. von Wahl. Semilinear elliptic and parabolic equations of arbitrary order. *Proc. Royal Soc. Edinburgh*, A 78:193–207, 1978.

407. W. Walter. Ganze Lösungen der Differentialgleichung $\Delta^p u = f(u)$. *Math. Z.*, 67:32–37, 1957.

408. G. Wang. σ_k-scalar curvature and eigenvalues of the Dirac operator. *Ann. Global Anal. Geom.*, 30:65–71, 2006.

409. X. Wang. On the Cauchy problem for reaction-diffusion equations. *Trans. Amer. Math. Soc.*, 337:549–590, 1993.

410. J. Wei and X. Xu. On conformal deformations of metrics on S^n. *J. Funct. Anal.*, 157:292–325, 1998.

411. J. Wei and X. Xu. Classification of solutions of higher order conformally invariant equations. *Math. Ann.*, 313:207–228, 1999.

412. K.-O. Widman. Inequalities for the Green function and boundary continuity of the gradient of solutions of elliptic differential equations. *Math. Scand.*, 21:17–37 (1968), 1967.

413. Ch. Wieners. A numerical existence proof of nodal lines for the first eigenfunction of the plate equation. *Arch. Math.*, 66:420–427, 1996.

414. T.J. Willmore. *Total curvature in Riemannian geometry.* Ellis Horwood Ltd., Chichester, 1982.

415. T.J. Willmore. *Riemannian Geometry.* Oxford Science Publications. The Clarendon Press, Oxford University Press, Oxford, 1993.

416. B. Willms. An isoperimetric inequality for the buckling of a clamped plate. Lecture at the Oberwolfach meeting on "Qualitative properties of PDE", organised by H. Berestycki, B. Kawohl, G. Talenti, 1995.

417. J. Wloka. *Partial Differential Equations.* Cambridge University Press, Cambridge, 1987. Transl. from the German by C. B. and M. J. Thomas. German original: *Partielle Differentialgleichungen.* Sobolevräume und Randwertaufgaben. B. G. Teubner, Stuttgart, 1982.

418. H. Yamabe. On the deformation of Riemannian structures on compact manifolds. *Osaka Math. J.*, 12:21–37, 1960.

419. D. Zhang. On multiple solutions of $\Delta u + \lambda u + |u|^{4/(n-2)} u = 0$. *Nonlin. Anal., Theory Methods Appl.*, 13:353–372, 1989.

420. Z. Zhao. Uniform boundedness of conditional gauge and Schrödinger equations. *Comm. Math. Phys.*, 93:19–31, 1984.

421. Z. Zhao. Green function for Schrödinger operator and conditioned Feynman-Kac gauge. *J. Math. Anal. Appl.*, 116:309–334, 1986.

422. Z. Zhao. Green functions and conditioned gauge theorem for a 2-dimensional domain. In *Seminar on Stochastic Processes, 1987 (Princeton, NJ, 1987)*, volume 15 of *Progr. Probab. Statist.*, pages 283–294. Birkhäuser Boston, Boston, 1988.

423. V.V. Žikov, S.M. Kozlov, O.A. Oleĭnik, and Ha T'en Ngoan. Averaging and G-convergence of differential operators. *Russ. Math. Surv.*, 34:69–148, 1979. Russian original: *Uspekhi Mat. Nauk*, 34: 65–133 and 256, 1979.

Author–Index

Subject–Index

Lecture Notes in Mathematics

For information about earlier volumes
please contact your bookseller or Springer
LNM Online archive: springerlink.com

Vol. 1816: S. Albeverio, W. Schachermayer, M. Talagrand, Lectures on Probability Theory and Statistics. Ecole d'Eté de Probabilités de Saint-Flour XXX-2000. Editor: P. Bernard (2003)

Vol. 1817: E. Koelink, W. Van Assche (Eds.), Orthogonal Polynomials and Special Functions. Leuven 2002 (2003)

Vol. 1818: M. Bildhauer, Convex Variational Problems with Linear, nearly Linear and/or Anisotropic Growth Conditions (2003)

Vol. 1819: D. Masser, Yu. V. Nesterenko, H. P. Schlickewei, W. M. Schmidt, M. Waldschmidt, Diophantine Approximation. Cetraro, Italy 2000. Editors: F. Amoroso, U. Zannier (2003)

Vol. 1820: F. Hiai, H. Kosaki, Means of Hilbert Space Operators (2003)

Vol. 1821: S. Teufel, Adiabatic Perturbation Theory in Quantum Dynamics (2003)

Vol. 1822: S.-N. Chow, R. Conti, R. Johnson, J. Mallet-Paret, R. Nussbaum, Dynamical Systems. Cetraro, Italy 2000. Editors: J. W. Macki, P. Zecca (2003)

Vol. 1823: A. M. Anile, W. Allegretto, C. Ringhofer, Mathematical Problems in Semiconductor Physics. Cetraro, Italy 1998. Editor: A. M. Anile (2003)

Vol. 1824: J. A. Navarro González, J. B. Sancho de Salas, $\mathscr{C}^\infty$ – Differentiable Spaces (2003)

Vol. 1825: J. H. Bramble, A. Cohen, W. Dahmen, Multiscale Problems and Methods in Numerical Simulations, Martina Franca, Italy 2001. Editor: C. Canuto (2003)

Vol. 1826: K. Dohmen, Improved Bonferroni Inequalities via Abstract Tubes. Inequalities and Identities of Inclusion-Exclusion Type. VIII, 113 p, 2003.

Vol. 1827: K. M. Pilgrim, Combinations of Complex Dynamical Systems. IX, 118 p, 2003.

Vol. 1828: D. J. Green, Gröbner Bases and the Computation of Group Cohomology. XII, 138 p, 2003.

Vol. 1829: E. Altman, B. Gaujal, A. Hordijk, Discrete-Event Control of Stochastic Networks: Multimodularity and Regularity. XIV, 313 p, 2003.

Vol. 1830: M. I. Gil', Operator Functions and Localization of Spectra. XIV, 256 p, 2003.

Vol. 1831: A. Connes, J. Cuntz, E. Guentner, N. Higson, J. E. Kaminker, Noncommutative Geometry, Martina Franca, Italy 2002. Editors: S. Doplicher, L. Longo (2004)

Vol. 1832: J. Azéma, M. Émery, M. Ledoux, M. Yor (Eds.), Séminaire de Probabilités XXXVII (2003)

Vol. 1833: D.-Q. Jiang, M. Qian, M.-P. Qian, Mathematical Theory of Nonequilibrium Steady States. On the Frontier of Probability and Dynamical Systems. IX, 280 p, 2004.

Vol. 1834: Yo. Yomdin, G. Comte, Tame Geometry with Application in Smooth Analysis. VIII, 186 p, 2004.

Vol. 1835: O.T. Izhboldin, B. Kahn, N.A. Karpenko, A. Vishik, Geometric Methods in the Algebraic Theory of Quadratic Forms. Summer School, Lens, 2000. Editor: J.-P. Tignol (2004)

Vol. 1836: C. Năstăsescu, F. Van Oystaeyen, Methods of Graded Rings. XIII, 304 p, 2004.

Vol. 1837: S. Tavaré, O. Zeitouni, Lectures on Probability Theory and Statistics. Ecole d'Eté de Probabilités de Saint-Flour XXXI-2001. Editor: J. Picard (2004)

Vol. 1838: A.J. Ganesh, N.W. O'Connell, D.J. Wischik, Big Queues. XII, 254 p, 2004.

Vol. 1839: R. Gohm, Noncommutative Stationary Processes. VIII, 170 p, 2004.

Vol. 1840: B. Tsirelson, W. Werner, Lectures on Probability Theory and Statistics. Ecole d'Eté de Probabilités de Saint-Flour XXXII-2002. Editor: J. Picard (2004)

Vol. 1841: W. Reichel, Uniqueness Theorems for Variational Problems by the Method of Transformation Groups (2004)

Vol. 1842: T. Johnsen, A. L. Knutsen, K_3 Projective Models in Scrolls (2004)

Vol. 1843: B. Jefferies, Spectral Properties of Noncommuting Operators (2004)

Vol. 1844: K.F. Siburg, The Principle of Least Action in Geometry and Dynamics (2004)

Vol. 1845: Min Ho Lee, Mixed Automorphic Forms, Torus Bundles, and Jacobi Forms (2004)

Vol. 1846: H. Ammari, H. Kang, Reconstruction of Small Inhomogeneities from Boundary Measurements (2004)

Vol. 1847: T.R. Bielecki, T. Björk, M. Jeanblanc, M. Rutkowski, J.A. Scheinkman, W. Xiong, Paris-Princeton Lectures on Mathematical Finance 2003 (2004)

Vol. 1848: M. Abate, J. E. Fornaess, X. Huang, J. P. Rosay, A. Tumanov, Real Methods in Complex and CR Geometry, Martina Franca, Italy 2002. Editors: D. Zaitsev, G. Zampieri (2004)

Vol. 1849: Martin L. Brown, Heegner Modules and Elliptic Curves (2004)

Vol. 1850: V. D. Milman, G. Schechtman (Eds.), Geometric Aspects of Functional Analysis. Israel Seminar 2002-2003 (2004)

Vol. 1851: O. Catoni, Statistical Learning Theory and Stochastic Optimization (2004)

Vol. 1852: A.S. Kechris, B.D. Miller, Topics in Orbit Equivalence (2004)

Vol. 1853: Ch. Favre, M. Jonsson, The Valuative Tree (2004)

Vol. 1854: O. Saeki, Topology of Singular Fibers of Differential Maps (2004)

Vol. 1855: G. Da Prato, P.C. Kunstmann, I. Lasiecka, A. Lunardi, R. Schnaubelt, L. Weis, Functional Analytic Methods for Evolution Equations. Editors: M. Iannelli, R. Nagel, S. Piazzera (2004)

Vol. 1856: K. Back, T.R. Bielecki, C. Hipp, S. Peng, W. Schachermayer, Stochastic Methods in Finance, Bressanone/Brixen, Italy, 2003. Editors: M. Fritelli, W. Runggaldier (2004)

Vol. 1909: H. Akiyoshi, M. Sakuma, M. Wada, Y. Yamashita, Punctured Torus Groups and 2-Bridge Knot Groups (I) (2007)

Vol. 1910: V.D. Milman, G. Schechtman (Eds.), Geometric Aspects of Functional Analysis. Israel Seminar 2004-2005 (2007)

Vol. 1911: A. Bressan, D. Serre, M. Williams, K. Zumbrun, Hyperbolic Systems of Balance Laws. Cetraro, Italy 2003. Editor: P. Marcati (2007)

Vol. 1912: V. Berinde, Iterative Approximation of Fixed Points (2007)

Vol. 1913: J.E. Marsden, G. Misiołek, J.-P. Ortega, M. Perlmutter, T.S. Ratiu, Hamiltonian Reduction by Stages (2007)

Vol. 1914: G. Kutyniok, Affine Density in Wavelet Analysis (2007)

Vol. 1915: T. Bıyıkoğlu, J. Leydold, P.F. Stadler, Laplacian Eigenvectors of Graphs. Perron-Frobenius and Faber-Krahn Type Theorems (2007)

Vol. 1916: C. Villani, F. Rezakhanlou, Entropy Methods for the Boltzmann Equation. Editors: F. Golse, S. Olla (2008)

Vol. 1917: I. Veselić, Existence and Regularity Properties of the Integrated Density of States of Random Schrödinger (2008)

Vol. 1918: B. Roberts, R. Schmidt, Local Newforms for GSp(4) (2007)

Vol. 1919: R.A. Carmona, I. Ekeland, A. Kohatsu-Higa, J.-M. Lasry, P.-L. Lions, H. Pham, E. Taflin, Paris-Princeton Lectures on Mathematical Finance 2004. Editors: R.A. Carmona, E. Çinlar, I. Ekeland, E. Jouini, J.A. Scheinkman, N. Touzi (2007)

Vol. 1920: S.N. Evans, Probability and Real Trees. Ecole d'Été de Probabilités de Saint-Flour XXXV-2005 (2008)

Vol. 1921: J.P. Tian, Evolution Algebras and their Applications (2008)

Vol. 1922: A. Friedman (Ed.), Tutorials in Mathematical BioSciences IV. Evolution and Ecology (2008)

Vol. 1923: J.P.N. Bishwal, Parameter Estimation in Stochastic Differential Equations (2008)

Vol. 1924: M. Wilson, Littlewood-Paley Theory and Exponential-Square Integrability (2008)

Vol. 1925: M. du Sautoy, L. Woodward, Zeta Functions of Groups and Rings (2008)

Vol. 1926: L. Barreira, V. Claudia, Stability of Nonautonomous Differential Equations (2008)

Vol. 1927: L. Ambrosio, L. Caffarelli, M.G. Crandall, L.C. Evans, N. Fusco, Calculus of Variations and Non-Linear Partial Differential Equations. Cetraro, Italy 2005. Editors: B. Dacorogna, P. Marcellini (2008)

Vol. 1928: J. Jonsson, Simplicial Complexes of Graphs (2008)

Vol. 1929: Y. Mishura, Stochastic Calculus for Fractional Brownian Motion and Related Processes (2008)

Vol. 1930: J.M. Urbano, The Method of Intrinsic Scaling. A Systematic Approach to Regularity for Degenerate and Singular PDEs (2008)

Vol. 1931: M. Cowling, E. Frenkel, M. Kashiwara, A. Valette, D.A. Vogan, Jr., N.R. Wallach, Representation Theory and Complex Analysis. Venice, Italy 2004. Editors: E.C. Tarabusi, A. D'Agnolo, M. Picardello (2008)

Vol. 1932: A.A. Agrachev, A.S. Morse, E.D. Sontag, H.J. Sussmann, V.I. Utkin, Nonlinear and Optimal Control Theory. Cetraro, Italy 2004. Editors: P. Nistri, G. Stefani (2008)

Vol. 1933: M. Petkovic, Point Estimation of Root Finding Methods (2008)

Vol. 1934: C. Donati-Martin, M. Émery, A. Rouault, C. Stricker (Eds.), Séminaire de Probabilités XLI (2008)

Vol. 1935: A. Unterberger, Alternative Pseudodifferential Analysis (2008)

Vol. 1936: P. Magal, S. Ruan (Eds.), Structured Population Models in Biology and Epidemiology (2008)

Vol. 1937: G. Capriz, P. Giovine, P.M. Mariano (Eds.), Mathematical Models of Granular Matter (2008)

Vol. 1938: D. Auroux, F. Catanese, M. Manetti, P. Seidel, B. Siebert, I. Smith, G. Tian, Symplectic 4-Manifolds and Algebraic Surfaces. Cetraro, Italy 2003. Editors: F. Catanese, G. Tian (2008)

Vol. 1939: D. Boffi, F. Brezzi, L. Demkowicz, R.G. Durán, R.S. Falk, M. Fortin, Mixed Finite Elements, Compatibility Conditions, and Applications. Cetraro, Italy 2006. Editors: D. Boffi, L. Gastaldi (2008)

Vol. 1940: J. Banasiak, V. Capasso, M.A.J. Chaplain, M. Lachowicz, J. Miȩkisz, Multiscale Problems in the Life Sciences. From Microscopic to Macroscopic. Bȩdlewo, Poland 2006. Editors: V. Capasso, M. Lachowicz (2008)

Vol. 1941: S.M.J. Haran, Arithmetical Investigations. Representation Theory, Orthogonal Polynomials, and Quantum Interpolations (2008)

Vol. 1942: S. Albeverio, F. Flandoli, Y.G. Sinai, SPDE in Hydrodynamic. Recent Progress and Prospects. Cetraro, Italy 2005. Editors: G. Da Prato, M. Röckner (2008)

Vol. 1943: L.L. Bonilla (Ed.), Inverse Problems and Imaging. Martina Franca, Italy 2002 (2008)

Vol. 1944: A. Di Bartolo, G. Falcone, P. Plaumann, K. Strambach, Algebraic Groups and Lie Groups with Few Factors (2008)

Vol. 1945: F. Brauer, P. van den Driessche, J. Wu (Eds.), Mathematical Epidemiology (2008)

Vol. 1946: G. Allaire, A. Arnold, P. Degond, T.Y. Hou, Quantum Transport. Modelling, Analysis and Asymptotics. Cetraro, Italy 2006. Editors: N.B. Abdallah, G. Frosali (2008)

Vol. 1947: D. Abramovich, M. Mariño, M. Thaddeus, R. Vakil, Enumerative Invariants in Algebraic Geometry and String Theory. Cetraro, Italy 2005. Editors: K. Behrend, M. Manetti (2008)

Vol. 1948: F. Cao, J-L. Lisani, J-M. Morel, P. Musé, F. Sur, A Theory of Shape Identification (2008)

Vol. 1949: H.G. Feichtinger, B. Helffer, M.P. Lamoureux, N. Lerner, J. Toft, Pseudo-Differential Operators. Quantization and Signals. Cetraro, Italy 2006. Editors: L. Rodino, M.W. Wong (2008)

Vol. 1950: M. Bramson, Stability of Queueing Networks, Ecole d'Eté de Probabilités de Saint-Flour XXXVI-2006 (2008)

Vol. 1951: A. Moltó, J. Orihuela, S. Troyanski, M. Valdivia, A Non Linear Transfer Technique for Renorming (2009)

Vol. 1952: R. Mikhailov, I.B.S. Passi, Lower Central and Dimension Series of Groups (2009)

Vol. 1953: K. Arwini, C.T.J. Dodson, Information Geometry (2008)

Vol. 1954: P. Biane, L. Bouten, F. Cipriani, N. Konno, N. Privault, Q. Xu, Quantum Potential Theory. Editors: U. Franz, M. Schuermann (2008)

Vol. 1955: M. Bernot, V. Caselles, J.-M. Morel, Optimal Transportation Networks (2008)

Vol. 1956: C.H. Chu, Matrix Convolution Operators on Groups (2008)

Vol. 1957: A. Guionnet, On Random Matrices: Macroscopic Asymptotics, Ecole d'Eté de Probabilités de Saint-Flour XXXVI-2006 (2009)

Vol. 1958: M.C. Olsson, Compactifying Moduli Spaces for Abelian Varieties (2008)

Vol. 1959: Y. Nakkajima, A. Shiho, Weight Filtrations on Log Crystalline Cohomologies of Families of Open Smooth Varieties (2008)

Vol. 1960: J. Lipman, M. Hashimoto, Foundations of Grothendieck Duality for Diagrams of Schemes (2009)

Vol. 1961: G. Buttazzo, A. Pratelli, S. Solimini, E. Stepanov, Optimal Urban Networks via Mass Transportation (2009)

Vol. 1962: R. Dalang, D. Khoshnevisan, C. Mueller, D. Nualart, Y. Xiao, A Minicourse on Stochastic Partial Differential Equations (2009)

Vol. 1963: W. Siegert, Local Lyapunov Exponents (2009)

Vol. 1964: W. Roth, Operator-valued Measures and Integrals for Cone-valued Functions and Integrals for Cone-valued Functions (2009)

Vol. 1965: C. Chidume, Geometric Properties of Banach Spaces and Nonlinear Iterations (2009)

Vol. 1966: D. Deng, Y. Han, Harmonic Analysis on Spaces of Homogeneous Type (2009)

Vol. 1967: B. Fresse, Modules over Operads and Functors (2009)

Vol. 1968: R. Weissauer, Endoscopy for GSP(4) and the Cohomology of Siegel Modular Threefolds (2009)

Vol. 1969: B. Roynette, M. Yor, Penalising Brownian Paths (2009)

Vol. 1970: M. Biskup, A. Bovier, F. den Hollander, D. Ioffe, F. Martinelli, K. Netočný, F. Toninelli, Methods of Contemporary Mathematical Statistical Physics. Editor: R. Kotecký (2009)

Vol. 1971: L. Saint-Raymond, Hydrodynamic Limits of the Boltzmann Equation (2009)

Vol. 1972: T. Mochizuki, Donaldson Type Invariants for Algebraic Surfaces (2009)

Vol. 1973: M.A. Berger, L.H. Kauffmann, B. Khesin, H.K. Moffatt, R.L. Ricca, De W. Sumners, Lectures on Topological Fluid Mechanics. Cetraro, Italy 2001. Editor: R.L. Ricca (2009)

Vol. 1974: F. den Hollander, Random Polymers: École d'Été de Probabilités de Saint-Flour XXXVII – 2007 (2009)

Vol. 1975: J.C. Rohde, Cyclic Coverings, Calabi-Yau Manifolds and Complex Multiplication (2009)

Vol. 1976: N. Ginoux, The Dirac Spectrum (2009)

Vol. 1977: M.J. Gursky, E. Lanconelli, A. Malchiodi, G. Tarantello, X.-J. Wang, P.C. Yang, Geometric Analysis and PDEs. Cetraro, Italy 2001. Editors: A. Ambrosetti, S.-Y.A. Chang, A. Malchiodi (2009)

Vol. 1978: M. Qian, J.-S. Xie, S. Zhu, Smooth Ergodic Theory for Endomorphisms (2009)

Vol. 1979: C. Donati-Martin, M. Émery, A. Rouault, C. Stricker (Eds.), Séminaire de Probablitiés XLII (2009)

Vol. 1980: P. Graczyk, A. Stos (Eds.), Potential Analysis of Stable Processes and its Extensions (2009)

Vol. 1981: M. Chlouveraki, Blocks and Families for Cyclotomic Hecke Algebras (2009)

Vol. 1982: N. Privault, Stochastic Analysis in Discrete and Continuous Settings. With Normal Martingales (2009)

Vol. 1983: H. Ammari (Ed.), Mathematical Modeling in Biomedical Imaging I. Electrical and Ultrasound Tomographies, Anomaly Detection, and Brain Imaging (2009)

Vol. 1984: V. Caselles, P. Monasse, Geometric Description of Images as Topographic Maps (2010)

Vol. 1985: T. Linß, Layer-Adapted Meshes for Reaction-Convection-Diffusion Problems (2010)

Vol. 1986: J.-P. Antoine, C. Trapani, Partial Inner Product Spaces. Theory and Applications (2009)

Vol. 1987: J.-P. Brasselet, J. Seade, T. Suwa, Vector Fields on Singular Varieties (2010)

Vol. 1988: M. Broué, Introduction to Complex Reflection Groups and Their Braid Groups (2010)

Vol. 1989: I.M. Bomze, V. Demyanov, Nonlinear Optimization. Cetraro, Italy 2007. Editors: G. di Pillo, F. Schoen (2010)

Vol. 1990: S. Bouc, Biset Functors for Finite Groups (2010)

Vol. 1991: F. Gazzola, H.-C. Grunau, G. Sweers, Polyharmonic Boundary Value Problems (2010)

Vol. 1992: A. Parmeggiani, Spectral Theory of Non-Commutative Harmonic Oscillators: An Introduction (2010)

Vol. 1993: P. Dodos, Banach Spaces and Descriptive Set Theory: Selected Topics (2010)

Vol. 1994: A. Baricz, Generalized Bessel Functions of the First Kind (2010)

Vol. 1995: A.Y. Khapalov, Controllability of Partial Differential Equations Governed by Multiplicative Controls (2010)

Vol. 1996: T. Lorenz, Mutational Analysis. A Joint Framework for Cauchy Problems *In* and *Beyond* Vector Spaces (2010)

Vol. 1997: M. Banagl, Intersection Spaces, Spatial Homology Truncation, and String Theory (2010)

Vol. 1998: M. Abate, E. Bedford, M. Brunella, T.-C. Dinh, P. Giorgio, D. Schleicher, N. Sibony, Holomorphic Dynamical Systems. Editors: G. Gentili, J. Guenot (2010)

Vol. 1999: H. Schoutens, The Use of Ultraproducts in Commutative Algebra (2010)

Vol. 2000: H. Yserentant, Regularity and Approximability of Electronic Wave Functions (2010)

Recent Reprints and New Editions

Vol. 1702: J. Ma, J. Yong, Forward-Backward Stochastic Differential Equations and their Applications. 1999 – Corr. 3rd printing (2007)

Vol. 830: J.A. Green, Polynomial Representations of GL_n, with an Appendix on Schensted Correspondence and Littelmann Paths by K. Erdmann, J.A. Green and M. Schoker¯ 1980 – 2nd corr. and augmented edition (2007)

Vol. 1693: S. Simons, From Hahn-Banach to Monotonicity (Minimax and Monotonicity 1998) – 2nd exp. edition (2008)

Vol. 470: R.E. Bowen, Equilibrium States and the Ergodic Theory of Anosov Diffeomorphisms. With a preface by D. Ruelle. Edited by J.-R. Chazottes. 1975 – 2nd rev. edition (2008)

Vol. 523: S.A. Albeverio, R.J. Høegh-Krohn, S. Mazzucchi, Mathematical Theory of Feynman Path Integral. 1976 – 2nd corr. and enlarged edition (2008)

Vol. 1764: A. Cannas da Silva, Lectures on Symplectic Geometry 2001 – Corr. 2nd printing (2008)

LECTURE NOTES IN MATHEMATICS   Springer

Edited by J.-M. Morel, F. Takens, B. Teissier, P.K. Maini

Editorial Policy (for the publication of monographs)

1. Lecture Notes aim to report new developments in all areas of mathematics and their applications - quickly, informally and at a high level. Mathematical texts analysing new developments in modelling and numerical simulation are welcome.

    Monograph manuscripts should be reasonably self-contained and rounded off. Thus they may, and often will, present not only results of the author but also related work by other people. They may be based on specialised lecture courses. Furthermore, the manuscripts should provide sufficient motivation, examples and applications. This clearly distinguishes Lecture Notes from journal articles or technical reports which normally are very concise. Articles intended for a journal but too long to be accepted by most journals, usually do not have this "lecture notes" character. For similar reasons it is unusual for doctoral theses to be accepted for the Lecture Notes series, though habilitation theses may be appropriate.

2. Manuscripts should be submitted either to Springer's mathematics editorial in Heidelberg, or to one of the series editors. In general, manuscripts will be sent out to 2 external referees for evaluation. If a decision cannot yet be reached on the basis of the first 2 reports, further referees may be contacted: The author will be informed of this. A final decision to publish can be made only on the basis of the complete manuscript, however a refereeing process leading to a preliminary decision can be based on a pre-final or incomplete manuscript. The strict minimum amount of material that will be considered should include a detailed outline describing the planned contents of each chapter, a bibliography and several sample chapters.

    Authors should be aware that incomplete or insufficiently close to final manuscripts almost always result in longer refereeing times and nevertheless unclear referees' recommendations, making further refereeing of a final draft necessary.

    Authors should also be aware that parallel submission of their manuscript to another publisher while under consideration for LNM will in general lead to immediate rejection.

3. Manuscripts should in general be submitted in English. Final manuscripts should contain at least 100 pages of mathematical text and should always include

    - a table of contents;
    - an informative introduction, with adequate motivation and perhaps some historical remarks: it should be accessible to a reader not intimately familiar with the topic treated;
    - a subject index: as a rule this is genuinely helpful for the reader.

    For evaluation purposes, manuscripts may be submitted in print or electronic form, in the latter case preferably as pdf- or zipped ps-files. Lecture Notes volumes are, as a rule, printed digitally from the authors' files. To ensure best results, authors are asked to use the LaTeX2e style files available from Springer's web-server at:

    ftp://ftp.springer.de/pub/tex/latex/svmonot1/ (for monographs).

Additional technical instructions, if necessary, are available on request from:
lnm@springer.com.

4. Careful preparation of the manuscripts will help keep production time short besides ensuring satisfactory appearance of the finished book in print and online. After acceptance of the manuscript authors will be asked to prepare the final LaTeX source files (and also the corresponding dvi-, pdf- or zipped ps-file) together with the final printout made from these files. The LaTeX source files are essential for producing the full-text online version of the book (see www.springerlink.com/content/110312 for the existing online volumes of LNM).

    The actual production of a Lecture Notes volume takes approximately 12 weeks.

5. Authors receive a total of 50 free copies of their volume, but no royalties. They are entitled to a discount of 33.3% on the price of Springer books purchased for their personal use, if ordering directly from Springer.

6. Commitment to publish is made by letter of intent rather than by signing a formal contract. Springer-Verlag secures the copyright for each volume. Authors are free to reuse material contained in their LNM volumes in later publications: a brief written (or e-mail) request for formal permission is sufficient.

Addresses:
Professor J.-M. Morel, CMLA,
École Normale Supérieure de Cachan,
61 Avenue du Président Wilson, 94235 Cachan Cedex, France
E-mail: Jean-Michel.Morel@cmla.ens-cachan.fr

Professor F. Takens, Mathematisch Instituut,
Rijksuniversiteit Groningen, Postbus 800,
9700 AV Groningen, The Netherlands
E-mail: F.Takens@math.rug.nl

Professor B. Teissier, Institut Mathématique de Jussieu,
UMR 7586 du CNRS, Équipe "Géométrie et Dynamique",
175 rue du Chevaleret
75013 Paris, France
E-mail: teissier@math.jussieu.fr

For the "Mathematical Biosciences Subseries" of LNM:

Professor P.K. Maini, Center for Mathematical Biology,
Mathematical Institute, 24-29 St Giles,
Oxford OX1 3LP, UK
E-mail: maini@maths.ox.ac.uk

Springer, Mathematics Editorial I, Tiergartenstr. 17
69121 Heidelberg, Germany,
Tel.: +49 (6221) 487-8259
Fax: +49 (6221) 4876-8259
E-mail: lnm@springer.com

If you have any concerns about our products,
you can contact us on
ProductSafety@springernature.com

In case Publisher is established outside the EU,
the EU authorized representative is:
Springer Nature Customer Service Center GmbH
Europaplatz 3, 69115 Heidelberg, Germany

Printed by Libri Plureos GmbH
in Hamburg, Germany